FUNDAMENTALS OF SOLAR CELLS

Photovoltaic Solar Energy Conversion

FUNDAMENTALS OF SOLAR CELLS

Photovoltaic Solar Energy Conversion

Alan L. Fahrenbruch
Richard H. Bube
Department of Materials Science
and Engineering
Stanford University
Stanford, California

1983

ACADEMIC PRESS
A Subsidiary of Harcourt Brace Jovanovich, Publishers
New York London
Paris San Diego San Francisco São Paulo Sydney Tokyo Toronto

ACADEMIC PRESS, INC.
111 Fifth Avenue, New York, New York 10003

United Kingdom Edition published by
ACADEMIC PRESS, INC. (LONDON) LTD.
24/28 Oval Road, London NW1 7DX

Library of Congress Cataloging in Publication Data

Fahrenbruch, Alan.
Fundamentals of solar cells.

Includes index.
1. Solar cells. I. Bube, Richard H., Date
II. Title.
TK2960.F33 1983 621.31'244 82-13919
ISBN 0-12-247680-8

PRINTED IN THE UNITED STATES OF AMERICA

83 84 85 86 9 8 7 6 5 4 3 2 1

To Phyllis, and to my parents, John and Lillian

ALF

The heavens are telling the glory of God;
and the firmament proclaims His handiwork. . . .
In [the heavens] He has set a tent for the sun . . .
Its rising is from the end of the heavens . . .
and there is nothing hid from its heat. Psalm 19:1,4,6

RHB

CONTENTS

CHAPTER 8 Heterojunction and Heteroface Structure Cells

CHAPTER 9 Polycrystalline Thin Films for Solar Cells

CHAPTER 10 The Cu_xS/CdS Cell: A Case History of an All-Thin-Film Cell

CHAPTER 11 Other Photovoltaic Cells

CHAPTER 12 Concentrators, Concentrator Systems, and Photoelectrochemical Cells

PREFACE

In the 1960s and 1970s there was a significant change in people's concept of the earth. An awareness developed that the fossil energy resources, which we all had come to take for granted, were not inexhaustible. New, and preferably renewable, energy resources had to be found. It is probable that a combination of sources will fill this need, including coal, solar-thermal, solar-photovoltaic, wind, nuclear, ocean-thermal, biomass, and others. It is our hope that this book will aid the expansion of solar photovoltaics from relatively small and specialized use to a large-scale contribution to our energy supply.

This book is intended for upper-level graduate students who have a reasonably good understanding of solid state physics and for scientists and engineers involved in research and development of solar cells. It is the outcome of a course taught by one of the authors (ALF) beginning in 1977 at Stanford University. The field of photovoltaics is changing rapidly, and the increasing interest of the scientific, engineering, and business communities has produced an avalanche of literature that attempts to keep up with the state of art. We could not hope to aim at such a moving target in a book of this kind, and fortunately it is not our purpose to do so. This book focuses instead on the fundamental physical principles of solar cells rather than on the details of particular devices. Specific devices are introduced principally as examples of basic phenomena and to illustrate the possibilities for innovative design. Although a solar cell is conceptually simple, it requires the application of many disciplines to produce efficient and economical devices. It is relatively easy to make a solar cell with less than 5% efficiency from a variety of materials. Indeed, the photovoltaic response of a cell based on spinach has been measured.† But to increase the solar efficiency figure to 10–20% requires great understanding and care.

The book begins with a procedure like that used by children who are confronted with clocks and other mysterious mechanisms: the cell is taken apart to expose the fundamental processes taking place. Then each of the basic aspects of its nature is explored in the most intuitive way possible. We begin this examination in Chapter 2 with the solar input. In Chapter 3 semiconductor basics, optical absorption and recombination, drift, and diffusion of carriers are reviewed both to establish a common language and for a convenient resource. Recombination and the flow of photogenerated carriers are related through the transport equation in Chapter 4, enabling us to calculate the total light-generated current and the spectral response. Charge separation and the characteristics of junction barriers are examined in detail in Chapter 5. In Chapter 6 the cell is "reassembled," the overall solar efficiency is calculated, series and parallel resistance losses are considered, and these are related to the semiconductor properties of the materials involved.

Chapters 7, 8, and 10 are devoted to examples of specific solar cell devices where the practicalities of fabrication are faced and the potential for innovative modification is introduced. We consider a single-crystal homojunction (Si), a single-crystal-heterojunction/buried-homojunction (AlGaAs/GaAs), and a polycrystalline, thin-film cell (Cu_xS/CdS). Chapter 9 describes the complexities that the use of thin-film polycrystalline material brings to the design of solar cells, including the modification of transport and recombination by grain boundary potential barriers. The book concludes with discussions of concentrator devices and systems and of photoelectrochemical cells.

The last parts of Chapters 5 and 10 relate to two areas that might be considered to be more in the realm of pure science: heterojunction interfaces and single-crystal Cu_xS/CdS cells. However, this background is essential for experimentalists dealing with cells that involve active interfaces and cells in which photoactive impurity energy levels are important. In addition, these complexities exist in all cells, even if only of second order in magnitude.

Finally we should like to express our appreciation to the people who reviewed particular chapters: Chapters 1 and 6, Dr. Kim W. Mitchell; Chapter 2, Dr. Arlon J. Hunt; Chapter 3, Dr. Julio Aranovich; Chapter 5, Dr. A. Rose; Chapter 7 and the concentrator section of Chapter 12, Pro-

† *New York Times*, July 31, 1979.

fessor Richard Swanson; Chapter 8, Professor G. R. Pearson; Chapter 9, Dr. Lewis Fraas; Chapter 10, Professor Allen Rothwarf and Dr. William Haines; Chapter 11, Professor Sigurd Wagner; and Chapter 12 (the section on PEC cells), Dr. Bruce Parkinson.

In addition, we wish to thank all the graduate students who gave valuable feedback and caught many errors in the manuscript, especially Chris Eberspacher. Finally we wish to give credit to NSF-RANN, ERDA, and the Department of Energy for financial support for a portion of the research discussed here.

LIST OF SYMBOLS

A	Ideality factor in diode equations
A^*	Richardson constant modified by the effective mass $[=120(m^*/m_0)]$ (A cm^{-2} °K^{-2})
A^{**}	Richardson constant modified by tunneling, image forces, etc.
A_0	Avogadro's number $[=6.023 \times 10^{23}]$ (mole^{-1})
$\mathscr{A}_A, \mathscr{A}_D$	Active illuminated area and total junction diode area
a	Lattice constant (Å)
B	Constant
$\mathbf{B}$	Magnetic induction (G)
C	Capacitance (F)
$\mathcal{C}$	Solar concentration ratio
D	Atomic diffusion coefficient (cm^2 sec^{-1})
D, D_n, D_p	Diffusion coefficients: general, for electrons, and for holes, respectively (cm^2 sec^{-1})
E	Energy (usually in eV)†
E_A, E_D	Acceptor- and donor-level energies (eV)
E_c	Energy of conduction band edge (eV)
E_F	Fermi-level energy (eV)
E_{Fn}, E_{Fp}	Quasi-Fermi-level energies for electrons and holes
E_g	Band-gap energy (eV)
E_i	Intrinsic-level energy (eV)
E_r	Recombination-level energy (eV)
E_v	Energy of valence band edge (eV)
E_0, E_{00}	Characteristic energies for thermionic-field emission junction transport (eV) [Section 5.5.6]
$\mathscr{E}$	Electric field (V cm^{-1})
$\mathscr{F}$	Faraday constant [=96486 C g-equiv^{-1}]
f	Fermi function
ff	Fill factor
G	Generation rate (cm^{-3} sec^{-1})

† In this book energies (e.g., E_F, E_i, E_c) are given with respect to an arbitrary reference. Energy differences are denoted explicitly, i.e., $(E_c - E_F)$, or by $\triangle E$, E_g, δ, E_0, and E_{00}.

g	Collection function, bulk (see Section 6.3.1)
H	Collection function, total
h	Collection function, interfacial
h	Planck's constant [$\hbar = h/2\pi = 1.053 \times 10^{-34}$ J sec $= 0.66 \times 10^{-15}$ eV sec]
I	Current (A)
I_{sc}	Short-circuit current (A)
J	Current density (A cm^{-2})
J_L, J_{Ln}, J_{Lp}	Light-generated current density: total, for electrons, and for holes, respectively (A cm^{-2})
J_m	Current density at maximum-power point (A cm^{-2})
J_n, J_p	Electron and hole current density (A cm^{-2})
J_{sc}	Short-circuit current density (A cm^{-2})
J_0	Preexponential factor in diode equation (A cm^{-2})
J_{00}	Preexponential factor for J_0 [e.g., $J_0 = J_{00} \exp(-\Delta E/kT)$]
K_0, K_L	Radiation damage coefficients [Section 7.5.1]
$\mathcal{K}$	Thermal conductivity (W cm^{-1} $°C^{-1}$)
k	Boltzmann constant [$= 1.38 \times 10^{-23}$ J $°K^{-1}$ $= 0.864 \times 10^{-4}$ eV $°K^{-1}$]
k_0	Distribution coefficient
L, L_n, L_p	Diffusion length: general, for electrons, and for holes, respectively (cm)
L_{dr}	Drift length [$= \mu \mathscr{E} \tau$] (cm) [Eq. (11.1)]
$\mathscr{L}_s, \mathscr{L}_p$	Loss fraction for series and parallel resistance
ln	Natural logarithm
M	Atomic mass
m_e^*, m_h^*	Effective masses of electrons and holes
m_r	Optical air-mass number
m_t^*	Tunneling effective mass
m_0	Rest mass of electron [$= 9.11 \times 10^{-31}$ kg]
N	Number density (cm^{-3})
N_A, N_D	Density of acceptors and donors (cm^{-3})
N_c, N_v	Effective density of states in the conduction and valence bands (cm^{-3})
N_{gb}	Density of grain boundary states (cm^{-2})
N_I	Density of imperfection centers (cm^{-3})
N_I^*	Density of imperfection centers available for capture of minority carriers (cm^{-3})
N_i	Density of interface states (cm^{-2})
N_r	Density of recombination centers (cm^{-3})
N_{ss}	Density of interface states (cm^{-2} eV^{-1})
n	Electron density (cm^{-3})
n_i	Intrinsic carrier density (cm^{-3})
n_n, n_p	Nonequilibrium carrier density of electrons in n-type and p-type materials
n_{n0}, n_{p0}	Equilibrium carrier density of electrons in n- and p-type materials
n_r	Index of refraction
P	Power density (W cm^{-2})
P_m	Power density at the maximum-power point (W cm^{-2})
P_s	Total solar input power (W cm^{-2})
p	Hole density (cm^{-3})
p_p, p_n	Nonequilibrium hole densities in p- and n-type materials (cm^{-3})
p_{p0}, p_{n0}	Thermal equilibrium hole densities in p- and n-type materials (cm^{-3})
Q	Charge density (C cm^{-2} or C cm^{-3})
Q_I	Insulator bulk charge (C cm^{-3})
Q_{ss}	Interface or surface state charge (C cm^{-2})
q	Electron charge [$= \lvert q \rvert = 1.6 \times 10^{-19}$ C]
R	Reflectance
R	Resistance (Ω)
R_s, R_p, R_c	Series, parallel (shunt), and contact resistance (Ω)

S	Index of interface behavior [Section 5.7.2]
S, S_i	Surface and interfacial recombination velocities (cm sec^{-1})
S_{gb}	Grain boundary recombination velocity (cm sec^{-1})
T	Temperature (°C or °K)
$\mathcal{T}$	Tunneling coefficient
t	Time (sec)
t	Layer thickness (cm)
U	Bulk recombination rate (cm^{-3} sec^{-1})
U_{gb}	Grain boundary recombination rate (cm^{-2} sec^{-1})
V	Voltage (V)
V_d	Diffusion (or "built-in") voltage (V)
V_{dgb}	Diffusion voltage of grain boundary (used where V_d would be confusing) (V)
V_m	Voltage at maximum-power point (V)
V_{oc}	Open-circuit voltage (V)
v_d	Diffusion velocity [= D/L] (cm sec^{-1})
v_{th}	Average thermal velocity of electrons or holes (cm sec^{-1}) [$=(8kT/\pi m^*)^{1/2}$; at 300°K, $v_{th} = (1.91 \times 10^7)(m_0/m^*)^{1/2}$ cm sec^{-1}]
W_d	Depletion layer width (cm)
W_{dgb}	Depletion layer width on one side of a grain boundary (cm)
W_i	Insulating layer width (cm)
W_p	Peak watts (W)
x	Distance coordinates (cm)
x_n, x_p	Distance coordinates of depletion layer edge for *n*-QNR and *p*-QNR (cm) [Fig. 5.1]
$x_{n'}$	Distance coordinate such that $(x_{n'} - x_n)$ is thickness of *n*-QNR (cm) [Fig. 5.1]
$x_{p'}$	$\lvert x_{p'} - x_p \rvert$ is thickness of *p*-QNR (cm)
y	Distance coordinate difference (e.g., $y_n = \lvert x_{n'} - x_n \rvert$) (cm)
z	Zenith angle of sun
z	Distance coordinate (cm)
α	Optical absorption coefficient (cm^{-1})
α	Tunneling exponential factor (V^{-1})
α_Σ	Minimum angle subtended by sun at earth's radius [= 0.533°]
β	Recombination coefficient (cm^3 sec^{-1})
Γ	Photon flux (photons cm^{-2} sec^{-1})
Γ_0	Photon flux incident on front surface of solar cell (cm^{-2} sec^{-1})
γ	Grain size (cm)
Δ	Increment
ΔE	Energy difference (eV)
$\Delta E_c, \Delta E_v$	Conduction and valence band discontinuities (eV) [Fig. 5.14]
δ	Thickness of insulating layer in MIS structure (cm or Å)
δ_n, δ_p	Energy difference between E_F and conduction or valence band edges [$\delta_n = E_c - E_F$; $\delta_p = E_F - E_v$] (eV)
ϵ_r	Dielectric constant [$\epsilon_s = \epsilon_r \epsilon_0$]
$\epsilon_s, \epsilon_{s0}, \epsilon_{shf}$	Permittivity of semiconductor, dc, and high frequency (F cm^{-1})
ϵ_0	Permittivity of free space [$= 8.85 \times 10^{-14}$ F cm^{-1}]
η_c	Concentrator efficiency (%)
η_Q	Quantum efficiency (%)
η_s	Solar efficiency (%)
θ	Angle
λ	Wavelength (μm)
μ_n, μ_p	Electron and hole mobilities (cm^2 V^{-1} sec^{-1})
ν	Frequency of light (sec^{-1})
Π	Product
ρ	Charge density (C cm^{-3})
ρ	Bulk resistivity (Ω cm)
ρ_c, ρ_{c0}	Contact resistivity: general and at zero-bias voltage, respectively (Ω cm^2)
ρ_s	Sheet resistivity (Ω $\square^{-1}$)
Σ	Sum
σ	Conductivity (Ω^{-1} cm^{-1})
$\sigma, \sigma_n, \sigma_p$	Recombination cross sections:

Symbol	Description
	general, for electrons, and for holes, respectively (cm^2)
τ, τ_n, τ_p	Minority carrier lifetimes: general, for electrons, and for holes, respectively (sec)
τ_{n0}, τ_{p0}	Extreme lifetime for electrons and holes (sec) [Section 3.5.2]
τ_r	Carrier scattering relaxation time (sec) [Eq. (3.20)]
Φ	Radiation fluence (particles cm^{-2})
ϕ	Work function (V)
ϕ	Electrostatic potential (V)
ϕ_b	Potential barrier height, measured from the Fermi level (V)
ϕ_d	Potential difference induced by electric dipole at junction (V)
ϕ_{gb}	Potential barrier height at grain boundary, measured from the Fermi level (V)
χ	Electron affinity (V)
Ω	Ohms
ω	Angular frequency (radians sec^{-1})

LIST OF ACRONYMS†

AM1, AMm_r	Air-mass numbers [Section 2.2]
BSF	Back-surface field (Section 4.5.1)
BSR	Back-surface reflection (optical, Section 7.6.3)
CNR	Comsat nonreflective (cell) [Section 7.4.4]
CVD	Chemical vapor deposition
CZ	Czochralski crystal growth (e.g., Si)
DLTS	Deep-level transient spectroscopy
EFG	Edge-fed growth (e.g., of Si ribbon)
FZ	Float zone crystal growth (e.g., Si)
I/D	Indicating the injection/diffusion component of diode transport ("id" when used as subscript, e.g., J_{id})
ITO	Indium tin oxide
LPE	Liquid-phase epitaxial (growth)
MIS	Metal/insulator/semiconductor junction
MOS	Metal/oxide/semiconductor junction
OS	Oxide semiconductor
PEC	Photoelectrochemical cell
PIN	*p*-type/insulator/*n*-type junction
PX	Polycrystalline
QNR	Quasi-neutral region
R/G	Indicating the recombination/generation component of junction transport (or as subscript "rg," e.g., J_{rg})
SIS	Semiconductor/insulator/semiconductor junction
VE	Vacuum evaporation
VMJ	Vertical multijunction cell

† Table 9.1 lists acronyms for the deposition methods.

CHAPTER 1

SURVEY OF BASIC CONCEPTS

Photovoltaic conversion of solar energy appears to be one of the most promising ways of meeting the increasing energy demands of the future in a time when conventional sources of energy are being depleted. In terms of human history, the life of the sun is effectively infinite, its energy is being radiated to the earth whether it is directly used or not, and the use of photovoltaics for direct conversion of solar energy introduces no direct contamination of the environment. Furthermore, it appears that major advances in the understanding and fabrication of semiconductor devices over the past 30 years have prepared the way for the kind of massive effort required if photovoltaics are to make a significant contribution to the energy requirements of the future.

1.1 SOURCES OF ENERGY

The present growing interest in photovoltaic conversion is a consequence of the concern to identify future sources of energy that will be inexpensive as well as consistent with the maintenance and safety of the environment. Traditional sources of energy—fossil fuels such as coal, petroleum, and natural gas—are running out as presently foreseeable rates of use are projected into the near future. Just 100 years ago, the principal source of energy in the United States was wood. Subsequent industrial development was built on the large exploitation of our coal resources. About the time of World War I, oil began to become a major contributor to our energy consumption. Finally, about 1950, natural gas took over a significant role in our economy. The large utilization of these fossil fuels is therefore a rather recent development against the history of the planet itself. The evanescent nature of our oil and gas supplies, as viewed from a historical perspective, is illustrated in Fig. 1.1 from Hubbert's 1972 study of oil and gas reserves in the United States.

Figure 1.2 diagrams the normalized energy flow in the United States in 1978, illustrating the inputs and outputs. As of that date, our energy was supplied 49% from oil, 19% from coal, 25% from natural gas, and 4% from nuclear, with almost negligible inputs from geothermal and hydroelectric. Of the 26% of the total energy input used for the generation of electrical energy, 35% came from oil and natural gas, 14% from nuclear, 5% from hydroelectric, and 50% from coal. It is in this area that photovoltaic conversion of solar energy can make a contribution. Figure 1.2 also illustrates the relevance of conservation of energy since only 45% of the total input energy is finally obtained in useful output.

Table 1.I gives a brief overview of the types of energy sources that might be called on in the future with some of their major advantages and disadvantages. Even if coal is utilized in many different ways not presently used (e.g., solvent refining, pyrolysis, gasification, magnetohydrodynamic generators), its supply is hardly infinite, and we shall probably begin to run out of coal in 300 years or less. The attempt to use coal more widely also threatens air quality through volatile and particulate pollutants. Domestic petroleum production peaked in 1971. Although 89% of

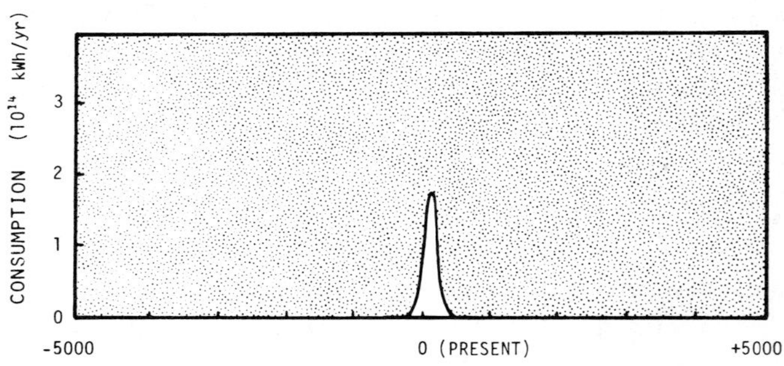

Fig. 1.1. The rise and fall of the world's rate of consumption of fossil fuel resources. [After "U.S. Energy Resources, a Review as of 1972, Part I," Serial No. 93-40 (92-75), prepared by M. King Hubbert, Office of Energy Resources, U.S. Geological Survey, at the request of Henry M. Jackson, Chairman, Committee on Interior and Insular Affairs, U.S. Senate, June 1974. U.S. Gov. Print. Off. Washington, D.C., 1974 (31-8200). [Figure only redrawn from *Phys. Today,* Dec., p. 11 (1976).]

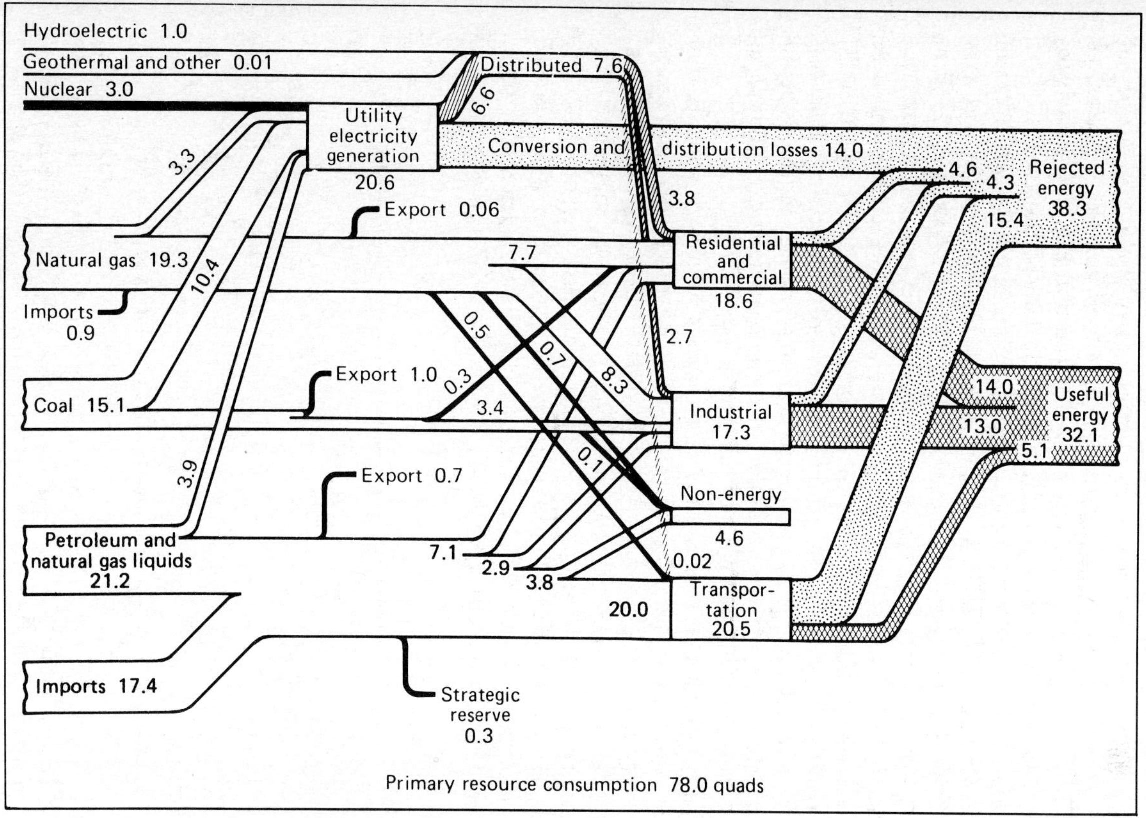

Fig. 1.2. Energy flow in the United States in quads (2.12 quads = 1 MBD, millions of barrels of oil per day-equivalent) in 1978. Energy inputs are on the left, energy use categories are on the right. Transportation absorbs more than 50% of the total supply of petroleum and rejects over 75% of what it consumes. [After A. L. Austin, Suppose the oil stops . . . *Energy Technol. Rev., Lawrence Livermore Natl. Lab.* June, p. 19 (1980). Flow chart, courtesy of Dr. Austin, its originator.]

Table 1.I

Present and Future Energy Sources[a]

Energy source	Types of processes	Advantages	Disadvantages
Petroleum	Drilling, oil shale	Convenient, low pollution	Limited supply, nonrenewable (20-year supply)
Natural gas	Drilling	Convenient, low pollution	Very limited supply, nonrenewable (5–10-year supply)
Coal	Deep or strip mining, solvent refining, pyrolysis, gasification, magnetohydrodynamic usage	Easy to handle, provides source of hydrocarbons in gaseous or liquid form as well as normal solid form	Atmospheric pollution, nonrenewable, much research needed for production of power by unconventional means, nonrenewable (200–300-year supply)
Nuclear fission	Light-water reactors	Existing technology, compact	Limited fuel supply, thermal pollution, operating and transport hazards, storage of wastes
Breeder reactors	Liquid-metal-cooled reactors	Reduces limitation of fuel availability	As for nuclear fission, but generally more severe
Nuclear fusion	Magnetic containment, laser initiated	Unlimited fuel supply	Much materials research needed, radiation leakage problems
Solar thermal	Direct heating and cooling via absorption of solar radiation	Pollution-free unlimited energy source	Need for extensive architectural changes, storage required
Solar thermal: electric	Use of solar energy to operate a steam turbine	Pollution-free unlimited energy source	Research needed on collection, focusing and storage

Solar photovoltaic	Photovoltaic effect in semiconductor junction devices	Pollution-free unlimited energy source	Present high cost of cells, energy storage, and operating lifetime
Hydroelectric	Fall of water under gravity used to generate electricity	Renewable, inexpensive, can be used as storage process, large potential in Third World	Limited to special locations, most of which in U.S. have been developed
Tidal energy	Motion of water under gravitational pull of moon used to generate electricity	Pollution-free, renewable	Limited number of exploitable sites
Wind	Force of the wind used to generate electricity	Pollution-free, renewable	Limited to special locations, large-scale effect on weather unknown, storage required
Ocean thermal	Uses thermal gradients to drive heat engine and generate electrical energy	Pollution-free, renewable	Special materials problems in resisting corrosion, costly transmission
Geothermal	Decay of radioactive materials within the earth produce heat	Inexpensive	Limited to special locations, corrosion problems, nonrenewable
Biomass	Conversion of solid organic matter into synthetic fuel	Ready supply, could utilize wastes	Uses arable land if deliberately planted, cost uncertain, sludge disposal

[a] An excellent summary from which these statements are derived is the book by A. L. Hammond, W. D. Metz, and T. H. Maugh, II, "*Energy and the Future*." Am. Assoc. Adv. Sci., Washington, D.C., 1973.

all fossil fuels remaining today are in the form of coal, 77% of U.S. consumption involves the use of oil and natural gas (Hammond *et al.*, 1973). Of all known petroleum reserves in the world, 75% are in the Middle East, where they will continue to be constantly threatened by international politics. It is estimated that we shall begin actually to run out of petroleum in about 25 years (Hammond *et al.*, 1973). Natural gas is an ideal fuel, but its supply is so limited that the United States may see its last cubic foot of natural gas enshrined in a museum within 20 years.

Alternative sources of energy may be singled out—nuclear fission, nuclear fusion, solar thermal, solar electric, wind, geothermal, hydroelectric, and fuel from wastes, to mention the most common—but no single one of these, or simple combination of them, is known today to be both an economically feasible and a socially acceptable means of providing the energy requirements of the year 2000. It is for this reason that continued investigation of all viable alternatives is essential. In recognition of the potential of solar energy in all its forms, U.S. government support of research and development in solar energy through the Department of Energy (DOE) has increased markedly from year to year.

1.2 THE ROLE OF PHOTOVOLTAIC CONVERSION

What is needed to make photovoltaic conversion of solar energy a useful contributor to our future energy needs? There are three main areas in which a commercially viable solar cell must excel: (1) cost, (2) efficiency, and (3) operating lifetime. Clearly, these are not independent, but are mutually related.

The first photovoltaic cells to have wide application were silicon single-crystal homojunction cells for utilization in the space program. In 1956 these cells cost $350 per peak watt (a peak watt is the maximum power generated by a cell in the course of an ideal day); by 1966 this cost had dropped to $100 per peak watt through improved processing and technology (Wolf, 1972), and by 1977 it was below $15 per peak watt for arrays. The 1978 goal of the Department of Energy was to have established the feasibility of an array cost of $2 per peak watt by 1980, with a goal of $0.15 to $0.50 by 1990. To put these costs into perspective, the cost of outfitting a private home with photovoltaic sources of electricity with solar cells at the current array cost of $15 per peak watt is $50,000 to $100,000. If this cost can be reduced by a factor of 30, the cost per home becomes about $3300, a reasonable capital outlay in view of increasing energy supply costs from other sources.

Photovoltaic cells for terrestrial application may be expected to assume one of two main forms: large-area, thin-film cells or concentrating systems using single-crystal cells. In the thin-film approach, reduction of cost is achieved by using small amounts of material and inexpensive processing. In the concentrator system approach, the higher cost of the single-crystal solar cell itself becomes small with respect to the overall cost of the concentrator system, and one cell operating efficiently under 100-suns illumination can economically cost many times more than a cell intended for use under 1 sun.

The cost goals previously described are estimated for a photovoltaic cell with an efficiency of 10%, i.e., a cell that can convert 10% of the incident sun's power into electrical power. Even if an extremely inexpensive cell were available with an appreciably lower efficiency, it is not likely that it would prove useful because of the fixed cost of support equipment and the need for larger areas. Although it is not universally agreed that 10% itself will be adequate, it is generally agreed that it is a reasonable minimum value to be considered seriously. Single-crystal silicon cells have efficiencies between 12 and 15% with maximum values of 15 to 20%. Single-crystal gallium arsenide based cells have efficiencies exceeding 20% without concentration, and efficiencies up to 24% with concentration. Most thin-film systems have efficiencies in the 8–12% range as of 1982.

Finally, to be effective, a photovoltaic cell must have a sufficiently long operation lifetime to repay both the financial costs and the energy costs required for its initial production. The generally recognized goal for operating lifetime is 20 years. Using 1977 fabrication methods, it is estimated that a single-crystal silicon cell can repay the energy costs of its production from SiO_2 in 3.8–6.4 years (Lindmayer, 1977). Another estimate of the energy payback time ranges from 12 years with present technology to 4 months with only slightly improved methods (Grant, 1976). The dominant processes that limit operating lifetime for terrestrial applications are quite different from space applications; in the latter, damage to the crystal lattice by the high-energy particle fluxes present in space can degrade the lifetime, whereas in the former, interdiffusion or corrosion processes dominate.

As the total generating capacity available from photovoltaic cells increases, so, of course, does the number of likely applications. Current photovoltaic systems, because of their high cost, are chiefly being used as power sources for microwave relays, corrosion protection, water pumping, and other needs for a remote and self-supporting power supply. With increased cell production and consequent lower cost, photovoltaic cells can be applied on a large scale to individual homes in remote areas, for agricultural water pumping, and for remote villages in Third World countries. Finally, full-scale commercial utilization would be associated with the capability of generating megawatts of power, and pho-

Fig. 1.3. (Top) Part of the solar cell arrays powering an experimental irrigation project near Mead, Nebraska. This Department of Energy project uses 120,000 individual cells to produce 25 peak kilowatts of electrical power to drive an irrigation pumping system. The system was turned on in early 1978. (DOE photo courtesy of Norman Avery, Solar Energy Research Institute.) (Bottom) This Solarex array provides power for internal lights, fog horn, and gas and weather data transmitting equipment in this gas monitoring platform in the Gulf of Mexico. (Photo courtesy of Solarex Corporation.)

tovoltaics would be used by public utilities to help out in periods of peak power demand. The Department of Energy is currently financing many trial systems to stimulate economic growth of the industry, two of which are shown in Fig. 1.3.

1.3 HISTORICAL SURVEY

The development of photovoltaic cells can be traced back 140 years to the discovery by Becquerel (1839) that a photovoltage resulted from the action of light on an electrode in an electrolyte solution. Some 40 years later, Adams and Day (1877) observed a similar effect in the solid material selenium shortly after Smith (1873) had demonstrated the phenomenon of photoconductivity in selenium. Subsequent work on photovoltaic effects in selenium and cuprous oxide led to the development of the selenium photovoltaic cell widely used for many years in photographic exposure meters. By 1914 solar conversion efficiencies of about 1% were achieved with the selenium cell after it was finally realized that an energy barrier was involved both in this cell and in the copper/copper oxide cell.

The modern era for photovoltaics began in 1954. In that year Chapin *et al.* (1954) reported a solar conversion efficiency of 6% for a silicon single-crystal cell. With improved technology, silicon cell efficiency under terrestrial sunlight had reached 14% by 1958. It was also in 1954 that a paper by Reynolds *et al.* (1954) reported 6% solar conversion efficiency in what later came to be understood as the cuprous sulfide/cadmium sulfide heterojunction. The silicon single-crystal photovoltaic cell has become the prototype of all homojunction cells and has been the prime focus of research and development for many years. The cuprous sulfide/cadmium sulfide heterojunction was the first all-thin-film photovoltaic system to receive significant attention; after more than 20 years of intermittent research and development its solar efficiency has been increased to about 10%. By taking advantage of new technology, researchers have rapidly raised the efficiency of the gallium arsenide based cell reported by Jenny *et al.* (1956) with 4% efficiency to present efficiencies approaching 24%.

1.4 SOLAR CELLS AND SOLAR SYSTEMS: A TECHNICAL OVERVIEW

In order to describe the properties and use of photovoltaic solar cells, it is helpful to consider the sequence of phenomena involved: from the radiation received from the sun, through the processes of absorption of this radiation, generation and transport of charge carriers in the semiconductor, separation of charge carriers by the junction, collection of these carriers at the contacts to the device, and finally to the various types of power conditioning and storage that might be used, and the actual utilization of power generated. For eventual terrestrial use, each stage of this sequence depends critically on the others. For example, photovoltaic generation of electricity must be coupled with an effective storage process in order to compensate for the intermittent nature of solar radiation. All of the systems considerations involved in a full exposition of these steps is beyond the scope of this particular book, which focuses primarily on the phenomena involving the photovoltaic junction itself.

A solar cell consists of a potential energy barrier within a semiconductor material that is capable of separating the electrons and holes that are generated by the absorption of light within the semiconductor. The four most common types of barriers are (1) *homojunctions: p/n* junctions within the same semiconductor material; (2) *heteroface structures:* similar to homojunctions but with a window layer of a larger band-gap semiconductor added to reduce surface recombination loss; (3) *heterojunctions: p/n* junctions between two different semiconductor materials; and (4) *Schottky barriers:* metal/semiconductor junctions. As illustrated in the schematic diagrams of Fig. 1.4, each type is a layered structure. In homojunctions the light is incident through a contact grid on a thin layer of one conductivity type (usually *n* type). In heterojunctions, two configurations are possible: light incident through a grid on the larger band-gap material (called the *backwall* configuration) and through a grid on a thin layer of the smaller band-gap material (called the *frontwall* mode). In Schottky barriers it is possible to have the light incident either through the semitransparent metal forming the barrier (frontwall) or through the semiconductor (backwall).

The various electronic processes occurring in a solar cell are illustrated in Fig. 1.5, which is drawn specifically for a *p/n* homojunction. Six domains can be distinguished: (1) the metal contact to the *p*-type semiconductor, introducing a contact resistance R_{cp}; (2) the bulk of the *p*-type region, where most of the electron/hole pairs are generated by absorption of the light and where the minority carriers (electrons) are transported by diffusion and are partially lost by recombination; (3) the junction region itself with the associated depletion regions, where carriers are separated by the junction electric field; (4) the bulk of the *n*-type region, which contributes mainly a series resistance; (5) the contact to the *n*-type semiconductor, introducing a contact resistance R_{cn}, and (6) the front surface, where surface recombination loss of minority carriers can occur.

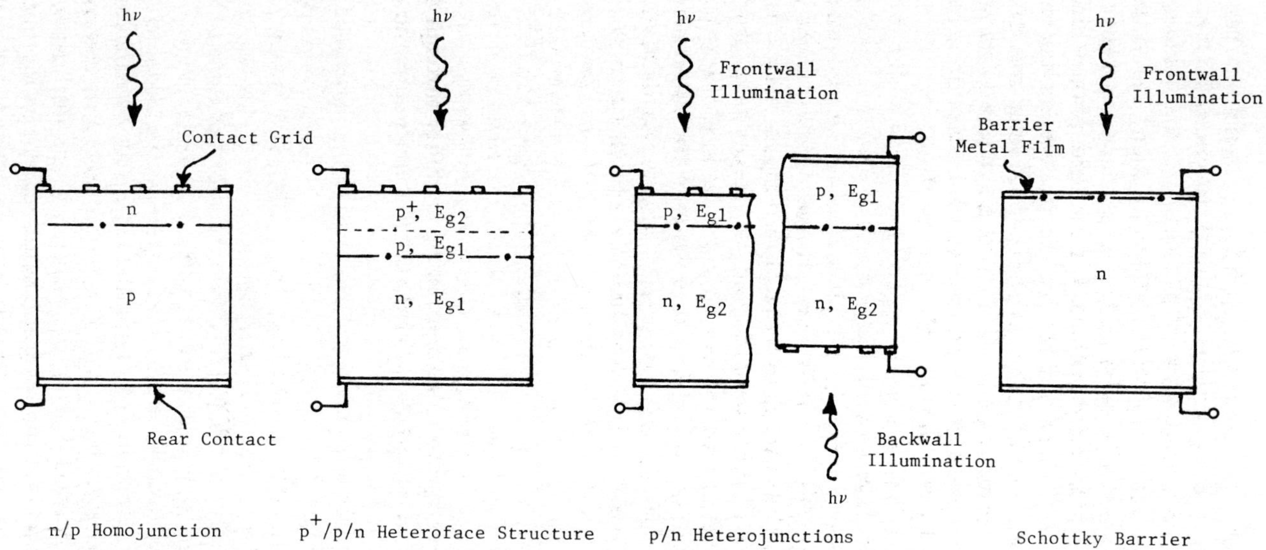

Fig. 1.4. Illustrations of different types of solar cells. In each case the junction barrier is indicated by—·—·—. For the heteroface and heterojunction structures, $E_{g1} < E_{g2}$.

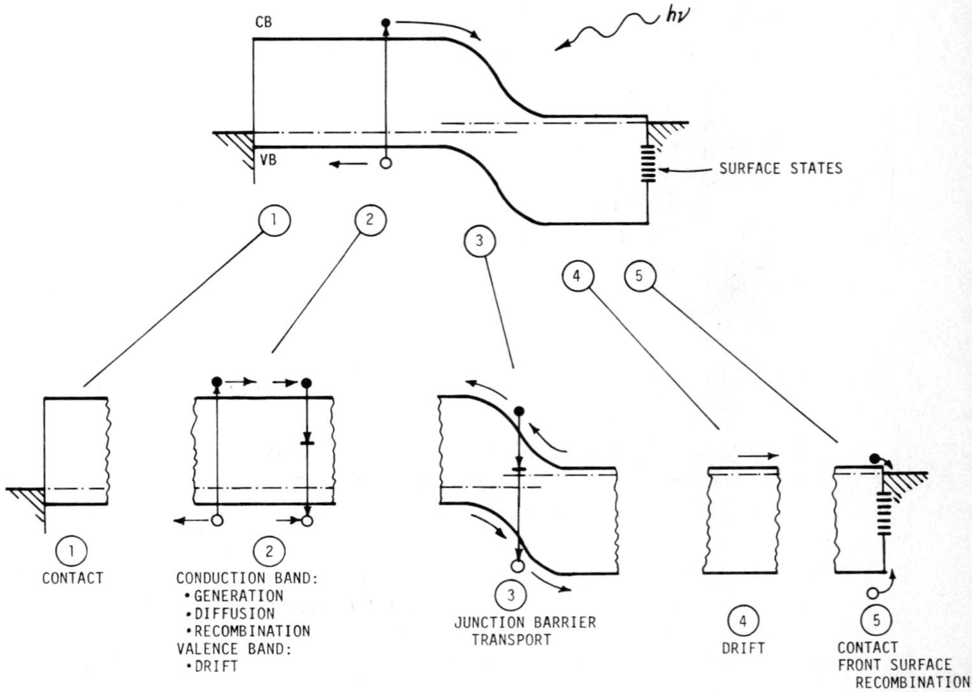

Fig. 1.5. The basic phenomena occurring in a p/n homojunction solar cell, shown at a small forward bias.

The contact resistances are determined by the work functions of the metals used for contacts, the electron affinity and carrier density of the contacted semiconductor, and the possible effects of interface states at the metal/semiconductor junction. The total series resistance of the cell can be expressed as

$$R_{\mathrm{s}} = R_{cp} + R_{\mathrm{b}p} + R_{\mathrm{b}n} + R_{cn}, \tag{1.1}$$

where $R_{\mathrm{b}p}$ and $R_{\mathrm{b}n}$ are the series resistances of the bulk p- and n-type regions in (2) and (4), respectively, in Fig. 1.5.

The generation rate for electron/hole pairs in the bulk region (2) is determined by the illumination flux and the optical absorption constant $\alpha(\lambda)$ of the semiconductor. The diffusion transport process is controlled by the mobility of the photogenerated minority electrons in the p-type region. The recombination loss can be described in terms of a lifetime τ for bulk recombination, as well as by a surface recombination velocity S for recombination of carriers generated near the free surfaces of the device.

The idealized junction current in the dark for the junction region (3) of Fig. 1.5 is given by the diode equation

$$I = I_0\{\exp[q(V - IR_s)/AkT] - 1\}, \tag{1.2}$$

where the reverse saturation current I_0 and the diode factor A are determined by the specific processes occurring at the junction. Localized states near the interface of a heterojunction and within the junction region may act as additional recombination loss pathways. Under illumination, Eq. (1.2) becomes

$$I = I_0\{\exp[q(V - IR_s)/AkT] - 1\} - I_L, \tag{1.3}$$

where I_L is the light-generated current determined by the processes in the bulk p-type region of Fig. 1.5. For an ideal cell, I_0 and A are not functions of illumination, and any parallel shunting resistance across the junction is effectively infinite. As the band gap E_g of the absorbing region is increased, the diode parameter I_0 generally decreases strongly, which in turn increases the voltage obtainable from the cell. Increasing E_g, however, also decreases the photogenerated current available since less and less of the solar spectrum is absorbed. The compromise between these two effects produces a maximum in the solar efficiency at band-gap energies of 1.4 to 1.5 eV, as illustrated in Fig. 1.6.

Figure 1.7 shows plots of Eqs. (1.2) and (1.3) for an ideal junction in order to introduce some additional quantities commonly used in describing solar cells. These include the open-circuit voltage V_{oc}, the short-circuit current I_{sc}, the voltage for maximum power V_m, and the current for maximum power I_m. The solar efficiency of the cell is defined as the ratio of the maximum power P_m to the solar power incident on the cell. A quantity called the fill factor ff is commonly introduced to relate P_m to the product $V_{oc}I_{sc}$,

$$\mathrm{ff} = I_m V_m / I_{sc} V_{oc} = P_m / I_{sc} V_{oc}. \tag{1.4}$$

Thus the solar efficiency of a cell can be written as

$$\eta_s = \mathrm{ff} I_{sc} V_{oc} / P_s, = P_m / P_s \tag{1.5}$$

where P_s is the solar power. Typical values of these quantities for a single cell are $V_{oc} = 0.5$–0.8 V, $I_{sc} = 10$–40 mA cm^{-2} of cell area, ff $= 0.6$–0.8, and $\eta_s = 6$–20%.

A schematic diagram of a photovoltaic power generation system is shown in Fig. 1.8. The output from an array of cells is controlled by a power conditioning unit to maintain the maximum power output from the cell array. The electrical energy produced usually must be stored in some way, e.g., by charging up a chemical storage battery, producing hydrogen

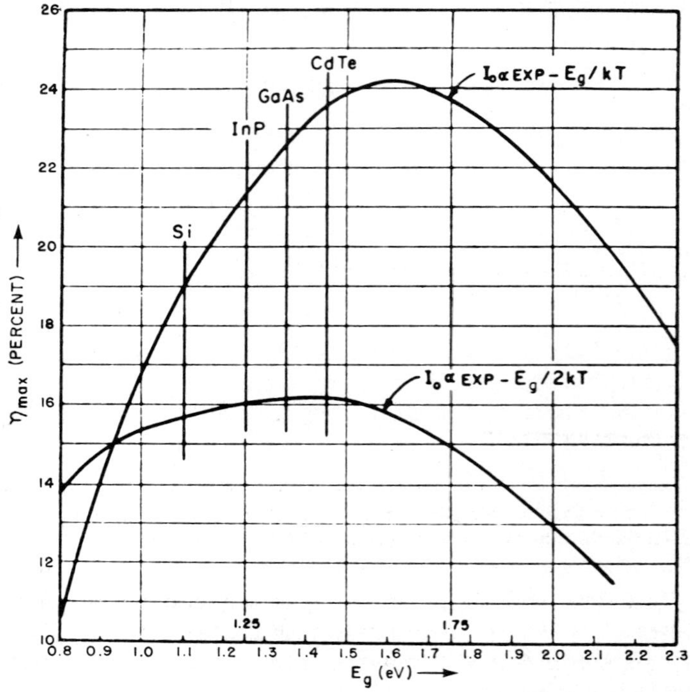

Fig. 1.6. Theoretical solar efficiency versus semiconductor band gap for ideal homojunction cells with no surface recombination loss. Curves for two different diode mechanisms are shown: (top) $A = 1$ for injection dominated current and (bottom) $A = 2$ for recombination in the depletion layer. [After J. J. Loferski, *J. Appl. Phys.* **27,** 777 (1956).]

by electrolysis of water, or pumping water to an elevated height. Finally, since most uses require alternating current, an inverter may be needed to convert the direct current output of the storage system into usable power. The efficiency of the total solar electric system is the product of the efficiency of the individual parts; since the storage element is 50–80% efficient and the inverter is about 95% efficient, an array of photovoltaic cells with an efficiency of 10% may result in a solar electric system with an overall solar electric power system efficiency of about 5–8%. An overall system efficiency of 10% would require solar cells with an efficiency in excess of 13%.

Photovoltaic solar cells are also used as elements in more complex systems designed for increased total conversion efficiency and/or lower system cost. Promising examples of these systems are

(1) tandem cell, spectrum splitting concentrator systems that use two

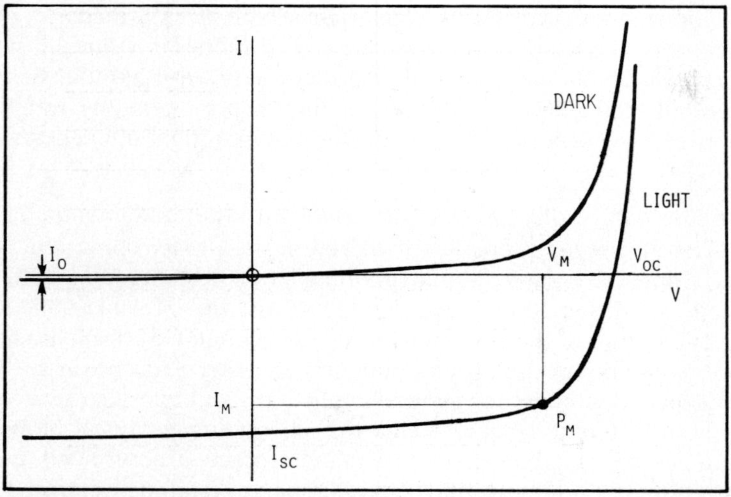

Fig. 1.7. Typical variation of the current I as a function of the voltage V for a solar cell in the dark and under illumination. The magnitude of I_0 is exaggerated for clarity.

or more solar cells with different band-gap semiconductors to intercept different parts of the solar spectrum to provide theoretical efficiencies in the 30–45% range [at this writing a laboratory system has provided a solar efficiency above 28% (Moon *et al.*, 1978)],

(2) photovoltaic/thermal hybrid systems in which solar cells are operated at a higher temperature so that thermal energy can also be extracted to operate a heat engine,

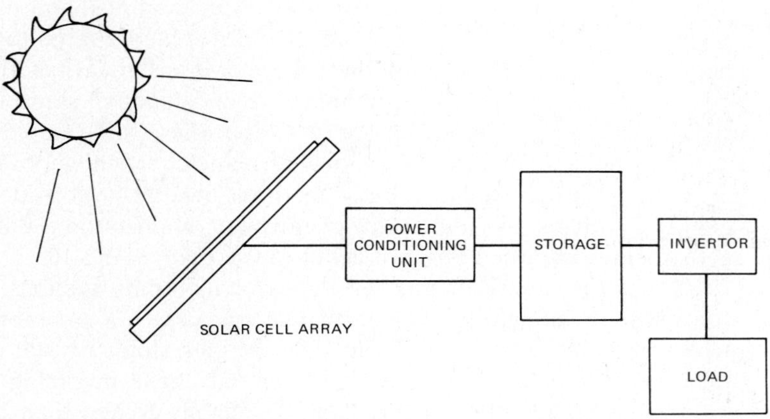

Fig. 1.8. The components of a photovoltaic solar electric system.

(3) thermophotovoltaic convertors in which an intermediate, refractory absorber is placed in the solar flux. The radiation from the hot absorber (1800–2500°C) is converted to electricity with a small band-gap photovoltaic cell. This conversion of high-energy photons into larger numbers of lower-energy photons promises theoretical efficiencies up to 45%.

The first half of this book is concerned with the fundamental description of the electronic processes involved in the photovoltaic junction in general. Some information about solar insolation is included as well, since solar cells are designed to fit specific inputs and the information is essential for evaluation of their performance. The second half of the book considers in some detail the specific junction systems that appear to be the most serious contenders for general application and that also serve as archetypes of different types of cells: the silicon single-crystal homojunction, the gallium arsenide single-crystal heteroface structure cell, and the cuprous sulfide/cadmium sulfide thin-film polycrystalline heterojunction. A survey is also included of other systems that appear promising from a scientific or practical perspective.

1.5 MATERIALS AND MATERIALS PROBLEMS

Unless total recourse is taken to the type of systems involving solar concentration that use small, relatively expensive single-crystal cells of high efficiency, it appears that thin-film technology is essential to the development of terrestrial solar cells. Thin films are necessary not only for the photoelectrically active layers but also for contacts, window materials, antireflection coatings, transparent conducting coatings, and passivating layers. Methods of film deposition, therefore, play a large role in developing useful solar cells. In addition to more or less standard methods such as vacuum evaporation (VE), nonreactive and reactive sputtering, electron beam evaporation, molecular beam evaporation (MBE), epitaxial layer growth from liquid or vapor (LPE or VPE), and chemical vapor deposition (CVD), other less well-developed methods need to be considered, such as spray pyrolysis (SP) and electrochemical crystallization. Some of these methods are described in more detail in Chapters 9 and 10.

To illustrate the variety of possible types of materials systems that might contribute meaningfully to photovoltaic solar cells, a summary of experimental systems is given in Table 1.II. Besides single-crystal cells involving silicon or gallium arsenide, to date only cells involving five other semiconductor absorbing layers give efficiencies greater than 10%, even in single-crystal form: InP [with either CdS, 15% (Shay *et al.*, 1977) or indium oxide, 14.4% (Sree Harsha *et al.*, 1977) as the large band-gap

Table 1.II
Illustrative Photovoltaic Systems and Parameters

Description of cell	Reference	V_{oc}, V	Solar efficiency, % (AMl)
Single-crystal Si homojunction	*a, b*	0.63	16.8
Polycrytalline Si	*c*	0.57	9.5
Epitaxial Si layer on edge-fed-grown Si	*d*	0.56	10
Diffused junction in edge-fed-grown Si	*d*	0.52	8
Cr/Sr metal–insulator–semiconductor (MIS) junction	*e*	0.50	8.1
Al/Si MIS junction	*f*	0.45	8
Au/Si MIS junction	*g*	0.55	9
Indium–tin oxide/Si junction	*h*	0.51	12
Pt Schottky barrier on amorphous Si	*i*	0.80	5.5
Single-crystal GaAs *p/n* homojunction	*j*	0.97	20
Single-crystal AlGaAs/GaAs junction	*k*	0.9	22
Au/GaAs MIS junction	*l*	0.73–0.83	15–17
Au/GaAs MIS junction on polycrystalline GaAs	*d*	0.62	3.5
Evaporated CdS, dipped Cu_2S to form Cu_2S/CdS junction	*m*	0.52	9.1
Evaporated ZnCdS, dipped Cu_2S to form Cu_2S/ZnCdS junction	*n*	0.61–0.67	7.6
Spray pyrolysis Cu_2S/CdS junction	*d*	0.40	5.2
Evaporated ZnCdS, evaporated CuCl to form Cu_2S/ZnCdS junction	*d*	0.68	4.0
CVD CdS on InP to form CdS/InP junction	*o*	0.79	15
Evaporated CdS on CVD InP to form CdS/InP junction	*p*	0.72	11.9
CVD CdS on polycrystalline InP to form CdS/InP junction	*q*	0.46	5.7
Ion-beam-deposited indium–tin oxide on InP to form ITO/InP junction	*r*	0.76	14.4
rf sputtered ITO on CVD InP to form ITO/InP junction	*s*	0.69	12.4
Evaporated CdS on CdTe to form CdS/CdTe junction	*t*	0.63	8

(*Continued*)

Table 1.II *(Continued)*

Description of cell	Reference	V_{oc}, V	Solar efficiency, % (AMl)
CVD CdS on CdTe to form CdS/CdTe junction	*u*	0.67	12
Spray pyrolysis CdS on CdTe to form CdS/CdTe junction	*v*	0.53–0.72	5.5–6.5
Spray pyrolysis ZnCdS on CdTe to form ZnCdS/CdTe junction	*w*	0.82	6–8
Totally screen-printed CdS/CdTe junction	*x*	0.69	8.1
rf sputtered indium–tin oxide on CdTe to form ITO/CdTe junction	*y*	0.82	8
Evaporated CdS on $CuInSe_2$ to form CdS/$CuInSe_2$ junction	*z*	0.49	12
All thin-film evaporated CdS/$CuInSe_2$ junction	*aa*	0.49	6.6
All thin-film evaporated CdS/$CuInS_2$ junction	*aa*	0.51	3.25
All thin-film evaporated $CuInSe_2$ homojunction	*aa*	0.31	3.0
All thin-film evaporated $CuInS_2$ homojunction	*aa*	0.41	3.6

[a] F. A. Lindholm, A. Neugroschel, S. C. Pao, J. G. Fossum, and C.-T. Sah, *Proc. 13th IEEE Photovoltaic Specialists Conf.*, p. 1302 (1978).

[b] J. G. Fossum, R. D. Nasby, and E. L. Burgess, *Proc. 13th IEEE Photovoltaic Specialists Conf.*, p. 1294 (1978).

[c] T. L. Chu, S. S. Chu, E. D. Stokes, C. L. Lin, and R. Abderrassoul, *Proc. 13th IEEE Photovoltaic Specialists Conf.*, p. 1106 (1978).

[d] As reported at the *ERDA Semiannu. Sol. Photovoltaic Program Rev. Meet., Orono, Maine, 1976.*

[e] W. A. Anderson, J. K. Kim, and A. E. Delahoy, *IEEE Trans. Electron Devices* **ED-24,** 453 (1977).

[f] E. J. Charlson and J. C. Lien, *J. Appl. Phys.* **46,** 3982 (1975).

[g] J. P. Pompon and P. Siffert, *J. Appl. Phys.* **47,** 3248 (1976).

[h] J. B. DuBow, D. E. Burk, and J. R. Sites, *Appl. Phys. Lett.* **29,** 494 (1976).

[i] D. E. Carlson, *IEEE Trans. Electron Devices* **ED-24,** 449 (1977).

[j] J. C.-C. Fan and C. O. Bozler, *Proc. 13th IEEE Photovoltaic Specialists Conf.*, p. 953 (1978).

[k] J. M. Woodall and H. J. Hovel, *Appl. Phys. Lett.* **30,** 492 (1977).

[l] R. J. Stirn, Y. C. M. Yeh, E. Y. Wang, F. P. Ernest, and C. J. Wu, *Tech Dig., 1977 Int. Electron Devices Meet.; IEEE Trans. Electron Devices* **ED-24,** 476 (1977).

[m] A. M. Barnett, J. A. Bragagnolo, R. B. Hall, J. E. Phillips, and J. D. Meakin, *Proc. 13th IEEE Photovoltaic Specialists Conf.*, p. 419 (1978).

Table 1.II (*Continued*)

[n] As reported in *Annu. Photovoltaic Res. Dev. Rev. Meet. Dep. Energy, 1978.*

[o] J. L. Shay, S. Wagner, M. Bettini, K. J. Bachman, and E. Buehler, *IEEE Trans. Electron Devices* **ED-24,** 483 (1977).

[p] H. M. Manasevit, K. L. Hess, P. D. Dapkus, R. P. Ruth, J. J. Yang, A. G. Campbell, R. E. Johnson, L. A. Moudy, R. H. Bube, L. B. Fabick, A. L. Fahrenbruch, and M.-J. Tsai, *Proc. 13th IEEE Photovoltaic Spec. Conf.*, p. 165 (1978).

[q] J. L. Shay, S. Wagner, M. Bettini, K. J. Bachmann, and E. Buehler, *IEEE Trans. Electron Devices* **ED-24,** 483 (1977).

[r] K. S. Sree Harsha, K. J. Bachmann, P. H. Schmidt, E. G. Spencer, and F. A. Thiel, *Appl. Phys. Lett.* **30,** 645 (1977).

[s] See reference *p* above.

[t] K. W. Mitchell, A. L. Fahrenbruch, and R. H. Bube, *J. Appl. Phys.* **48,** 4365 (1977).

[u] K. Yamaguchi, N. Nakayama, H. Matsumoto, and S. Ikegami, *Jpn. J. Appl. Phys.* **15,** 1575 (1976); **16,** 1203 (1977).

[v] Y. Y. Ma, A. L. Fahrenbruch, and R. H. Bube, *Appl. Phys. Lett.* **30,** 423 (1977).

[w] S.-Y. Yin, A. L. Fahrenbruch, and R. H. Bube, *J. Appl. Phys.* **49,** 1294 (1978).

[x] N. Nakayama, H. Matsumoto, K. Yamaguchi, S. Ikegami, and Y. Hioki, *Jpn. J. Appl. Phys.* **15,** 2281 (1976).

[y] R. H. Bube, *Proc. Soc. Photo-Opt. Instrum. Eng.* **114,** 7 (1977).

[z] S. Wagner, J. L. Shay, P. Migliorato, and H. M. Kasper, *Appl. Phys. Let.* **25,** 434 (1974).

[aa] L. L. Kazmerski, "Second Quarter Report to NSF-RANN and ERDA," AER-75-19576-A01, April 1977: NSF/AER-75-19576-2.

conducting layer], $CuInSe_2$ [with a CdS layer, 12% (Wagner *et al.*, 1974)], CdTe [with a CdS layer, 12% (Yamaguchi *et al.*, 1976)], Cu_xS [with a $Zn_xCd_{1-x}S$ layer (Hall *et al.*, 1981)], and amorphous Si, 10% (Catalano *et al.*, 1982).

Table 1.III summarizes some of the major criteria that must be considered in choosing the materials systems judged suitable for further development as solar cells using some form of heterojunction. It is commonly argued with some justification that an elemental system (e.g., Si homojunctions) is certain to be simpler to prepare and control and to be more stable under operation than a multielemental system. In the semiconductor world, for example, Si has held off challenges from GaAs at least partially for reasons based on such considerations. One might argue, therefore, that any attempt to go to a multielemental complex heterojunction structure is destined to be impractical, no matter how felicitous its behavior might seem to be in the scientific laboratory. Heterojunctions such as CdS/InP and CdS/CdTe should, however, not be considered to be three- or four-element systems as though all parts were equally electrically active. In each of these cases the major useful optical absorption is limited to just one of the compounds involved (i.e., InP or CdTe), with the other compounds playing a rather passive role, in many cases approaching the equivalent in device complexity to transparent conducting coatings and antireflection coatings. Even the simplest elemental solar

Table 1.III

Criteria for Choice of a Heterojunction System for Photovoltaic Solar Cells[a]

Property	Criteria
Band gap of smaller band-gap material	Band gap near 1.4 eV to maximize absorption of solar radiation, while minimizing diode current that limits V_{oc} Direct optical absorption so that carriers are generated close to the junction Long minority-carrier diffusion length
Band gap of larger band-gap material	As large as possible while maintaining low series resistance
Conductivity type	Smaller band-gap material should usually be p-type because of longer electron diffusion lengths
Electron affinities	Materials should be chosen such that no potential spike occurs at the junction for the minority photoexcited carriers
Diffusion voltage	As large as possible, since the maximum V_{oc} is proportional to the diffusion voltage
Lattice mismatch	As little mismatch in lattice constant between the two materials as possible (this appears to minimize interface state density and recombination losses through such states)
Deposition methods	Suitable deposition methods for thin-film formation and control should be available
Electrical contacts	It should be possible to form low-resistance electrical contacts to both n- and p-type materials
Material availability	Supplies of the material should be sufficient to allow large-area cell production
Material cost	Cost of the material should be competitive with alternative systems
Material toxicity	Materials should be nontoxic, or control of toxicity should be possible
Cell stability and lifetime	Cell must have an operating lifetime sufficient to pay back economic and energy costs required to produce it

[a] Some of the basic semiconductor terms used in this summary are defined in more detail in Chapter 3.

cell must include the latter elements. Whether investigation of ternary compounds as the photoelectrically active element, such as in the CdS/$CuInSe_2$ system, or even more complex systems designed to achieve optimal band-gap and lattice match through the use of multicomponent solid solutions, can be defended on the same grounds remains to be seen. It is hoped that the information presented in the remaining chapters of this book will both illuminate some of these suggestions and provide the reader with the background necessary to make these decisions.

1.6 A GENERAL PERSPECTIVE ON ENERGY AND THE FUTURE

In comparing various solar conversion systems among themselves and with conventional systems there is a tendency to focus on peak efficiency values and cost per peak watt. Although this is a natural approach for a competitive and analytical society, other factors may well be more crucial. The cost and efficiency comparisons invoked today are almost wholly based on straight-line extrapolation of conditions in the past. Such comparisons are based largely on a philosophy of continued growth (because this is the standard model in the Western world for a healthy society) and on an implicit assumption that natural resources are unlimited.

Although we have known for some time that natural resources are not unlimited, we have a difficult time putting this knowledge into practice. We know that the curve of rate of use versus time for gas, oil, coal, and even ^{235}U is not an ever-increasing one, but instead is a bell-shaped curve that eventually returns to zero as the fuel becomes more and more expensive (Fig. 1.1). The question is, How long do we have before the situation becomes critical? Hubbert contends that we passed the peak for natural gas in 1967. To speak in this way would have been almost unthinkable 10 to 20 years ago. We can expect that extrapolation 20 years into the future is going to stretch our imagination even further. Our thinking and projections must include the possibility of a different way of living, more in harmony with the limited nature of our resources. An extrapolation based on the number of electrical appliances major industries would like to see in each home in 1990 has become quite dangerous.

A typical modern house today uses occasional electric heat and cooling and many rooms use artificial light during the day. Electricity, some 20–40 kW h per day, arrives through the power lines at the $\simeq 99.9\%$ reliability that we have come to expect. Such a house is hardly suitable (although not impossibly so) for the application of solar electric power. In contrast, consider a future home in which all rooms use natural light during the day. The house is well insulated, and any air conditioning uses solar heat cycles directly. Water heating and clothes drying also use solar heat directly. Electrical reliability is cut to 97.5%. Such a home uses 4–8 kW h day^{-1}, and retains the essence of our present standard of living. Such an analysis predicts that solar electricity can make an early and sizable contribution without sacrificing our standard of living substantially.

Another example of these other crucial factors that must influence our overall planning for the future follows from Rose's (1976) perspective on global energy usage [see also the analysis by Wolf (1973)]. Rose uses solar units, the solar power density falling on the earth averaged throughout the day and year, to make his point. From purely thermodynamic considerations of the earth's energy balance with outer space, he shows that if we consume (and thus dissipate) an additional solar unit in the form of fossil or nuclear fuels, the temperature of the earth must increase to that of boiling water. The use of only 0.01 solar units of fuels outside the solar energy balance will raise the temperature by an average of 1°C, producing large-scale climatic conditions and considerable melting of the polar ice caps. The U.S. level of energy consumption is already 2×10^{-3} solar units, and urban levels are considerably above that—the heat being carried off by convection to the outlying regions. The average world level is currently about 10^{-4} solar units. There is clearly an absolute limit to energy use; at present growth rates, we shall reach that limit within 50 years. Only solar energy sources (including wind, water, and biomass) are exempt from this limiting calculation since the net energy balance is unchanged by their use.

Clifford (1976) has argued that the problems facing photovoltaics are not so much technological as they are economic and political. He points out, for example, that several of the Department of Energy goals for polycrystalline silicon cells have been met as much as 8 years ahead of schedule. The photovoltaics industry is proceeding on a highly promising 75% learning curve (price per unit versus accumulated volume). The question is whether or not investment capital and public acceptance are available to maintain the industry volume growth required for continued price reductions. The United States government, through the Department of Energy and other agencies, initially stimulated the market; in 1977, for example, 400 kW of terrestrial cells were purchased. Clifford goes on to say that many of the cost factors affecting the cost of electrical power and in turn the potential position of photovoltaics in that scheme are also largely political: substantial subsidies for construction of both fossil- and nuclear-powered plants, subsidies for many phases of existing nuclear generation (e.g., fuel and storage of radioactive wastes), prices of oil being determined by OPEC nations, large congressional lobbies for energy interest groups, and finally, the manner in which the public itself uses energy.

The ancient author of Ecclesiastes wrote, "What has been is what will be, and what has been done is what will be done; and there is nothing new under the sun." The future appears to demand that there be something new under the sun—and it is precisely the sun itself that can help to provide it.

GENERAL REFERENCES

Solar Photovoltaics

Backus, C. E., ed. (1976). "Solar Cells," IEEE Selected Reprint Series. Wiley (Interscience), New York. A collection of relevant papers from 1954 through 1973.

Brandhorst, H. W., and Godlewski, M. P., eds. (1977). *IEEE Trans. Electron Devices* **ED-24** No. 4. Special issue devoted to photovoltaics.

Ehrenreich, H., chm. (1979). "Principal Conclusions of the American Physical Society Study Group on Solar Photovoltaic Energy Conversion." Am. Phy. Soc., New York.

Fonash, S. J. (1981). "Solar Cell Device Physics." Academic Press, New York.

Hovel, H. J. "Semiconductors and Semimetals, Vol. 11, Solar Cells." Academic Press, New York. A book with considerable depth, written for the scientist already familiar with photovoltaics.

Lindholm, F. A., and Rappaport, P., eds. (1980). *IEEE Trans. Electron Devices* **ED-27** No. 4. Special issue devoted to photovoltaics.

Merrigan, J. A. (1975). "Sunlight to Electricity." MIT Press, Cambridge, Massachusetts. Perspective on energy usage and economic impact of solar cells.

Neville, R. C. (1978). "Solar Energy Conversion: The Solar Cell." Elsevier, Amsterdam. Electrical engineering point of view, sound basic text.

Palz, W. (1978). "Solar Electricity." Butterworth, London. Good sections on photovoltaics, direct use of solar thermal energy, and thermodynamic conversion. Interesting because of European perspective.

Pulfrey, D. L. (1978). "Photovoltaic Power Generation." Van Nostrand-Reinhold, New York. Covers photovoltaics at an intermediate level, with large sections on solar electric systems and economics.

General Energy

Angrist, S. W. (1971). "Direct Energy Conversion," 2nd ed. Allyn & Bacon, Boston, Massachusetts. General, intermediate level text.

Daniels, F. (1964). "Direct Uses of the Sun's Energy." Yale Univ. Press, New Haven, Connecticut. A general text written for the layman.

Stein, C., ed. (1976). "Critical Materials Problems in Energy Production." Academic Press, New York. A series of distinguished lectures on a wide variety of energy related topics given at the Joint Center for Materials Science of New Mexico.

Semiconductor Physics

Bube, R. H. (1974). "Electronic Properties of Crystalline Solids." Academic Press, New York. Comprehensive, basic text on electronic aspects of solid state physics.

Bube, R. H. (1978). "Photoconductivity of Solids." Krieger, Huntington, New York. Comprehensive basic text on photoconductivity. New printing of the Wiley 1967 edition.

Maissel, L., and Glang, R. (1970). "Handbook of Thin Film Technology." McGraw-Hill, New York. Very comprehensive book on thin-film deposition techniques and properties.

Milnes, A., and Feucht, D. (1972). "Heterojunctions and Metal-Semiconductor Junctions." Academic Press, New York. Includes an overview of solar cell theory and details of efficiency and series resistance calculations.

Mort, J., and Pai, D., eds. (1976). "Photoconductivity and Related Phenomena." Elsevier North-Holland, Amsterdam. A collection of papers.

Sharma, B. L., and Purohit, R. K. (1974). "Semiconductor Heterojunctions." Pergamon, Oxford.

Smith, R. A. (1968). "Semiconductors." Cambridge Univ. Press, London and New York. Excellent text on basic semiconductor properties.

Sze, S. (1969). "Physics of Semiconductor Devices." Wiley, New York. Includes good sections on diode theory and solar cells plus a concise, clear overview of semiconductor basics.

Solar Politics

Behrman, D. (1976). "Solar Energy—The Awakening Science." Little, Brown, Boston, Massachusetts. A good perspective on who's doing what in solar science.

Clifford, A. (1976). Some economic and political aspects of photovoltaic development. *Proc. 12th IEEE Photovoltaic Specialists Conf.*, p. 826. Mr. Clifford is an employee of the Solarex Corp.

Cowan, E. (1976). Who should develop solar energy. *N.Y. Times* Nov. 12.

Commoner, B. (1976). A reporter at large (Energy I, II, and III). *The New Yorker* Feb. 2, 9, and 16. One man's view of the energy problem.

Lovins, A. (1976). Energy strategy: The road not taken. *Foreign Affairs*, October.

REFERENCES

Adams, W. G., and Day, R. E. (1877). The action of light on selenium. *Proc. R. Soc. London, Ser. A* **25,** 113.

Becquerel, E. (1839). On electric effects under the influence of solar radiation. *C. R. Hebd. Seances Acad. Sci.* **9,** 561.

Catalano, A., D'Aiello, R., Dresner, J., Faughnan, B., Firester, A., Kane, J., Smith, Z. E., Schade, H., Swartz, G., and Triano, A. (1982). *Proc. 16th IEEE Photovoltaic Specialists Conf.*, to be published.

Chapin, D. M., Fuller, C. S., and Pearson, G. L. (1954). A new silicon *p-n* junction photocell for converting solar radiation into electrical power. *J. Appl. Phys.* **25,** 676.

Clifford, A. (1976). *Proc. 12th IEEE Photovoltaic Specialists Conf.*, p. 826.

Grant, L. P. (1976). *Proc. 12th IEEE Photovoltaic Specialists Conf.*, p. 347.

Hammond, A. L., Metz, W. D., and Maugh, T. H., II (1973). "Energy and the Future" Am. Assoc. Adv. Sci., Washington, D.C.

Hall, R. B., Birkmire, R. W., Phillips, J. E., and Meakin, J. D. (1981). *Appl. Phys. Lett.* **38,** 925.

Jenny, D. A., Loferski, J. J., and Rappaport, P. (1956). *Phys. Rev.* **101,** 1208.

Lindmayer, J. (1977). *CEC Photovoltaic Sol. Energy Conf., Luxembourg* p. 825.

Moon, R. L., James, L. W., Vander Plas, H. A., Yep, T. O., Antypus, G. A., and Chai, Y. (1978). *Proc. 13th IEEE Photovoltaic Spec. Conf.,* p. 859.

Reynolds, D. C., Leies, G., Antes, L. L., and Marburger, R. E. (1954). Photovoltaic effect in cadmium sulfide. *Phys. Rev.* **96,** 533.

Rose, A. (1976). A global view of solar energy in rational units. Unpublished.

Shay, J. L., Wagner, S., Bettini, M., Bachmann, K. J., and Buehler, E. (1977). *IEEE Trans. Electron Devices* **ED-24,** 483.

Smith, W. (1873). Effect of light on selenium during the passage of an electric current. *Nature (London)* **7,** 303.

Sree Harsha, K. S., Bachmann, K. J., Schmidt, P. H., Spencer, E. G., and Thiel, F. A. (1977). *Appl. Phys. Lett.* **30,** 645.

Wagner, S., Shay, J. L., Migliorato, P., and Kasper, H. M. (1974). *Appl. Phys. Lett.* **25,** 434.

Wolf, M. (1972). Historical development of solar cells. *Proc. 25th Annu. Power Sources Symp., Atlantic City, N.J.* PSC Public Comm., Red Bank, New Jersey.

Wolf, M. (1973). Solar energy, an endless supply. *Consult. Eng. E* **40**(3), 171.

Yamaguchi, K., Matsumoto, H., Nakayama, N., and Ikegami, S. (1976). *Jpn. J. Appl. Phys.* **15,** 1575.

CHAPTER 2

SOLAR INSOLATION

The solar efficiency of a photovoltaic system depends critically on the spectral distribution of the radiation coming to us from the sun. The study of solar insolation is a complex and involved one that is a field all by itself. Here we present a brief survey of the information currently available about solar insolation and of the way in which this information can be interpreted, before beginning to consider the details of the photovoltaic systems themselves.

2.1 SOLAR SPECTRUM

The sun is a complex radiator whose spectrum can be aproximated by a 6050°K blackbody. This blackbody spectrum is modified by the variation in temperature across the sun's disk, the effect of the solar atmosphere, and the Fraunhofer absorption lines. Figure 2.1 shows the solar spectrum outside the atmosphere, and Fig. 2.2 compares a smoothed solar spectrum to various blackbody curves.

In outer space, 98% of the total energy radiated by the sun lies between 0.25 and 3.0 μm. A quantity called the *solar constant* is defined as the rate at which energy is received on a unit surface, perpendicular to the sun's direction, in free space at the earth's mean distance from the sun; the most recently accepted value is 1.353 kW m^{-2}. The actual solar radiation in free space at the earth's mean distance from the sun differs from this value by $\pm 3.35\%$ because of changes in the earth–sun distance

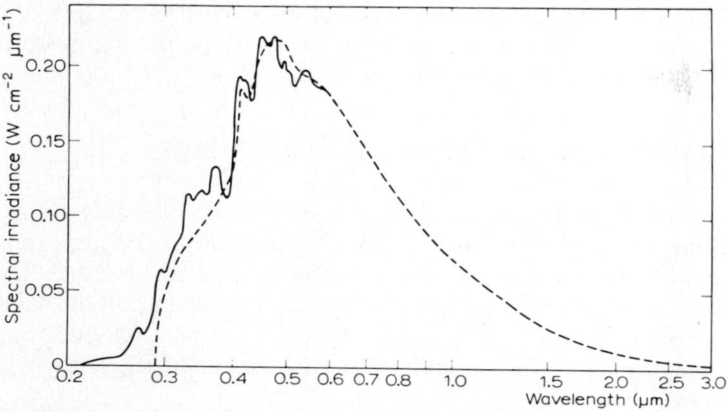

Fig. 2.1. Spectral distribution of the solar radiation outside of the earth's atmosphere. ———, Smithsonian (P. Moon); —, Naval Research Laboratory. (After N. Robinson, "Solar Radiation," p. 3. Elsevier, Amsterdam, 1966.)

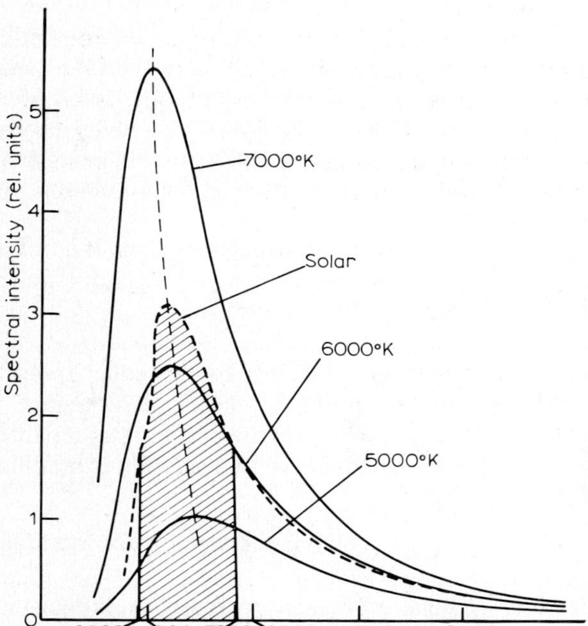

Fig. 2.2. Spectral distribution of the sun's radiation compared with blackbody distributions corresponding to different temperatures. (After N. Robinson, "Solar Radiation." Elsevier, Amsterdam, 1966.)

throughout the year, producing a radiation intensity of 1.40 kW m^{-2} on January 3 and 1.31 kW m^{-2} on July 6. This is the only appreciable variation outside the earth's atmosphere.

2.2 EFFECTS OF THE EARTH'S ATMOSPHERE

Both the intensity and the spectral distribution of the radiation arriving at the earth's surface depend on the composition of the atmosphere, as well as the path length of the radiation through the atmosphere. The most important parameters of the atmosphere are the water content w; the turbidity coefficient B, expressing the effect of haze and related scattering†; the ozone content; the cloudiness of the sky; and the effect of ground reflection.

Since the insolation reaching the periphery of the earth's atmosphere, normal to the sun's direction, depends only on the earth–sun distance, other geometrical effects, such as the sun's declination angle, are manifested primarily by varying the path length through the atmosphere. If we consider only direct radiation on a receiver pointing directly toward the sun, these geometric effects can be almost completely described by specifying a zenith angle of the sun z, the angle between the earth–sun radius and the normal of a plane containing the horizon circle, as illustrated in Fig. 2.3. This zenith angle z is a function of the time of day, the season, the latitude, and the longitude. At real solar noon, z is given by the latitude minus 23.5° cos(360N/365), where N is the number of days since the summer solstice, in degrees.

The path length through the atmosphere is conveniently described in terms of an equivalent relative *air mass* m_r. The path length for a zenith angle z is just sec z times the path length for $z = 0$, and this relative air mass m_r is defined as $m_r = \sec z$. Specific solar spectra are labeled AMm_r (read: air mass m_r). AM0 corresponds to the solar spectrum in outer space; the average solar spectrum at the earth's surface corresponds approximately to AM2. AM0 and AM2 spectra are compared in Fig. 2.4.

On its path through the atmosphere, the sunlight is modified by the following processes:

(1) Rayleigh scattering, which is responsible for the blue color of the sky.

(2) Electronic absorption bands, primarily in oxygen, nitrogen, and ozone. Nearly all the radiation with $\lambda < 0.29$ μm is absorbed by ozone.

† Sometimes scattering coefficients α and β are used rather than B.

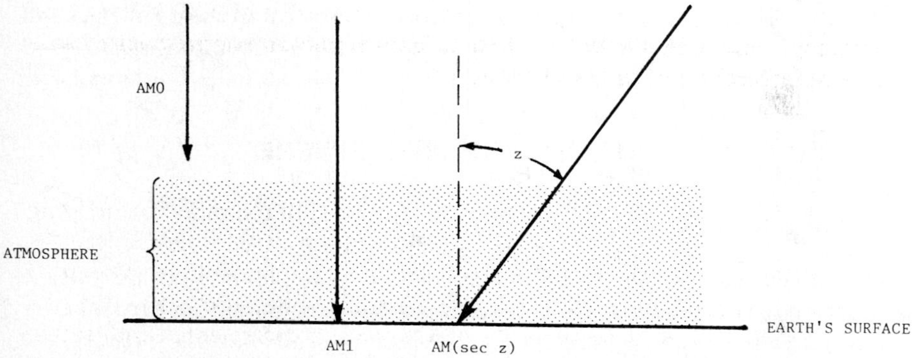

Fig. 2.3. Illustration of the solar radiation spectrum labeled AM0 in space, AM1 at the earth's surface for normal incidence, and AMm at the earth's surface with $m = \sec z$, where z is the deviation from normal incidence. The diagram is drawn in the plane including the zenith and the earth–sun radius.

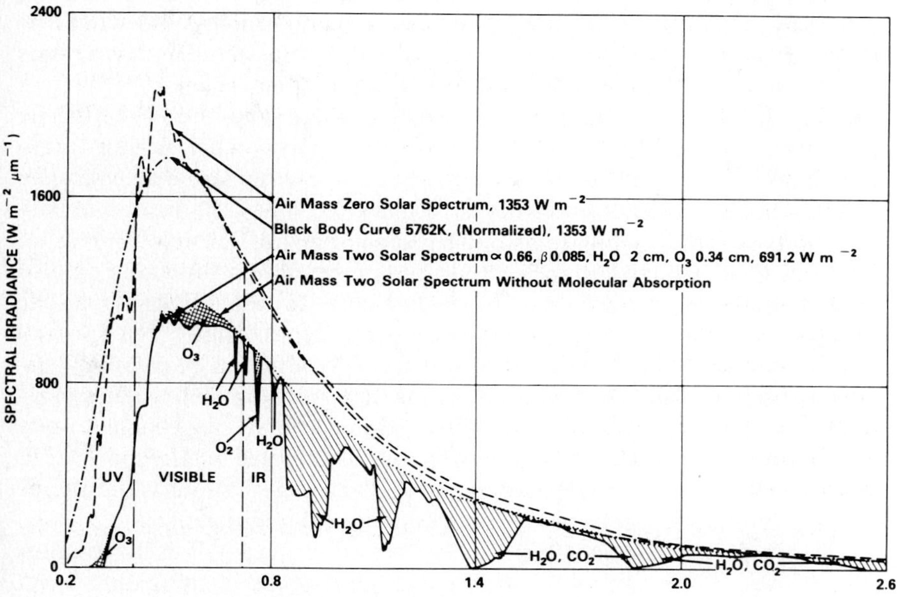

Fig. 2.4. Comparison of AM0 and AM2 solar spectra, showing the various atmospheric absorption bands in AM2. The α and β for the AM2 spectrum are turbidity coefficients. (After M. P. Thekaekara, Data on incident solar energy. *In* "The Energy Crisis and Energy From the Sun." Inst. Environ. Sci., Mt. Prospect, Illinois, 1974.)

(3) Molecular rotational and vibrational absorption bands in H_2O and CO_2 (see Fig. 2.4). Nearly all the radiation with $\lambda > 3.0\ \mu m$ is absorbed by H_2O and CO_2 (with the exception of the so-called atmospheric windows at much longer wavelengths).

(4) Scattering by aerosols and particulate matter, which is greater for the shorter wavelengths.

(5) Refraction and turbulence due to variations in the index of refraction with temperature and pressure—a small effect in this context.

Finally, the incident radiation is modified by the addition of the diffuse component of radiation from the sky, strongly dependent on aerosol concentration, cloudiness, and the local reflection of the earth's surface. The diffuse radiation is strongly peaked in the blue portion of the spectrum and contributes about 8–10% of the total radiation on a fairly clear day at noon and considerably more at greater zenith angles. The sum of the total direct and diffuse radiation is called the *global radiation*.

It is interesting to note that in northern Europe, the sunlight may be as much as 60% diffuse (Mallinson and Landsberg, 1977, 1978). Since the spectral content of this diffuse radiation is shifted toward the blue, a solar cell with a slightly higher band gap than indicated in Fig. 1.6 would be most efficient. For concentrator systems, lower solar concentration ratios would be most efficient because of their larger apertures.

During the diurnal variation, the solar zenith angle, and hence the air mass, is continuously changing. Therefore, a continuously varying intensity and spectral distribution is incident on the earth's surface. Averaged over 24 h, the radiation intensity is *roughly* 0.2 times the peak intensity (but this number is highly variable, depending on latitude and other factors). The geometrical relative-air-mass formulation functions only to approximate the reduction in beam intensity by the thickness of the atmosphere, which increases with increasing zenith angle. The spectral distribution of the radiation is also substantially affected by weather conditions, altitude, and the real air mass (as determined by barometric pressure and temperature). The real air mass, of course, approaches unity only under perfect conditions when $z = 0$. Therefore, AM1.5 or AM2 is used as a very approximate estimate of the "average" air mass during the daylight hours for the purpose of calculating cell efficiencies.

Because the calculation of solar insolation in detail is a complex endeavor, a variety of more economical approaches is commonly used. Computer calculations based on measured AM0 spectra as modified by semiempirical mathematical models for the effects of ozone, H_2O, CO_2, and scattering parameters are used to generate idealized "standard" spectra. Such a standard AM1.5 spectrum, shown in Fig. 2.5, is valuable

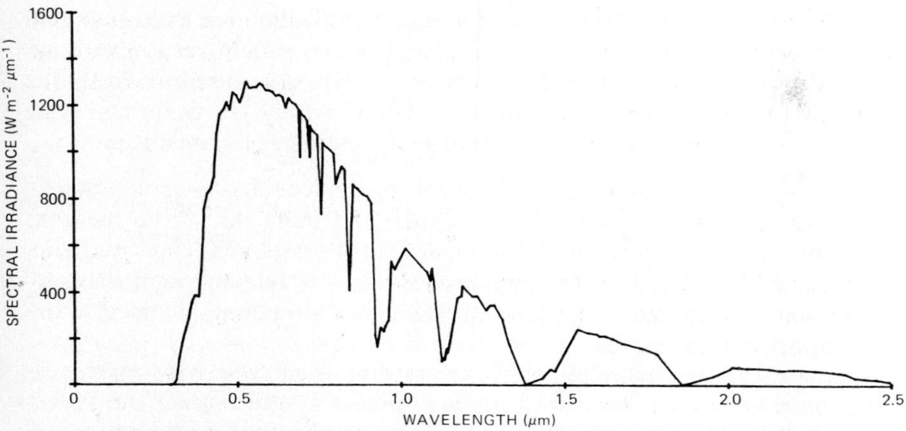

Fig. 2.5. Standard AM1.5 solar spectrum. Computer generated from AM0-measured data, assuming precipitable $H_2O = 2.0$ cm, ozone = 0.34 cm, air mass = 1.5, and aerosol scattering parameters, $\alpha = 1.3$ and $\beta = 0.12$. (From "Terrestrial Photovoltaic Measurement Procedures," ERDA/NASA/1022-77/16, NASA TM 73702. NASA-Lewis Res. Cent., Cleveland, Ohio, 1977.)

for calculation of theoretical solar cell efficiencies.† Direct, hour-by-hour measurements of insolation, which have been made throughout the year and in many parts of the world, are useful for computer simulation of solar photovoltaic systems, but unfortunately do not usually include spectral variations. For laboratory measurements on experimental devices, various solar simulators are available that provide the required constant intensity and approximate the solar spectrum.

2.3 MEASUREMENT OF SOLAR INSOLATION

A variety of instruments is used to measure solar insolation. The radiation detector components of these instruments are of two basic types:

(1) semiconductor junction devices very similar to solar cells (usually Si p/n junctions), which are essentially photon counters with spectral response ranges much narrower than the solar spectrum (e.g., see Fig. 7.17), and

† A more useful form for light-generated current calculations is the integrated photon flux versus photon-energy curve shown in Fig. 6.3, based on the data of Fig. 2.5 that is tabulated in Appendix 1.

(2) thermopiles and pyroelectric detectors, which are essentially energy integrators with generally wavelength-independent response (compared to the junction devices) that have a bandwidth, determined by the window material and the absorber coating, which is generally wide enough to encompass essentially all the terrestrial solar spectrum.

Pyrheliometers are insolation-measuring devices that generally have a small angular field of view (5–10° or less) and track the sun to measure only the direct component of the insolation (normal-incidence pyrheliometers, NIP). Usually a thermopile is used as a detector, and the total solar wavelength range is measured. Bandpass filters may be used to obtain spectral information.

A new class of pyrheliometers, referred to as active cavity devices, is gaining popularity. These instruments operate by measuring the power (supplied by the instrument) used to heat a cavity that is alternately exposed to and blocked from the sun by a chopper. These instruments have the advantage of being calibrated in an absolute sense and do not require secondary standards.

Pyranometers are solar radiometers with a 180° field of view used to measure global radiation (i.e., direct plus diffuse). They are generally used with a thermopile detector mounted in a horizontal position. A Si solar cell is sometimes used as a detector, but in this case the value of total insolation must be obtained from a calibration constant. Using such a Si solar cell can lead to errors because only a portion of the solar spectrum is measured; the instrument is insensitive to variations (often considerable) outside the passband of the detector and may lead to incorrect readings for the total insolation.

For higher-concentration applications it is of interest to know the angular intensity distribution of sunlight reaching an observer on the earth. This distribution influences the maximum useful concentration ratio used. A considerable proportion of the solar intensity appears to be incident from outside the solar disk, having been scattered by the earth's atmosphere. This is called *circumsolar radiation;* the ratio of circumsolar intensity to the intensity directly from the solar disk (C/S) varies from $<1\%$ to nearly totally circumsolar, depending on atmospheric conditions and the zenith angle. Circumsolar is actually the near-axis portion of the so-called diffuse component of solar radiation.

Figures 2.6a and 2.6b are examples of the circumsolar intensity patterns (unfiltered) for a clear day and a hazy day as reported by Grether *et al.* (1976). The data were taken by an automated circumsolar telescope, which scans solar and circumsolar regions of the sky with an aperture that has an extremely narrow field of view. An ideal, nonimaging concentrator

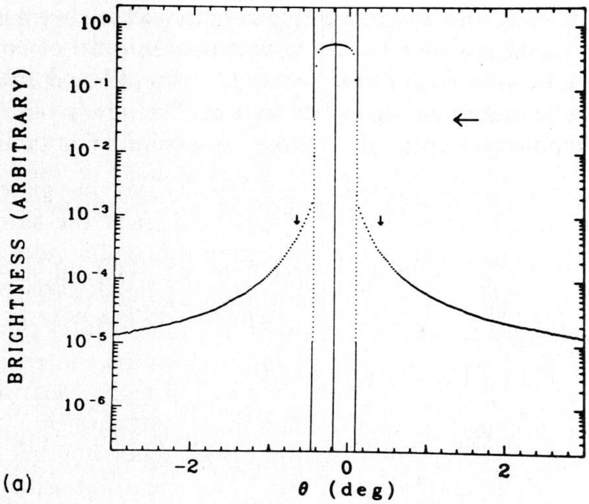

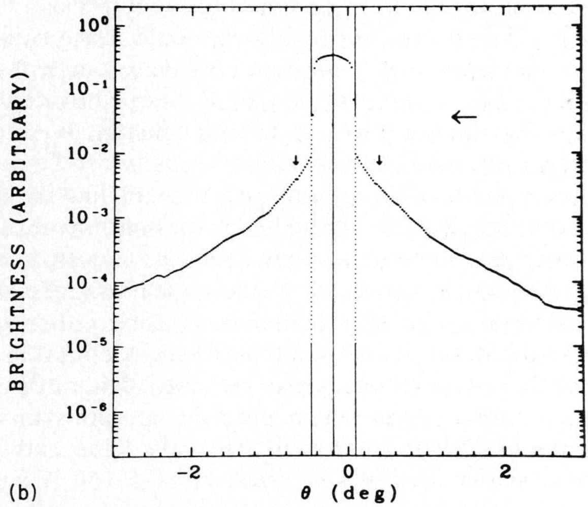

Fig. 2.6. Unfiltered circumsolar intensity patterns versus angle, taken at China Lake, California, on 30 July 1976: (a) fairly clear conditions (11:11 AM, $C/S = 1.7\%$, NIP intensity for 6° field of view is 783 W m^{-2}) and (b) hazy conditions (12:05 AM, $C/S = 20.1\%$, NIP intensity for 6° field of view is 599 W m^{-2}). (A. J. Hunt, Lawrence Berkeley Lab., personal communication, 1978.)

with a field of view encompassing the same angle as the plots shown (6° total included angle) would have a concentration ratio of 365.†

The distinction between what is called "direct" and "diffuse" is of course a relative matter subject to definition. The whole pattern shown in Fig. 2.6 would normally be called "direct." The direct component can vary greatly from day to day. The most efficient field of view for a high-concentration system therefore depends on the meteorological conditions encountered.

A wealth of circumsolar information exists; data tapes are available with similar data taken every 10 min of the day over periods of a year or more at four different locations around the United States (Grether *et al.*, 1976). This same data base contains information on the relative proportions of "direct" radiation (6° field of view) in eight spectral passbands of roughly equal energy content. Thus these data can be used to determine the amount and type of variation in cell efficiency because of both short- (hours) and long- (months) term solar spectral variations.

2.4 SOLAR SIMULATION

For long-term stability and reproducibility in the testing of solar cells, a solar simulator is required for the laboratory comparison of cells under standardized conditions. Such a simulator should, for example, approximate the solar spectrum over the spectral response range of the solar cell to within about ± 10% at any wavelength. A total equivalent intensity of 0.5 to 1.0 kW m^{-2} is required that can be held constant to better than ± 5% over long periods of time. The equivalent intensity can be determined by comparing the response of a thermopile or solar cell first to direct sunlight and then to the solar simulator at selected wavelengths (e.g., using narrow bandpass interference filters) chosen in spectral ranges where the effects of water vapor and haze variations would be least significant.

Solar simulators are usually constructed using either xenon arc or tungsten–iodine light sources with suitable filters to adjust the source spectrum to that of the sun. Simulators may be large, elaborate instruments or small laboratory installations which produce an approximate spectrum suitable for research testing on small-area cells. The extremely simple solar simulator shown in Fig. 2.7 consists of a 650-W quartz–iodine photo-spot lamp (color temperature of 3200°K), a 7.5-cm water filter, and a Corning 1-62 color conversion filter immersed in the water, which attenuates the long wavelengths to approximate the solar spectrum. A comparison of the output of this simple simulator with natural sunlight using

† The solar disk has an included angle of 0.535° as viewed from earth.

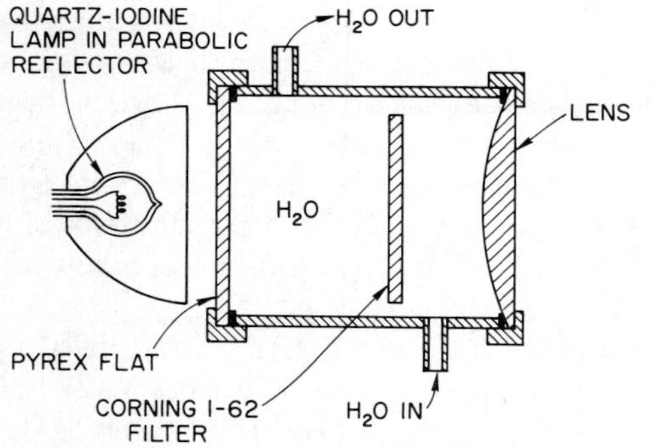

Fig. 2.7. Simple solar simulator which approximates the solar spectrum between 0.5 and 1.0 μm to $\pm 40\%$. Flowing H_2O is required to cool the Corning filter as well as to remove large IR component from tungsten source.

interference filters indicates average deviations of $\pm 40\%$ between 0.5 μm and 1.0 μm. Simple, unfiltered, quartz iodine lamps with cold mirror reflectors (which eliminate a large portion of the infrared) are also used as light sources approximating the solar spectrum (e.g., the General Electric "ELH" 300-W projector lamp). The spectral irradiance of eight different types of solar simulators and their effects on solar cell measurements are compared by Seaman *et al.* (1980).

For an absolute efficiency measurement, a solar cell should be exposed to direct sunlight, the total integrated intensity of which is measured using a thermopile with a National Bureau of Standards traceable calibration. Alternatively, such cells can be tested at laboratories such as the National Aeronautics and Space Administration (NASA) Lewis Research Center in Cleveland, Ohio, or the Solar Energy Research Institute (SERI) in Golden, Colorado, where careful attention is paid to establishing standards and testing procedures. To obtain AM0 comparisons between developmental cells, it is common practice to use high-altitude balloon or rocket flights to expose the test cells to the sun above the earth's atmosphere. This latter approach demonstrates that solar simulation and measurement can be a difficult endeavor. As another example of this difficulty, see Allison and Crabb's (1976) discussion of the difficulties with comparison of solar simulator results from different countries.

2.5 SOLAR CELL TESTING METHODS

The three methods that have been approved by NASA-Lewis under the direction of the Department of Energy for the testing of solar cells are summarized in Fig. 2.8 (NASA, 1977).

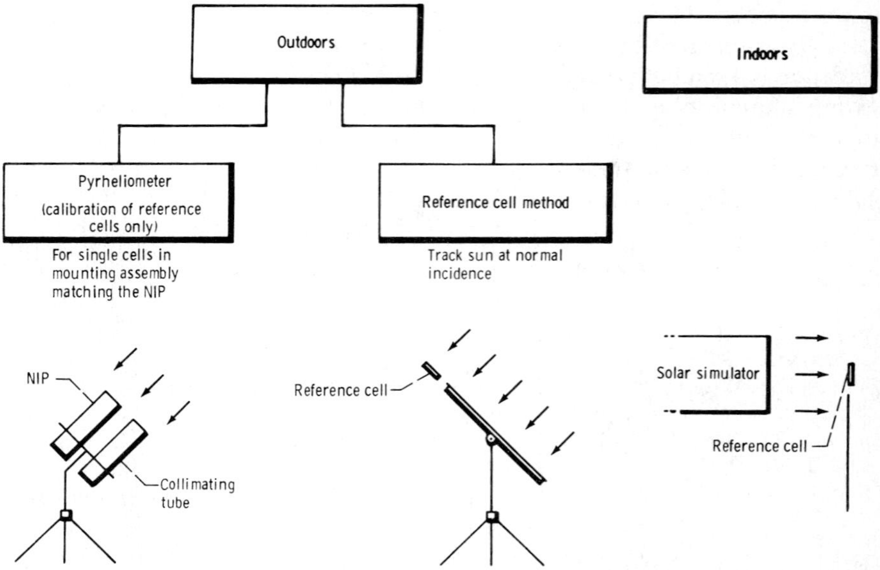

Fig. 2.8. DOE/NASA solar cell testing methods (NIP = normal incidence pyrheliometer). Reference cells are of the same type (i.e., Si, GaAs, Cu_2S/Cds, etc.) as the cell to be measured. (From "Terrestrial Photovoltaic Measurement Procedures," ERDA/NASA/1022-77/16, NASA TM 73702. NASA-Lewis Res. Cent., Cleveland, Ohio, 1977.)

A standard solar cell is a cell made from the same materials as the test cell (to ensure similar spectral response) and used to set the radiation intensity level of a solar simulator. It is calibrated in units of short-circuit current output per unit of radiant energy input in milliwatts per square centimeter. The NASA/DOE standard test conditions are a cell temperature of 28° ± 2°C, an irradiance of 100 mW cm^{-2}, a short-circuit current measured with a precision load resistor with no more than 20 mV drop across the resistor, and an open-circuit voltage measured across an unloaded cell with a voltmeter with internal resistance of at least $10^4\ \Omega\ V^{-1}$. The test cell area is the *entire* front surface area of the cell, including the area covered by grids and contacts.†

It must be emphasized here that many reported experimental solar efficiency results are calculated from insolation intensities measured with broad-band detectors such as thermopiles. Such efficiencies, especially for infrared-blind, high band-gap cells, are very sensitive to optical air

† Occasionally in the literature, it has been the practice to report efficiencies in terms of the *active* area of the cell in the case of small experimental cells.

mass number. The infrared radiation (which only the detector sees) is strongly attenuated by H_2O vapor in the atmosphere, whereas the visible radiation (which both detector and cell see) is not attenuated significantly, giving inflated solar efficiency values. An example of the effect is a series of measurements of an AlGaAs/GaAs cell whose measured efficiency (with respect to a thermopile pyranometer) varied from 19.1% ($\approx$ AM2) to 15.3% ($\approx$ AM1) (Woodall and Hovel, 1972).

GENERAL REFERENCES

Angrist, S. W. (1971). "Direct Energy Conversion," 2nd ed. Allyn & Bacon, Boston, Massachusetts.

Bird, R. E., Hulstrom, R. L., Kliman, A. W., and Eldering, H. G. (1982). Solar spectral measurements in the terrestrial environment. *Appl. Opt.* **21,** 1430.

Gates, D. M. (1966). Spectral distribution of solar radiation at the earth's surface. *Science* **151,** 523.

Hunt, A. J., Grether, D. F., and Wahlig, M. (1977). "Techniques for Measuring Circumsolar Radiation," LBL-8345. Lawrence Berkeley Lab., Berkeley, California.

Luft, W. (1962). Stationary Si solar cell converter calculations. *Sol. Energy* **6,** No. 1, 27.

Meinel, A. B., and Meinel, M. P. (1976). "Applied Solar Energy—An Introduction." Addison-Wesley, Reading, Massachusetts.

Robinson, N. (1966). "Solar Radiation." Elsevier, Amsterdam.

"Solar Energy Data Workshop Proceedings" (1973). NSF-RA-N-74-062. Silver Spring, Maryland.

"Terrestrial Photovoltaic Measurements" (1975). Workshop Proc., NASA TM X-71802. Cleveland, Ohio.

"Terrestrial Photovoltaic Measurements—II" (1976). Workshop Proc., NASA CP-2010 and ERDA/NASA-1022/76/10. Baton Rouge, Louisiana.

REFERENCES

Allison, J. F., and Crabb, R. L. (1976). *Proc. 12th IEEE Photovoltaic Specialists Conf.,* p. 554.

Grether, D. F., Hunt, A. J., and Wahlig, M. (1976). "Results from Circumsolar Radiation Measurements," LBL-5292. Lawrence Berkeley Lab., Berkeley, California.

Mallinson, J. R., and Landsberg, P. T. (1977). *Proc. R. Soc. London, Ser. A* **355,** 115.

Mallinson, J. R., and Landsberg, P. T. (1978). *Appl. Phys.* **16,** 399.

NASA (1977). "Terrestrial Photovoltaic Measurement Procedures," ERDA/NASA/1022-77/16, NASA TM 73702. National Aernautics and Space Administration–Lewis Research Center, Cleveland, Ohio.

Seaman, C. H., Anspaugh, B. E., Downing, R. G., and Estey, R. S. (1980). *Proc. 14th IEEE Photovoltaics Specialists Conf.,* p. 494.

Woodall, J. M., and Hovel, H. J. (1972). *Appl. Phys. Lett.* **21,** 379.

CHAPTER 3

PROPERTIES OF SEMICONDUCTORS

In this and the next several chapters we are concerned with the specific processes involved in each of the regions of a photovoltaic cell as summarized in Fig. 1.5. Since these various processes of generation of carriers, diffusion, recombination, and drift are fundamental to the operation of a solar cell, we consider in this chapter some of the relevant electrical, optical, and photoelectronic properties of semiconductors.

3.1 ENERGY LEVELS

A semiconductor is a material for which the allowed energy band of highest energy for electrons (the valence band) is almost totally occupied. An *intrinsic* semiconductor is one in which the free electrons in the conduction band and the free holes in the valence band arise solely from thermal excitation across the band gap. An *extrinsic* semiconductor is one in which additional electrons are present in the conduction band because of ionized donor imperfections, or one in which additional holes are present in the valence band because of ionized acceptor imperfections. A *degenerate* semiconductor is one in which so many donor (acceptor) imperfections are present that states at the bottom of the conduction band are almost totally occupied by electrons (states at the top of the valence band are almost totally occupied by holes) and the Fermi level lies within

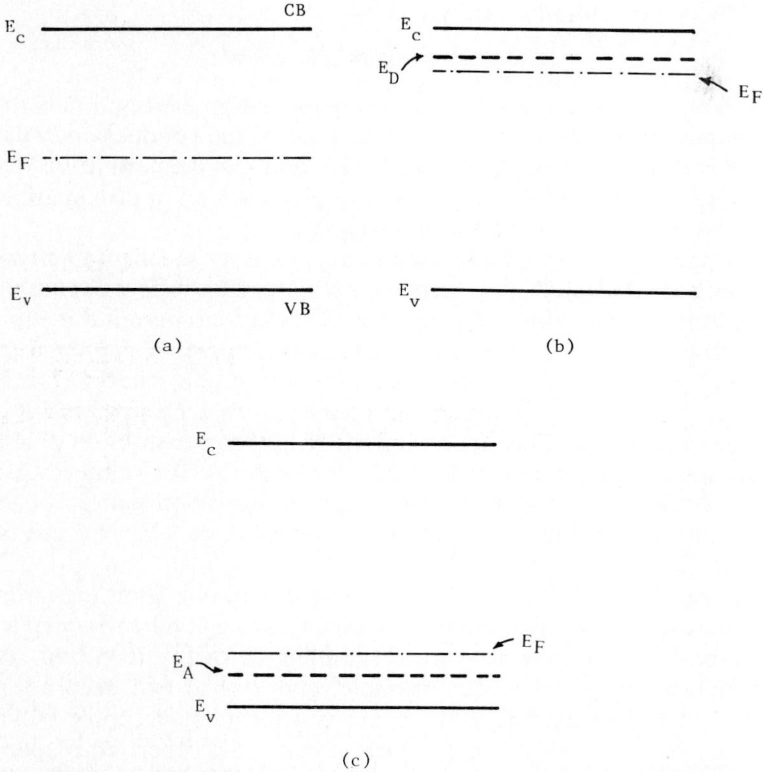

Fig. 3.1. Energy-level diagrams for semiconductors (E_c is the energy of the bottom of the conduction band CB, E_F is the energy of the Fermi level, and E_v is the energy of the top of the valence band VB): (a) An intrinsic semiconductor, (b) extrinsic *n*-type with electron donor imperfection levels E_D, and (c) extrinsic *p*-type with electron acceptor levels E_A.

the conduction (valence) band. The energy band diagrams for these cases are shown in Fig. 3.1.

The electrical conductivity of semiconductors can be controlled by the incorporation of suitable donor or acceptor imperfections. Such incorporation may be achieved by having a desired impurity added in the preparation of the material or by introducing the impurity at a later time by diffusion, ion implantation, or other methods.

In general, impurities that differ in valence by one from the atom for which they are substituted behave as shallow donors or acceptors with energy levels that can be approximately described in terms of a hydrogenic model (Kohn, 1957). In such a model, the energy levels for a donor (acceptor) impurity are given by

$$\Delta E = (1/n^2)(q^4 m^*/32\pi^2\hbar^2\epsilon_s^2), \tag{3.1}$$

where n is an integer equal to one for the ionization energy, m^* is the effective mass of an electron (hole) at the edge of the conduction (valence) band, $\hbar$ is Planck's constant divided by 2π, and ϵ_s is the permittivity of the semiconductor. Equation (3.1) is able to give only an approximation for the actual values of the ionization energies.

Phosphorus is a commonly used donor impurity in silicon, and boron is a commonly used acceptor impurity. Both can be diffused readily into Si, and both have a sufficiently high solubility in Si to permit the impurity concentrations desired. The measured donor ionization energy for P is $(E_c - E_D) = 0.044$ eV, and the acceptor ionization energy for B is $(E_A - E_v) = 0.045$ eV. Other dopant energy levels are shown in Fig. 7.8. Other examples of shallow donor impurities are the substitution of indium for cadmium in cadmium sulfide (CdS) or of chlorine for sulfur in CdS; in both cases donor levels are formed with an ionization energy of about 0.03 eV, in reasonable agreement with the value of 0.026 eV predicted from Eq. (3.1).

Impurities that differ in valence by more than one from the atom for which they are substituted commonly give rise to a number of deep levels with large ionization energies corresponding to highly localized states. Zinc impurity atoms in Si, for example, give rise to two deep acceptor levels. One, corresponding to the impurity after it has accepted two electrons, lies 0.52 eV below the conduction band; the other, corresponding to the impurity after it has accepted one electron, lies 0.31 eV above the valence band. Copper impurity in Si also produces two deep levels. One, corresponding to the impurity after it has accepted one electron, lies 0.49 eV above the valence band; the other, corresponding to the impurity after it has given up one electron, lies 0.24 eV above the valence band. In this case the former level is an acceptor, and the latter level is a donor. Impurities with deep levels are frequently deleterious for photovoltaic solar cell operation: they do not contribute to lowering the resistance of the material, several (Au, Cu, Ag) diffuse readily and contribute to instability, and they may serve as fast recombination sites that reduce the minority carrier diffusion length.

3.2 DETERMINATION OF THE FERMI ENERGY

In a nondegenerate semiconductor the Fermi level lies within the band gap of the material as shown in Fig. 3.1. For the condition of thermal equilibrium between the conduction and valence bands, the total density of free electrons in an n-type material can be written

$$n_0 = N_c \exp[-(E_c - E_F)/kT], \tag{3.2}$$

where the subscript for n_0 denotes a thermal equilibrium value, k is Boltzmann's constant, and T is the temperature in degrees Kelvin.† The term N_c is called the *effective density of states* in the conduction band and is given by

$$N_c = 2M_c[2\pi m_{de}^* kT/h^2]^{3/2}, \tag{3.3}$$

where M_c is the number of equivalent minima in the conduction band and h is Planck's constant. Similarly, the total density of free holes in a p-type material is given by

$$p_0 = N_v \exp[-(E_F - E_v)/kT], \tag{3.4}$$

where N_v is the effective density of states in the valence band (with M_v equivalent maxima) given by

$$N_v = 2M_v[2\pi m_{dh}^* kT/h^2]^{3/2}. \tag{3.5}$$

The density-of-states effective masses m_{de}^* and m_{dh}^* are for electrons at the bottom of the conduction band and holes at the top of the valence band, respectively.‡

The product of Eqs. (3.2) and (3.4),

$$n_0 p_0 = N_c N_v \exp[-(E_c - E_v)/kT], \tag{3.6}$$

shows that $n_0 p_0$ is a constant for a particular temperature depending on the band gap $E_g = E_c - E_v$ of the material. Equation (3.6) is an expression of the law of mass action, and it is always true for thermal equilibrium. In an intrinsic semiconductor $n_0 = p_0$, and an intrinsic density n_i can be defined such that

$$n_i^2 = n_0 p_0. \tag{3.7}$$

Since this is a restatement of Eq. (3.6), Eq. (3.7) remains true for extrinsic semiconductors, where $n_0 \neq p_0$.

† The energies E_F, E_c, E_v, and E_i are taken with respect to an arbitrary reference level.

‡ The density-of-states effective masses are given by $m_{de}^* = (m_1^* m_2^* m_3^*)^{1/3}$ where m_1^*, m_2^*, and m_3^* are the effective masses along the principal axes of an ellipsoidal energy surface. In the case of Si, m_2^* equals m_3^* and the effective mass values are commonly denoted as longitudinal m_L^*, and transverse $m_T^* = m_2^* = m_3^*$. Then $m_{de}^* = (m_L^* m_T^{*2})^{1/3}$ and $M_c = 6$. For holes, there may be a "light" hole band and "heavy" hole band, in which case $m_{dh}^* = (m_{lh}^{*3/2} + m_{hh}^{*3/2})^{2/3}$. See, e.g., Smith (1968).

It is sometimes convenient to regard the position of the Fermi level in an intrinsic semiconductor (by definition nondegenerate) as a reference level E_i. Then from Eqs. (3.2) and (3.4)

$$E_i = E_{F,\text{intrinsic}} = (E_c + E_v)/2 + (kT/2)\ln(N_v/N_c). \qquad (3.8)$$

The intrinsic level E_i usually lies quite close to the center of the band gap and is a weak function of temperature. The Fermi level in extrinsic semiconductors can then be described in terms of E_i:

$$n_0 = n_i \exp[(E_F - E_i)/kT] \quad \text{and} \quad p_0 = n_i \exp[(E_i - E_F)/kT]. \qquad (3.9)$$

Typical values for these various quantities are given in Table 3.I for several semiconductor materials.

For a degenerate semiconductor in which the Fermi level lies well within the band (more than several kT from the band edge), simple free-electron calculations can be used to locate the Fermi energy. In this case

$$(E_F - E_c) \simeq (\hbar^2/2m_e^*)(3\pi^2 n)^{2/3} \qquad (3.10)$$

for n-type degenerate material, and

$$E_v - E_F \simeq (\hbar^2/3m_h^*)(3\pi^2 p)^{2/3} \qquad (3.11)$$

Table 3.I
Electronic Quantities for Selected Semiconductors

Quantity[a]	Si	GaAs	CdTe	InP
E_g (eV)	1.12 Indirect	1.42 Direct	1.50 Direct	1.35 Direct
N_c (cm^{-3})	2.9×10^{19}	4.3×10^{17}	7.5×10^{17}	5.5×10^{17}
N_v (cm^{-3})	1.0×10^{19}	8.2×10^{18}	1.8×10^{19}	1.8×10^{19}
n_i (cm^{-3})	1.4×10^{10}	1.8×10^{6}	6.9×10^{5}	1.5×10^{7}
m^* (electrons)	—	0.068	0.11	0.078
m_L^*	0.98	—	—	—
m_T^*	0.19	—	—	—
m^* (holes)				
m_{lh}^*	0.16	0.087	0.13	—
m_{hh}^*	0.49	0.47	1.0	0.8
χ (eV)[b]	4.01	4.07	4.28	4.38
μ_e ($cm^2\ V^{-1}\ sec^{-1}$)[c]	1900	8800	1050	4700
μ_h ($cm^2\ V^{-1}\ sec^{-1}$)[c]	495	400	80	150
ϵ_s/ϵ_o, dielectric constant (dc)	11.7	13.1	9.4	12.35

[a] All quantities given at 300°K.
[b] Electron affinity.
[c] Highest experimental Hall mobilities for single crystals.

for p-type degenerate material. Degeneracy occurs when the density of free carriers becomes comparable to the effective density of states.

Given the location of the Fermi level, the occupancy of all other states in the material is also known. If a total density N_D of donor impurities with an ionization energy of ΔE_D is occupied by n_D electrons, for example,

$$n_D = N_D/\{1 + (1/g)\exp[(E_c - E_F - \Delta E_D)/kT]\}, \qquad (3.12)$$

where g is the degeneracy of the impurity. For the simple case where only spin degeneracy is involved, $g = 2$; the donor level can accept one electron of either spin. For occupation of acceptors by holes, a similar expression holds except that $g = 4$ for Si, Ge, and GaAs because the acceptor level is doubly degenerate as a result of the two valence bands at $k = 0$ in these materials.

Perhaps the most general condition to be satisfied in a semiconductor is that of charge neutrality. In the extrinsic case, charge neutrality can be expressed as

$$n_0 + N_A^- = p_0 + N_D^+, \qquad (3.13)$$

where $N_D^+ = N_D - n_D$ is the total density of ionized donors and N_A^- is the density of ionized acceptors. As donors (or acceptors) are added to the semiconductor, the Fermi level adjusts accordingly. The charge neutrality condition is used, together with Eqs. (3.2), (3.4), and (3.12) (for the donor impurity case), to obtain the position of the Fermi level at any temperature. The resulting electron density is shown in Fig. 3.2.

For semiconductors with dopant levels close to the band edges nearly all the donors or acceptors are ionized at room temperature (corresponding to the saturation range of Fig. 3.2) so that $N_D^+ \simeq N_D$ and $N_A^- \simeq N_A$. For a moderately doped, n-type material with $n_0 \gg n_i$, we have $n_0 \simeq N_D^+ \simeq N_D$, and, using Eq. (3.2), $N_D \simeq N_c \exp[-(E_c - E_F)/kT]$. The hole density can be obtained immediately from the mass action law:

$$p_0 \simeq n_i^2/N_D \qquad (3.14)$$

so that

$$(E_c - E_F) \simeq kT \ln(N_c/N_D). \qquad (3.15)$$

Under nonequilibrium conditions such as photoexcitation in a bulk semiconductor or forward-bias injection near a p/n junction, Eq. (3.7) must be reconsidered and

$$pn > n_i^2 = p_0 n_0, \qquad (3.16)$$

where p and n are the *nonequilibrium* carrier densities. On the other hand, in the case of a reverse biased p/n junction, $pn < n_i^2$. In these cases it is

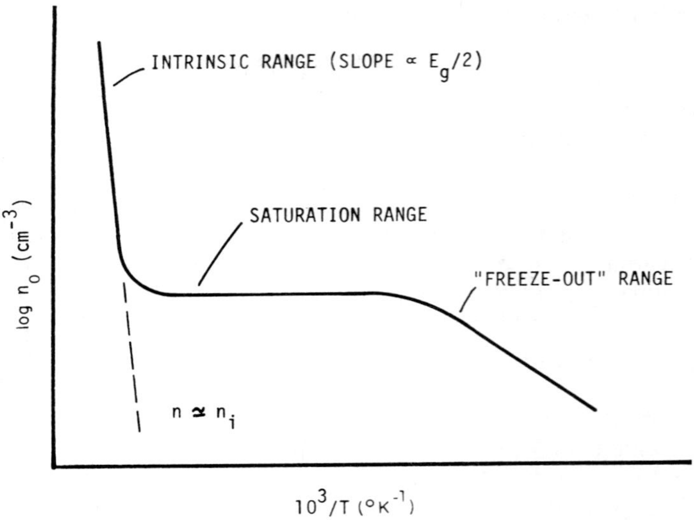

Fig. 3.2. Electron density versus inverse temperature for an *n*-type semiconductor. In the "freeze-out" range, the slope is proportional to ΔE_D, the donor energy level, when the semiconductor is partially compensated ($N_A \neq 0$). For $N_A = 0$, the slope is $\Delta E_D/2$.

convenient to introduce two *quasi-Fermi levels,* E_{Fn} for electrons and E_{Fp} for holes, rather than a single Fermi level. The positions of the quasi-Fermi levels are given by analogy with Eq. (3.9).

$$n = n_i \exp[(E_{Fn} - E_i)/kT] \quad \text{and} \quad p = n_i \exp[(E_i - E_{Fp})/kT]. \tag{3.17}$$

Note that as $n \to n_0$ and $p \to p_0$, the quasi-Fermi levels aproach each other and E_F. Figure 4.3 shows a solar cell in which the Fermi level is split into quasi-Fermi levels E_{Fn} and E_{Fp} by photogeneration of excess carriers and by forward bias. The quasi-Fermi level E_{Fn} can continue to be used to describe the occupancy of dopant levels in effective thermal equilibrium with the conduction band (and E_{Fp} for dopant levels in effective thermal equilibrium with the valence band).

3.3 ELECTRICAL CONDUCTIVITY

The general expression for the electrical conductivity of a semiconductor is

$$\sigma = q(n\mu_n + p\mu_p), \tag{3.18}$$

where μ_n and μ_p are the electron and hole mobilities, respectively. Thus the intrinsic conductivity is given by

$$\sigma_i = qn_i(\mu_n + \mu_p). \tag{3.19}$$

For Si at 300°K, $\sigma_i \approx 4.3 \times 10^{-6}$ $(\Omega\ \text{cm})^{-1}$, using the values of Table 3.I. In extrinsic semiconductors, however, such as are important in photovoltaic cells, either p_0 or $n_0 \gg n_i$, and a simplified expression for one-carrier conductivity can be used with $\sigma_n = nq\mu_n$ in n-type material and $\sigma_p = pq\mu_p$ in p-type material.

The temperature dependence of the carrier mobility $\mu(T)$ depends on the mechanism of scattering that dominates. If the scattering is *elastic*, i.e., if the change in energy on scattering is not greater than kT, then scattering can be described in terms of an average relaxation time between scattering events $\langle\tau_r(E)\rangle$,

$$\mu(T) = q\langle\tau_r(E)\rangle/m^*_{\text{cond}}, \tag{3.20}$$

where m^*_{cond} is the conductivity effective mass.† The most common types of elastic scattering in semiconductors are by acoustic lattice waves (for which $\tau_r(E) \propto E^{-1/2}$ and the acoustic mobility is approximately proportional to $T^{-3/2}$) and by charged impurity scattering (for which $\tau_r(E) \propto E^{3/2}$ and the charged impurity mobility is approximately proportional to $T^{3/2}/N_I$, where N_I is the charged impurity density in the material). For Si, these two mechanisms usually predominate. The most common type of *inelastic* scattering is by optical lattice waves. Here the probability of scattering is proportional to the density $\bar{n}$ of optical mode phonons, where

$$\bar{n} = 1/[\exp(h\nu_{\text{phonon}}/kT) - 1], \tag{3.21}$$

and the corresponding $\mu(T)$ increases approximately exponentially with decreasing temperature.

For polycrystalline semiconductor materials, grain boundary potentials may influence the overall mobility, introducing an apparent exponential increase in the effective mobility with temperature (see Chapter 9). The temperature dependences of the mobility are summarized in Fig. 3.3. Several scattering modes are almost always present and the total mobility can be obtained approximately by addition of the reciprocals, i.e., $1/\mu = 1/\mu_a + 1/\mu_I$.

To obtain the temperature dependence for the one-carrier conductivity of an n-type material, for example, Eq. (3.2) may be used:

$$\sigma_n = qN_c\mu_n(T)\exp[-(E_c - E_F)/kT]. \tag{3.22}$$

A plot of measured values of $\ln \sigma_n$ versus $1/T$ will give a straight line with

† In the case of a general equal energy surface, $1/m^*_{\text{cond}} = \frac{1}{3}\{1/m^*_1 + 1/m^*_2 + 1/m^*_3\}$.

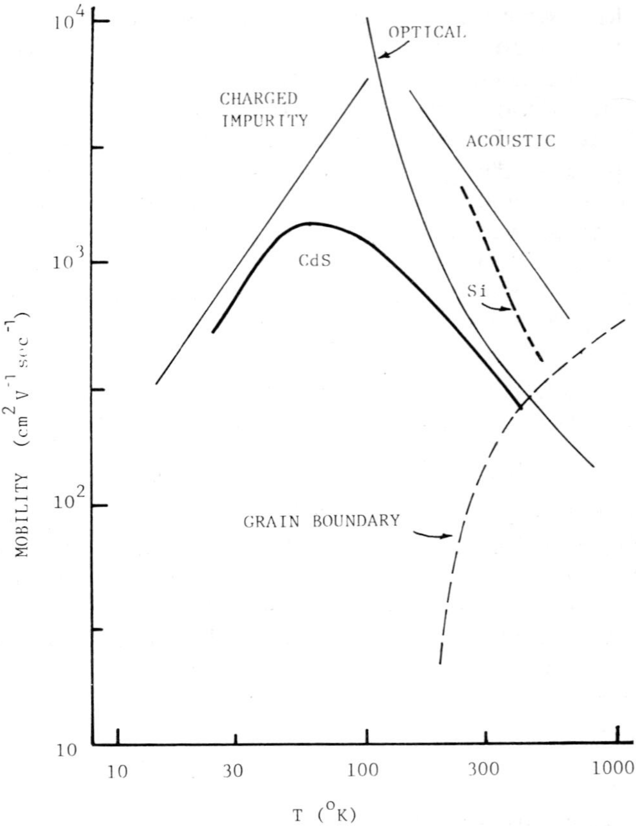

Fig. 3.3. Schematic plot of theoretical variation of mobility with temperature for various scattering mechanisms. Approximate experimental data are shown for CdS with $N_I = 2 \times 10^{16}$ cm^{-3} (predominantly optical and impurity scattering) and *n*-Si with $N_I = 10^{16}$ cm^{-3} (acoustic and impurity scattering).

a slope of $-(E_c - E_F)/k$ *only if* all other temperature dependences cancel out. In general, the mobility μ_n, the Fermi level, and the density of states are all functions of T. The simplest case is for scattering of electrons by acoustic lattice waves for which $\mu_n(T) \propto T^{-3/2}$ (which just cancels the temperature dependence of N_c) and for $E_c - E_F = (E_c - E_F)^0 + bT$, where b is a constant and the superscript 0 denotes the $T = 0$ value [e.g., Eq. (3.8)]. In this case a plot of $\ln \sigma_n$ versus $1/T$ gives a straight line with slope of $-(E_c - E_F)^0/k$. In general, however, it is necessary to obtain measured values of $n_0(T) = \sigma_n(T)/q\mu_n(T)$ by other means. For example, one might measure the Hall effect as a function of temperature and then plot $\ln(n_0T^{-3/2})$ versus $1/T$ to get a straight line with slope $-(E_c - E_F)^0/k$.

Once the carrier density has been separated from the mobility by experimental measurement, the variation of carrier density with temperature can be used to determine the characteristics of the donors and acceptors. The relationship between the measured slope of a plot of $\ln(n_0 T^{-3/2})$ versus $1/T$ and the ionization energy of the donor impurities involved depends on the specific model applicable. The dependence of the Fermi level on temperature can be calculated from the conditions required for charge neutrality. For example, for an intrinsic material where $n_0 = p_0$, see Eq. (3.8).

If only donors are present, charge neutrality requires that

$$n_0 = p_0 + N_D^+ \simeq N_D^+, \tag{3.23}$$

where N_D^+ is the density of ionized donors, which is large compared to p_0. Combining Eqs. (3.2) and (3.12) gives for $n_0 \ll N_D$ and $N_D^+ \gg p_0$

$$E_c - E_F = (E_c - E_D)/2 + (kT/2)\ln(gN_c/N_D). \tag{3.24}$$

The value of $E_c - E_F$ saturates at $kT\ln(N_c/N_D)$ when n_0 approaches N_D. Thus in this case the value of $(E_c - E_F)^0$ is one-half of the ionization energy of the donors, and this is the slope observed in a plot of $\ln(n_0 T^{-3/2})$ versus $1/T$ (i.e., the "freeze-out" region of Fig. 3.2).

It is quite common to have at least a small fraction of the donors compensated by some acceptor impurities because of unintentionally introduced impurities or crystal defects. Then the condition for charge neutrality becomes

$$n_0 - N_A^- = p_0 + N_D^+ \simeq N_D^+, \tag{3.25}$$

where N_A^- is the density of ionized acceptors and $p_0 \ll N_D^+ - N_A^-$. As long as $N_D \gg N_A$, as would be the case in a high-conductivity material, two cases are found:

$$E_c - E_F = E_c - E_D + kT\ln[gN_A/(N_D - N_A)],$$
$$\text{for} \quad n_0 \ll N_A, (N_D - N_A), \text{ and} \tag{3.26}$$

$$E_c - E_F = (E_c - E_D)/2 + (kT/2)\ln[gN_c/(N_D - N_A)]$$
$$\text{for} \quad N_A \ll n_0 \ll (N_D - N_A). \tag{3.27}$$

Finally, saturation at $E_c - E_F = kT\ln[N_c/(N_D - N_A)]$ occurs when $n_0 = N_D - N_A$. A typical example of such a situation is illustrated in Fig. 3.4. In this case a plot of $\ln(n_0 T^{-3/2})$ versus $1/T$ has a slope equal to the total ionization energy of the donors at low temperatures, changing to a slope equal to one-half the ionization energy of the donors in an intermediate temperature range before the donors are completely ionized. Finally, in the saturation range, the slope changes sign because of the temperature dependence of the mobility.

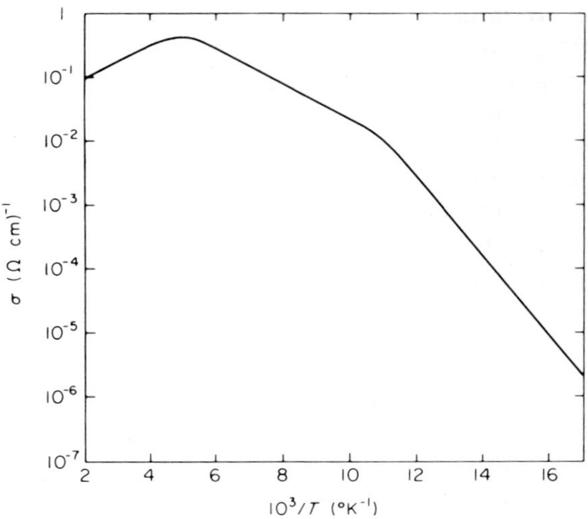

Fig. 3.4. Temperature dependence of conductivity in an extrinsic semiconductor with partially compensated acceptors. Parameters used in the calculation are $m_h^* = 0.2m$, $(E_A - E_v) = 0.105$ eV, $N_A = 3.9 \times 10^{15}$ cm^{-3}, $N_D = 9.6 \times 10^{12}$ cm^{-3}, $\mu_p = 1790\,(100/T)^{3/2}$ cm^2 V^{-1} sec^{-1}. (From R. H. Bube, "Electronic Properties of Crystalline Solids," Academic Press, New York, 1974.)

3.4 OPTICAL ABSORPTION

Since the generation of both free electrons and free holes is essential to produce a photovoltaic current, intrinsic optical absorption of a single photon across the band gap of the semiconductor is the dominant optical absorption process in a solar cell. Extrinsic absorption by deep, localized states in the band gap may act to affect charge transport (e.g., by charging states to change a junction profile or by emptying traps), but it produces negligible net current since the absorption coefficients associated with extrinsic absorption are many orders of magnitude smaller than for the intrinsic process.

The magnitude of absorption is described in terms of an absorption constant $\alpha(\lambda)$, which is a function of the wavelength of the light. If a photon flux $\Gamma_0(\lambda)$ is present at $x = 0$ in the absorbing medium, then

$$\Gamma(\lambda, x) = \Gamma_0(\lambda) \exp[-\alpha(\lambda)x]. \tag{3.28}$$

The flux Γ may be measured in units of photons cm^{-2}sec^{-1} or in mW cm^{-2}. The rate of free carrier generation $G(x)$ by intrinsic absorption is

$$G(\lambda, x)\, dx = -d\Gamma(\lambda) = \alpha(\lambda)\Gamma(\lambda, x)\, dx. \tag{3.29}$$

For wavelengths of interest, α is in the range of $10^3 - 10^5$ cm^{-1}.

There are two major types of intrinsic absorption processes involved in determining α; these are called *direct* and *indirect* absorption. Direct optical absorption is illustrated in Fig. 3.5. It is a first-order process involving only the absorbed photon and can be represented approximately as a vertical line on an energy versus wave vector plot. The absorption transition conserves energy so that $\Delta E = h\nu$ (where $h\nu$ is the photon energy), and there is no change in k between initial and final states except for the small momentum of the photon. The expected variation of α with photon energy near the minimum energy for absorption E_g^d is given by

$$\alpha(h\nu) = C(h\nu - E_g^d)^{1/2}, \tag{3.30}$$

where C is essentially constant. As indicated in Fig. 3.5b, a plot of α^2 as a function of $h\nu$ yields a straight line for a direct absorption, with an intercept on the photon energy axis equal to the direct band gap of the material. At the onset of a direct optical absorption, the absorption constant α increases rapidly with increasing photon energy to values in the range of $10^4-10^6\ \text{cm}^{-1}$. For higher photon energies, the absorption constant α remains high and varies only slightly with photon energy, depending on transitions to states well above the bottom of the conduction band or to other higher-lying bands.

For degenerate semiconductors, the states near the bottom of the conduction band are filled, permitting transitions only to states at or above the degenerate Fermi level. The apparent direct band gap is shifted to a higher energy $E_g^{d'}$ given, for n-type material, by

$$E_g^{d'} \simeq E_g^d + (\hbar^2/2m_e^*)(2\pi^2 n)^{2/3}. \tag{3.31}$$

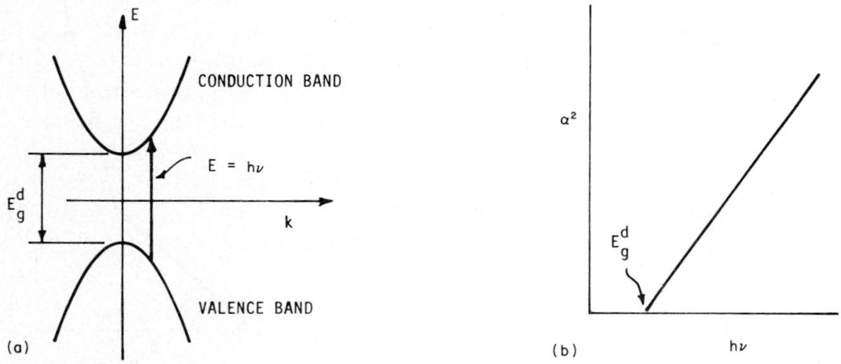

Fig. 3.5. Direct optical absorption. (a) A typical direct optical transition from the valence band to the conduction band. The minimum energy for the transition to occur is the direct band gap E_g^d. (b) Variation of optical absorption constant α with photon energy for a direct transition.

This effect, known as the Burstein shift (Burstein, 1954), occurs most strongly in semiconductors with small effective masses and thus small N_c. It is important in metal oxide semiconductors used as windows for heterojunction solar cells.

An indirect optical transition is illustrated in Fig. 3.6. It is a second-order process involving both the absorbed photon and a simultaneously absorbed or emitted phonon, which occurs when the minimum of the conduction band and the maximum of the valence band occur at different values of wave vector k. The absorption transition conserves energy by requiring that $\Delta E = h\nu \pm E_p$, where E_p is the phonon energy, and the change in k between the initial and final states Δk is just equal to the k value of the phonon involved. The expected variation of the indirect absorption coefficient α_i is given by (Smith, 1968)

$$\begin{aligned}
\alpha_i &= \frac{C_i}{(E_g^i)^2}\left\{\frac{(h\nu - E_g^i - E_p)^2}{1-\exp(-E_p/kT)} + \frac{(h\nu - E_g^i + E_p)^2}{\exp(E_p/kT)-1}\right\} \\
&\quad \text{for} \quad h\nu > E_g^i + E_p, \\
\alpha_i &= \frac{C_i}{(E_g^i)^2}\left\{\frac{(h\nu - E_g^i + E_p)^2}{\exp(E_p/kT)-1}\right\} \quad \text{for} \quad E_g^i - E_p < h\nu \le E_g^i + E_p, \\
\alpha_i &= 0 \qquad \text{for} \quad h\nu \le E_g^i - E_p,
\end{aligned} \tag{3.32}$$

and C_i is essentially constant.

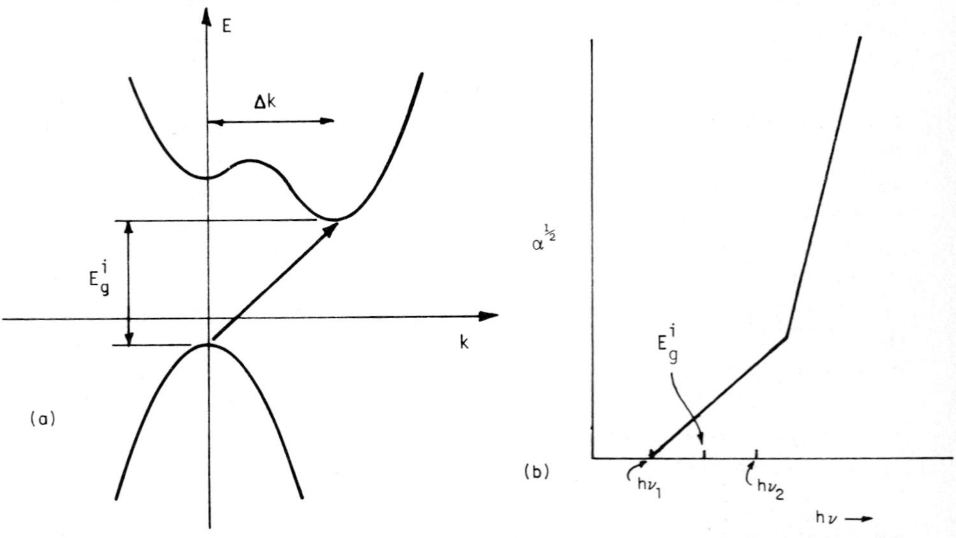

Fig. 3.6. Indirect optical absorption. (a) A typical optical indirect transition from the valence band to the conduction band. The minimum energy for the absorption to occur is the indirect band gap ($E_g^i - E_p$). (b) Variation of absorption constant α with photon energy for an indirect transition.

The photon energy is such that if $\alpha^{1/2}$ is plotted as a function of photon energy, two straight-line segments are found as indicated in Fig. 3.6b. One line segment with intercept of $h\nu_2$ on the photon energy axis corresponds to emission of a phonon, while the other, with intercept of $h\nu_1$, corresponds to absorption of a phonon. The indirect band gap E_g^i and the phonon energy involved can both be obtained from these intercepts:

$$E_g^i = \tfrac{1}{2}(h\nu_1 + h\nu_2) \qquad \text{and} \qquad E_p = \tfrac{1}{2}(h\nu_1 - h\nu_2). \tag{3.33}$$

This discussion of direct and indirect optical transitions can be illustrated for the case of single-crystal germanium (Ge) in which both types of transitions are observed, the direct band gap E_g^d being larger than the indirect band gap E_g^i. Figure 3.7 shows optical absorption spectra for Ge at 77 and 300°K, together with the results of analysis of the 300°K data in terms of an indirect transition dominating at lower photon energies and a direct transition dominating at higher photon energies. These results emphasize the fact that the range of magnitude for α for indirect transitions (10^{-1}–10^2 cm^{-1} in this case) is appreciably less than that for direct transitions (10^2–10^4 cm^{-1} in this case).

Optical absorption spectra for several materials are shown in Fig. 3.8. Considering the major materials of interest for photovoltaic solar cells, the indirect band gap is smallest in Si and AlAs, whereas the direct gap is smallest in GaAs, InP, CdS, and CdTe. Figure 3.9 illustrates the behavior of the direct and indirect band gaps of $Al_xGa_{1-x}As$ as x is varied.

When an indirect band-gap material is used as the absorbing component of a solar cell, the relatively low value of the absorption constant means that a relatively large thickness of the material must be used to absorb most of the light; furthermore, since many carriers are generated by light absorbed relatively far from the collecting junction, long minority carrier diffusion lengths are required for the generated carriers to reach the junction before recombination. A typical thickness for such an indirect band-gap material would be about 20–50 μm in order to absorb most of the available solar spectrum. By contrast, only 1–3 μm of a direct band-gap material are required to absorb most of the light. In the case of a direct band-gap material, however, a homojunction must have a very thin section on the side of illumination so that the light may penetrate to the junction region. These possibilities are illustrated in Fig. 3.10. In the case of such a thin junction region, the probability of carriers generated by the light diffusing to the surface, where recombination loss can occur, is large compared to the probability of diffusing to the junction to be collected. Thus direct band-gap material homojunctions are frequently limited by surface recombination losses. The use of a heterojunction structure with a

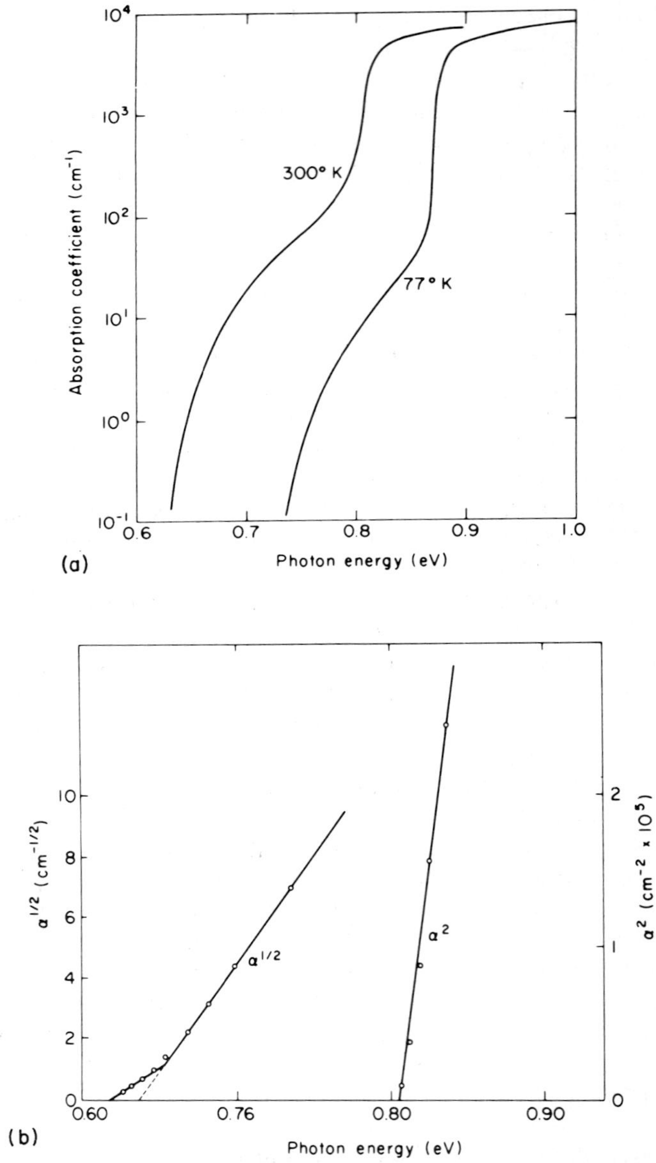

Fig. 3.7. (a) Optical absorption spectra for single-crystal germanium at 77°K and 300°K, after W. C. Dash and R. Newman, *Phys. Rev.* **99,** 1151 (1955). (b) Results of the analysis of the data of (a) in terms of an indirect optical absorption with $E_G^i = 0.63$ eV and $h\nu$ (phonon) ~ 0.01 eV, and of a direct optical transition with $E_G^d = 0.81$ eV. (From R. H. Bube, "Electronic Properties of Crystalline Solids." Academic Press, New York, 1974.)

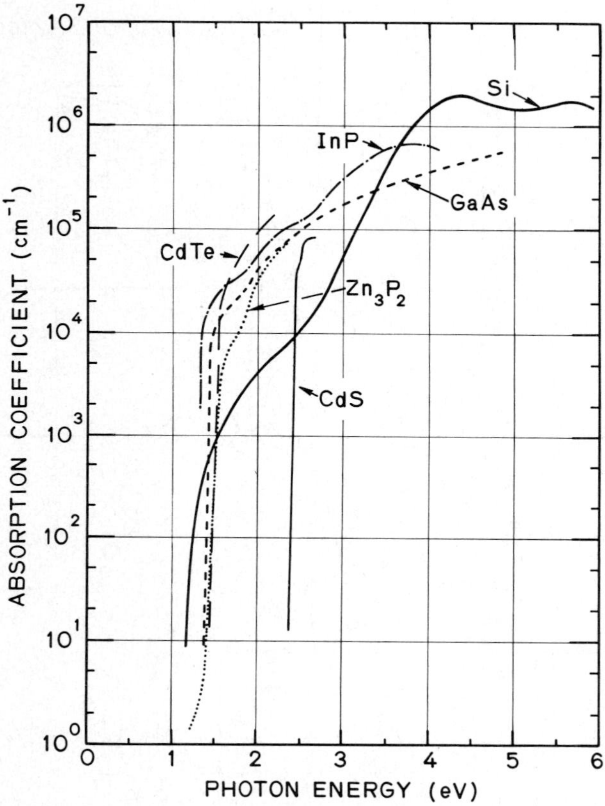

Fig. 3.8. Optical absorption coefficients of various single-crystal semiconductors.

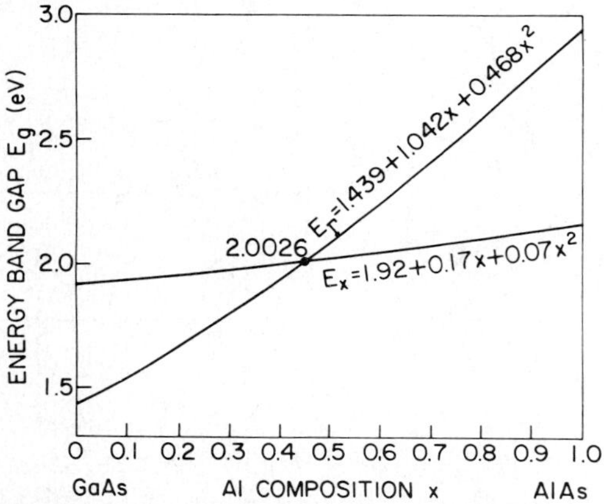

Fig. 3.9. The direct energy band gap E_Γ and indirect energy band gap E_x of $Al_xGa_{1-x}As$ as a function of composition x at 300°K. [After L. C.-C. Shen, Ph.D. Thesis, Stanford University Department of Electrical Engineering (1976).]

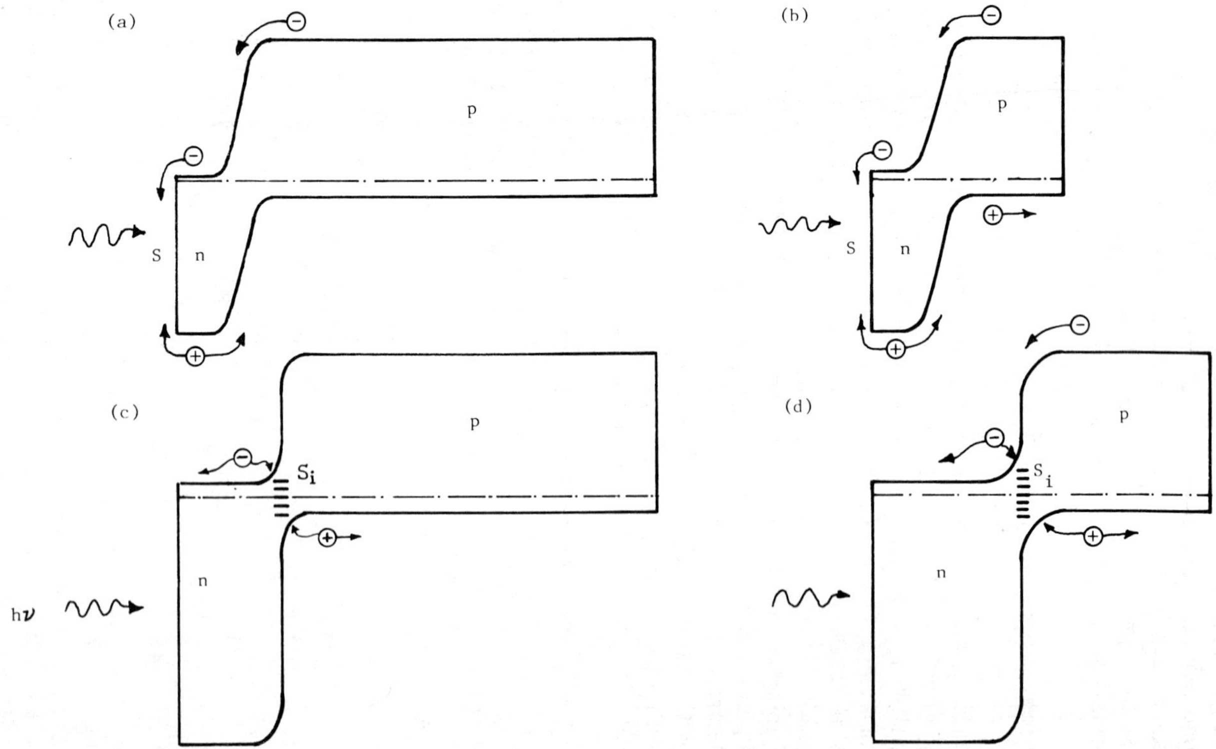

Fig. 3.10. Illustrations of cell configuration possibilities: (a) indirect band-gap material homojunction, (b) direct band-gap material homojunction, (c) indirect band-gap material heterojunction, and (d) direct band-gap material heterojunction. Surface and interface recombination losses are indicated by S and S_i, respectively.

larger band-gap material on the direct band-gap absorbing material helps to overcome some of these problems, although if there is appreciable lattice mismatch between the two different materials of the heterojunction, interface states or other interface interactions may exist that still limit the cell performance through interface recombination losses.

3.5 RECOMBINATION

After optical absorption has generated carriers in the bulk of the material in a photovoltaic cell, it is necessary for these carriers to pass across (be collected by) the junction barrier before they are lost by recombination (the inverse of the generation process). Recombination attempts to restore the nonequilibrium carrier densities generated by photoexcitation to their dark, thermal equilibrium values. A typical carrier generation/recombination model is pictured in Fig. 3.11. In this model, electron/hole pairs are being generated at a rate of G cm^{-3} sec^{-1}. These excited electrons and holes may recombine again either by an intrinsic process (with recombination coefficient β_i) or by a two-step, extrinsic process with recombination coefficients β_n (for capturing an electron when unoccupied) and β_p (for capturing a hole when occupied by an electron).

Since the rate of recombination U depends both on the recombination coefficients and on the densities of the recombining species, it is necessary to have information on both of these quantities in order to describe

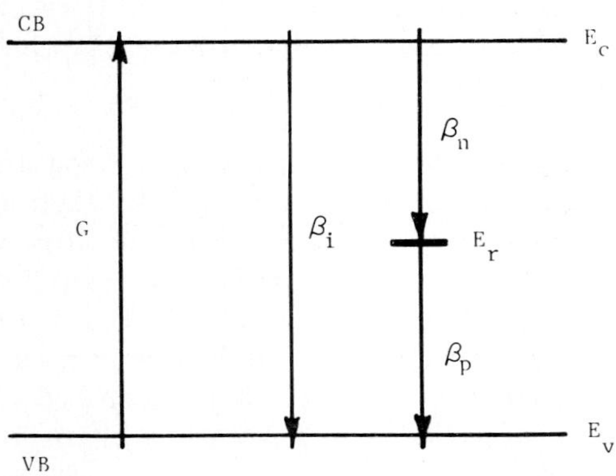

Fig. 3.11. Generation/recombination model. The rate of intrinsic generation of electron–hole pairs is G (cm^{-3} sec^{-1}). The coefficient for recombination between free electrons and free holes is β_i. The coefficient for recombination of a free electron with an unoccupied imperfection level is β_n. The coefficient for recombination of a free hole with an electron-occupied level is β_p.

recombination processes fully. We describe first the considerations that determine the magnitude of the recombination coefficient and then those that determine the densities of the recombining species.

For a generalized recombination process with recombination coefficient β_{12}, the rate of recombination U is given by $U = n_1 n_2 \beta_{12}$, where n_1 and n_2 are the densities of the two recombining species. Recombination processes are sometimes also expressed in terms of a recombination *cross section* σ. Since the probability of capture is a function of the carrier energy,

$$\beta = \langle \sigma(E) v_{\mathrm{th}} \rangle = \sigma \langle v_{\mathrm{th}} \rangle, \tag{3.34}$$

where v_{th} is the carrier's thermal velocity, $\sigma(E)$ is the cross section as a function of energy, and σ is a quantity defined as the cross section for the process.

3.5.1 Recombination Processes

The major processes for recombination can be separated into those that are *radiative* (the energy is dissipated by emission of photons), those that are *nonradiative and phonon assisted* (the energy is dissipated by emission of phonons), and those that are *nonradiative and carrier excitation assisted* (the energy is dissipated by excitation of other free carriers).

The magnitude of the recombination coefficient depends on the process by which the energy of the carrier is dissipated on recombination. If this energy is given off as a photon, a radiative process is involved, and it can be shown from a consideration of detailed balance that the cross section $\sigma_{\mathrm{intrinsic}}$ for the intrinsic radiative process is given by (Rose, 1963)

$$\sigma_{\mathrm{intrinsic}} = (5 \times 10^{-25}) \alpha n_{\mathrm{r}}^2 E_{\mathrm{g}}^2 (m_{\mathrm{e}}^* m_{\mathrm{h}}^*)^{-3/2} (300/T\,^\circ\mathrm{K})^{5/2}\ \mathrm{cm}^2, \tag{3.35}$$

where n_{r} is the refractive index of the material, α is the optical absorption constant for photons with energies slightly greater than the band-gap energy E_{g}, and m_{e}^* and m_{h}^* are the electron and hole effective mass *ratios*, respectively. Values of $\sigma_{\mathrm{intrinsic}}$ for typical materials range from $2.4 \times 10^{-19}\ \mathrm{cm}^2$ for Ge to $9.3 \times 10^{-18}\ \mathrm{cm}^2$ for GaP. It is expected that the radiative cross section for extrinsic recombination (σ_n or σ_p) would have a similar or slightly smaller magnitude. Thus, compared to a "geometric" atomic cross section of $\sim 10^{-15}$ cm, the cross section for a radiative recombination process is relatively small.

If the energy of the photoexcited carrier is dissipated by the emission of one or more phonons, a *nonradiative phonon-assisted* recombination is involved. Since the energy to be dissipated is usually on the order of several tenths of an electron volt and a phonon energy is on the order of a few millielectron volts, such a process requires the involvement of many phonons. If it is required that these phonons be emitted simultaneously in one burst of phonons, the probability is relatively small. The recombination cross section calculated for even the small ionization energy of a hydrogenic impurity in Si, for example, is $\sim 10^{-15}$ cm^2, or about the same as the geometric cross section (Gummel and Lax, 1955). For a deeper recombination center, the corresponding cross section would be expected to be orders of magnitude smaller. In the case of a *Coulomb-attractive center,* however, a new possibility exists. Such an imperfection center has a number of hydrogenlike excited states above the ground state. If a carrier is captured into a large excited-state orbit of such a center, the possibility exists for it to make a number of transitions down an "excited-state ladder" to the ground state, emitting only a few phonons with each transition and with very small probability of being reexcited thermally to escape from the center (Lax, 1960). Capture cross sections associated with such centers have been observed at low temperatures for As and Sb impurity in Ge with values of 10^{-11} cm^2 (Ascarelli and Rodriguez, 1961, 1962). A very simple approximation to the magnitude and temperature dependence of the cross section for a Coulomb-attractive center can be obtained by considering the cross section $\sigma_{\text{att}} = \pi r^2$, where r is the radial distance from the center at which the Coulomb-attractive potential energy is just equal to the mean thermal energy kT of the carrier. This gives

$$\sigma_{\text{att}} = q^4/16\pi\epsilon_s^2 k^2T^2, \tag{3.36}$$

where ϵ_s is the permittivity of the material. For typical values of ϵ_s, Eq. (3.36) gives $\sigma_{\text{att}} \sim 10^{-12}(300/T)^2$ cm^2.

If the charge on a recombination center is the same as the charge of the free carrier, a *Coulomb-repulsive* cross section σ_{rep} will be observed. In this case the center with approximately geometric dimensions σ_{geo} can be considered to be surrounded by a potential barrier of height ΔE. If capture requires the free carrier to be thermally excited over this barrier, then the cross section for the repulsive center can be expressed as

$$\sigma_{\text{rep}} = \sigma_{\text{geo}} \exp(-\Delta E/kT). \tag{3.37}$$

If $\Delta E \simeq 0.4$ eV, then at 300°K, $\sigma_{\text{rep}} \simeq 10^{-6}\sigma_{\text{geo}} = 10^{-21}$ cm^2. An alternative route to capture, in addition to thermal excitation over the barrier, would be tunneling through the barrier with or without thermal assistance. An example of such an impurity is Zn in Si; when Zn has accepted two electrons, it is effectively doubly negative with respect to the crystal. Even after capturing a photoexcited hole, it is still singly negative, and it is with this singly negative center that a photoexcited electron must recombine. Measured values give $\sigma_{\text{rep}} = 10^{-20}$ cm^2 without temperature dependence at low temperatures for Zn in Si, indicating that tunneling through the barrier dominates over this range (Sklensky and Bube, 1972).

If the energy of the photoexcited carrier is dissipated by increasing the energy of another free carrier, a *nonradiative Auger* recombination is involved. This kind of Auger recombination process is the inverse of the process of impact ionization, in which an energetic free carrier loses energy in the process of ionizing an imperfection. Auger recombination is a three-body process, and thus there are many possible interactions involving free carriers and charged imperfection centers (e.g., nnN_r^*, npN_r^*, n^2p, and np^2 (Beck and Conradt, 1973; Dziewior and Schmid, 1977).† in general, the recombination rate increases with decreasing band-gap energy, so the Auger processes are most important for small band-gap semiconductors. As an example, for p-Si the recombination rate for the dominant one of these processes is given by

$$U = Bnp^2, \tag{3.38}$$

so that the Auger lifetime is $\tau = n/U = 1/Bp^2 \simeq 1/BN_A^2$, with the Auger recombination coefficient $B \simeq 1.2 \times 10^{-31}$ cm^6 sec^{-1} (Burtscher *et al.*, 1975). Thus Auger recombination is important in p-Si for $N_A \gtrsim 2 \times 10^{17}$ cm^{-3} ($\rho \lesssim 0.1\ \Omega$ cm) and sets a rather abrupt limit on minority carrier lifetime for high-conductivity material.‡

By calculating the rate of impact ionization and setting it equal to the rate of Auger recombination, involving a charged imperfection, it is possible to calculate a value for B (Burstein *et al.*, 1956). This has been done for a hydrogenic imperfection, with the result that $B \simeq 10^{-24}$ cm^6 sec^{-1}. Because of the dependence of the rate of Auger recombination on the free carrier density, this mode of recombination also becomes important only when the carrier density is reasonably high. For Auger recombination to compete with a Coulomb-attractive recombination process with $\sigma_{\text{att}} = 10^{-12}$ cm^2, for example, the carrier density would have to exceed 10^{19} cm^{-3} using $B = 10^{-24}$ cm^6 sec^{-1}.

Actually, there is reason to believe that when deep levels are involved in recombination, the effective value of B may be much smaller than this hydrogenic value; experimental data suggest that B may be less than 10^{-28} cm^6 sec^{-1} for a deep center associated with Cu impurity in GaAs, lying 0.45 eV above the valence band (Saura and Bube, 1965).

In general, for direct band-gap semiconductors, direct, intrinsic, radia-

† A chapter in Blakemore (1962) is devoted to Auger processes.

‡ For n-Si, $\tau = 1/B'N_D^2$, where $B' = 2.8 \times 10^{-31}$ cm^6 sec^{-1} (Dziewior and Schmid, 1977). Also see Redfield (1980).

The second term is the rate of thermal excitation of electrons from the center to the conduction band. It can be written in the form $n\nu \exp\{-(E_c - E_r)/kT\}$, where ν is an "attempt-to-escape" frequency; consideration of the thermal equilibrium balance of the centers with the conduction band shows that $\nu = N_c\beta_n$. Similarly, the net rate of hole capture by the centers is

$$d(N_r - n_r)/dt = n_r p\beta_p - (N_r - n_r)N_v\beta_p \exp\{-(E_r - E_v)/kT\}. \quad (3.54)$$

For steady-state conditions

$$U = dn_r/dt = d(N_r - n_r)/dt, \quad (3.55)$$

and the electron and hole lifetimes can be calculated. In order to accomplish this, values of n_r must be determined as they are affected by the location of the thermal equilibrium Fermi level, by temperature, and by light intensity. This can be done by setting Eqs. (3.53) and (3.54) equal, to obtain

$$U = \frac{pn - n_i^2}{\tau_{p0}(n + n') + \tau_{n0}(p + p')}, \quad (3.56)$$

where

$$n' = n_i \exp\{(E_r - E_i)/kT\} \quad (3.57)$$

$$p' = n_i \exp\{-(E_r - E_i)/kT\} \quad (3.58)$$

$$\tau_{n0} = 1/\beta_n N_r = 1/\sigma_n N_r v_{th} \quad (3.59)$$

$$\tau_{p0} = 1/\beta_p N_r = 1/\sigma_p N_r v_{th}. \quad (3.60)$$

The quantities n' (and p') correspond to the free electron (hole) density when the Fermi level is located at the recombination center level. The quantities τ_{n0} and τ_{p0} correspond to the extreme *minimum* lifetime for electrons and holes that are obtained for each type of carrier when *all* the recombination centers are filled with carriers of the opposite sign, i.e., $U_n = (n_p - n_{p0})/\tau_{n0}$ when all the centers are occupied by holes and $U_p = (p_n - p_{n0})/\tau_{p0}$ when all the centers are occupied by electrons. For steady-state conservation of charge, $U_n = U_p$.

In the small-signal case with $\tau_n = \tau_p = \tau$, $(n - n_0) = (p - p_0) \ll n_0, p_0$, it is relatively simply to obtain an explicit expression for the lifetime from Eq. (3.56).

$$\tau = \frac{n - n_0}{U} = \tau_{p0}\left(\frac{n_0 + n'}{n_0 + p_0}\right) + \tau_{n0}\left(\frac{p_0 + p'}{n_0 + p_0}\right). \quad (3.61)$$

The small-signal lifetime is therefore largest for a particular recombination center level when the Fermi level lies at the middle of the band gap and is largest for a particular Fermi level position when the level lies near the band edge, i.e., for shallow imperfections.

The general solution for the large-signal case with $(n - n_0)$, $(p - p_0) \gg n_0, p_0$ and where $\tau_n \neq \tau_p$, i.e., $n - n_0 \neq p - p_0$, cannot be obtained in closed form exactly, but usually a very good approximation can be obtained (Blakemore, 1962). This approximation takes a fairly simple form for sufficiently high light intensities that $n - n_0$ and $p - p_0$ are both much larger than n_0, p_0, n', and p':

$$\tau_n^\infty \simeq \frac{(n - n_0)(\tau_{n0} + \tau_{p0}) + \tau_{n0}N_r}{n - n_0 + N_r(1 - f)} \tag{3.62}$$

$$\tau_p^\infty \simeq \frac{(p - p_0)(\tau_{n0} + \tau_{p0}) + \tau_{p0}N_r}{p - p_0 + N_r f}, \tag{3.63}$$

where f is the Fermi distribution function, $f = p'/(p_0 + p')$.

Figure 3.14 shows the variation of τ_n with the location of the recombination centers in the band gap for typical values of the parameters at constant illumination level. A p-type material controlled by acceptors N_A (like the absorbing layer in a Si solar cell) is assumed, together with donorlike recombination centers (positively charged when empty, neutral when electron occupied). A fixed value of G is chosen to produce an n_p of the order of magnitude found under solar illumination.† The hole quasi-Fermi level is fixed by the acceptors since $N_A \gg N_r$, except at the highest n_p when $p_p \simeq n_p$. The plot was made by solving Eq. (3.56) numerically for $n(U)$ and then calculating $\tau_n = U/n$, using the relation $p = n + N_A$ to define the relationship between p and n. The plot of Fig. 3.14 illustrates the fact that the longest lifetime occurs when the centers lie close to the band edges. As E_r approaches the conduction band edge, τ_n increases strongly because of thermal emptying of the recombination centers into the conduction band.

A plot of electron lifetime and recombination rate as a function of the quasi-Fermi level position E_{Fn} is given in Fig. 3.15 for the same parameters as above except that E_r is fixed at the intrinsic level and n is allowed to vary. Note the broad plateau of constant τ_n for which the lifetime is controlled by the shorter of τ_{n0} or τ_{p0}. As the value of n exceeds the acceptor concentration (as would obtain under solar concentration conditions), $n \simeq p$ and the lifetime is controlled by the longer of τ_{n0} or τ_{p0}. As pn approaches n_i^2, the recombination rate approaches zero, as it must for

† We have neglected diffusion of minority carriers out of the volume of interest, and so have decreased G to give values of n approximating solar illumination when $E_r = 0.5$ eV.

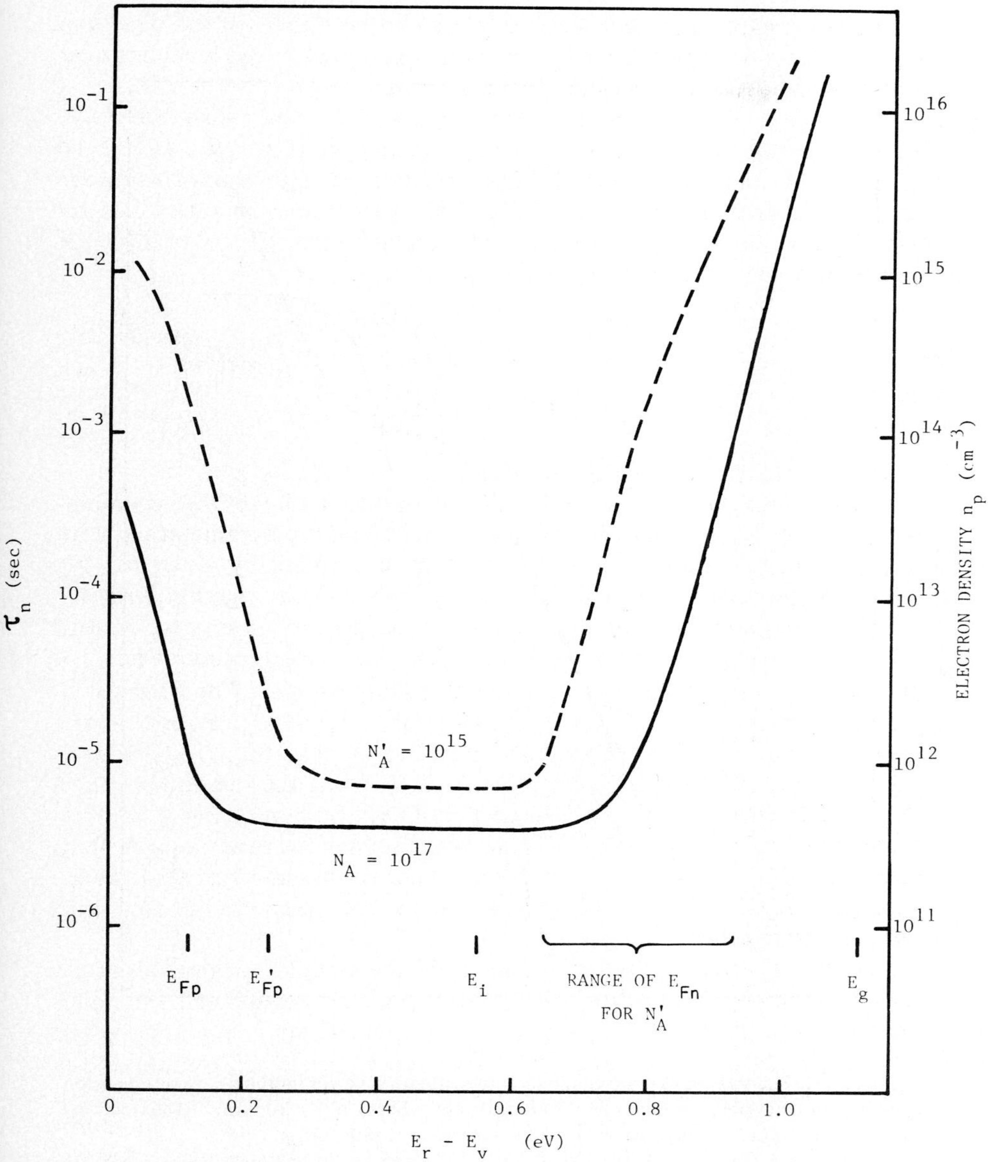

Fig. 3.14. Plot of effective minority carrier lifetime τ_n and electron density versus recombination center energy level E_r in p-Si for acceptor densities $N'_A = 10^{15}$ and $N_A = 10^{17}$ cm^{-3}. Assumed parameters: $N_r = 10^{10}$ cm^{-3}, $\sigma_n = 10^{-12}$ cm^2, and $\sigma_p = 10^{-15}$ cm^2. The range of E_{Fn} values shown in the plot is obtained by variation of the recombination center energy level for constant generation rate and $N'_A = 10^{15}$ cm^{-3}.

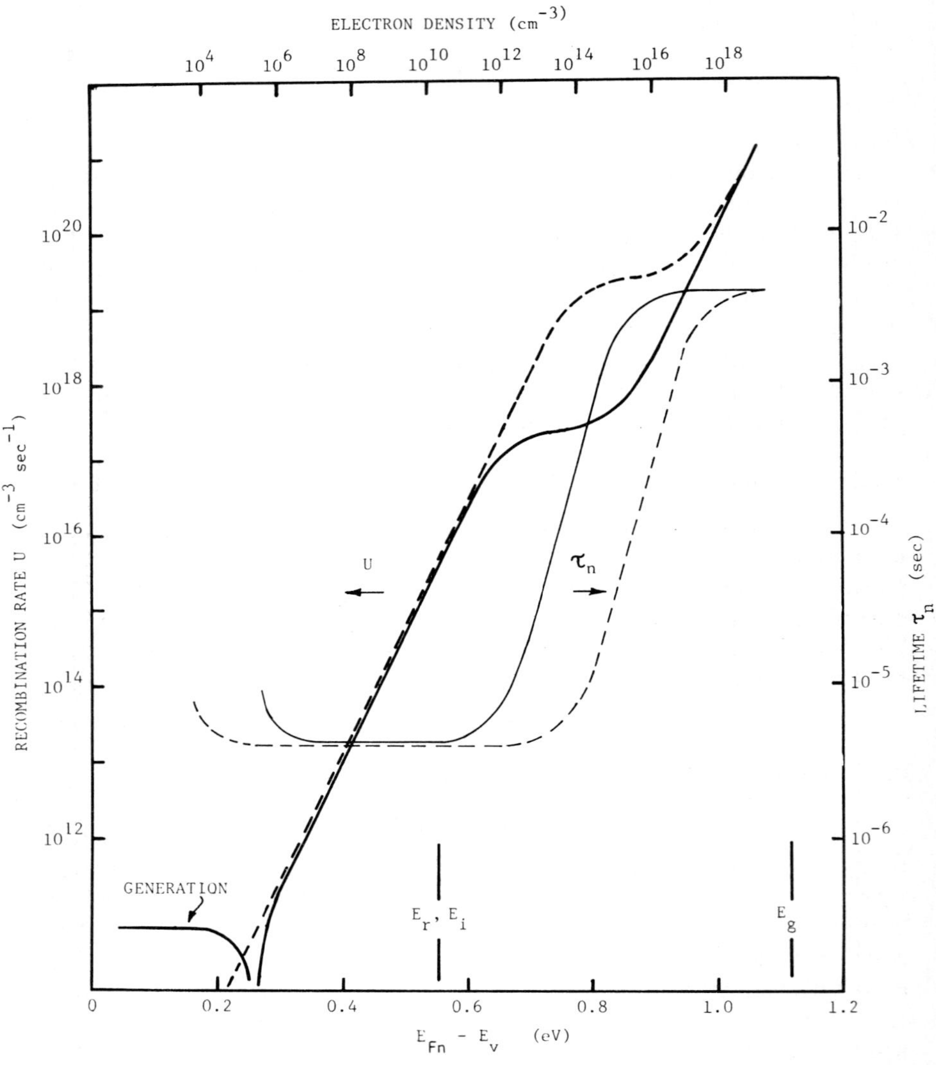

Fig. 3.15. Recombination rate U and effective minority carrier lifetime τ_n versus electron quasi-Fermi level position (and electron density) in p-Si for $N_A' = 10^{15}$ cm^{-3} (solid curves) and $N_A = 10^{17}$ cm^{-3} (dashed curves). Note generation region for $pn < n_i^2$.

thermal equilibrium. For $pn \ll n_i^2$, carriers are *generated* at the imperfection centers at a rate

$$U \simeq -n_i^2/N_A\,\tau_{n0} = -n_{p0}/\tau_{n0} \tag{3.64}$$

for this case where $p_p \simeq N_A \gg n_i$. If n and p are both smaller than n_i (as might occur in a junction depletion region), the generation rate is given by

$$U \simeq -n_i/(\tau_{n0} + \tau_{p0}). \qquad (3.65)$$

These types of situations are found in reverse-biased semiconductor-based junctions.

As seen previously, the kinetics of recombination are complex even when only one recombination center is involved. Fortunately, for most solar cells without solar concentration (e.g., at AM1 in p-Si, $n_p \simeq 10^{12}$ cm^{-3}), the lifetime is a constant over a considerable range of light intensities, as evidenced by a linear relationship between light-generated current and photon flux over many orders of magnitude. Practical values of τ_n range from 10^{-9} to 6×10^{-8} sec in p-GaAs and from 10^{-6} to 5×10^{-4} sec in p-Si crystals. Recombination in degenerate semiconductors is discussed by Hu and Oldham (1979) and by Von Roos (1978).

3.5.3 Surface Recombination

In addition to recombination at imperfections in the bulk of the semiconductor, there is also recombination at surface or interface states. The same approaches discussed earlier apply although the situation is usually described more simply. If the density of such states effective for recombination is N_s cm^{-2}, the recombination rate at the surface or interface can be described in terms of a *surface recombination velocity* S,

$$S = \beta_s N_s = \sigma_s v_{th} N_s, \qquad (3.66)$$

and the surface or interface recombination current density J_s is

$$J_s = (n_p - n_{p0})qS \qquad (3.67)$$

in a p-type material. Since the particle flux normal to the surface is given by $nv_{th}/4$, when S approaches one-fourth of the thermal velocity v_{th} of the carriers, the surface recombination loss is controlled only by diffusion from the bulk, with the surface acting as an infinite sink. The value of S is extremely sensitive to surface preparation, and S ranges from 10^2 cm sec^{-1} for a freshly cleaved or etched Si surface to greater than 10^6 cm sec^{-1} for a heavily damaged surface (e.g., a sandblasted surface).

The application of the Shockley–Read recombination model (and others) to surface recombination is treated by Many *et al.* (1971).

GENERAL REFERENCES

Blakemore, J. S. (1962). "Semiconductor Statistics" (H. K. Henisch, ed.), Semiconductors, Vol. 3. Pergamon, New York.

Blatt, F. J. (1963). "Physics of Electronic Conduction in Solids." McGraw-Hill, New York.

Bube, R. H., (1974). "Electronic Properties of Crystalline Solids." Academic Press, New York.
Bube, R. H. (1978). "Photoconductivity of Solids." Krieger, Huntington, New York.
Grove, A. S. (1967). "Physics and Technology of Semiconductor Devices." Wiley, New York. A very clear discussion on recombination.
Harbeke, G. (1972), Optical properties of semiconductors. *In* "Optical Properties of Solids" (B. Abeles, ed.), p. 21. North-Holland Publ., Amsterdam.
Moll, J. L. (1964). "Physics of Semiconductors." McGraw-Hill, New York.
Rose, R. (1963). "Concepts in Photoconductivity and Allied Problems." Wiley (Interscience), New York.
Smith, R. A. (1968). "Semiconductors." Cambridge Univ. Press, London and New York.
Spenke, E. (1958). "Electronic Semiconductors." McGraw-Hill, New York.
Sze, S. (1969). "Physics of Semiconductor Devices." Wiley, New York.
Ziman, J. M. (1964). "Principles of the Theory of Solids." Cambridge Univ. Press, London and New York.

REFERENCES

Ascarelli, G., and Rodriguez, S. (1961). *Phys. Rev.* **124,** 1321.
Ascarelli, G., and Rodriguez, S. (1962). *Phys. Rev.* **127,** 167.
Beck, J. D., and Conradt, R. (1973). *Solid State Commun.* **13,** 93.
Blakemore, J. S. (1962). "Semiconductor Statistics" (H. K. Henisch, ed.), Semiconductors, Vol. 3, p. 280. Pergamon, New York.
Burstein, E. (1954). *Phys. Rev.* **93,** 632.
Burstein, E., Picus, G., and Sclar, N. (1956). "Photoconductivity Conference," p. 353. Wiley, New York.
Burtscher, J., Dannhäuser, F., and Krausse, J. (1975). *Solid-State Electron.* **18,** 35.
Dziewior, J., and Schmid, W. (1977). *Appl. Phys. Lett.* **31,** 346.
Gummel, H., and Lax, M. (1955). *Phys. Rev.* **97,** 1469.
Hu, C., and Oldham, W. G. (1979). *Appl. Phys. Lett.* **35,** 636.
Kohn, W. (1957). Shallow impurity states in germanium and silicon. *Solid State Phys.* **5,** 257.
Lax, M. (1960). *Phys. Rev.* **119,** 1502.
Many, A., Goldstein, Y., and Grover, N. B. (1971). "Semiconductor Surfaces," p. 194. North-Holland Publ., Amsterdam.
Redfield, D. (1980). *Proc. 14th IEEE Photovoltaic Specialists Conf.,* p. 580.
Rose, A. (1963). "Concepts in Photoconductivity and Allied Problems," p. 120. Wiley (Interscience), New York.
Saura, J., and Bube, R. H. (1965). *J. Appl. Phys.* **36,** 3660.
Shockley, W., and Read, W. T. (1952). *Phys. Rev.* **87,** 835.
Sklensky, A., and Bube, R. H. (1972). *Phys. Rev.* **6,** 1328.
Smith, R. A. (1968). "Semiconductors." Cambridge Univ. Press, London and New York.
von Roos, O. (1978). *Solid-State Electron.* **21,** 633.

CHAPTER 4

APPLICATION OF THE TRANSPORT EQUATION

4.1 OVERVIEW

The absorption of photons in the active, absorber layer of a solar cell produces electron/hole pairs causing an excess minority carrier concentration to build up there. The processes of diffusion and bulk and surface recombination attempt to restore the conduction-band/valence-band system to its dark, thermal equilibrium condition under the constraint of charge neutrality. In an efficient solar cell most of these excess carriers must diffuse to the junction region, where they can be usefully collected. The transport equation relates the processes of generation, recombination, diffusion, and field-induced drift of the photoexcited carriers. Finally, the solutions of the transport equation enable us to write theoretical expressions for the total light-generated current to be obtained from a solar cell.

Since the solutions of the transport equation are often so complex as to obscure the physics behind them and are sometimes impossible to obtain at all except by numerical methods, we first focus on several simple but important solutions to build physical intuition. These solutions apply to systems with simple boundary conditions in which the law of superposition holds.† From these solutions we can obtain the theoretical quantum

† The law of superposition holds when the differential equations and the boundary conditions of the system are linear with respect to the carrier densities and their derivatives; the carrier densities (and their derivatives) can then be summed directly; e.g., $dn_1/dx + dn_2/dx = d(n_1 + n_2)/dx$, in the relevant equations. The linearity of the equation then implies that the light and dark carrier densities and currents can be summed or superposed.

Table 4.I

Typical Semiconductor Parameters for Solar Cells

	Si	GaAs
Majority carrier (hole) density p_{p0} (cm^{-3})	$10^{15}-10^{18}$	$10^{15}-10^{18}$
Minority carrier (electron) density (cm^{-3})		
dark, thermal equilibrium n_{p0}	10^5	10
under AM1.5 illumination n_p	10^{12}	$\simeq 10^9$
Hole mobility μ_p (cm^2 V^{-1} sec^{-1})[a]	400	400
Electron mobility μ_n (cm^2 V^{-1} sec^{-1})[a]	1200	4000
Minority carrier (electron) lifetime τ_n (sec)	$\simeq 10^{-4}$	$\simeq 10^{-8}$
Minority carrier (electron) diffusion length L_n (μm)	$\simeq 300$	$\simeq 3$
Absorption constant α (cm^{-1} at 1.9 eV)	4×10^3	8×10^4

[a] Highest experimental Hall mobilities.

efficiency and spectral response of the solar cell. Similar theoretical approaches are offered by Smith (1968), Landsberg (1975), Wolf (1960), and Wolf and Prince (1958), for example.

We next focus on more detailed considerations: the effect of an electric field, the situation for high injection levels, situations where the law of superposition does not apply, and, finally, the boundary condition assumptions. In the last section the measurement of minority carrier lifetime and of diffusion length is discussed, not only to examine the measurement per se, but to illustrate the application of transport equation principles to real systems.

Our examples are written in terms of a p-type material that is the usual major absorbing element in a solar cell (because the minority carrier diffusion length for electrons is usually greater than that for holes). The typical parameters of such a p-layer are given in Table 4.I to describe the domain in which we apply the transport equation.

Most of the considerations of this chapter are equally applicable to homojunctions and to heterojunctions or Schottky devices since we focus on the quasi-neutral regions and the adjoining depletion layer on one side of the junction only.

4.2 BASIC INGREDIENTS FOR THE TRANSPORT EQUATION

Consider an isolated, thin volume of absorber under uniform, monochromatic illumination, shown in Fig. 4.1. From Eq. (3.29), the bulk generation rate per unit volume is given by $G(x, \lambda) = \alpha(\lambda)\Gamma(x)$. This assumes

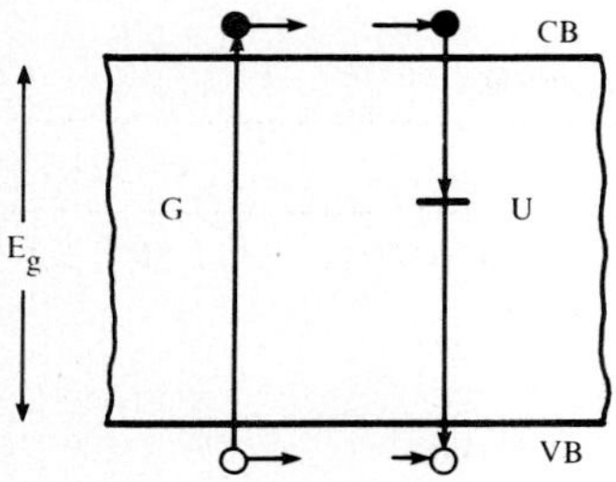

Fig. 4.1. Generation and recombination in a thin region of semiconductor.

that single photon processes dominate and that the quantum yield for the generation process is unity—both good assumptions for this domain. $G(x, \lambda)$ is a function of photon energy through $\alpha(\lambda)$ with a maximum on the order of $10^{21}-10^{22}$ cm^{-3} sec^{-1} when integrated over the AM1.5 spectrum.

The recombination of excess carriers is described by the Shockley–Read formulation discussed in Chapter 3. The bulk recombination rate can be approximated by using an effective minority carrier lifetime τ_n, which is constant with respect to carrier density and position. For $n_p \ll p_p$ the recombination rate $U(x)$ is given by

$$U(x) = (n_p - n_{p0})/\tau_n, \tag{4.1}$$

where n_{p0} is the thermal equilibrium density of minority carriers. The time rate of increase of n_p per unit volume is then

$$dn_p/dt = G(x) - U(x) = \alpha(\lambda)\Gamma - (n_p - n_{p0})/\tau_n \tag{4.2}$$

and, in steady state and without loss at the boundaries, the density of excess carriers is $n_p - n_{p0} = \alpha(\lambda)\Gamma\tau_n$. The meaning of τ and its measurement are discussed in more detail at the end of this chapter. Photoexcited carriers with energies in excess of the band-gap energy E_g very quickly come to thermal equilibrium with the lattice so that most of them have energies within a few kT of the band edge. The carrier scattering relaxation time $\tau_r \simeq \mu m^*/q \simeq 10^{-13}$ sec within the bands is much smaller than the minority carrier lifetime τ so that carriers equilibrate *within* the bands in times much shorter than that required for thermal equilibrium *between* bands. Thermal equilibrium within the bands is one of the basic assumptions for the application of the transport equation.

The concept of carrier diffusion is illustrated nicely by the Shockley–Haynes (1951) experiment in which a very thin slab of excess minority carriers is generated in a semiconductor bar at time $t = 0$ (Fig. 4.2) by the absorption of a short pulse of photons (an electron beam might

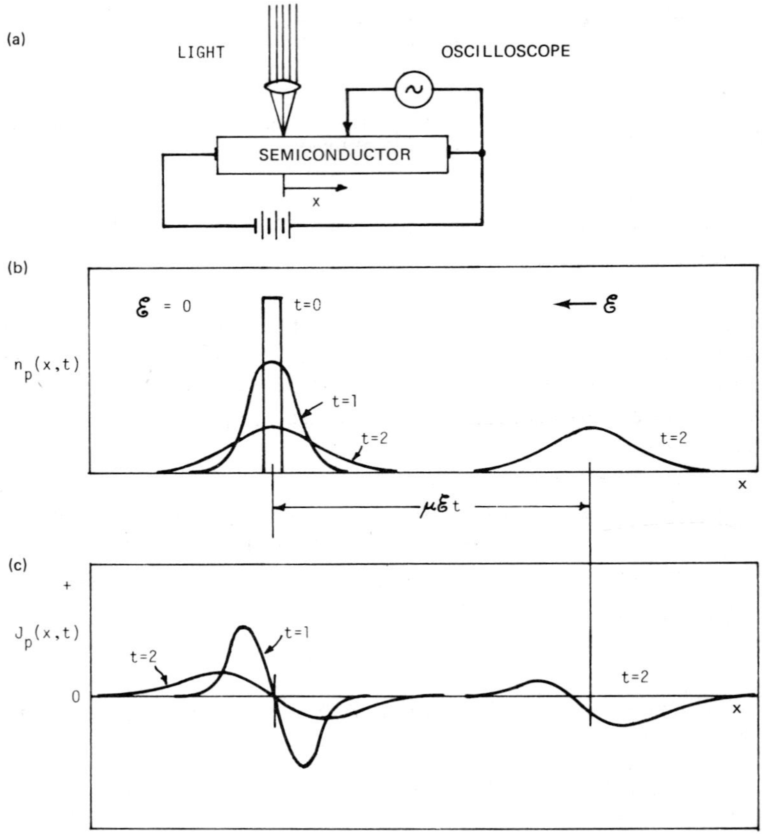

Fig. 4.2. (a) Schematic diagram of Haynes–Shockley experiment. Diffusion and drift of a slab of minority carriers generated at $t = 0$ for the infinite lifetime case. Carrier density (b) and current density (c) versus distance are shown for $\mathscr{E} = 0$ and $\mathscr{E} < 0$.

also be used to excite the slab of carriers). The electrons rapidly diffuse out of the volume, and their behavior is described by the diffusion equation

$$J_n = qD_n[\partial(n_p - n_{p0})/\partial x], \tag{4.3}$$

where D_n is the diffusion constant, which is related to the conductivity or drift mobility μ_n by the Einstein relation $D = kT\mu_n/q$. The continuity equation

$$\partial n_p/dt = \nabla \cdot J_n \tag{4.4}$$

must also hold since at this point we have assumed no sources or sinks within the semiconductor (after the excitation is turned off). In the absence of electric fields and for infinite lifetime, the one-dimensional solutions are error functions (see, e.g., Sze, 1969):

$$n_p(x, t) = [N/(4\pi D_n t)^{1/2}] \exp(-x^2/4D_n t) + n_{p0}, \qquad (4.5)$$

where N is the total number of photoexcited carriers initially generated per unit area and $n_p(x)$ and the corresponding electron current, using Eq. (4.3), are shown in Fig. 4.2.

When a small electric field $\mathscr{E}$ is applied, the whole packet of minority carriers moves in response, and a field-induced drift velocity $\mu\mathscr{E}$ is superimposed linearly on the random thermal motion. Then x in Eq. (4.5) is replaced by $(x - \mu\mathscr{E}t)$. The minority carrier mobility measured in this way is called the drift mobility and is identical to the conductivity mobility when $p \gg n$ or $n \gg p$, provided that trapping effects are absent. To consider finite carrier lifetimes, we must use the transport equation derived in the next section.

4.3 DERIVATION OF THE TRANSPORT EQUATION

In general, the current density is given by the sum of drift and diffusion components:

$$\mathbf{J}_n = nq\mu\vec{\mathscr{E}} + qD_n\nabla n \qquad (4.6)$$

$$\mathbf{J}_p = pq\mu\vec{\mathscr{E}} - qD_p\nabla p \qquad (4.7)$$

for electrons and holes, respectively, and the total current is $\mathbf{J} = \mathbf{J}_n + \mathbf{J}_p$. With the addition of a continuity condition these equations are sufficient to assemble the transport equation. For an incremental volume of semiconductor, the condition for conservation of charge is

$$\partial\rho/\partial t + \nabla \cdot \mathbf{J} = 0, \qquad (4.8)$$

$$\partial n/\partial t - \nabla \cdot \mathbf{J}_n/q = G_n - U_n, \qquad (4.9)$$

$$\partial p/\partial t + \nabla \cdot \mathbf{J}_p/q = G_p - U_p. \qquad (4.10)$$

Since the holes and electrons are generated in pairs by band-to-band excitation in our model, $G_n = G_p$. Since we assume that net charge neutrality must hold† ($\rho = 0$ and $\partial\rho/\partial t = 0$), we can show, using Eqs. (4.8)–(4.10), that

$$U_n = U_p = (n - n_0)/\tau_n = (p - p_0)/\tau_p, \qquad (4.11)$$

† Small deviations from charge neutrality may occur in transient cases or for strong gradients in the generation rate. The latter situation gives rise to the so-called Dember voltage, discussed in Section 4.5.3. We have also neglected trapping effects.

where n_0 and p_0 are the thermal equilibrium values. Using Eqs. (4.6) and (4.7) we obtain the one-dimensional transport equations:

$$\partial n/\partial t = D_n \partial^2 n/\partial x^2 + \mu_n \mathscr{E}\, \partial n/\partial x + \mu_n n\, \partial \mathscr{E}/\partial x - (n - n_0)/\tau_n + G_n(x), \tag{4.12}$$

$$\partial p/\partial t = D_p\, \partial^2 p/\partial x^2 - \mu_p \mathscr{E}\, \partial p/\partial x - \mu_p p\, \partial \mathscr{E}/\partial x - (p - p_0)/\tau_p + G_p(x). \tag{4.13}$$

In general, these equations must be solved simultaneously with Poisson's equation. In solar cells, however, we can usually uncouple the transport equations for holes and for electrons because $G_n = G_p$ and either $p \gg n$ or $n \gg p$. Also, in most cases, we are dealing with steady-state conditions (compared with carrier lifetimes) so that $dn/dt = dp/dt = 0$ and a single transport equation suffices to describe the excess minority carrier density,

$$D_n\, d^2 n_p/dx^2 + \mu_n \mathscr{E}\, dn_p/dx - (n_p - n_{p0})/\tau_n + G(x) = 0, \tag{4.14}$$

for p-type material with $p \gg n$, for example. Note that the term involving $d\mathscr{E}/dx$ drops out when the equations are uncoupled. Smith (1968, Chap. 8) and Jonscher (1960) are good general references on this topic. The time-dependent and steady-state transport expressions for minority carriers for several ideal source and sink geometries are given by Van Roosbroeck (1955).

When the injected minority carrier density becomes appreciable compared with the majority carrier density, this approximation fails (e.g., under high-intensity light, $\gtrsim 100$ suns, or at high forward bias). These conditions are discussed in Section 4.5.3.

4.4 SOLUTION OF THE TRANSPORT EQUATION

4.4.1 Boundary Conditions

Considering the n/p homojunction solar cell shown in Fig. 4.3, we assume for the moment that all the useful absorption is in the p-type, quasi-neutral region ($x_p \leq x \leq x_p'$). At the edge of the depletion layer $x = x_p$, the excess minority carriers are swept out rapidly by the junction field so that n_p approaches n_{p0} as $x \to x_p$ for zero applied bias. For forward bias V, the carrier density is given by

$$n_p = n_{p0} \exp(qV/kT) \qquad \text{at} \quad x = x_p, \tag{4.15}$$

which is a consequence of the constancy of the electron quasi-Fermi level across the junction depletion layer under forward bias (Section 5.2.3), a basic assumption in Shockley's (1949) treatment of the p/n homojunction.

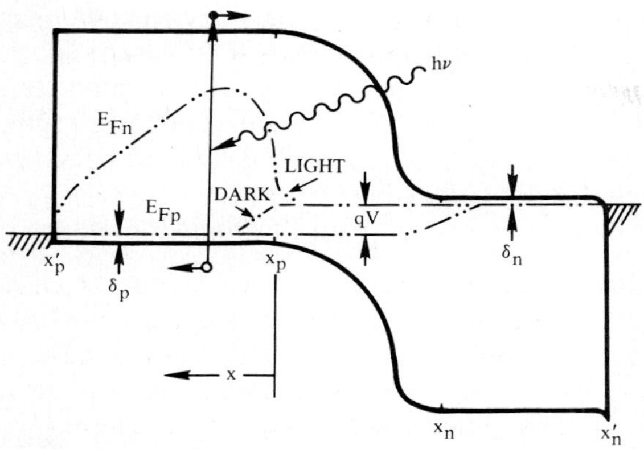

Fig. 4.3. Schematic band diagram of a simple homojunction solar cell at forward-bias voltage V showing quasi-Fermi levels with $n_p = n_{p0} \exp(qV/kT)$ at $x = x_p$ assumed. The width of the depletion layer is exaggerated. In this example it is assumed that no photogeneration takes place there.

These commonly used assumptions are quite good for most situations and greatly simplify the solution. This point is discussed further in Section 4.5.4.

4.4.2 The Semi-Infinite Absorber Case

For this example we assume that the thickness of the absorber layer is much larger than both the minority carrier diffusion length and the optical absorption length at the wavelength considered$[(x_p' - x_p) \gg 1/\alpha, L_n]$. It is further assumed that $\mathscr{E} \simeq 0$ in the quasi-neutral p-type layer, $x_p < x < x_p'$. The transport equation reduces to

$$d^2n_p/dx^2 - (n_p - n_{p0})/D_n\tau_n = -G(x)/D_n = -(\alpha\Gamma/D_n)\exp[-\alpha(x - x_p)], \tag{4.16}$$

where $\Gamma = \Gamma(\lambda)$ is the monochromatic illumination photon flux density at $x = x_p$. The boundary conditions are

$$n_p = n_{p0}\exp(qV/kT) \quad \text{at} \quad x = x_p \qquad \text{and} \qquad n_p \to n_{p0} \quad \text{as} \quad x \to \infty.$$

The solution is straightforward, and the nonequilibrium carrier density is

$$n_p = n_{p0}[\exp\frac{qV}{kT} - 1]\exp\left[\frac{-(x - x_p)}{(D_n\tau_n)^{1/2}}\right] + \frac{\alpha\Gamma(\lambda)}{D_n(\alpha^2 - 1/D_n\tau_n)}\left\{\exp\left[\frac{-(x - x_p)}{(D_n\tau_n)^{1/2}}\right] - \exp[-\alpha(x - x_p)]\right\} + n_{p0}. \tag{4.17}$$

The first term is the diffusion-induced minority carrier density due to forward bias of this half of a Shockley diode in the dark. The second term is the light-induced minority carrier density. The quantity $(D_n\tau_n)^{1/2}$ is called the minority carrier diffusion length L_n. In a uniform semiconductor, an electron excited a distance L_n from the collection point will have a $1/e$ probability of reaching that point.†

The electron current caused by diffusion is

$$\begin{aligned} J_n(x) &= qD_n dn_p/dx \\ &= \frac{-(qD_n n_{p0})}{L_n} \exp\left[\frac{-(x - x_p)}{L_n}\right]\left[\exp\left(\frac{qV}{kT}\right) - 1\right] \\ &+ \frac{q\alpha\Gamma(\lambda)}{(\alpha^2 - 1/L_n^2)}\left\{\alpha \exp[-\alpha(x - x_p)] - \frac{1}{L_n}\exp\left[\frac{-(x - x_p)}{L_n}\right]\right\}. \end{aligned} \tag{4.18}$$

The first term on the right-hand side of Eq. (4.18) is the diffusion current in *half* of a forward-biased Shockley diode in the dark. Note that the monochromatic photogenerated current given by the second term is not affected by applied bias and similarly the dark, forward-bias current is not affected by light. This is the situation for a so-called *ideal* solar cell, and it is a consequence of the principle of superposition. Since the differential equation (4.16) and the boundary conditions are linear in n_p, the resulting current is the sum of the currents arising from dark, forward-bias injection over the barrier and from photogeneration in the bulk. When the injection mechanisms interact (e.g., when the lifetime is a function of n_p or when generation and recombination in the depletion layer depends on J_n), the equations are no longer linear, and we cannot make these simple analyses. Lindholm *et al.* (1976), Tarr and Pulfrey (1979), and Rothwarf (1978) discuss the application of the law of superposition to solar cells.

At the depletion layer edge $x = x_p$ and with zero applied bias, Eq. (4.18) reduces to

$$J_n(x_p) = q\Gamma(\lambda)/(1 + 1/\alpha L_n) = J_\mathrm{L}, \tag{4.19}$$

where J_L is the monochromatic, light-generated current. The total current $J = J_n + J_p$ must be continuous throughout the device and at $x = x_p$, $J_p = 0$ at zero applied bias because of the junction barrier to hole current passing to the right. Thus we find

$$J_p(x) = J(x_p) - J_n(x) = J_\mathrm{L} - J_n(x). \tag{4.20}$$

These results are plotted in Figs. 4.4 and 4.5. Note that at $x = x_p$ the

† For purposes of approximation, the peak carrier density due to photoexcitation in the absorber layer is given by $n_{p\max} \simeq \Gamma(\lambda)/\alpha D_n$ in layers where the loss due to bulk recombination is small (i.e., high quantum efficiency).

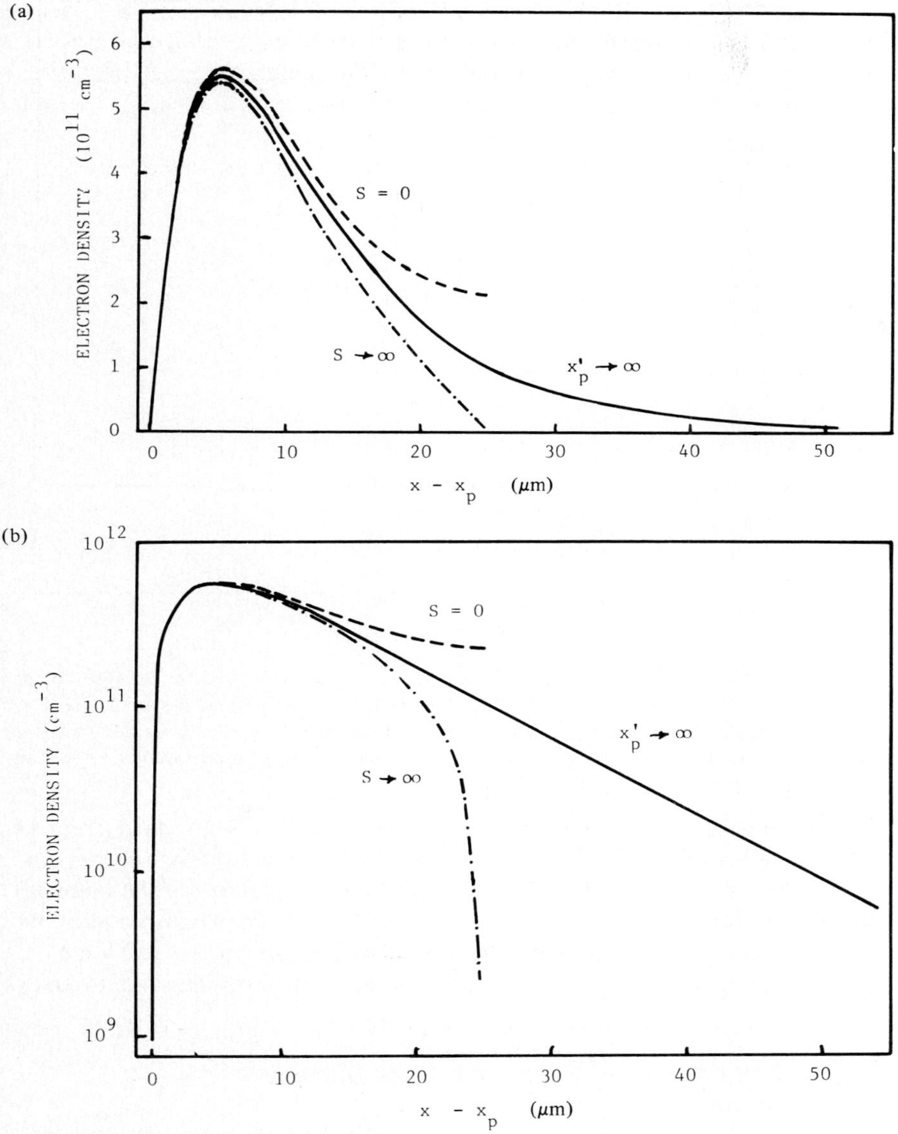

Fig. 4.4 (a) Linear and (b) log plots of carrier density versus x for a semi-infinite absorber and for an absorber of length $|x_p' - x_p| = 25\ \mu m$ with $S = 0$ and $S \to \infty$. Parameters used are $1/\alpha = 3\ \mu m$, $L_n = 10\ \mu m$, $D = 25\ cm^2\ sec^{-1}$, and $\Gamma = 10^{17}$ photons $cm^{-2}\ sec^{-1}$. Larger values of L_n would result in considerably greater differences between the peak carrier densities for the $S = 0$ and $S \to \infty$ cases.

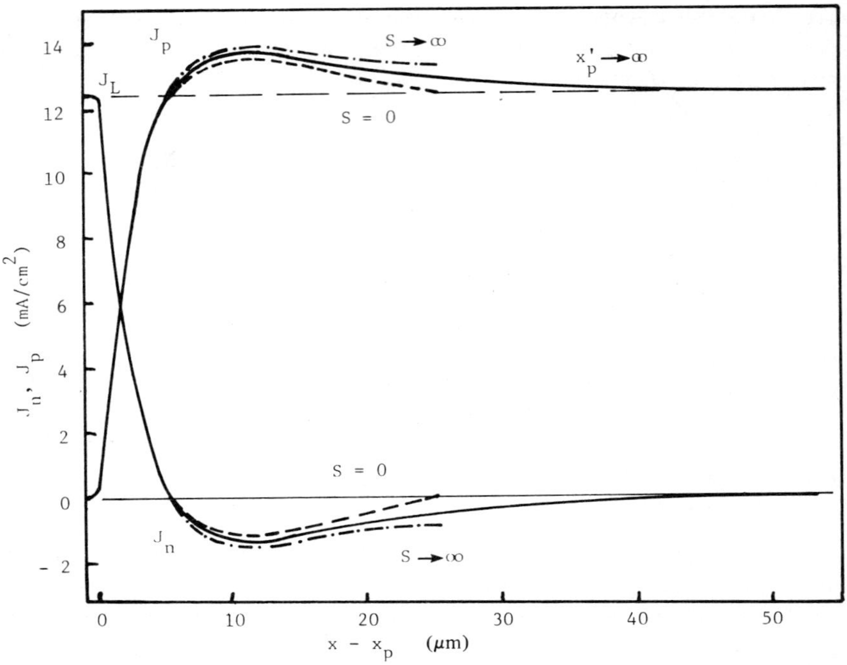

Fig. 4.5. Current density versus x for a semi-infinite absorber and for an absorber of thickness $|x'_p - x_p| = 25\ \mu$m with $S = 0$ and $S \rightarrow \infty$ at x'_p. Same parameters as for Fig. 4.4. Quantum efficiencies are $x'_p \rightarrow \infty$, 76.9%, $S = 0$, 76.4%; and $S \rightarrow \infty$, 77.3%. Larger values of L_n would result in considerably greater differences between the quantum efficiencies for the $S = 0$ and $S \rightarrow \infty$ cases.

current is carried by electrons only, whereas for $x \rightarrow x'_p$ the current is carried by holes. The major part of this hole current is a drift current induced by a field $\mathscr{E} \simeq \rho_p J_p$, where ρ_p is the bulk resistivity of the material for holes. This field ($\sim$0.1 V cm^{-1}) is generally very small compared to the junction field ($\sim 10^4$ V cm^{-1}) or to the diffusion field ($kT/qL_n \simeq 250$ V cm^{-1}). Thus the assumption that $\mathscr{E} \simeq 0$ in the quasi-neutral region is quite justifiable.

4.4.3 Finite Absorber Layer with Back Surface Recombination

Usually, the thickness of the absorber is finite, and there is a nonzero surface recombination velocity S at the rear contact. Since the electron current to the contact is primarily due to diffusion, the boundary condition at $x = x'_p$ is then

$$J_n = (n_p - n_{p0})qS = -qD_n\, dn_p/dx. \tag{4.21}$$

With these two modifications, the transport equation (4.16) can again be solved to obtain the expression for the monochromatic light current for zero applied bias (letting $y = |x_p' - x_p|$)

$$J_L = J_n(x_p)$$

$$= \frac{q\Gamma(\lambda)}{(1 + 1/\alpha L_n)(1 - 1/\alpha L_n)}$$

$$\times \left\{ 1 - \frac{(S/\alpha D_n)[\cosh(y/L_n) - \exp(-\alpha y)] + (1/\alpha L_n)\sinh(y/L_n) + \exp(-\alpha y)}{(SL_n/D_n)\sinh(y/L_n) + \cosh(y/L_n)} \right\}, \quad (4.22)$$

and the resulting carrier density and current density are plotted in Figs. 4.4 and 4.5 for limiting values of S. Expressions for $n(x)$ and $J_n(x)$ for various geometries are given in Appendix 2.

4.4.4 Current Generation in a Complete *p*/*n* Junction

In real solar cells there are nonnegligible contributions to the photocurrent from absorption in the depletion layer and from the n-type layer (usually quite thin). In the depletion layer most of the carriers are swept out before they can be lost by recombination, so the effective L_n is very large. The photocurrent from the depletion layer is approximately equal to the number of photons absorbed there per second multiplied by the electronic charge,

$$J_{dl} \simeq q\Gamma_0\{1 - \exp(-\alpha|x_p - x_n|)\}\exp(-\alpha|x_n' - x_n|), \quad (4.23)$$

where Γ_0 is the photon flux at the light-incident side of the n-type layer. Note that J_{dl} is a slowly varying function of bias voltage (since the depletion layer width $|x_p - x_n|$ changes with bias). This represents a departure from ideality for the device.

All the carriers generated in the depletion layer are assumed to be collected. If superposition holds, the total light current J_L is the sum of the currents from each of these three layers, each adjusted for the intensity of illumination reaching the respective layer. For such a cell

$$J_L = J_{p\text{-layer}} + J_{n\text{-layer}} + J_{dl}, \quad (4.24)$$

where the first term on the right-hand side of Eq. (4.24) is like Eq. (4.22), but with the photon flux Γ at the light-incident side of the p-layer.

The common configurations for solar cells are shown in Fig. 4.6, and the total J_L for each can be evaluated using expressions similar to Eqs. (4.22)–(4.24) or those given in Appendix 2.

In most cases and especially for cells utilizing indirect band-gap,

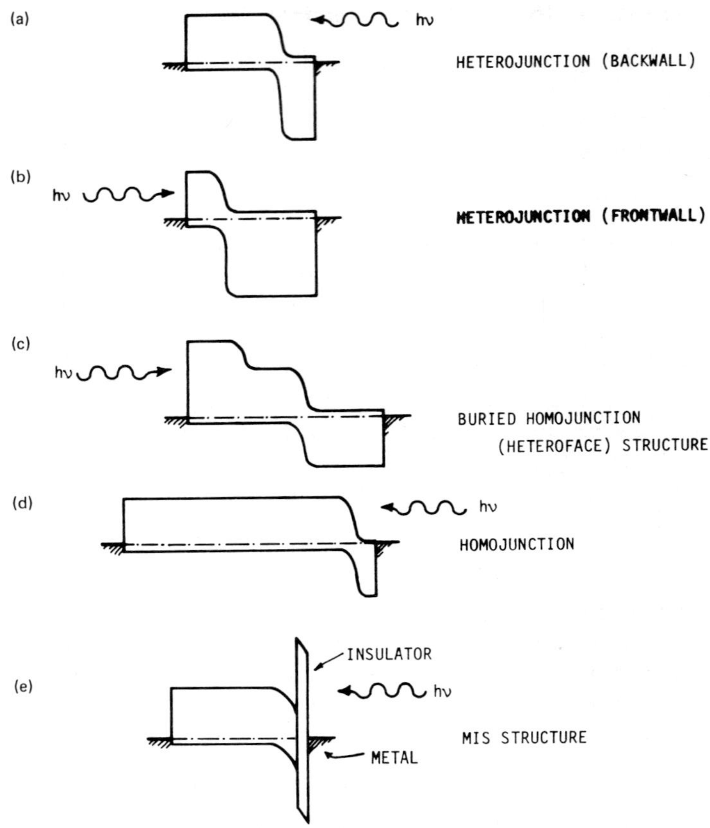

Fig. 4.6. Common configurations for solar cells. In cases (a), (b), and (c) the contribution to J_L from the large band-gap material is usually negligible.

low-absorption-coefficient semiconductors, the majority of the useful absorption takes place in the absorber layer, and the contributions from the depletion region and the light-entrant layer (*n*-type in our previous example) are small. Therefore, the approximations used in Eq. (4.23) are quite justified.

When lifetime, mobility, and electric field are not constant in the absorbing region, the solution of the transport equation becomes much more complex, and numerical methods must be used, such as have been used by Tsaur *et al.* (1972) for GaAs cells and by Fossum (1976) for Si cells.

4.4.5 Quantum Efficiency and Spectral Response

The internal quantum efficiency η_Q of an absorber layer is the ratio of the current of photogenerated carriers to the photon flux incident just inside the light-incident boundary of that particular layer (i.e., after accounting for reflection):

$$\eta_Q(\lambda) = J_L(\lambda)/q\Gamma(\lambda). \tag{4.25}$$

When the appropriate dependence of α on λ is inserted, Eq. (4.25), with (4.19) or Eq. (4.22) gives the *internal* quantum efficiency of a layer. One can also speak of the η_Q of a device with several layers, each of which contributes to the total current. In this case, Eq. (4.24) is used in Eq. (4.25), carefully accounting for the photon flux incident on each of the layers. To obtain the *external* quantum efficiency, these theoretical expressions must be multiplied by the term $(1 - R)$, where the reflectivity R is generally a function of λ. The external quantum efficiency is written with respect to the photon flux just outside the light-incident boundary; for example,

$$\eta_Q(\lambda) = [1 - R(\lambda)]/[1 + 1/\alpha(\lambda)L_n]$$

for the example of Eq. (4.19).

Another, less common, expression for η_Q includes the effect of nonuseful absorption and scattering within the semiconductor layers. If this loss is represented by A and J_L is given by a theoretical model, for example, the internal quantum efficiency might be written

$$\eta_Q(\lambda) = (1 - A)J_L(\lambda)/q\Gamma(\lambda).$$

For solar cells operating in a linear fashion, the experimental measurement of quantum efficiency can be done at any illumination intensity (provided the voltage drop across the current meter is negligible compared with V_{oc}). For nonlinear solar cells, such as Cu_xS/CdS devices or even Si cells operated at high intensity, the quantum efficiency depends on illumination intensity and spectral content.† In these cases measurement using bias light is usually required. In the bias-light method, a constant, white bias light with intensity and spectral content close to that of the sun is superimposed on a chopped monochromatic illumination of smaller intensity. The response ΔJ_L to the monochromatic beam is measured with a synchronous detector, and the quantum efficiency is then

† For Cu_xS/CdS cells this is because of wavelength-dependent trapping effects. For cells operated at high intensity it is a result of the change in lifetime at high minority carrier densities.

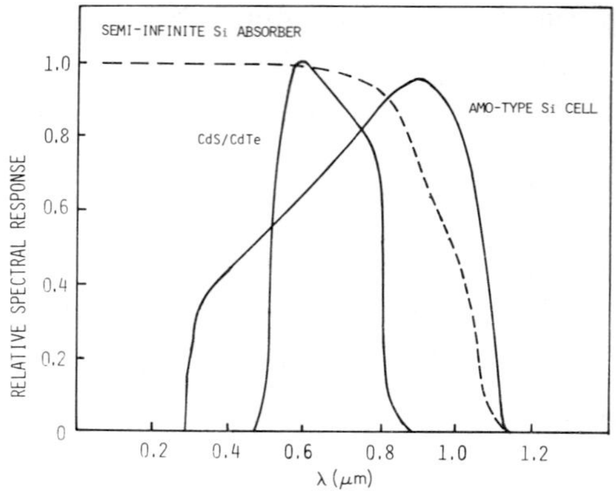

Fig. 4.7. Spectral response curves for a semi-infinite absorber (Si) with no front surface losses, an AM0-type Si *p/n* junction, and an experimental CdTe/CdS heterojunction cell.

given by

$$\eta_Q = (1/q)[\Delta J_L/\Delta\Gamma(\boldsymbol{\lambda})].$$

The term *spectral response* is usually reserved for the relation between the light-generated current J_L and the wavelength or photon energy hc/λ of the incident light. The spectral response can be written

$$dJ_L(\lambda)/d\lambda = q\eta_Q d\Gamma/d\lambda,$$

and if the photon flux is constant as λ is varied, then the spectral response and the quantum efficiency are proportional.†

The total light-generated current J_L in sunlight is then obtained by integration of the product of the monochromatic quantum efficiency $\eta_Q(\lambda)$ and the solar photon flux spectrum $d\Gamma(\lambda)/d\lambda$:

$$J_{L,\mathrm{total}} = q\int_0^\infty \eta_Q(\lambda)\frac{d\lambda}{d\Gamma(\lambda)}\,d\lambda. \tag{4.26}$$

Several spectral response curves are shown in Fig. 4.7, including a calculated curve for a Si semi-infinite absorber without front surface loss as given by Eq. (4.19), an experimental Si cell curve that shows losses at the front (x_n') surface and in the *n*-type and depletion layers, and, finally, a calculated heterojunction cell curve where the response is cut off at higher wavelengths by nonuseful absorption in the large band-gap window component. A relative spectral response curve calculated for a complete Si

† This differential expression for the spectral response and its integral form Eq. (4.26) are written for cells with a linear response; for cells in which J_L and Γ are not linearly related (e.g., the Cu_2S/CdS cell), the equations may not hold.

cell is shown in Fig. 7.16 with the contributions from the p-type (base), depletion, and n-type (diffused) layers shown separately as per Eqs. (4.22),† (4.23), and (4.24), respectively.

4.5 SPECIAL TOPICS

4.5.1 The Effect of Electric Fields

Electric fields may be used to decrease the losses due to surface and bulk recombination in practical solar cells. In the bulk case the carrier diffusion length L_n is effectively increased by an electric field $\mathscr{E}$. The general solution for the transport equation (4.14) with constant electric field can be written

$$n_p - n_{p0} = A \exp(x/L_+) + B \exp(-x/L_-) + \mathscr{F}[G(x)],$$

where A and B are constants to be defined by the boundary conditions and $\mathscr{F}[G(x)]$ is a functional depending on the form of the generation rate and the boundary conditions. The "upstream" and "downstream" diffusion lengths L_+ and L_- respectively, are given by

$$1/L_\pm = (1/2L_n)\,\{[(\mathscr{E}/\mathscr{E}_c)^2 + 4]^{1/2} \pm (\mathscr{E}/\mathscr{E}_c)\}, \tag{4.27}$$

where L_n is the zero field diffusion length. The term $\mathscr{E}_c = kT/qL_n$ is called the critical field (Smith, 1968, pp. 243, 247). The diffusion lengths L_+ and L_- can be aproximated quite well by

$$L_+ \simeq L_n[(\mathscr{E}/\mathscr{E}_c) + 1]^{-1}, \qquad L_- \simeq L_n[(\mathscr{E}/\mathscr{E}_c) + 1]. \tag{4.28}$$

(The approximations of Eq. (4.28) are for positive $\mathscr{E}$; for negative $\mathscr{E}$ the + and − subscripts must be interchanged.) These relationships, illustrated in Fig. 4.8 for injection at a point, show that the effective diffusion length is increased when carriers are swept away from the point of injection by the field.

The form of the functional $\mathscr{F}$ depends on the generation term, and in the situation for which Eq. (4.17) was derived, for example, it has the form

$$\mathscr{F}[G(x)] = [-(\alpha\Gamma_0/D_n)\exp(-\alpha x)]/[\alpha^2 - (\alpha\mathscr{E}/L_n\mathscr{E}_c) - 1/L_n^2]. \tag{4.29}$$

Solutions for constant $\mathscr{E}$ can be written explicitly for many situations, as, for example, those given by Hovel (1975, p. 52), Wolf (1963), and Ellis and Moss (1970).

† Equation (4.22) must be modified for the n-type layer because of the direction of illumination.

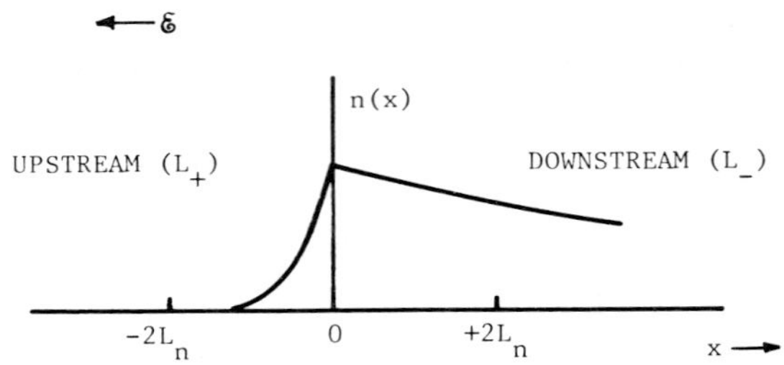

Fig. 4.8. Carrier density in a uniform bar with electrons injected at the point $x = 0$. $\mathscr{E} \gg \mathscr{E}_c$.

The electric field can be built into the absorber layer by variation of the acceptor concentration N_A over its thickness. The field is given approximately by

$$\mathscr{E} \simeq (kT/q)(1/N_A)\, dN_A/dx,$$

and the effective minority carrier diffusion length is given by

$$L_- \simeq L_n[(\mathscr{E}/\mathscr{E}_c) + 1] \simeq L_n[(L_n/N_A)(dN_A/dx) + 1].$$

Although a built-in electric field spanning the bulk absorber layer is seemingly an attractive way to increase L_n and hence the J_L, such fields have rarely been used. This may be due to the difficulty of producing the required doping gradient without reducing lifetime as well. Further, such a field necessarily decreases the obtainable diffusion voltage at the junction, essentially using up a portion of it to produce the potential gradient. This decreases the V_{oc} of the cell and so some compromise must be reached between achieving high J_L and V_{oc}. Drift fields in low lifetime materials are discussed theoretically by Urli *et al.* (1975).

Some attention has been given to electric field effects on recombination in the space charge region for Schottky barrier cells. A computer calculation for such a cell including the spatial variation of $\mathscr{E}$ is given by McQuat and Pulfrey (1976). Van Overstraeten and Nyuts (1969) give numerical solutions for the transport equation with spatial variation of $\mathscr{E}$, τ, and μ. A treatment of recombination in the space charge layer is given by Mallinson and Landsberg (1978) for p/n Si cells. In amorphous Si cells that have characteristically short L, enhancement of the effective L by the depletion layer field is generally *required* to produce reasonable quantum efficiency (see Chapter 11).

So-called back-surface fields (BSFs) have been used quite successfully in modern Si cells to reduce surface recombination losses at the back ohmic contact of the absorber layer. Such a field, produced by a short diffusion of additional dopant at the back surface, acts to "reflect" the photoexcited carriers and reduce their density there. Back-surface fields are discussed in more detail in Section 7.5.2.

4.5.2 On the Constancy of the Photogenerated Current across the Junction

Implicit in our derivation for J_L was the assumption that the photogenerated electron current was constant across the depletion layer and that once the carriers passed the depletion layer edge they were fully collected. The current of photogenerated carriers raises the quasi-Fermi level within the depletion layer above its equilibrium value (as in Fig. 5.7), implying that more recombination takes place there and that some of the carriers are lost in the depletion layer, even considering the increased effective diffusion length due to the high junction field.

A simplified, approximate treatment is given here that assumes that J_L can be superimposed on the dark, forward-bias current and that the important depletion layer recombination occurs within a thin layer at the interface between the p- and n-type layers (appropriate for heterojunction cells). The recombination loss is lumped into an effective interface recombination velocity S_i that is assumed to be constant with respect to applied bias. The continuity of photogenerated current across the interface is expressed by $J_{in} = J_{out} + qnS_i$ and the carrier density at the interface can be approximated by $n(x_i) = J_{out}/q\mu\mathscr{E}$, where $\mathscr{E} \simeq 2(V_d - V)/W_d(V)$. The depletion layer width is $W_d(V)$ and x_i is the interface position. Combining we have

$$J_{out}/J_{in} = 1/[1 + S_i/\mu\mathscr{E}(V)]. \tag{4.30}$$

In this way it is possible to treat the recombination loss of J_L in the junction region rather independently of the dark junction current. This approach has been used by Mitchell *et al.* (1977) and Rothwarf *et al.* (1978).

An alternative approach assumes that $S_i = \delta_i U_i(V)$, where δ_i is the thickness of the interfacial layer and $U_i(V)$ is the recombination rate there that depends on the applied bias voltage. For recombination centers with a single energy level midway between E_{Fn} and E_{Fp} in a symmetrical junction, the recombination rate U_i might be given by $U_i(V) \simeq U_{i0}[\exp(qV/2kT) - 1]$ for example. Since the electric field at the junction is high ($\simeq 10^4$ V cm^{-1}), the carrier velocity $v(\mathscr{E})$ is no longer given by $\mu\mathscr{E}$, but is more appropriately expressed by the scattering-limited velocity, i.e., $v(\mathscr{E}) \rightarrow v_{sl} \simeq 10^7$ cm sec^{-1}. Then Eq. (4.30) can be rewritten

$$J_{out}/J_{in} = 1/[1 + S_i(V)/v_{sl}].$$

Loss of photogenerated carriers in the depletion layer is discussed further in Section 6.3.1.

4.5.3 High Injection Effects

When the minority carrier density is increased substantially by operation at high light intensities ($\gtrsim 100$ suns) and/or high forward bias ($V \gtrsim 0.7\ E_g/q$), several assumptions used in the calculation of J_L must be reconsidered. Consider a p-type absorber for simplicity.

1. Since the condition $p_p \gg n_p$ no longer holds, the bulk recombination is no longer controlled solely by the photogenerated excess minority carrier density. A more complete recombination model, such as Eq. (3.56) must be used. The dependence of the recombination rate may no longer be linear in n_p, and longer lifetimes are generally obtained (Mataré and Wolff, 1978).

2. The criteria for neglecting the majority carrier transport equation, Eq. (4.13) in our previous example, may no longer be valid, and then both electron and hole transport equations must be solved simultaneously.

3. The boundary condition $n_p = n_{p0} \exp(qV/kT)$ may no longer be valid, requiring that the transport problem be expanded to include the depletion layer.

4. At high photogenerated carrier densities the Dember voltage (Dember, 1931, 1932), due to different mobility values for electrons and for holes in the presence of a large spatial variation in generation rate, can be as large as $\simeq 0.05$ V for 500 suns in Si (Lindholm and Fossumi, 1977). Proper choice of n- or p-type absorber and light-incidence direction will add the Dember voltage to the junction voltage rather than subtracting from it.†

These complications dictate the use of numerical methods to solve the transport equation. Other effects of high injection are evident with respect to the junction behavior and series resistance losses as discussed in Chapter 6.

High injection effects in solar cell design are discussed by Dalal and Moore (1977) and Fossum and Burgess (1976).

4.5.4 Critique of Assumptions

In this section we examine several of the assumptions underlying our derivation of the photogenerated current J_L. We wish to concentrate our attention on those assumptions that concern transport in the depletion

† A very readable account of the Dember potential is given by von Roos (1979).

layer, thinking of the quasi-neutral absorber more as a source that is modified only by conditions at the boundary of the depletion layer.

The Recombination Rate Expression The recombination rate expression, Eq. (4.1), is based on the Shockley–Read model for a single-level recombination center when $p_p \gg n_p$ (assuming a *p*-type absorber) and τ_n can be considered constant with respect to n_p and position. The approximation is probably valid in the quasi-neutral regions of most systems of interest at AM0 illumination levels, providing that the τ used is an effective lifetime that is measured under similar conditions of carrier density, illumination, and temperature.

In the depletion layer, and under high illumination, $p_p \gg n_p$ no longer holds, and the full Shockley–Read expression must be used. The recombination rate in the depletion layer then depends on the position of the recombination levels and the quasi-Fermi levels for electrons and holes in a complex way, generally reaching a maximum value when the electron quasi-Fermi level is in the vicinity of the intrinsic level E_i. This consideration is particularly relevent for direct band-gap cells, where the fraction of the light absorbed in the depletion region may be appreciable. The total recombination current in the depletion region is given by integration of the recombination rate across the junction:

$$J_r = \int_{x_n}^{x_p} qU(x)\,dx. \tag{4.31}$$

The fact that the rate reaches a strong maximum within a limited region of the depletion layer allows some simplifying assumptions, e.g., allowing the replacement of $U(x)$ by its maximum value within $x_n < x < x_p$ so that the integral can be replaced by a simple product.

The Boundary Condition at the Depletion Layer Edge For a boundary condition we previously assumed that *all* the photogenerated carriers are swept away by the junction field, driving the minority carrier concentration at the depletion layer edge to its dark, thermal equilibrium value. Although this is a great convenience in calculation, we know intuitively that this cannot be true. The density of minority carriers at $x = x_p$ must be influenced to some degree by the properties $[\mu, \mathscr{E}(x)]$ of the depletion layer into which they flow; the depletion layer is not quite a perfect sink for carriers as was previously assumed. If we assume that the photogenerated carriers traversing the junction are traveling at a scattering-limited velocity v_{sl}, the limiting value under high fields, then the minimum carrier density in the depletion layer is J_L/qv_{sl}, which is considerably greater than the dark, equilibrium value for most of the depletion layer. For example, for GaAs, μn begins to saturate at $\mathscr{E} \sim 2 \times 10^3\ \mathrm{V\ cm^{-1}}$ and

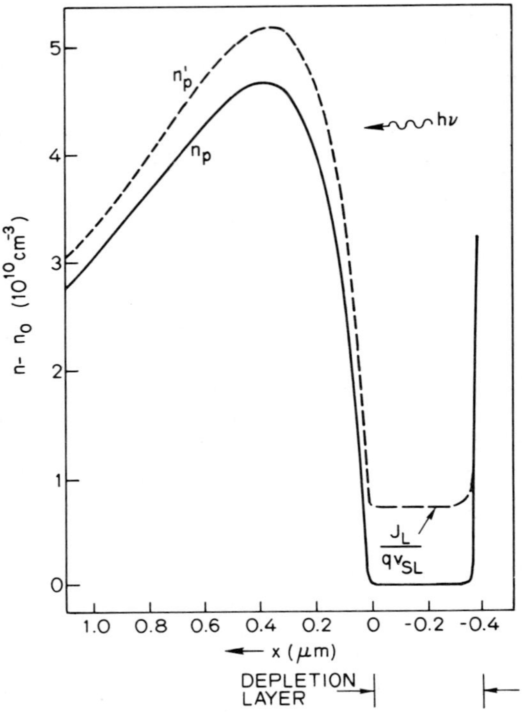

Fig. 4.9. Calculated carrier density in a CdTe p/n junction at zero bias with (-----) and without (—) scattering-limited-velocity limitation. The solid $n_p(x)$ curve is "forced" to fit the boundary condition $n_p = n_{p0}$ at $x = x_p$.

$v_{sl} \sim 10^7$ cm sec^{-1}. Thus the carrier density in these junctions must be approximately as shown in Fig. 4.9. For a Si homojunction the minimum carrier density as limited by v_{sl} is much smaller than the light-generated carrier density so the effect is not important and $n_p = n_{p0}$ at $x = x_p$ is still a good assumption. However, for direct band-gap materials, which commonly have a much shorter τ, the maximum photogenerated carrier density in the quasi-neutral region is much smaller, and the effect of the scattering-limited-velocity carrier density is relatively larger. If we assume a v_{sl} throughout the depletion layer to obtain an approximate electron density there, then the reduction factor for J_L is approximately $(1 - D_n/L_n v_{sl})$ with D_n and L_n taken in the quasi-neutral region. This is a *lower* bound on the reduction of J_L and in the example of Fig. 4.9 for CdTe, J_L is reduced by a factor of $\approx 2\%$ assuming $L_n = 1$ μm and $v_{sl} = 1.3 \times 10^7$ cm sec^{-1}.

Diffusion of Hot Carriers within the Depletion Layer Consideration of the carrier density in the junction region is further complicated by the fact that the minority carrier density in the depletion layer is not in equilibrium when light current is flowing. The excess carriers are "hot" with respect to the dropping conduction band edge, and they are thermalizing to approach equilibrium with the carriers on the other side of the depletion layer. Equilibrium should be reached within several electron mean free paths (electron mean free path $\sim 0.02\ \mu m$), which may be an appreciable fraction of the depletion layer width. In fact, the thermalization of hot carriers in the depletion layer is a major but unavoidable loss mechanism for solar cells, reducing the output power by approximately $[1 - (V_m/V_d)] \simeq 20\%$, where V_m is the voltage at the maximum power point and V_d is the junction diffusion voltage.† (See Section 6.5.)

The diffusion of hot electrons in a junction field is discussed by Seeger (1973), Stratton (1962, 1969), and Persky and Bartelink (1970). The usual treatment is to use the so-called diffusive approximation for the electron distribution function [$f = f_0 - \tau_m \mathscr{E}_z(q/m^*)\ \partial f_0/\partial v_z$, for example], where τ_m is the scattering relaxation time and v_z and $\mathscr{E}_z$ are the components of the velocity and electric field in the z direction. The Boltzmann approximation for the electron distribution function f_0 is still used but with an electron temperature T_e which is higher than that of the lattice, $f_0 \sim \exp(-mv_z^2/2kT_e)$. Equivalent T_e values of $\sim 1000°K$ are predicted (Sigmon and Gibbons, 1969) and observed (Davies, 1964) for junction fields of $\sim 10^4 V\ cm^{-1}$. The general effects of these high fields are as follows:

1. the Einstein relation in its simple form $(D = kT\mu/q)$ no longer holds,
2. the carrier velocity saturates toward the thermal velocity at high fields, and
3. the current–voltage relation for the diode is changed somewhat and must include T_e, which is itself a function of the electron current (Seeger, 1973; *Proc. Int. Conf.*, 1978).

This consideration introduces the realization that thermal equilibrium may not hold in the dark depletion layer under forward bias and that it certainly does not hold in the light. The principle of superposition used in the previous calculations is not strictly applicable, although it appears to give satisfactory results (see Lindholm *et al.*, 1976a). A question that remains to be answered by the solar cell theorist is how to describe the den-

† $qV_d = (E_g - \delta_n - \delta_p)$, where E_g is the absorber band-gap energy and δ_n and δ_p are the energy differences between the Fermi levels and the conduction and valence band edges in the n- and p-type quasi-neutral regions.

sity of hot carriers thermalizing in the depletion layer and, further, how to describe the recombination of the photoexcited barriers traversing the junction in the nonequilibrium situation.

Despite the preceding complications, the assumption that $n_p = n_{p0} \exp(qV/kT)$ at $x = x_p$ is almost universally used since it gives experimentally verifiable results and is consistent with the approximations used in Shockley diode theory.

4.6 MEASUREMENT OF MINORITY CARRIER LIFETIME AND DIFFUSION LENGTH

Since the magnitude of the optical absorption coefficient $\alpha(\lambda)$ is an inherent property of a semiconductor material that can be varied only within relatively narrow limits, the minority carrier lifetime τ remains as one of the major connections between materials processing and the final efficiency of a solar cell. The principal effect of increasing τ is to decrease the diode parameter J_0 by reducing recombination of forward-bias-injected carriers and ultimately to increase V_{oc}. For example, consider the V_{oc} of an ideal Si p/n junction solar cell where the diode current is dominated by diffusion and recombination in the base region. For AM1 illumination, V_{oc} is increased by about 0.03 V (or $\sim 5\%$) when τ is increased by a factor of 10. The accompanying percentage increase in J_{sc} is about an order of magnitude smaller. Although these changes seem small, it is common for τ to vary between 10^{-5} and 10^{-3} sec during Si processing [see e.g., Fischer and Pschunder (1975)].

The techniques that we discuss here measure the effective minority carrier lifetime τ or an equivalent quantity, the minority carrier diffusion length $L = (D\tau)^{1/2}$. As discussed in Chapter 3, τ is related to the extremum electron and hole lifetimes τ_{n0} and τ_{p0} and is also a function of the position of the electron and hole quasi-Fermi levels (i.e., the doping density and the excitation intensity). The response time of a sample or device is influenced by trapping and surface recombination effects in many cases so that one must be cautious in relating this response time to a minority carrier lifetime (e.g., a p/n junction device).

The methods of lifetime measurement can be separated into two broad categories. The first uses a sample of the material, often of appreciable size, on which measurements such as Haynes–Shockley drift, photoconductivity decay, or surface photovoltage are done. The second category makes use of the properties of a device in its final state. Examples of such measurements on a p/n junction or Schottky diode device include electron or light-beam-induced current, V_{oc} decay, and capacitance transients. Since processing, even a simple low-temperature heat treatment,

Table 4.II

Summary of Lifetime and Diffusion Length Measurements

Method	Mode	Independent variable	Excitation	Result	Range
Haynes–Shockley	Transient	Electric field, collector position	Point	τ, μ	$>10^{-7}$ sec
Photoconductivity decay	Transient	Time	Uniform	τ	$>10^{-8}$ sec
Photoelectromagnetic effect	Steady state	Magnetic field	Uniform	τ	10^{-8}–10^{-5} sec
V_{oc} decay	Transient	Time	V_F[a] or $\exp[-\alpha(\lambda)x]$	τ	10^{-7}–10^{-3} sec
MIS capacitance decay	Transient	Time	V_F[a]	τ	10^{-9}–10^{-6} sec
Surface photovoltage	Steady state	Wavelength and intensity of light	$\exp[-\alpha(\lambda)x]$	L	>0.1 μm (10^{-8}–10^{-3} sec)
Light-generated current	Steady state	Wavelength	$\exp[-\alpha(\lambda)x]$	L	>0.1 μm
Beam-induced current	Steady state	Source–collector distance	Point	L	>0.2 μm

[a] V_F is forward-bias voltage.

can change τ so much, the latter methods seem preferable if they can be applied. A number of techniques are listed in Table 4.II. The measurement of carrier lifetime in Si for solar cells is the subject of a review by Graff and Fischer (1979).

4.6.1 Haynes–Shockley or Drift Measurement of τ

The Haynes and Shockley (1951) method, one of the oldest techniques, was devised to measure both mobility and lifetime. A thin, short-duration pulse of carriers is injected into a rod of the semiconductor using near-band-gap light or x-rays, as shown in Fig. 4.2. The pulse of excess carriers is swept down the rod by an electric field and its magnitude is measured with a contact probe. The pulse height is decreased both by diffusion and by recombination, with the shape of the carrier density pulse being given by (Haynes and Shockley, 1951)

$$n_p(x, t) = [N_0/(4\pi D_n t)^{1/2}] \exp\{-[(x - \mu\mathscr{E}t)^2/4D_n t] - t/\tau\} + n_{p0}, \qquad (4.32)$$

where N_0 is the total initial number of injected carriers per unit area. By observing the decay in pulse height with time, lifetimes of $\sim 10^{-7}$ sec or larger can be determined. Use of the technique is discussed by Durrant (1955) and Spitzer *et al.* (1955).

In this and certain other methods to be mentioned, surface recombination can greatly lower the observed lifetime in samples with large surface-to-volume ratios or when the injecting radiation is absorbed close to the surface. For this reason photons of energy just above the band gap are preferable; in measurements on Si, for example, a thin Si sheet is used to filter the incident radiation.

4.6.2 Photoconductivity Decay

In this technique the entire sample is illuminated as uniformly as possible by a short pulse of penetrating light. The excess carrier density is given by

$$\Delta n_p = n_p - n_{p0} = \Delta n_{pi} \exp(-t/\tau_n),$$

where Δn_{pi} is the excess carrier density at $t = 0$. A similar relation holds for holes with a lifetime τ_p and the conductivity is given by

$$\sigma = (n_{p0} + \Delta n_p)\mu_n + (p_{p0} + \Delta p_p)\mu_p.$$

Usually $\Delta n_p = \Delta p_p$, $\Delta p_p \ll p_{p0}$, and $n_{p0} \ll p_{p0}$ for p-type material so that $\tau_n = \tau_p = \tau$ in the absence of trapping and the change in conductivity induced by the light pulse is

$$\Delta\sigma = \Delta n_p(\mu_n + \mu_p) = \Delta n_{pi}(\mu_n + \mu_p) \exp(-t/\tau). \qquad (4.33)$$

With a constant electric field across the sample a measurement of the photocurrent versus time gives the lifetime. Since $\Delta\sigma$ is usually much smaller than the dark conductivity for solar cell material, intense light pulses are needed and ac methods may be required. Lifetimes of $\geq 10^{-6}$ sec are commonly measured by this technique.

In both of the techniques discussed so far the decay time of the light source must be small compared to the measured lifetime. Xenon flash tubes, rotating mirror arrangements, and flash x-ray sources have been used. Sinusoidal excitation of frequency ω using a Kerr cell or modulated laser can be used for which the phase shift angle θ is given by $\tan\theta = \omega\tau$. Using this last method, lifetimes of $\geq 10^{-8}$ sec have been measured.

The photoconductivity decay can be detected directly using ohmic contacts or indirectly by using microwave absorption techniques (Larrabee, 1960; Atwater, 1961), by using capacitive coupling (Weingarten and Rothberg, 1961), by monitoring the free carrier optical absorption (Harrick, 1956), or by observing photoluminescence (Vilms and Spicer, 1965).

In this and other measurement techniques the presence of carrier trapping can have a strong effect on the measured response time, causing it to be much larger than the actual lifetime. For band-to-band recombination or when a carrier resides on a mid-band-gap center for a time short compared to the lifetime, electron and hole lifetimes are equal and the measurement should be valid. However, when a carrier is trapped on a center for a time longer than the lifetime, the time required to empty the traps dominates the measured response time. The phenomena associated with trapping are complex, and the situation is best avoided by choosing appropriate experimental conditions if possible. A qualitative description of the effects of trapping and surface recombination is shown in Fig. 4.10. Generally, for higher injection levels, trapping effects become small because the density of free carriers exceeds the density of trapped carriers, and the trapping population can become saturated. Van Roosbroeck (1960) and Rose (1963) give a theoretical treatment of the effects of trapping on photoconductivity.

4.6.3 Surface Photovoltage

In this technique the surface of the bulk semiconductor is illuminated by strongly absorbed light while the steady-state surface potential is monitored using a transparent, capacitive probe with transient or ac techniques. If $L_n \gg W_d$, the excess carrier density at the surface is given by

$$\Delta n_p \simeq q\Gamma(1 - R)/\{[(D_n/L_n) + S][1 + 1/\alpha(\lambda)L_n]\}, \tag{4.34}$$

where R is the reflectivity of the surface and Γ is the incident photon flux.

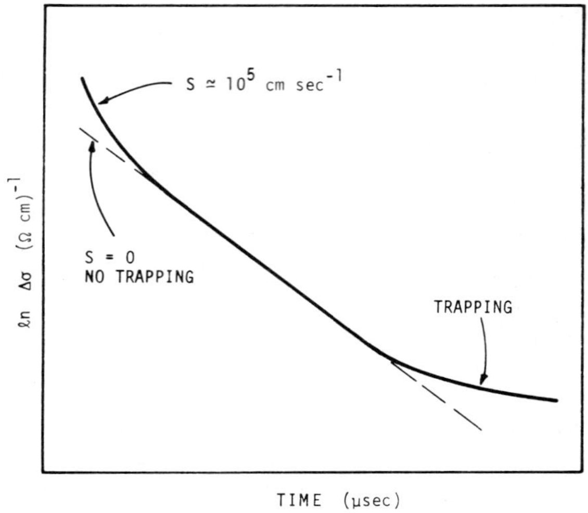

Fig. 4.10. Photoconductivity versus time showing the effects of surface recombination (with fairly strongly absorbed light) and trapping.

The potential at the surface V_s is a function of Δn_p given approximately as $V_s \sim (kT/q)\ln(\Delta n_p/n_{p0})$. If the magnitude of Γ required to maintain a constant V_s (i.e., a constant Δn_p) is plotted against $1/\alpha(\lambda)$ for various wavelengths, the value of L_n can be obtained from the resulting curve since D_n, L_n, and S are constant. The relation of V_s to Δn_p need not be known, but values for $\alpha(\lambda)$ must be available. The method is discussed by Goodman (1961), Johnson (1957a,b), and Choo and Sanderson (1970).

For already fabricated p/n junctions or Schottky diodes, the open-circuit voltage can be used rather than V_s for this measurement if the diode parameters do not depend on illumination (Waldner, 1959; Subhashiev *et al.*, 1961; Loferski and Wysocki, 1961). Wang *et al.* (1974) give an extension of the surface photovoltage (SPV) technique to shallow p/n junctions.

Another variation on this sort of analysis utilizes the variation in the steady-state light current J_L with wavelength λ in an already fabricated solar cell. In this case we must also have $L_n \gg W_d$; for a heterojunction in backwall mode, Eq. (4.19) is used to obtain

$$J_L = q\Gamma h[1 - R(\lambda)]/[1 + 1/\alpha(\lambda)L_n], \qquad (4.35)$$

where h is a constant representing the junction collection efficiency for photoexcited carriers and $R(\lambda)$ is the optical reflectivity. A plot of $1/J_L$ versus $1/\alpha(\lambda)$ at constant Γ gives a straight line, and L_n can be determined from the slope (Stokes and Chu, 1977).

These methods are useful for values of L on the order of the optical absorption length $1/\alpha$ so that extremely small lifetimes can be measured in direct band-gap materials. Particular advantages of these techniques are that they are insensitive to surface recombination, majority carrier trapping, and moderate levels of minority carrier trapping (Choo and Mazur, 1970).

A related method, in which variations of reverse-bias voltage change the effective thickness of the absorbing layer to determine diffusion length and surface recombination velocity is described by Kim *et al.* (1980).

4.6.4 Photoelectromagnetic (PEM) Effect

The Hall effect on a photoinduced diffusion current can be used to measure very small values of τ ($\sim 10^{-8}$ sec) (Subhashiev, 1963). Although the measurement is relatively independent of trapping effects, surface recombination can have a strong influence in thin samples. If the PEM current I_{PEM} and the steady-state, zero-field photoconductivity $\Delta\sigma$ are measured with the same illumination, then if $\tau_n = \tau_p = \tau$

$$\tau = B^2 D (\Delta\sigma)^2 / I_{PEM}^2 ,$$

where $D = kT\mu/q$ and B is the magnetic field, and the result is independent of S and light intensity, assuming L is much smaller than the sample thickness (Smith, 1968, p. 314).

4.6.5 Electron- and Photon-Beam-Induced Current

In this method a very small region of the material is excited by a photon or electron beam at a distance x from a p/n junction or Schottky barrier where the excited carriers are collected, as shown in Fig. 4.11. The collected current I_{sc} is a complex function of geometry and materials properties (Berz and Kuiken, 1976), but for moderate to large values of x/L the relationship reduces to

$$I_{sc} \simeq A \exp(-x/L), \tag{4.36}$$

where A is a constant. The measurement is quite insensitive to trapping since the excitation is essentially steady state, but may be influenced strongly by surface recombination as indicated by the schematic plot in Fig. 4.12a. A typical I_{sc} versus x plot for a p/n junction is shown in Fig. 4.12b.

A theoretical treatment of the currents induced by such point or line excitation at the surface of a semiconductor, including the effects of surface recombination, has been formulated by Van Roosbroeck (1955). The theory was extended by Berz and Kuiken (1976) for the case of excitation

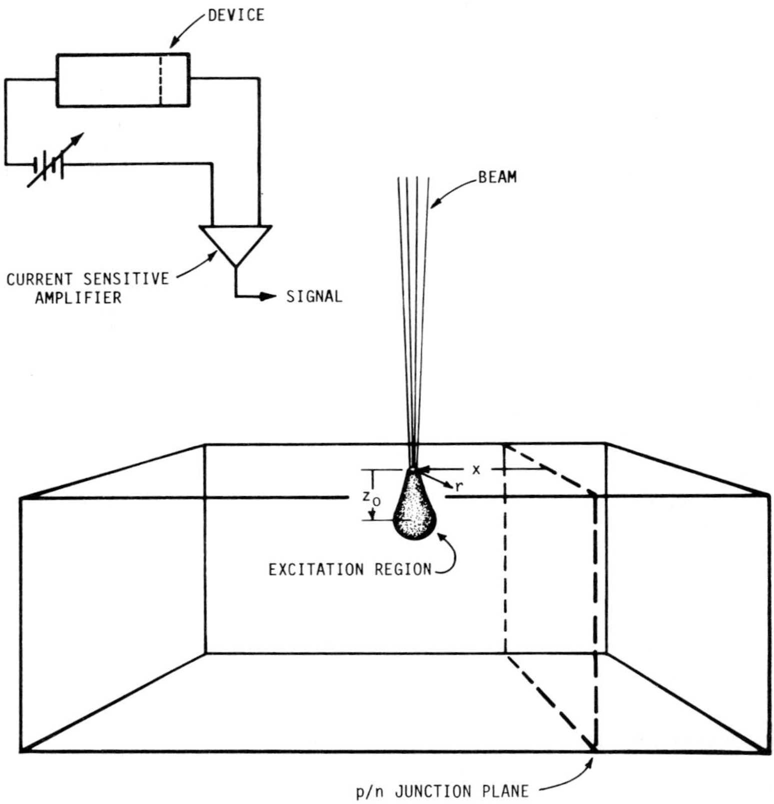

Fig. 4.11. Beam-induced-current geometry. Circuit is shown at upper left.

by an electron beam such as that used in a scanning electron microscope (SEM). An excellent tutorial introduction and review of the SEM technique is given by Leamy *et al.* (1979).

The region of excitation by an electron beam is a roughly teardrop-shaped distribution described by Hackett (1972), for example, as

$$\Delta n(r, z) = f(r) \exp[-a(z - z_0)^2],$$

with $a \simeq 5.4/z_0$, the radial distance from the beam axis is r, and z is measured from the surface of the semiconductor. The range parameter z_0 depends on beam voltage and is given by Everhart and Hoff (1971):

$$z_0 = (B/\rho_m)V^{1.7}, \tag{4.37}$$

where B is a constant depending on the average atomic number of the

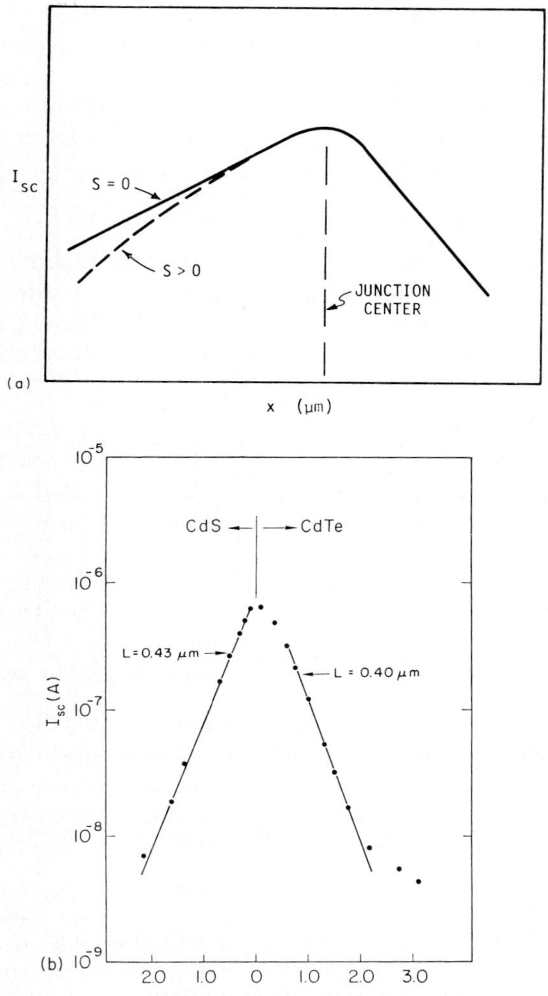

Fig. 4.12. (a) Schematic plot of beam-induced current versus distance from the junction. (b) Electron-beam-induced current as a function of beam position for an experimental CdTe/CdS device. [After K. W. Mitchell, A. L. Fahrenbruch, and R. H. Bube (1977).].

semiconductor and ρ_m is the mass density. Shea *et al.* (1978) give a semiempirical form for the lateral range function $f(r)$. They show that $f(r)$ is reduced to approximately one-tenth of its $r = 0$ value when $r \simeq 0.25z_0$ and that for $L > 0.25z_0$ the measurement should be independent of the excitation volume. Shea *et al.* used the relation $z_0 = 0.021\ (V, \text{kV})^{1.62}$

μm for CdS, for example, giving an effective excitation-volume radius of ~0.42 μm at 15 kV.

In the presence of surface recombination the effective value of the diffusion length is decreased. Jastrzebski *et al.* (1975) give a relation for the decreased diffusion length L_{eff}:

$$L_{\text{eff}}^2 = L_0^2\{1 - [s/(s + 1)]\} \exp(-z_0/L_0) \tag{4.38}$$

for a point source, in terms of a reduced surface recombination velocity $s = SL_0/D$. Variation of the range z_0 by changing the beam energy allowed these authors to measure S as well as the bulk diffusion length L_0.

The electron-beam-induced current (EBIC) method offers several advantages:

1. It can conveniently be done in a scanning electron microscope using a line scan mode that can be superimposed on the secondary electron image of the sample.
2. The junction plane can be located with respect to the metallurgical junction.
3. Variations of L with position in both x and y directions and with the junction field can be conveniently measured and the effect of visible defects such as grain boundaries determined.
4. Values of L as small as 0.1–0.2 μm can readily be measured.

In a variation of the preceding technique, Hackett (1972) used a somewhat larger beam (~10 μm) incident on an angle-lapped junction whose lapped surface makes a small angle ϕ with the junction plane. The range z_0 is small compared with L and the beam-generated carriers must diffuse a distance $x \sin \phi$ to the junction. The current is given by

$$I_{\text{sc}} \simeq \{A/[1 + (SL/D)]\} \exp[-(x \sin \phi)/L] \tag{4.39}$$

so that S can be eliminated from the determination of L.

Kamm and Bernt (1978) discuss EBIC methods for separate determination of the diffusion constant, the bulk lifetime, and the surface recombination velocity in a single Schottky-barriered specimen. An extensive theoretical examination of the use of EBIC analysis of solar cells and other devices has been made by von Roos (1978, 1979).

4.6.6 Open-Circuit Voltage Decay

The V_{oc} decay method involves injection of excess minority carriers, by either forward bias or photoexcitation, into the base region† of a p/n

† Following transistor terminology, the thick, absorber layer is commonly called the *base*, while the thin front layer (usually highly doped) is called the diffused region or the emitter.

junction device and observation of the V_{oc} transient when the excitation is abruptly terminated. The method is capable of measuring moderately short lifetimes ($\tau \gtrsim 10^{-7}$ sec) and has the advantage of measurement under conditions closely approximating those of actual use of the device. It requires relatively simple apparatus. The method was apparently originally introduced by Gossick (1953, 1955). Davies (1963) used forward-bias injection in $p^+/$"L"$/n^+$ structures to determine both low and high injection lifetimes in the lightly doped "L" layer. Wilson (1967) and Bassett (1969) also discuss the method for the case of forward-bias injection. Mahan *et al.* (1979) compared V_{oc} decay using both forward-bias and photoexcitation modes with beam-induced current methods (Section 4.6.5) and found that photoexcitation gave much more reliable results in the Si cells measured.

A simplified derivation, following that by Davies (1963), which neglects diffusion effects is given here. Assuming thermal equilibrium across the junction, the minority carrier density at the depletion layer edge x_p (see Fig. 4.1) of the absorber (base) region (assumed here to be p type) is related to the junction voltage by

$$qV_{oc} = E_{Fn} - E_{Fp} = kT\ln(n_p/n_i) - kT\ln(p_p/n_i) = kT\ln(n_p p_p/n_i^2)$$

(as discussed in Section 5.2.3). Since $p_p \simeq N_A$, we can write

$$V_{oc}(t) \simeq (kT/q)\ln[n_p(t)N_A/n_i^2]. \tag{4.40}$$

For this derivation we are interested in a region with thickness smaller than a diffusion length, i.e., $|x_p' - x_p| < L_n$, so that the injection level is almost uniform throughout the region and diffusion can be neglected. In this case the recombination rate for the entire region can be approximated by that for an incremental volume if τ is not a function of position:

$$\partial n_p/\partial t \simeq -(n_p - n_{p0})/\tau. \tag{4.41}$$

Using Eqs. (4.40) and (4.41), the lifetime is given by

$$\tau \simeq -kT/(\partial V_{oc}/\partial t). \tag{4.42}$$

For the high injection condition, $n_p \simeq p_p$ and the result becomes

$$\tau \simeq -2kT/(\partial V_{oc}/\partial t). \tag{4.43}$$

If the excited region is thicker than L_n, diffusion as well as the Dember voltage (arising from different $\mu\tau$ products for electrons and holes) must be taken into account and the analysis becomes much more complex [see, e.g., the analysis of Choo and Mazur (1970)].

In the foregoing we have also neglected the effect of the n-type (emitter) side of the junction. This can be valid if $N_D \gg N_A$ or if $|x_n' - x_n| \ll |x_p' - x_p|$ so that the injected charge on the p-type side dominates

(Gossick, 1953, 1955; Davies, 1963). Recombination in the depletion layer can also affect the results, particularly if $|x_p - x_n| \simeq |x_p' - x_p|$.

In summary, open-circuit voltage decay gives a sort of "average" time constant for the system (including n-type, p-type, and depletion regions) to relax to its dark, equilibrium state and thus represents the dominant relaxation mode for the system. One must ascertain whether the region of interest does in fact dominate the relaxation. Lindholm and Sah (1976) discuss these considerations rather elegantly. The method is expected to be most accurate in systems where the photogeneration region is thick compared with the depletion or window layers, where the other side of the junction has a much higher carrier density, and when recombination in the depletion layer can be neglected.

4.6.7 MIS Capacitance Decay

Minority carrier lifetimes as small as 10^{-10} sec can be determined by observation of the transient capacitance response of a metal/insulator/semiconductor (MIS) junction after the application of a large reverse-bias step. The method, variously called MOS capacitance, transient capacitance, or the *C-t* method, is particularly convenient for Si since a thin SiO_2 insulating layer can easily be grown on the surface. Using a matrix of small metal dots, lifetime maps of Si wafers can easily be made.

When a reverse-bias step is applied to an MIS junction, initially a large depletion layer is formed. As the voltage is kept constant, electron-hole pairs are generated in the depletion layer because the carrier density is now below the zero-bias equilibrium value. The generation rate G is given by the Shockley–Read formulation, Eq. (3.56), and, for $\sigma_n \gg \sigma_p$ or $\sigma_p \gg \sigma_n$, $G \simeq n_i/\tau$. For a junction on p-type material the newly generated holes drift to the edge of the depletion layer where they counterbalance the ionized acceptors, reducing the depletion layer width W_d and increasing the capacitance. The electrons drift to the insulator/semiconductor (I/S) interface to form an inversion layer that partially supports the applied voltage (Fig. 4.13). In essence, the capacitor relaxes under the constant applied voltage by exchange of charge from the unpaired acceptors at the depletion layer edge to form a similarly charged inversion layer closer to the metal layer. The capacitance is increased by this process, and generation at centers within the gap is the rate-controlling step.

To a first approximation the entire depletion layer width could be neutralized in a time t so that $GtW_d \simeq N_A W_d$. This yields an approximate relation between the observed relaxation time constant t and the carrier lifetime

$$t \simeq N_A/G \simeq \tau N_A/n_i .$$

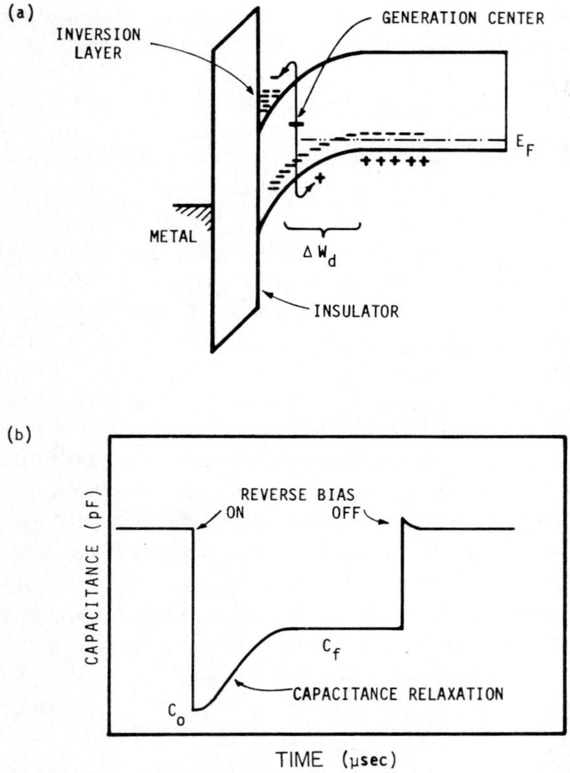

Fig. 4.13. (a) Reverse-bias MIS capacitor during relaxation, and (b) capacitance versus time. The change in depletion layer width ΔW_d is usually very small compared with the minority carrier diffusion length.

Since N_A/n_i is typically 10^5-10^8, exceedingly small lifetimes can measured. A more accurate derivation including the effect of the insulator capacitance is given by Heiman (1967) for example, which shows that the relaxation is a simple exponential for heavily doped semiconductors or thick insulator layers but nonexponential otherwise.

For long lifetimes, surface recombination loss of the inversion layer charge and trapping effects at the I/S interface must be taken into account. Indeed surface recombination velocities at the I/S interface can be extracted from the data using a method given by Zerbst (1966) and later discussed by Schroder and Nathanson (1970) and by Schroder and Guldberg (1971). Baliga and Adler (1978) were able to extend the above methods to measure lifetime as a function of depth in diffused layers. A variation on the method outlined above, observing instead charge transients, was used by Viswanethan and Takino (1978), enabling them to separate lifetime and surface recombination velocity.

4.6.8 Lifetime Measurement Summary

Many other techniques for measurement of lifetime and diffusion length exist [see, e.g., Runyan (1975) and Sah (1976)]. The methods of measurement of L generally involve variation of the distance between the effective excitation source and the collection point by moving an electron or photon beam or by changing the wavelength of illumination [as in the BIC, surface photovoltage, or $J_L(\lambda)$ methods]. For direct measurement of τ, the methods observe the dominant relaxation time for the bulk material (photoconductivity decay or PEM) or for a system (C-t method, diode recovery, or junction admittance versus frequency) from which the desired lifetime is inferred. In many cases, especially in less-well-known materials and devices, this inference can be a weak point in the determination since a model for the system must be postulated. In these cases, several different techniques must be used together to obtain unambiguous values of τ. For example, Neugroschel *et al.* (1977; see also Lindholm *et al.*, 1976b) point out that in some cases the emitter of a p/n junction solar cell may dominate the relaxation properties of the device. In this excellent paper they discuss methods by which several techniques may be combined to give both emitter and base lifetimes.

REFERENCES

Atwater, H. A. (1961). *Proc. IRE* **49,** 1440.
Baliga, B. J., and Adler, M. S. (1978). *IEEE Trans. Electron Devices* **ED-25,** 817.
Bassett, R. J. (1969). *Solid-State Electron.* **12,** 385.
Berz, F., and Kuiken, H. K. (1976). *Solid-State Electron.* **19,** 437.
Choo, S. C., and Mazur, R. G. (1970). *Solid-State Electron.* **13,** 553.
Choo, S. C., and Sanderson, A. C. (1970). *Solid-State Electron.* **13,** 609.
Dalal, V. L., and Moore, A. R. (1977). *J. Appl. Phys.* **48,** 1244.
Davies, E. A. (1964). *J. Phys. Chem. Solids* **25,** 201.
Davies, L. W. (1963). *Proc. IEEE* **51,** 1637.
Dember, H. (1931). *Phys. Z.* **32,** 554, 856.
Dember, H. (1932). *Phys. Z.* **33,** 209.
Durrant, N. F. (1955). *Proc. Phys. Soc., B* **68,** 562.
Ellis, B., and Moss, T. S. (1970). *Solid-State Electron.* **13,** 1.
Everhart, T. E., and Hoff, P. H. (1971). *J. Appl. Phys.* **42,** 5837.
Fischer, H., and Pschunder, W. (1975). *Proc. 11th IEEE Photovoltaic Specialists Conf.*, p. 25.
Fossum, J. G. (1976). *Solid-State Electron.* **19,** 269.
Fossum, J. G., and Burgess, E. L. (1976). *Proc. 12th IEEE Photovoltaic Spec. Conf.*, p. 737.
Goodman, A. M. (1961). *J. Appl. Phys.* **32,** 2550.

Gossick, B. R. (1953). *Phys. Rev.* **91,** 1012.
Gossick, B. R. (1955). *J. Appl. Phys.* **26,** 1356.
Graff, K., and Fischer, H. (1979). Carrier lifetime in Si and its impact on solar cell parameters. *In* "Solar Energy Conversion" (B. O. Seraphin, ed.), Topics in Applied Physics, Vol. 31, p. 173. Springer-Verlag, Berlin and New York.
Hackett, W. H., Jr. (1972). *J. Appl. Phys.* **43,** 1649.
Harrick, N. J. (1956). *J. Appl. Phys.* **27,** 1439.
Haynes, J. R., and Shockley, W. (1951). *Phys. Rev.* **81,** 835.
Heiman, F. P. (1967). *IEEE Trans. Electron Devices* **ED-14,** 781.
Hovel, H. J. (1975). "Solar Cells" (R. K. Willardson and A. C. Beer, eds.), Semiconductors and Semimetals, Vol. 11. Academic Press, New York.
Jastrzebski, L., Logowski, J., and Gates, H. C. (1975). *Appl. Phys. Lett.* **27,** 537.
Johnson, E. O. (1957a). *J. Appl. Phys.* **28,** 1349.
Johnson, E. O. (1957b). *RCA Rev.* **18,** 556.
Jonscher, A. K. (1960). "Principles of Semiconductor Device Operation." Wiley, New York.
Kamm, J. D., and Bernt, H. (1978). *Solid-State Electron.* **21,** 957.
Kim, Y. S., Drowley, C. I., and Hu, C. (1980). *Proc. 14th IEEE Photovoltaic Specialists Conf.,* p. 596.
Landsberg, P. T. (1975). Solid-State Electron. **18,** 1043.
Larrabee, R. D. (1960). *RCA Rev.* **21,** 124.
Leamy, H. J., Kimmerling, L. C., and Ferris, S. D. (1979). *Semicond. Int.* May, p. 75. Reprinted from *Scanning Electron Microsc.* **1** (1978).
Lindholm, F. A., and Fossum, J. G. (1977). *IEEE Trans. Electron Devices* **ED-24,** 325.
Lindholm, F. A. and Sah. C.-T. (1976). *J. Appl. Phys.* **47,** 4203.
Lindholm, F. A., Fossum, J. G., and Burgess, E. L. (1976a). *Proc. 12th IEEE Photovoltaic Specialists Conf.,* p. 33.
Lindholm, F. A., Neugroschel, A., Sah, C.-T., Godlewski, M. P., and Brandhorst, H. W. (1976b). *Proc. 12th IEEE Photovoltaic Specialists Conf.,* p. 1.
Loferski, J. J., and Wysocki, J. J. (1961). *RCA Rev.* **22,** 38.
McQuat, R. F., and Pulfrey, D. L. (1976). *J. Appl. Phys.* **47,** 2113.
Mahan, J. E., Ekstedt, T. W., Frank, R. I., and Kaplow, R. (1979). *IEEE Trans. Electron Devices* **ED-26,** 733.
Mallinson, J. R., and Landsberg, P. T. (1978). Proc. *CEC Photovoltaic Sol. Energy Conf., Luxembourg, 1977* p. 1027. Reidel Publ., Boston, Massachusetts.
Mataré, H. F., and Wolff, G. A. (1978). *Appl. Phys.* **17,** 335.
Mitchell, K. W., Fahrenbruch, A. L., and Bube, R. H. (1977). *J. Appl. Phys.* **48,** 4365.
Neugroschel, A., Lindholm, F. A., and Sah, C.-T. (1977). *IEEE Trans. Electron Devices* **ED-24,** 622.
Persky, G., and Bartelink, D. J. (1970). *Phys. Rev. B* **1,** 1640.
Proc. Int. Conf. Hot Electrons Semicond. (1978). *Denton, Tex., 1977; Solid-State Electron.* **21.**
Rose, A. (1963). "Concepts in Photoconductivity and Allied Problems." Wiley (Interscience), New York.
Rothwarf, A. (1978). *Proc. 13th IEEE Photovoltaic Specialists Conf.,* p. 1312.
Rothwarf, A., Phillips, J., and Wyeth, N. C. (1978). *Proc. 13th IEEE Photovoltaics Specialists Conf.* p. 399.
Runyan, W. R. (1975). "Semiconductor Measurements and Instrumentation," Texas Instruments Electronics Series. McGraw-Hill, New York.
Sah, C.-T. (1976). *Solid-State Electron.* **19,** 975.

Schroder, D. K., and Guldberg, J. (1971). *Solid-State Electron.* **14,** 1285.
Schroder, D. K., and Nathanson, H. C. (1970). *Solid-State Electron.* **13,** 577.
Seeger, K. (1973). "Semiconductor Physics." Springer-Verlag, Berlin and New York.
Shea, S. P., Partain, L. D., and Warter, P. J. (1978). *Scanning Electron Microsc.* **1,** 435.
Shockley, W. (1949). *Bell Tech.* **28,** 335.
Sigmon, T. W., and Gibbons, J. F. (1969). *Appl. Phys. Lett.* **15,** 320.
Smith, R. A. (1968). "Semiconductors." Cambridge Univ. Press, London and New York.
Spitzer, W. G., Firle, T. E., Cutler, M. Schulman, R. G., and Becker, M. (1955). *J. Appl. Phys.* **26,** 414.
Stokes, E. D., and Chu, T. L. (1977). *Appl. Phys. Lett.* **30,** 425.
Stratton, R. (1962). *Phys. Rev.* **126,** 2002.
Stratton, R. (1969). *J. Appl. Phys.* **40,** 4582.
Subhashiev, V. K. (1963). *Sov. Phys.–Solid State* **5,** 405.
Subhashiev, V. K., Dubrovskii, G. B., and Petrusivich, V. A. (1961). *Sov. Phys.—Solid State* **2,** 1781.
Sze, S. M. (1969). "Physics of Semiconductor Devices," p. 70. Wiley (Interscience), New York.
Tarr, N. G., and Pulfrey, D. L. (1979). *Solid-State Electron.* **22,** 265.
Tsaur, S. C., Milnes, A. G., Sahai, R., and Feucht, D. L. (1972). Gallium arsenide and related compounds. *Proc. 4th Int. Symp., Boulder, Colo.* (C. Hilsum, ed.), p. 156. Inst. Phys. Conf. Series, No. 17. Institute of Physics, London and Bristol, 1973.
Urli, N. B., Desnica, U. V., and Coffou, E. (1975). *IEEE Trans. Electron Devices* **ED-22,** 1077.
Van Overstraeten, R., and Nyuts, W. (1969). *IEEE Trans. Electron Devices* **ED-16,** 632.
Van Roosbroeck, W. (1955). *J. Appl. Phys.* **26,** 380.
Van Roosbroeck, W. (1960). *Phys. Rev.* **119,** 636.
von Roos, O. (1978). *Solid-State Electron.* **21,** 1063, 1069, and 1101.
von Roos, O. (1979a). *Solid-State Electron.* **22,** 229.
von Roos, O. (1979b). *Solid-State Electron.* **23,** 177.
Vilms, J., and Spicer, W. E. (1965). *J. Appl. Phys.* **36,** 2815.
Viswanethan, C. R., and Takino, T. (1978). *IEEE Trans. Electron Devices* **ED-25,** 817.
Waldner, M. (1959). *Proc. IRE* **47,** 1004.
Wang, E. Y., Baroana, C. R., and Brandhorst, H. W., Jr. (1974). *J. Electrochem. Soc.* **121,** 273.
Weingarten, I. R., and Rothberg, M. (1961). *J. Electrochem. Soc.* **108,** 167.
Wilson, P. G. (1967). *Solid-State Electron.* **10,** 145.
Wolf, M. (1960). *Proc. IRE* **48,** 1246.
Wolf, M. (1963). *Proc. IEEE* **51,** 674.
Wolf, M., and Prince, M. (1958). *J. Br. Inst. Radio Eng.* **18,** 583.
Zerbst, M. (1966). *Z. Angew. Phys.* **22,** 30.

CHAPTER 5

JUNCTIONS

5.1 INTRODUCTION

The previous chapter enabled us to calculate the light-generated current J_L for a variety of situations. The other basic requirement for calculation of the solar efficiency η_s is knowledge of the current–voltage relationship of the rectifying junction that determines the operating voltage of the cell.

This chapter includes descriptions of the basic homojunction (a junction between *n*- and *p*-type layers of the same semiconductor material), heterojunction and heteroface structures (junctions between two different semiconductor materials), and other junction devices such as Schottky barriers [metal/semiconductor junction (MS)] and MIS and SIS structures that include an intervening insulating layer. The chapter is not meant as a definitive treatise on junction devices but rather as an introduction to the basic principles, with an overview of the considerations relevant to photovoltaic cells.

Homojunctions have been studied in great detail because of their applications in rectifiers and transistors, and the theory is well developed. Although heterojunctions have also received considerable attention, chiefly because of their promise as solid-state lasers and light-emitting diodes, they cannot be said to be well understood.

A general overview of the current transport through a junction device is presented here. Only the charge carried by electrons is considered explicitly; the situation for holes is essentially analogous. In a forward-biased junction in the dark (Fig. 5.1), electrons are injected from the *n*-type region, where they are majority carriers, through the depletion

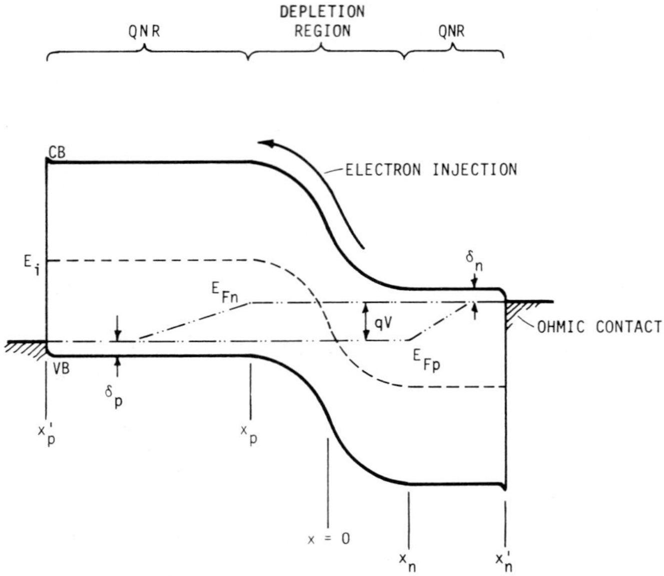

Fig. 5.1. Shockley homojunction at forward-bias voltage V in the dark.

layer into the p-type quasi-neutral region (QNR).† Here, as minority carriers, they recombine with holes. The circuit is completed by a current of holes from the p-type ohmic contact, supplied through the external circuit. A general expression for the current through the junction is

$$J = q \int_{x'_p}^{x'_n} U(x, V)\, dx, \tag{5.1}$$

where $q = |q|$ is the electronic charge, $U(x, V)$ is the recombination rate, and the integral is taken across the entire device, including the boundary surfaces at x'_p and x'_n. To obtain $U(x, V)$, the carrier densities must be known as a function of x and the bias voltage V.

In the Shockley homojunction diode case, where the transport is controlled by diffusion of minority carriers into the QNRs followed by recombination there, the effect of the depletion layer is neglected. The depletion layer is assumed to act as a perfect "transmitter" for the injected electrons, and electrons at x_p and x_n are in thermal equilibrium with respect to the carrier densities within each band. Solution of the transport equation (Eq. 4.16) throughout the p-type QNR, with the appropriate boundary

† In the QNR (the region $x'_p < x < x_p$, in this case) the bands are assumed to be almost flat, i.e., negligible electric field exists, and zero net charge density is assumed.

conditions, is used to obtain the electron current at the depletion layer edge ($x = x_p$). Since this must be equal to the total electron current injected into the p-type QNR, the integration of Eq. (5.1) is thus already accomplished. A similar treatment is used for hole injection into the n-type QNR. Unless the thicknesses of the QNRs are at least several times as large as the minority carrier diffusion length, the effects of surface recombination at the external boundaries of the device (x'_p and x'_n) must be included in the general summation of recombination current (e.g., as boundary conditions for the transport equations for the QNRs).

If recombination in the depletion region is an important contribution to the junction current, the usual approximation is to neglect the effect of diffusion in the depletion region in order to obtain $U(x, V)$ as a function of applied bias. $U(x, V)$ is then just a function of the carrier density and lifetime at each point in the depletion layer. Integration of Eq. (5.1) throughout the depletion region ($x_p < x < x_n$) gives the portion of the diode current recombining in the depletion region, which can be added to the QNR currents to obtain the total current through the device.

In the case of heterojunctions, there is almost always appreciable recombination current through energy states at the interface between the two semiconductors. This must also be included in the integration of Eq. (5.1), perhaps by the introduction of an effective interfacial recombination velocity. In some cases, for depletion layers that are effectively very thin, a portion of the recombination pathway includes tunneling that may even dominate the characteristics of the interfacial region pathway. The presence of the tunneling mechanism is reflected in the form of the current–voltage relationship of the diode (especially with regard to its temperature dependence). All of these cases are examined in more detail in the next sections.

5.2 HOMOJUNCTIONS

Junctions between n- and p-type portions of the same material are the simplest to treat theoretically, and the theory of such homojunctions is well established. Such junctions are usually formed by the diffusion of a dopant of the opposite type into a single crystal to overcompensate the existing doping, e.g., P diffused into B-doped p-type Si to produce a thin layer of n-type material.

First, we examine the formation and properties of an idealized step junction barrier. Then we use the transport equation to calculate the dark junction current as a function of applied bias for infinite and finite layer thicknesses. Next we consider the additional current through the junction when recombination/generation occurs in the junction depletion layer. After consideration of some of the details of our development, such as the constancy of the quasi-Fermi levels across the depletion layer and the physical meaning of the diode quality factor A, we finish the section by briefly discussing some special junction models: "high–low" junctions and the effect of back and front surface fields.

5.2.1 Properties of the Junction Barrier

When two layers of semiconductor material of opposite carrier types are intimately joined, an exchange of charge takes place so that the Fermi level becomes the same in both layers. Electrons from the n-type mate-

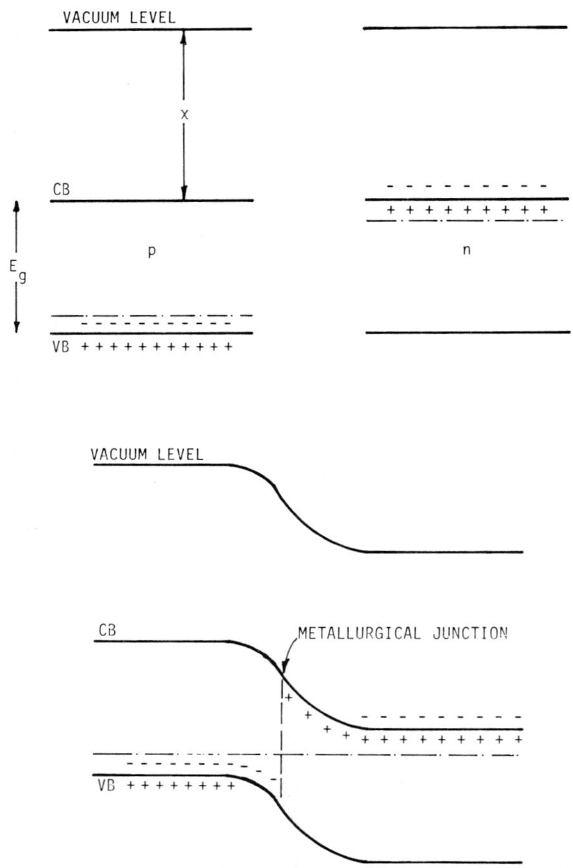

Fig. 5.2. Joining of p-type and n-type semiconductors, before and after contact. The electron affinity is denoted by χ.

rial adjacent to the junction flow into the p-type material and holes flow from the p-type side to the n-type side. This leaves uncompensated donors and acceptors on either side of the junction, as indicated in Fig. 5.2. The two layers of uncompensated charge form a dipole that supports the barrier formed by equalizing the Fermi levels and prevents further net flow of charge. To a good approximation, the edges of these depletion layers at x_n and x_p are abrupt and well defined.† The region between x_n and x_p, the space charge region, is almost completely depleted of mobile carriers, and n and p are negligibly small compared to the donor and acceptor densities there. This is called the depletion approximation. Overall charge neutrality must still hold, and if $x = 0$ at the center of the junction (Fig. 5.3),

$$N_\mathrm{D} x_n = -N_\mathrm{A} x_p. \tag{5.2}$$

This assumes that all the donors and acceptors are ionized within the depletion layer. The electric field in the junction region is given by integration of Poisson's equation, $d\mathscr{E}/dx = \rho/\epsilon_\mathrm{s}$, where ρ is the charge density and ϵ_s is the permittivity of the semiconductor. Assuming that N_D and N_A are constant with respect to position, the electric field is given by‡

$$\begin{aligned} \mathscr{E}(x) &= -qN_\mathrm{A}(x - x_p)/\epsilon_\mathrm{s} \quad \text{for} \quad -x_p < x < 0, \\ \mathscr{E}(x) &= qN_\mathrm{D}(x - x_n)/\epsilon_\mathrm{s} \quad \text{for} \quad 0 < x < x_n, \end{aligned} \tag{5.3}$$

which is obtained by setting $\mathscr{E} = 0$ at x_p and x_n. Note that the maximum electric field is at $x = 0$ and is given by

$$\mathscr{E}_\mathrm{max} = qN_\mathrm{D} x_n/\epsilon_\mathrm{s} = -qN_\mathrm{A} x_p/\epsilon_\mathrm{s}, \tag{5.4}$$

These relations are illustrated in Fig. 5.3.

The electrostatic potential profile is given by a second integration of Eq. (5.3), $dV/dx = -\mathscr{E}$, setting $V = 0$ at x_p to obtain

$$\begin{aligned} V(x) &= qN_\mathrm{A}(x + x_p)^2/2\epsilon_\mathrm{s} && \text{for} \quad -x_p < x < 0 \\ V(x) &= qN_\mathrm{D}(2xx_n - x^2)/2\epsilon_\mathrm{s} + qN_\mathrm{A} x_p^2/2\epsilon_\mathrm{s} && \text{for} \quad 0 < x < x_n. \end{aligned} \tag{5.5}$$

At x_n, $V(x) = V_\mathrm{d}$, the junction built-in voltage (or diffusion voltage), V_d is given by§

$$\begin{aligned} V_\mathrm{d} &= q(N_\mathrm{D} x_n^2 + N_\mathrm{A} x_p^2)/2\epsilon_\mathrm{s} = \mathscr{E}_\mathrm{max}(x_n - x_p)/2\epsilon_\mathrm{s} \\ &= qW_\mathrm{d}^2[N_\mathrm{A}N_\mathrm{D}/(N_\mathrm{D} + N_\mathrm{A})]/2\epsilon_\mathrm{s}, \end{aligned} \tag{5.6}$$

† The carrier density profile at the edges of the depletion layer is treated by Gückel *et al.* (1977).

‡ We follow the convention that $q = |q|$.

§ The diffusion voltage V_d is related to the band gap and donor and acceptor densities by $qV_\mathrm{d} = E_\mathrm{g} - (E_\mathrm{c} - E_\mathrm{F}) - (E_\mathrm{F} - E_\mathrm{v}) = E_\mathrm{g} - \delta_n - \delta_p = kT \ln(N_\mathrm{A}N_\mathrm{D}/n_\mathrm{i}^2)$.

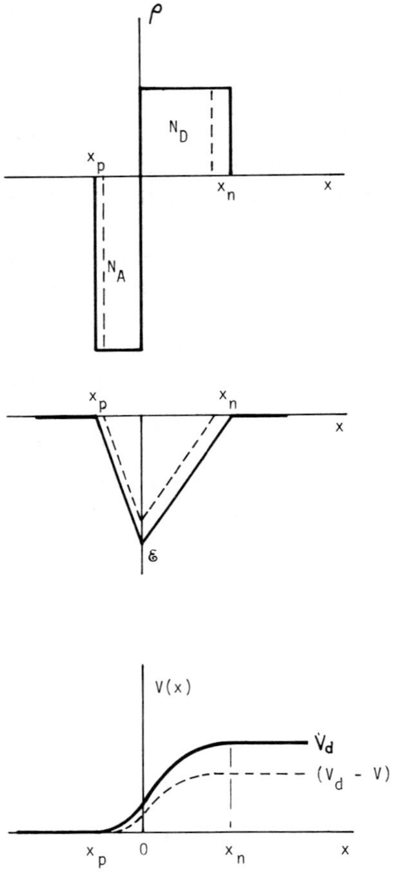

Fig. 5.3. Charge density, electric field, and electrostatic potential versus distance at an abrupt homojunction. Shown for zero applied bias and for forward-bias voltage V (dashed lines).

where $W_d = x_n - x_p$ is the total depletion layer width. To a good approximation, application of an external forward bias merely reduces the junction potential so that the potential across the junction is just $(V_d - V)$, as shown in Fig. 5.3, and the change in charge takes place at the depletion layer edges x_n and x_p.

When the applied voltage is zero, the potential profile is such that the currents due to drift in the junction electric field plus the diffusion flux sum to zero so that the net current is zero, as must be the case in equilibrium.

The capacitance of the junction is given by $C = A\epsilon_s/W_d$ where A is the junction area. The above relations can be rewritten in a more familiar form

$$1/C^2 = (W_d/\epsilon_s A)^2 = [2(N_A + N_D)/qA^2\epsilon_s N_A N_D](V_d - V). \quad (5.7)$$

Usually, the distributions of acceptors and donors are not constant with distance as in the example above, but rather are graded in some way. A similar analysis for a linearly graded junction, for example, gives a linear relationship between V and $1/C^3$ rather than $1/C^2$.

The abrupt step results are compared with a more exact error function dopant diffusion profile calculation by Lawrence and Warner (1960) for diffused junctions in Fig. 5.4. Generally, the step approximation is quite accurate for diffused junctions in which $W_d > 2(Dt)^{1/2}$, where D is the atomic diffusion constant of the dopant and t is the diffusion time (i.e., W_d is large compared to the width of the region in which the doping density is varying with x). For junctions in which $W_d < (Dt)^{1/2}$, the linearly graded approximation gives the best results.

The charge transfer required for a change of depletion layer width involves just the charge at the edges of the depletion layer. Thus for any ar-

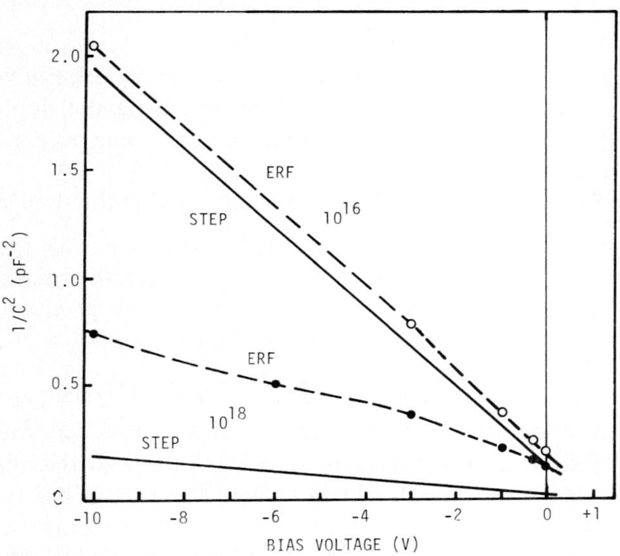

Fig. 5.4. $1/C^2$ versus V plots for two substrate carrier densities, $N_A = 10^{16}$ and 10^{18} cm^{-3}. Both the abrupt step approximation and a calculation for an error function diffusion profile (surface density, 10^{20} cm^{-3}, junction depth 1 μm) are shown. $N_D \gg N_A$ in both cases. (Redrawn from data of Lawrence and Warner presented in A. S. Grove, "Physics and Technology of Semiconductor Devices," p. 168. Wiley, New York, 1967.)

bitrary charge distribution, the variation of capacitance with an incremental change in bias voltage can be used to obtain the carrier density distribution at the depletion layer edges. The carrier density $N_D(x)$ [or $N_A(x)$] can be determined from the slope of a $1/C^2$ versus V curve for a junction with either N_A (or N_D) known or $N_A \gg N_D$ (or $N_D \gg N_A$). For $N_A \gg N_D$:

$$N_D(W_d) = -(2/qA^2\epsilon_s)[d(1/C^2)/dV]^{-1}, \tag{5.8}$$

where $W_d = W_d(V)$. In addition, the extrapolation of the curve to $1/C^2 = 0$ ideally gives the value of V_d, although in practice the value obtained can be strongly influenced by charged states, dipoles, and/or insulating layers within the junction. Goodman (1963), Gossick (1964), and Grove (1967) give good discussions on junction capacitance and the quantities that can be inferred from its measurement.

5.2.2 Diffusion Transport: Infinite and Finite Layer Cases

In deriving the current–voltage relation for transport by injection and diffusion in the QNRs it is usual to make the following assumptions.

1. The Boltzmann approximation holds.
2. The junction is abrupt with well-localized depletion layer edges.
3. There is complete thermal equilibrium across the depletion layer within each band, i.e., the carrier densities at x_n and x_p are in thermal equilibrium.
4. No generation or recombination occurs within the depletion layers.

In this case, E_{Fn} and E_{Fp} can be assumed constant across the depletion layer for forward bias V and $qV = (E_{Fn} - E_{Fp})$. Then the boundary condition at x_p, for example, can be obtained from the np product, using Eq. (3.17): $n_p p_p = n_i^2 \exp(E_{Fn} - E_{Fp})/kT$, and since $p_p \simeq p_{p0} \simeq N_A$, we have

$$n_p \simeq (n_i^2/N_A)\exp(qV/kT) = n_{p0}\exp(qV/kT). \tag{5.9}$$

The electron transport is assumed to be controlled solely by diffusion into the p-type QNR and recombination there. To obtain the current–voltage relation, the transport equation Eq. (4.14) is solved in the p-type QNR using the boundary condition given by Eq. (5.9). Taking the boundary condition at x_p' as $n_p \to n_{p0}$ as $x \to x_p'$, which we can do if $|x_p - x_p'| > L_n$,† we solve the equation in much the same way as we did for the

† A diode for which the thickness of the base is much greater than L_n (or L_p) is called a "long-base diode."

photogenerated current to obtain

$$(n_p - n_{p0}) = n_{p0}[\exp(qV/kT) - 1] \exp[-(x_p - x)/(D_n\tau_n)^{1/2}] \quad (5.10)$$

(x is taken as positive on the n-type side here). Since the carrier transport is controlled by diffusion,

$$J_n(x) = qD_n dn_p/dx = [qn_{p0}(D_n/\tau_n)^{1/2}][\exp(qV/kT) - 1] \times \exp\{-(x_p - x)/L_n\}. \quad (5.11)$$

Since all the current flowing into the p-QNR must pass x_p, the calculation of $J_n(x_p)$ is equivalent to the integration of Eq. (5.1):

$$J_n = J_n(x_p) = [q(D_n/\tau_n)^{1/2}\, n_i^2/N_A][\exp(qV/kT) - 1]. \quad (5.12)$$

The treatment for hole transport into the n-QNR by diffusion is completely analogous, and the total junction current is given by

$$J = qn_i^2\,\{(D_n/\tau_n)^{1/2}/N_A + (D_p/\tau_p)^{1/2}/N_D\}\,[\exp(qV/kT) - 1]. \quad (5.13)$$

Plots of n, p, J_n, and J_p for this diode model are shown in Fig. 5.5.†

Two of the assumptions made earlier—that $\mathscr{E}$ and net charge density are zero in the QNR—are discussed by Sah *et al.* (1957). A much more detailed solution of the p/n junction, without many of the assumptions used here, has been obtained numerically by De Mari (1968), who presents his results in the form of plots of E_{Fn}, E_{Fp}, n, and p throughout the device.

For QNRs with thicknesses of the order of the minority carrier diffusion length, surface recombination at the front and back surfaces of the device influences the current transport. At x'_p, for example, the current density is determined by the surface recombination velocity S_n. To obtain the minority carrier current, we again use the transport equation, this time with the boundary condition

$$J(x'_p) = qD_n(dn/dx)|_{x=x'_p} = q[n_p(x'_p) - n_{p0}]S_n, \quad (5.14)$$

just as we did for the calculation of light-generated current in a finite absorber layer. The solution is straightforward, although somewhat involved algebraically, and we obtain for the electron current injected into the p-type layer of thickness $y = |x'_p - x_p|$:

$$J_n = J_{0n}[\exp(qV/kT) - 1],$$

with

$$J_{0n} = q\left(\frac{D_n}{\tau_n}\right)^{1/2}\left(\frac{n_i^2}{N_A}\right)\left[\frac{(S_nL_n/D_n)\cosh(y/L_n) + \sinh(y/L_n)}{(S_nL_n/D_n)\sinh(y/L_n) + \cosh(y/L_n)}\right] \quad (5.15)$$

† Note that $(D_n/\tau_n)^{1/2} = D_n/L_n = L_n/\tau_n$.

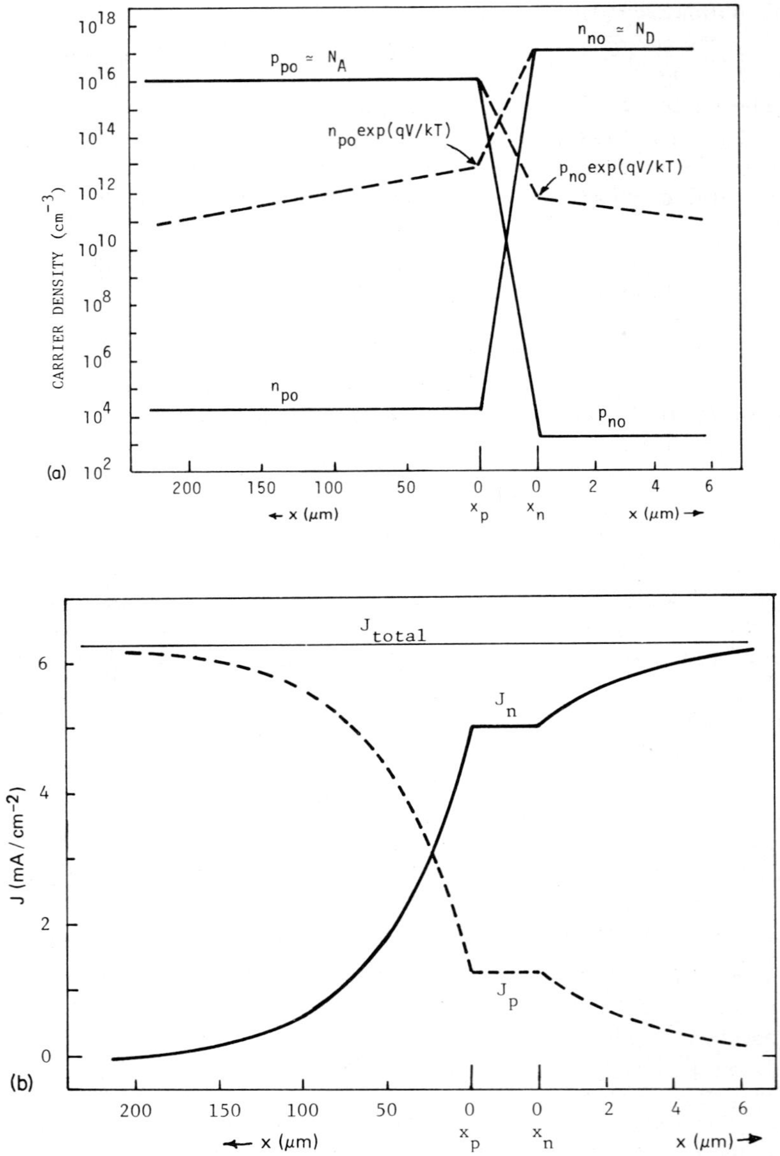

Fig. 5.5. (a) Log n and p versus distance for a Shockley diode at forward biases of zero and 0.5 V. Parameters used are $D_n = 25$ cm^2 sec^{-1}, $L_n = 50$ μm, $D_p = 4$ cm^2 sec^{-1}, and $L_p = 3.2$ μm. Note change of scale for n and p QNRs. The depletion layer thickness is exaggerated for clarity. Surface effects are neglected here. (b) Current density versus distance for the junction of (a) at a forward bias of 0.5 V. Semi-infinite slab approximation.

and a similar equation for hole transport [cf., Eq. (4.22)]. Note that Eq. (5.15) is the same as Eq. (5.12) except for the factor in brackets, which is plotted in Fig. 5.6.

The influence of the surface extends only a few L_n into the QNR and for $y/L_n \gtrsim 1.5$, the term in brackets is approximately unity, and the influence of the surface is negligible no matter what the value of S_n. For smaller values of y/L_n there are three regions. For $S_nL_n/D_n > 100$:

$$J_{0n} \simeq qn_i^2D_n/N_Ay \tag{5.16}$$

and transport is controlled by the mobility in the layer. For $100 > S_nL_n/D_n > 0.01$:

$$J_{0n} \simeq qn_i^2S_n/N_A \tag{5.17}$$

and transport is controlled by the surface recombination velocity. For $S_nL_n/D_n < 0.01$:

$$J_{0n} \simeq qn_i^2D_ny/N_AL_n^2 = qn_i^2y/N_A\tau_n, \tag{5.18}$$

which is equivalent to transport into a region of uniform recombination rate n_p/τ_n and thickness y, so that transport is controlled by the lifetime in the layer. This fact can be used to advantage, for example, by making the diffused front layer of a Si cell thin enough so that the hole current can be reduced to reasonable values even though the hole lifetime is very small there.

In reverse bias, for the Shockley diode model, the defects in the QNR act as generation centers with a rate

$$U = -n_{p0}/\tau_{n0} = -n_i^2/N_A\tau_{n0}. \tag{5.19}$$

The preceding development, Eq. (5.11), still gives the correct form for the reverse bias current except that the lifetime is now given exactly by τ_{n0} rather than the lifetime τ_n discussed in Section 3.5.2. For this model the reverse bias saturation current is constant with bias until junction breakdown by avalanche multiplication occurs for large V_r.

For the case when electric fields due to doping gradients exist in the QNRs, the expressions for J_0 become more complex; Ellis and Moss (1970) give such expressions for the constant electric field situation.

At this point it is interesting to examine the energy balance of such a diode system at a forward-bias voltage V. We neglect I^2R power losses for the moment. Since higher-energy majority carriers are "skimmed off" of the electron energy distribution in the n-type QNR by injection into the p-type side, there is a net heat flow from the n-type QNR to the p-type QNR. When the carriers recombine in the p-QNR to complete the circuit, each electron/hole pair gives up an energy equal to E_g, either by emitting

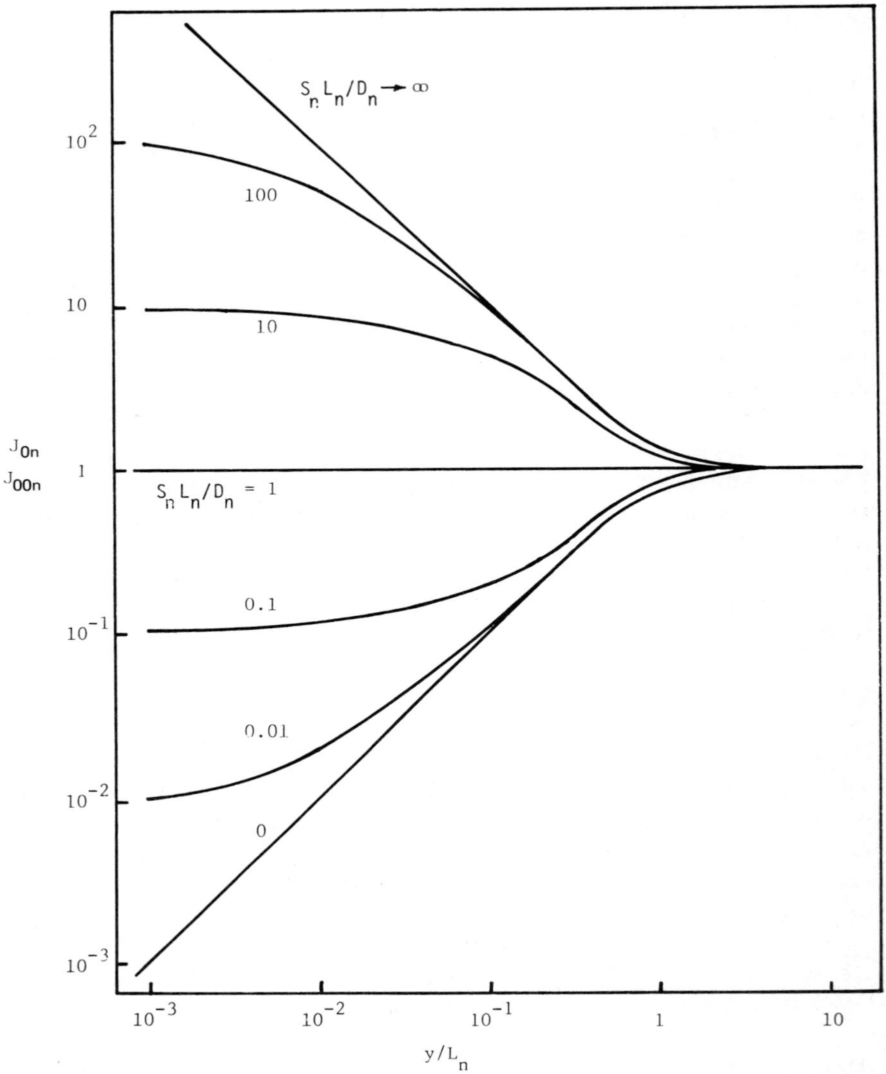

Fig. 5.6. The effect of surface recombination on the diode saturation current J_{0n}. The value of J_{0n} is normalized to the infinite layer case by division by $J_{00n} = q(D_n/\tau_n)^{1/2} n_i^2/N_A$ and the quotient is plotted versus layer thickness in terms of y/L_n for various values of the ratio S_nL_n/D_n. The thickness is $y = |x_p' - x_p|$.

photons or as heat. Since the power input to the diode is $P = JV$, the heat flux across the junction is

$$\text{heat flux} = JE_g/q - JV = J[(E_g/q) - V]. \tag{5.20}$$

5.2.3 Position of the Quasi-Fermi Levels within the Depletion Layer

Fundamental to the calculation of forward-bias current transport (and the light-current transport discussed in Section 4.3) is the assumption that the quasi-Fermi levels are constant across the depletion layer. Thus, in our simplified treatment for the calculation of the light-generated current J_L and the forward-bias, dark diode current we assumed that $n_p = n_{p0} \exp(qV/kT)$ at $x = x_p$ to obtain the first-order boundary condition. For the homojunction case we can show that the quasi-Fermi level is almost constant within the depletion layer (i.e., $dE_{Fn}/dx \simeq 0$) for forward bias in the dark. The demonstration follows from that outlined by Sah *et al.* (1957). If thermodynamic equilibrium within the conduction and valence bands is assumed (as is commonly done), then for each value of bias voltage the electron current is given in terms of the electron quasi-Fermi level by

$$J_n(x) = \mu_n n(x)[dE_{Fn}/dx] \tag{5.21}$$

with

$$n(x) = n_i \exp[E_{Fn}(x) - E_i(x)]/kT, \tag{5.22}$$

where $E_i(x)$ is the intrinsic level. These relations lead directly to a differential equation with $\exp(E_{Fn} - E_i)/kT$ as the variable and with the solution

$$\exp\frac{E_{Fn}(x)}{kT} = C + \int \frac{J_n(x)}{n_i\mu_n kT} \exp\frac{E_i(x)}{kT}\,dx. \tag{5.23}$$

The constant C is eliminated by the choice of the limits of integration and, with the zero of energy at $x = x_n$ (i.e., $E_{Fn}(x_n) = 0$), we obtain

$$E_{Fn}(x_p) = kT \ln\left\{1 + \int_{x_n}^{x_p} \frac{J_n(x)}{n_i\mu_n kT} \exp\frac{E_i(x)}{kT}\,dx\right\}. \tag{5.24}$$

Note that $J_n(x)$ is negative for forward bias.

This integral can be evaluated approximately by using the value of $J_n(V)$ for a real diode characteristic, letting J_n be constant throughout the junction, and using a linear approximation for $E_i(x, V)$ across the junction. For this example, consider a rather poor Si p/n junction with $J_0 = 10^{-9}$ A cm^{-2}, $A = 1$, a depletion layer width W_d of 1 μm, and $\mu_n = 10^3$

cm^2 V^{-1} sec^{-1}. For this example we find that the electron quasi-Fermi level changes by less than $0.4kT$ from one side of the junction to the other. In general, for the Shockley diode in forward bias, the depletion layer behaves like an infinite source of carriers and E_{Fn} is almost constant as long as $(kT/qL_n) \ll (V_d - V)/W_d$, where V_d is the diffusion voltage and L_n is the minority carrier diffusion length in the p-type QNR (i.e., the diffusion field in the p-QNR is much less than the junction field). For a recombination/generation diode, the variation in E_{Fn} within the depletion layer is even smaller because of the change in J_n with x. Middlebrook (1957) and DeMari (1968) also discuss the constancy of the quasi-Fermi levels across the depletion layer. These results should also hold in the heterojunction case.

The quasi-Fermi levels are *not* constant across the depletion layer in the light, for very high forward-bias voltages when resistance drops in the depletion layer must be accounted for, or for appreciable reverse bias in the recombination/generation (R/G) diode (Sah *et al.*, 1957).

Under illumination, the current is opposite in sign, and Eq. (5.24) can again be used to calculate the position of the quasi-Fermi level. This is done for the case of the forward-biased illuminated diode described in Fig. 4.9, where the carrier density is assumed to be scattering-velocity limited. The result is shown in Fig. 5.7. Since the photogenerated carriers

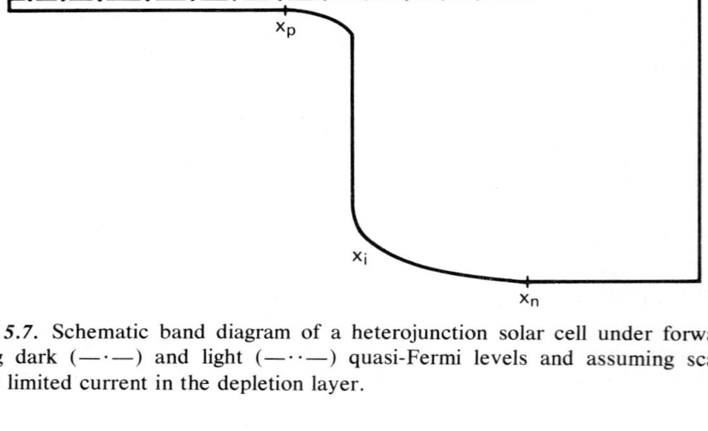

Fig. 5.7. Schematic band diagram of a heterojunction solar cell under forward bias showing dark (—·—) and light (—··—) quasi-Fermi levels and assuming scattering-velocity limited current in the depletion layer.

are "hot" with respect to the lattice, E_{Fn} is somewhat lower than that shown and the "true E_{Fn}" should lie between E_{Fn} and E_{Fn0} within the depletion layer.

5.2.4 Recombination/Generation in the Depletion Layer

In many junctions recombination in the depletion layer cannot be neglected. This is especially true at low forward bias and for heterojunctions. After outlining an approximate treatment to get an overview of the situation, we comment briefly on the Sah, Noyce, and Shockley (SNS) treatment of the symmetrical homojunction and the Choo extension for asymmetrical junctions.

The approximate treatment of the recombination/generation (R/G) transport mechanism within the depletion layer is based on the assumption that the quasi-Fermi levels are constant there. Examination of the Shockley–Read recombination expression, Eq. (3.56), shows that the maximum recombination rate occurs when the intrinsic level lies approximately halfway between the electron and hole quasi-Fermi levels. The rate drops rapidly as the intrinsic level differs appreciably from $(E_{Fn} + E_{Fp})/2$, and thus the major recombination occurs for a small portion of the depletion layer width in forward bias. The recombination rate $U(x)$ can thus be replaced by its maximum value in the integration of Eq. (5.1). If we assume that the recombination center level lies at E_i and that $\tau_{n0} = \tau_{p0}$,† then

$$U_{max} \simeq \frac{n_i^2\{\exp[(E_{Fn} - E_{Fp})/kT] - 1\}}{\tau_{p0}n_i\{\exp[(E_{Fn} - E_i)/kT] + 1\} + \tau_{n0}n_i\{\exp[(E_i - E_{Fp})/kT] + 1\}}$$

$$\simeq \frac{n_i[\exp(qV/kT) - 1]}{2\tau_{n0}[\exp(qV/2kT) + 1]} \simeq \frac{n_i}{2\tau_{n0}}\left[\exp\left(\frac{qV}{2kT}\right) - 1\right]. \tag{5.25}$$

The recombination rate falls exponentially on either side of the maximum with a characteristic length $kT/q\mathscr{E}$, where $\mathscr{E}$ is the junction field given by $\mathscr{E} = 2(V_d - V)W_d$, so that the effective thickness of the recombination region can be taken as approximately

$$2(kT/q\mathscr{E}) = kTW_d/q(V_d - V). \tag{5.26}$$

The junction R/G current for forward bias is given approximately by

$$J_{rg}^+ = q\int_{x_p}^{x_n} U_{max}\,dx \simeq \frac{n_i W_d kT}{2(V_d - V)\tau_{n0}}\left[\exp\left(\frac{qV}{2kT}\right) - 1\right]. \tag{5.27}$$

This approximation for J_{rg}^+ is fairly good for forward bias, although the

† If τ_{n0} does not equal τ_{p0} or if E_r is not at E_i, then the position of the maximum recombination rate is just shifted to one side or the other within the depletion layer.

more exact results that follow predict that the A factor will *always be less than 2* for a symmetrically doped junction with a uniform spatial distribution of recombination centers.

For reverse bias, generation at the centers involved occurs with a maximum rate given by Eq. (3.65):

$$U_{\max} = -n_{\mathrm{i}}/(\tau_{n0} + \tau_{p0}). \tag{5.28}$$

In contrast to the forward-bias case where the region of maximum recombination rate is small with respect to W_{d} and rather constant in width, the region of maximum rate in reverse bias increases in width for increasing bias voltage, finally becoming constant over almost the whole depletion layer. The resulting current is

$$J_{\mathrm{rg}}^{-} \simeq -qW_{\mathrm{d}}n_{\mathrm{i}}/(\tau_{n0} + \tau_{p0}). \tag{5.29}$$

Since W_{d} increases approximately as $(V_{\mathrm{d}} + V)^{1/2}$, the current J_{rg}^{-} increases slowly with increasing reverse bias. Although the electron and hole quasi-Fermi levels deviate considerably from being constant across the depletion layer for appreciable reverse bias, this has little effect on the current since U rapidly saturates to the value given by Eq. (5.28), for the condition $E_{\mathrm{F}p} > E_{\mathrm{i}} > E_{\mathrm{F}n}$.

Sah *et al.* (1957) (SNS) attacked this problem more carefully, first determining that $E_{\mathrm{F}n}$ and $E_{\mathrm{F}p}$ positions are almost constant throughout the depletion layer of a symmetrical homojunction and then using the Shockley–Read formulation to obtain $U(x, V)$ within the depletion layer. They integrated $U(x, V)$ across the junction to obtain the total current. The result cannot be obtained in closed form, but for moderate forward bias it is given by

$$J_{\mathrm{rg}}^{+} = \frac{qn_{\mathrm{i}}W_{\mathrm{d}}(V)}{(\tau_{n0}\tau_{p0})^{1/2}} \left\{\frac{2\sinh(qV/2kT)}{q(V_{\mathrm{d}} - V)/kT}\right\} f(b), \tag{5.30}$$

where

$$f(b) \simeq \int_{z_1}^{z_2} \frac{dz}{[z^2 + 2bz + 1]},$$

$$b = \exp(-qV/kT)\cosh\{[(E_{\mathrm{r}} - E_{\mathrm{i}})/kT] + \ln(\tau_{p0}/\tau_{n0})^{1/2}\}, \tag{5.31}$$

$$z_{1,2} = (\tau_{n0}/\tau_{p0})^{1/2}\exp[\mp(qV_{\mathrm{d}} - E_{\mathrm{F}n} + E_{\mathrm{F}p})/2kT].$$

For symmetrical junctions and V at least several kT less than V_{d}, the integration limits z_1 and z_2 can be extended to zero and ∞, respectively, with small error. The function $f(b)$, plotted in Fig. 5.8, approaches $\pi/2$ for small values of b (i.e., large V). The diode quality factor A is a function of E_{r}, T, and V as indicated by Fig. 5.9, where A is plotted against V. *Note*

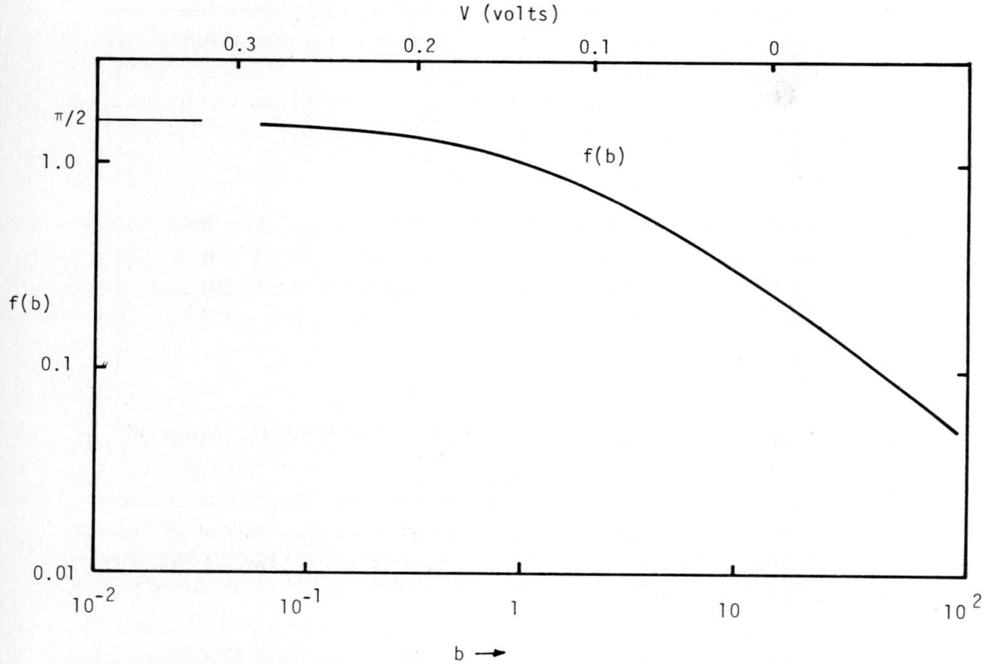

Fig. 5.8. The function $f(b)$ versus b. A forward-bias voltage scale is included for the conditions: $E_r = E_i$, $\tau_{p0}/\tau_{n0} = 10^3$, and $T = 300°K$. (Redrawn from Sah *et al.*, 1957. ©1957 IRE, now IEEE.)

that A reaches a maximum value of 1.8 for deep recombination centers [$\epsilon_r \simeq 0$, where $\epsilon_r = (E_r - E_i)/kT$], and that A approaches 1.0 for centers that are shallow ($\epsilon_r \simeq 10$). The results of SNS agreed very well with their experimental measurements of Si homojunction diodes.

For reverse bias, Eq. (5.30) is not quite accurate because of the constant E_{Fn}, E_{Fp} approximation. However, for reverse bias, Eq. (5.30) reduces to

$$J_{rg}^{-} = -qW_d(V)n_i\,(2(\tau_{n0}\tau_{p0})^{1/2}\cosh\{[(E_r - E_i)/kT] + \ln(\tau_{n0}/\tau_{p0})^{1/2}\})^{-1}. \tag{5.32}$$

Choo (1968) extended the SNS theory to junctions that are asymmetrically doped and in which τ_{n0} and τ_{p0} are allowed to vary more widely. These junctions again had a uniform distribution of recombination centers of a single energy throughout the depletion layer with properties similar to those in the bulk QNRs. In particular, Choo used a more accurate expression for the limits of the integral determining $f(b)$, making $f(b)$ a function

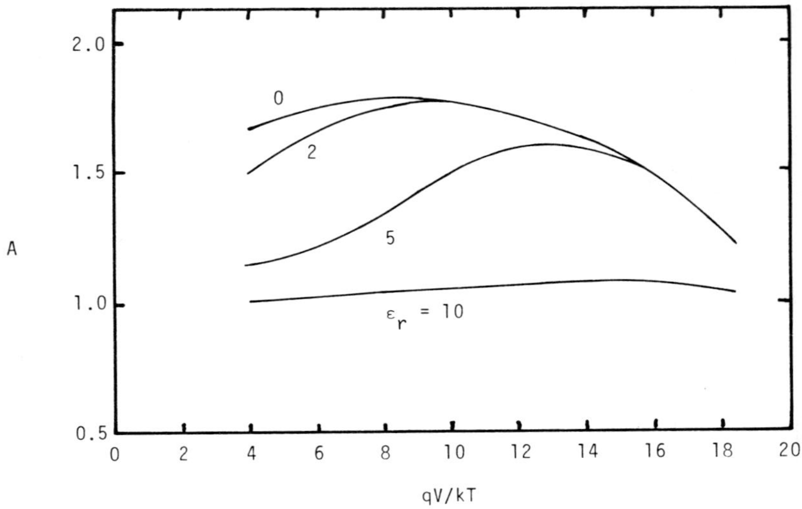

Fig. 5.9. Diode quality factor A versus normalized forward-bias voltage for a symmetrically doped homojunction with R/G transport. The recombination center energy level is $\epsilon_r = (E_r - E_i)/kT$, $\tau_{p0}/\tau_{n0} \simeq 20$, and the diffusion voltage is $qV_d/kT = 20$. (Redrawn from Sah *et al.*, 1957. ©1957 IRE, now IEEE.)

of N_D and N_A as well. Choo found that when donorlike centers were active in a n^+/p junction, the asymmetry produced an increasingly large reduction in $f(b)$ for high forward bias, as shown in Fig. 5.10. The amount of this reduction is greater when the ratio of τ_{p0}/τ_{n0} is large. For acceptorlike centers in a n^+/p junction the effect is almost negligible. The opposite relation holds for p^+/n junctions. The result of the asymmetry dependence is to reduce the R/G currents for increasing forward bias even to the point of causing saturation for the case of low bulk carrier concentration.

Although the SNS and Choo theories are successful in predicting the form of the current voltage characteristics, the R/G currents in real diodes are usually much larger, particularly for junctions involving materials with higher band gaps than Si. Although the Choo theory predicts currents that are even smaller than those predicted by SNS, the Choo theory does provide a mechanism for the A factors equal to or larger than 2, as commonly seen in heterojunctions.

Hovel (1973) has pointed out that it may be necessary to include the effects of several intercommunicating sets of recombination centers in an expanded theory to explain successfully both the magnitude and form of the J–V curves.

Up to this point we have considered only junctions with a distribution

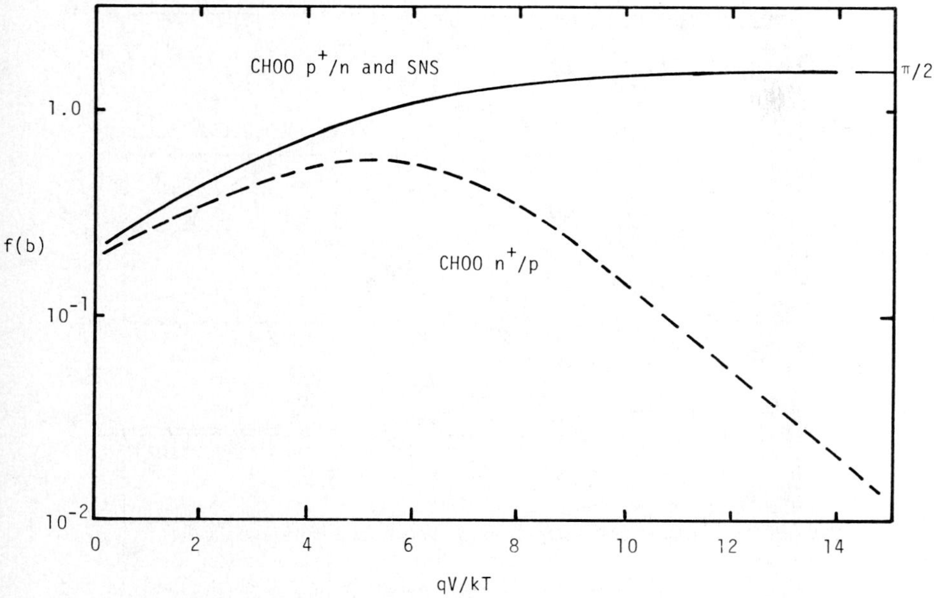

Fig. 5.10. The function $f(b)$ versus normalized bias voltage according to Choo and Sah, Noyce, and Shockley. Recombination center parameters: $\tau_{n0} = 8 \times 10^{-10}$ sec and $\tau_{p0} = 4 \times 10^{-5}$ sec; $E_r - E_i \simeq -8.9\,kT$. Choo's calculation for the p^+/n junction ($N_D = 8 \times 10^{13}$ cm^{-3} and $N_A = 8 \times 10^{18}$ cm^{-3}) is nearly the same as that of Sah, Noyce, and Shockley. Lower curve: Choo's n^+/p junction calculation ($N_D = 8 \times 10^{18}$ cm^{-3} and $N_A = 8 \times 10^{13}$ cm^{-3}). (Redrawn from Choo, 1968. Copyright 1968, Pergamon Press, Ltd.)

of single-energy recombination centers that is uniform throughout the device. If we now consider an asymmetrically doped homojunction with a high recombination rate region at the interface between *p*- and *n*-type regions (perhaps formed when an *n*-layer was grown epitaxially on a "dirty" *p*-type substrate), we see that the positions of E_{Fn} and E_{Fp} at the interface change considerably with respect to E_i, depending on the magnitude and direction of the asymmetry as shown in Fig. 5.11. The relative positions of E_{Fn}, E_{Fp}, E_r, and E_i determine the recombination rate $U(x)$, and hence the total junction current. This introduces an additional mode of dependence of the diode factor A on junction asymmetry. The treatment is readily extended, at least in principle, to heterojunctions.

We could also extend the SNS or Choo theories to a nonuniform spatial distribution of recombination centers in the depletion layer, perhaps with several energy levels. As the narrow region of maximum recombination moves through the depletion region it involves recombination from

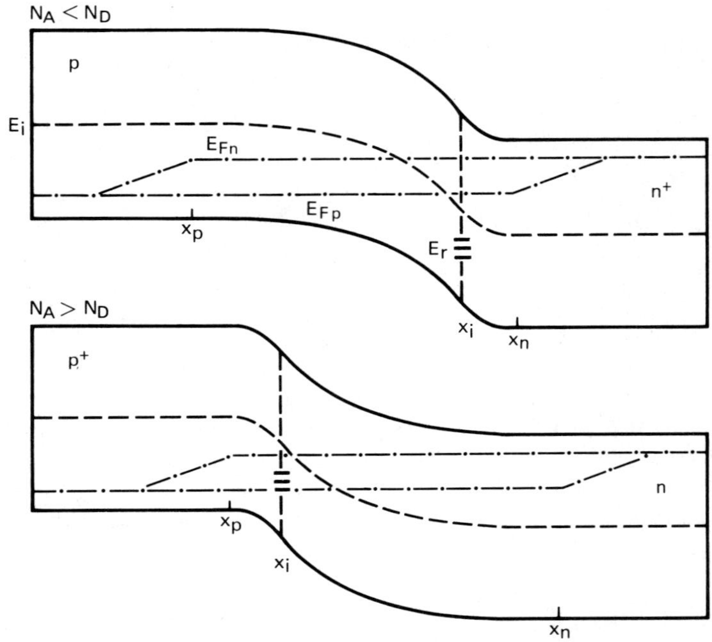

Fig. 5.11. Asymmetrical homojunctions with a nonuniform distribution of recombination centers, as for example from interface states.

different centers.† Both of the models above can lead to $A > 2$. A nonuniform distribution of charged recombination centers within the depletion layer is likely considering the possibility of field-aided atomic diffusion in the high electric field present there.

5.2.5 *J* versus *V* Relations for Homojunctions with Several Transport Mechanisms

Several other phenomena affect junction transport more or less strongly: series and parallel resistance, changes in transport at high forward bias, the structure of the dopant diffusion profile, and edge effects.

Series resistance arises from the bulk resistivities of the layers comprising the device and from the contacts. To a good approximation the resistance of the depletion layer can be neglected if $W_d < L_d$. In most cases, the resistances can be lumped to give the equivalent circuit shown

† This situation was discussed by Shockley and Queisser (1961, p. 518).

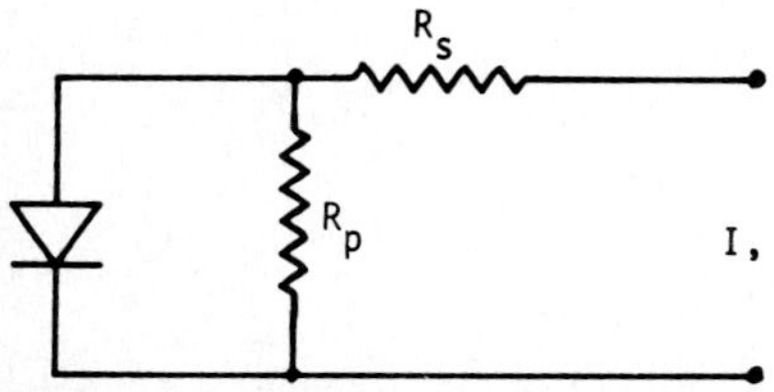

Fig. 5.12. Equivalent circuit of a diode with series (R_s) and parallel (R_p) resistances.

in Fig. 5.12 and the following current–voltage relationship for a simple diode device:

$$I = I_0\{\exp[q(V - IR_s)/kT] - 1\} + (V - IR_s)/R_p. \qquad (5.33)$$

From the form of the equation, one can see that the major effect of R_s occurs at high currents (high bias), whereas the effect of R_p occurs at small voltages such that the diode current is small compared with the shunt current.

The total current through the junction is the sum of the diffusion, R/G, and shunt resistance currents, all of which are usually considered independent. The diffusion component varies as n_i^2; it tends to dominate for small band-gap semiconductors ($E_g \lesssim 1.1$ eV) and for large applied bias voltages, whereas the R/G component usually dominates for larger band-gap material and small bias.† Figure 5.13 summarizes the modes of junction transport for an idealized Si homojunction.

Experimentally, it is often quite difficult to distinguish the shunt resistance current from the R/G current, particularly since the factor A can vary with applied bias voltage. Indeed, there is some evidence that the so-called R_p currents are in reality shunting currents arising from other modes of transport, such as surface conduction by R/G or tunneling at the periphery of the junction plane, or tunneling currents at three-dimensional imperfections in the junction plane.

At high forward bias the diffusion current generally dominates. When the injected carrier density in the QNRs becomes larger than the majority carrier density there, the "high-injection" mode of transport exists. By charge conservation $p_p = N_A + n_p \simeq n_p$ and the quasi-Fermi levels at x_p are given by the usual expressions. Multiplication gives

$$n_p p_p = n_i^2 \exp(qV/kT) \simeq n_p^2,$$

† The ratio of the R/G current, given by Eq. (5.27), to the diffusion current J_{id}, given by Eq. (5.12), is $J_{rg}/J_{id} = [W_d kTN_A/2(V_d - V)qn_i L_n]\exp(-qV/2kT)$. Thus, for large E_g (i.e., small n_i) and/or small V, J_{rg} dominates.

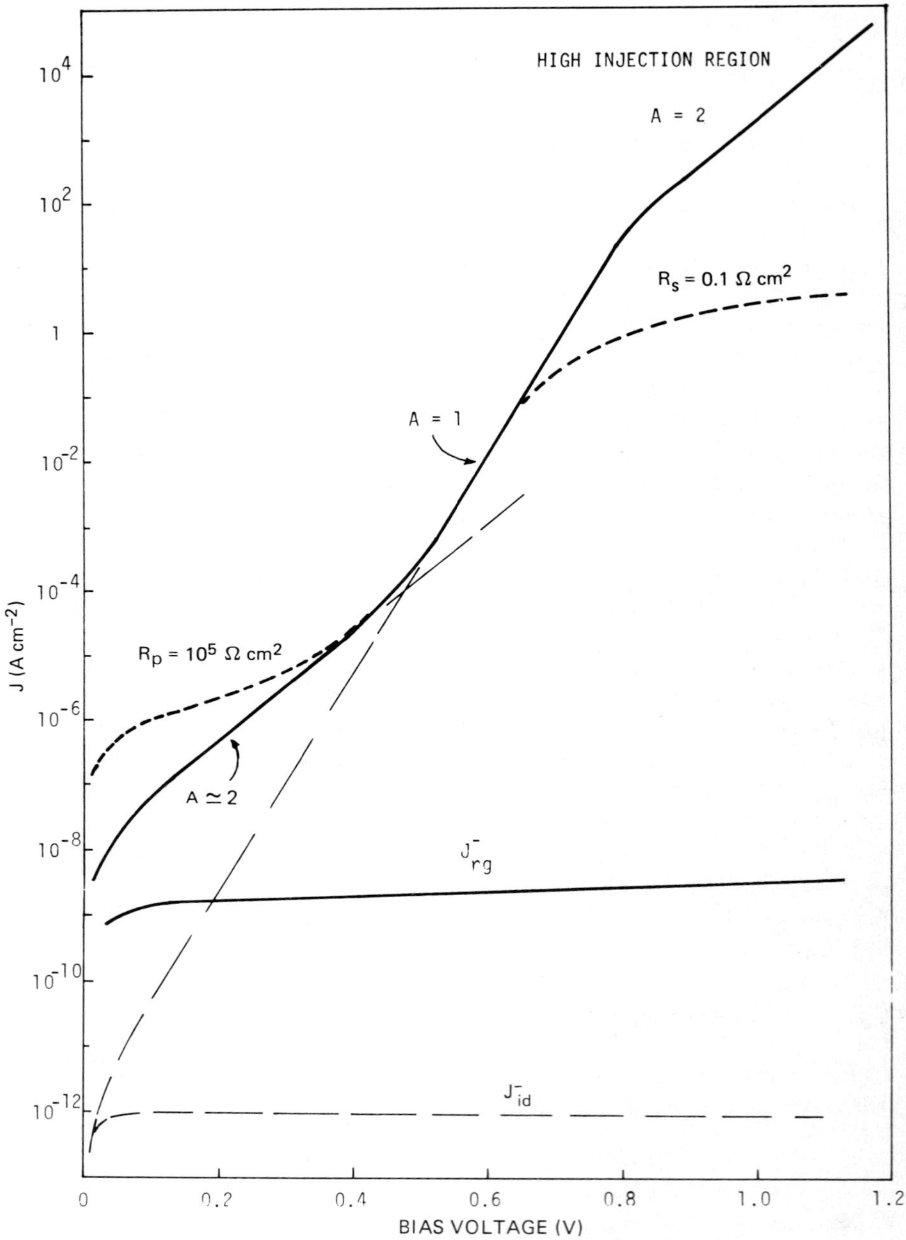

Fig. 5.13. Plot of log J versus V for a simple homojunction diode with injection/diffusion I/D and recombination/generation R/G modes of transport. The A factors and the influence of R_s and R_p are shown. The bias voltage scale is to be multiplied by 10 for the reverse-bias curves for the two cases.

so that the boundary condition at x_p is now

$$n_p(x_p) \simeq n_i \exp(qV/2kT). \tag{5.34}$$

Using the same arguments as in Section 5.2.2, the injected current is controlled by an A factor *equal* to 2 for high-injection currents, as illustrated in Fig. 5.13. Note that the effective recombination centers are now approximately midway between E_{Fn} and E_{Fp}, as in the recombination/generation diode case. For high-injection conditions the effective lifetime is the *larger* of τ_{n0} or τ_{p0} (for recombination center energy levels near midgap) as described in Section 3.5.2. The situation is further complicated by the fact that the transport in the p-QNR, for example, is now also affected by the diffusive transport of holes toward the depletion layer edge. Thus the pertinent diffusion length is a function of L_n, L_p, and position. The current density near the transition to the high-injection mode is about qD_nN_A/L_n for the p-QNR (or about 5 A cm^{-2} for good Si with N_A = 10^{16} cm^{-3}). Such current densities can be reached in concentrator cells. At these currents the resistivities of the bulk QNRs and even the depletion layer can become appreciable, reducing the current at high bias. High-level injection conditions in homojunctions are discussed by Nussbaum (1962), Van Vliet (1966), Gückel *et al.* (1977), and Lindholm and Fossum (1977).

In many cases, particularly for thin junctions, where W_d is thinner than the dopant diffusion profile, the junction profile is not abrupt, as has been assumed, but rather results from an error function type of atomic diffusion profile, as discussed in Section 5.2.1. When the major control of the current transport resides in the QNRs, the overall J–V characteristics are not changed significantly unless extremely thin diffused layers are present. The results for recombination–generation transport are altered more, particularly for the reverse bias case. Transport tends toward a variation of $J_{rg}^{-} \sim V^{1/3}$ (as in the linearly graded doping profile case) rather than $J_{rg}^{-} \sim V^{1/2}$ (as in the step junction case).

5.2.6 Three-Dimensional Variations in Homojunctions

Much of our modeling of junctions is based on the picture of a junction with one spatial dimension and one energy dimension, as shown by the usual band diagram. However, there are several considerations that require a description in three spatial dimensions. On a microscopic scale, doping profiles have statistical variations that may be aggravated by agglomeration around point defects. Patches of unintentional impurities or defects, introduced during processing, can produce areas with lifetimes orders of magnitude smaller than the surrounding, undegraded area, for example. These defects can, in some cases, dominate the $J-V$ characteristics of the device. Schwuttke (1974) demonstrated the effects of such nonuniformities with a two-dimensional map of carrier lifetime in single-crystal Si wafers, obtained using an array of tiny MOS capacitors. Through structural investigations on the same wafers by SEM and TEM, he found that many of the low-lifetime areas were associated with microcracks up to 100 μm deep introduced during the wafer sawing operation. Similarly, grain boundaries penetrating the junction plane and/or interfacial defects can be charged, producing high field regions and consequently large tunneling currents.

Another example, particularly crucial for solar cells, is the three-dimensional alteration in surface recombination velocity due to grid contacts. In the area between grid lines, the surface recombination velocity S can be greatly reduced by oxide passivation. Under the grid contacts, however, the value of S remains large (perhaps 10^6-10^7 cm sec^{-1}). If the properties of the diffused layers are such that J_0 is dominated by S (i.e., $0.1 < SL/D < 100$ and $y/L < 0.1$, from Fig. 5.6), then the $J-V$ characteristics can be considerably degraded even though the grids cover only 5–10% of the cell area.

It has been found that the "$A = 2$" current in many junctions, particularly those with small areas, is controlled by interfacial and surface recombination at the periphery of the junction plane. Stringfellow (1976) found that the recombination/generation current in GaAsP and GaP heterojunction LEDs was directly related to the area-to-perimeter ratio and that it could be reduced temporarily by CF_4 plasma etching. Although Sah *et al.* (1957) found negligible perimeter current in the Si p/n junctions used to substantiate their theory, later work by Sah (1962) showed that the perimeter current in Si p/n diodes *could* be substantial unless the proper surface oxide was present. Henry *et al.* (1978), in a very thorough piece of work, used luminescence techniques to show that the $A \simeq 2$ current in AlGaAs p/n heterojunctions is due to surface recombination at the junction perimeter for a wide variety of surface treatments. Henry attributed the "$A = 2$" behavior to a nearly constant ratio of electron to hole densities that is required to maintain overall charge neutrality at the surface in the presence of Fermi level pinning. Although these measurements are for small-area devices (≈ 0.04 mm^2), they clearly indicate that perimeter currents cannot be ignored in experimental solar cells with small areas.† These concepts can also be extended to polycrystalline solar cells in which such perimeter currents can exist at the grain boundaries that penetrate the electrical junction plane.

† In test devices a guard ring configuration can be used to eliminate the effects of peripheral currents. Alternatively, diodes with a range of area-to-perimeter ratios can be used to determine the peripheral current contribution.

5.2.7 Modifications of the Homojunction

High/Low Band Profiles A highly doped p^+ region next to the p-QNR (i.e., a $p^+/p/n$ structure) was apparently first introduced to modify the boundary condition at the contact/QNR interface and effectively decrease S there. In solar cells the additional region appears to have a greater effect on V_{oc} than would be accounted for by just decreasing S. This point is discussed further in Section 7.5.2.

p/i/n Junctions A structure consisting of p-type, insulator, and n-type layers, all of the same semiconductor material, is called a $p/i/n$ junction. The depletion layer width is increased according to

$$W_d^2 = W_i^2 + W_{d0}^2 \tag{5.35}$$

where W_i is the insulator thickness and W_{d0} is the depletion layer width without the insulator. For $W_i < L$, either the Shockley diode equation [Eq. (5.13), for negligible recombination within W_d], or the recombination/generation diode equation (Eq. (5.27)) apply. When $W_i > L$, the description of transport is considerably more complex, as discussed by Rose (1965). These $p/i/n$ devices have been used for high-intensity Si solar cells, where the large diffusion lengths that can be obtained in undoped materials are particularly beneficial (see Section 12.3.1), and in amorphous Si cells.

5.2.8 Summary

In the discussion of the preceding sections, the theories of homojunction transport have been centered around idealized models (e.g., step junctions, single-level recombination centers distributed uniformly, etc.). Nevertheless, the agreement with the experiment is remarkably good for Si and GaAs p/n junctions and predictions of these theories for solar efficiencies are quite accurate (except for the discrepancy between the predicted and realized V_{oc} in Si cells that is discussed in Chapter 7). The theories form a foundation and guide for the treatment of heterojunctions in the next sections. At this point the student may go on in this chapter to heterojunctions or return after completing Chapters 6 and 7.

An interesting and useful perspective on junction transport is provided by the "charge control" model of Rose (1954). In this general, qualitative model, the junction current is given by

$$\boxed{J = Q/\tau = qnx/\tau = qnv} \tag{5.36}$$

where Q is the available mobile charge, n is the average excess carrier density, τ is an effective lifetime, x is a relevant length, and v is an appropriate velocity.

Consider, as an example, a simple homojunction. The carrier density $n(x)$ in the p-QNR under forward bias is

$$n_{p0} \exp(qV/kT) \exp(-|x_p - x|/L_n),$$

where $n_{p0} \simeq n_i^2/N_A$. An estimate of the effective excess charge Q involved in recombination of the p-QNR is the product of the charge density at the depletion layer edge (x_p) and the effective length of the recombination region, that is just the minority carrier diffusion length L_n. Thus $Q = qn(x_p)L_n = [(qn_i^2/N_A)\exp(qV/kT)]L_n$, and the total junction current injected into the p-QNR is $J = Q/\tau_n$. In this case the estimate turns out to be exactly the result of the solution of the transport equation. The values and correspondence for various modes of junction transport are shown in Table 5.I.

For bulk-dominated processes the calculation of junction current is just the quotient of a generalized excess charge and an effective lifetime. For surface or interface processes, multiplication of an excess charge density by the relevant velocity gives $J(V)$. The J_0 term in a diode equation always contains a factor or N_c, N_v, or $(N_cN_v)^{1/2}$ and another factor of $\exp(-q\phi/kT)$, where ϕ is the relevant potential barrier, measured from the majority carrier Fermi level (i.e., $V_d + \delta_n$, $E_g/2q$, or ϕ_b).

5.3 HETEROJUNCTIONS

The use of a heterojunction (HJ) with a large band-gap window material and a small band-gap absorber material is a means to minimize the surface recombination loss that might otherwise dominate in direct band-gap materials. The use of HJs expands the semiconductor material possibilities for solar photovoltaics enormously. Heteroface cells (in which a p/n homojunction is interfaced with a lattice-matched material of larger band gap) have achieved extremely high solar efficiencies. Such heteroface structures are discussed briefly in Section 5.5.7 and in more detail in Chapter 8.

Unlike the situation for homojunctions, the carrier transport properties of HJs are generally dominated by phenomena in the interface region. The current transport in the depletion layer is variously attributed to recombination, to tunneling, or to combinations of tunneling *and* recombination involving energy levels near the interface. In very few HJ solar cells is transport controlled by injection into the QNRs outside the junction

Table 5.1

Charge Control Model Parameters for Various Junction Transport Mechanisms[a]

Model	$n_p\ (V)$	x	Q	τ	v	Additional factor involved
Bulk control						
Shockley I/D Eq. (5.12)	$(n_i^2/N_A)\exp(qV/kT)$ $= N_c \exp\{[-q(V_d - V) - \delta_n]/kT\}$	L_n	$(qL_n n_i^2/N_A)\exp(qV/kT)$	τ_n	Diffusion velocity $= L_n/\tau_n$	—
R/G Eq. (5.27)	$\simeq n_i \exp(qV/2kT)$ $= (N_c N_v)^{1/2} \exp[-(E_g - qV)/2kT]$	$W_d/2$	$\simeq (qW_d n_i/2)\exp(qV/2kT)$	τ	W_d/τ	—
Interface control						
Interface recombination velocity control Eq. (5.48)	$N_c \exp[-q(\phi_b - V)/kT]$	—	—	—	S_i[b]	—
Thermionic emission (Bethe) Eq. (5.62)	$N_c \exp[-q(\phi_b - V)/kT]$	—	—	—	$v_{th}/4$	—
Thermionic emission (Schottky) Eq. (5.68)	$N_c \exp[-q(\phi_b - V)/kT]$	—	—	—	$\mu \mathscr{E}_{max}$	—
Thermionic-field emission Eq. (5.54)	$\simeq N_c \exp[-q(\phi_b - V)/E_o]$	—	—	—	$v_{th}/4$	$\mathscr{T}(\phi_b, V)$

[a] v_{th} is the thermal velocity of the electrons. $\mathscr{E}_{max}$ is the maximum electric field in the junction. $\mathscr{T}(\phi_b, V)$ is the tunneling transmission factor as a function of barrier height and bias voltage. The equations referenced in the text may be written in slightly different form.

[b] Note that S_i may itself be a function of bias voltage.

region. (This is due in part to the larger band-gap absorbers that are generally used, so that the Shockley injection component is extremely small.) The theoretical treatment of the HJ is complicated by the presence of discontinuities in the conduction and valence bands and possibly interface dipole layers. In addition, the distribution of interface states may be electrically charged, which further distorts the junction profile. The theory of HJ transport is not yet as firmly based as that for homojunctions.

This section describes and contrasts several of the present HJ theories. Since these transport models are presented in numerous papers, in excellent review articles by Tansley (1971) and Van Ruyven (1972), and in the book by Milnes and Feucht (1972), our discussion is aimed at a rather general level, to seek out the essence of the models with particular emphasis on the implications for solar cells. An agenda for the discussion follows:

(1) extension of the Shockley diode theory to HJ with no interfaces states (the Anderson model),

(2) introduction of charged interface states and dipoles that act to distort the band profiles and as recombination centers,

(3) discussion of specific models of HJ transport, including mechanisms of recombination through interface states, tunneling, and combinations of the two.

5.3.1 The Basic Model

The model of Anderson (1960) forms the basis and starting point for most other HJ theories. The essence of the model is to incorporate the discontinuities in material properties, ϵ_s, χ, and E_g across an abrupt metallurgical interface into the Shockley diode theory. The discontinuity in the relative dielectric constant ϵ_s/ϵ_0 is easily dealt with by requiring that the electric displacement be continuous across the interface; i.e., $\epsilon_{s1}\mathscr{E}_1 = \epsilon_{s2}\mathscr{E}_2$. The differences in E_g and the electron affinities χ require discontinuities in the conduction and valence band edges ΔE_c and ΔE_v, which are easy to visualize but rather discomforting theoretically. The result of bringing the two semiconductors together and lining up the Fermi levels is shown in Fig. 5.14. Anderson's model assumes that no interface states are present and that current transport is via injection into the quasi-neutral regions or by recombination/generation in the depletion layer.

Consideration of the constancy of the electron and hole quasi-Fermi levels across the depletion layer follows the same arguments as those for the homojunction case in Section 5.2.3. The relation between the various quantities is given here:

$$\Delta E_c = (\chi_1 - \chi_2)q, \qquad \Delta E_v = (\chi_2 - \chi_1)q + E_{g2} - E_{g1}, \tag{5.37}$$

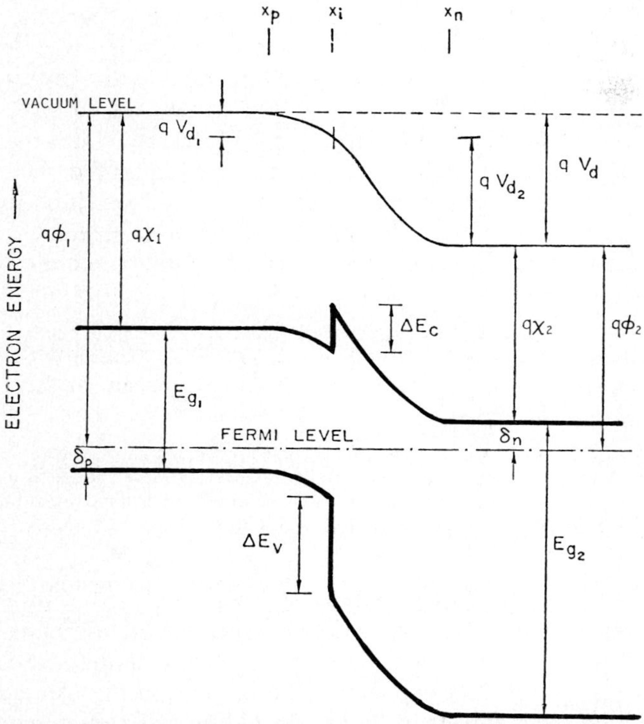

Fig. 5.14. Schematic Anderson heterojunction with $\Delta E_c > 0$ and $\Delta E_v > 0$.

$$qV_d = qV_{d1} + qV_{d2} = E_{g1} - \delta_n - \delta_p + \Delta E_c, \tag{5.38}$$
$$V = V_1 + V_2,$$

$$\begin{aligned} V_{d2}/V_{d1} &= V_2/V_1 = (V_{d2} - V_2)/(V_{d1} - V_1) \\ &= \epsilon_{s1}W_{d2}/\epsilon_{s2}W_{d1} = \epsilon_{s1}N_{A1}/\epsilon_{s2}N_{D2}, \end{aligned} \tag{5.39}$$

$$\begin{aligned} W_{d1}^2 &= (x_i - x_p)^2 = 2N_{D2}\epsilon_{s1}\epsilon_{s2}(V_d - V)/[qN_{A1}(\epsilon_{s1}N_{A1} + \epsilon_{s2}N_{D2})], \\ W_{d2}^2 &= (x_n - x_i)^2 = 2N_{A1}\epsilon_{s1}\epsilon_{s2}(V_d - V)/[qN_{D2}(\epsilon_{s1}N_{A1} + \epsilon_{s2}N_{D2})]. \end{aligned} \tag{5.40}$$

Here V_{d1} and V_{d2} are the diffusion voltages and V_1 and V_2 are the portions of the applied voltage dropped on either side of the junction. Of course other values of δ_n, δ_p, χ_1, χ_2, permit a wide variety of other profiles to be obtained, some of which are shown schematically in Fig. 5.15. For the HJ of Fig. 5.15a, which represents the usual configuration for a solar cell, we

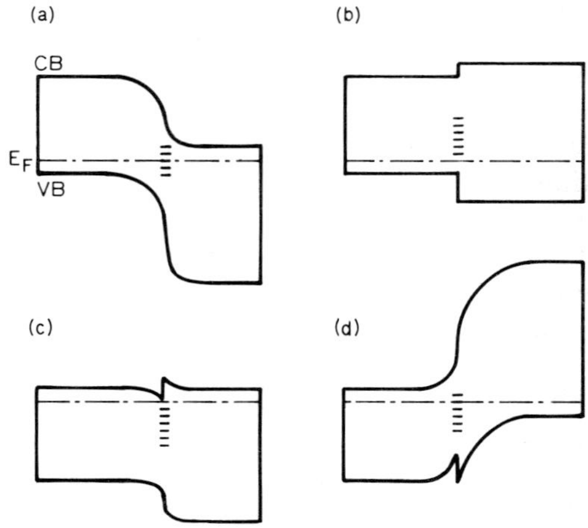

Fig. 5.15. Various heterojunction band configurations: (a) anisotype junction with no CB or VB spike, (b) isotype junction with no band bending, (c) isotype junction with band bending and CB spike, and (d) anisotype junction with VB spike.

want $E_{g2} > E_{g1}$ to maximize the band-gap window so that in most cases the transport of one of the carrier types *across* the metallurgical interface dominates. In the case shown, hole transport across the interface to recombine in the n-QNR is negligible with respect to the electron current because of the large barrier for holes. In addition, the photogeneration in the wide band-gap material is usually neglected because its thickness is much greater than both $1/\alpha$ and L_{p2} so that hole transport to the left may also be neglected in this example. For diffusion-dominated transport, the J versus V relationship for this diode is very similar to that for a homojunction, and, assuming that the electron and hole quasi-Fermi levels are constant across the depletion layer, the $J-V$ relation may be written out of hand:

$$J \simeq q(D_{n1}/L_{n1})(n_{i1}^2/N_{A1})[\exp(qV/kT) - 1], \qquad (5.41)$$

where V is the applied bias, $V = V_1 + V_2$.† From Eq. (5.41) we note that the magnitude of ΔE_c has no effect on transport until $V \simeq V_{d1} + V_{d2}$, the flat band condition. The Anderson model is somewhat successful in predicting the configuration of the bands at the junction but in most cases the

† This can be rewritten in its more usual form in terms of the densities of states in the conduction bands:

$$J \simeq q(D_{n1}/L_{n1})[N_{c1}N_{D1}/N_{c2}] \exp(-qV_b/kT) [\exp(qV/kT) - 1],$$

where $V_b = V_{d1} + V_{d2} - \Delta E_c$.

measured J–V relations are much different, both in quality and in magnitude, from those predicted. The measured values of J_0 obtained by extrapolation of log J versus V curves to $V = 0$ are generally much higher than expected. For example, in the Ge/GaAs HJ studied by Riben and Feucht (1966a) J_0 was six decades higher than the value predicted theoretically at 300°K. In addition, the observed slope at low forward bias was constant with temperature, whereas the model of Eq. (5.41) predicts a slope change by a factor of 4 over the temperature range used. The values of ΔE_c and ΔE_v, as measured by capacitance and other techniques, are not always as predicted by published χ values. For these and other reasons the Anderson model has been modified as described in the following sections.

Experimentally, most HJ admit one of several general formulations. Many fit the relation

$$J = J_{00} \exp(-\Delta E/kT)[\exp(qV/AkT) - 1], \tag{5.42}$$

where J_{00} and A may be slowly varying functions of T and V, and ΔE is a measured activation energy for the zero bias extrapolation of the ln J versus V curves. This may also be written in another general form:

$$J = J'_{00}\{\exp[-q(\phi_b - V)/AkT] - 1\}, \tag{5.43}$$

where $\phi_b = V_d + \delta_n/q$. For many HJs the slope of the log J versus V relationship is relatively temperature independent and thus cannot be characterized by Eq. (5.42) or (5.43). It is found that many such junction characteristics appear to fit the relation

$$J = J''_{00} \exp(\beta T) \exp(-\alpha\phi_b) \exp(\alpha V), \tag{5.44}$$

where β and α are constants that are relatively temperature independent. In many cases, the curves show well-defined branches with differing slopes. Thermally assisted tunneling transport appears to fit the relation of Eq. (5.44).

It should be pointed out here that for the purpose of solar cell design, the J–V relation must be known only over a rather small interval near V_{oc}. Thus, empirically, either Eq. (5.42) or (5.44) are very adequate descriptions. However, if one is unraveling the junction transport mechanisms, much more involved descriptions over as wide a range of voltages as possible are needed.

5.3.2 The Electron Affinity and Band Discontinuities

The Anderson model predicts band edge discontinuities, depending only on the electron affinities and the band gaps of the two materials. The presence of such discontinuities is clear from experimental measurements

(Riben and Feucht, 1966a; Anderson, 1960), although their magnitude, as well as their direct dependence on the electron affinities, is uncertain because of our inability to separate out interface effects. A Schottky barrier is the simplest example of such a discontinuity. When $\Delta E_c < 0$ as shown in Fig. 5.15a, the effect is merely to reduce V_d and ultimately to reduce the obtainable V_{oc} from a HJ solar cell. On the other hand, $\Delta E_c > 0$ implies a conduction band spike or notch (Fig. 5.14) that tends to impede photogenerated carrier transport and that, in some cases, may change the dark $J-V$ characteristics considerably.

By definition, the electron affinity χ is the energy required to raise an electron from the conduction band edge to the vacuum level (Kroemer, 1975). Intuitively we feel that it is a bulk property, and in many cases it has been treated as such in the calculation of ΔE_c and ΔE_v. The value of χ is usually inferred from measurements of the electron work function ϕ. All such measurements, however, involve a surface and unfortunately the value of χ is very surface sensitive. The principal means for measuring the work function are the following.

(1) In the Kelvin method of contact potential measurement a reference surface is brought into close proximity with the semiconductor surface and their spacing is varied periodically. The contact difference of potential across the vibrating capacitor so formed produces an ac voltage that may be nulled to measure $\Delta\phi$.

(2) In photoemission, electrons within the valence band are photoexcited with enough energy to be ejected through the surface of the material into the vacuum. The kinetic energy spectrum of the ejected electrons is measured and the data reduced to obtain the work function and hence the value of χ. A similar sort of experiment may be done utilizing thermionic emission. When a junction is made with another semiconductor or a metal, measurement of internal photoemission can be used to determine barrier heights (on both sides of the junction for a HJ) so that the effective $\Delta\chi$ can be inferred.

(3) The capacitance–voltage characteristics of a metal/semiconductor junction or of a HJ can in principle be used to obtain V_d, from which ΔE_c can be inferred with certain assumptions about interface states.

(4) The measurement of flat-band potential in an electrolytic cell, with respect to a standard hydrogen electrode, can be used to obtain ϕ (Scaife, 1980).

In all of these methods χ is measured with respect to a particular interface (vacuum in 1 or 2, a metal or semiconductor in 3, and an electrolyte in case 4). There is ample evidence that interactions between each pair of materials results in a unique case and that there is considerable influence on the effective electron affinity difference by surface states, pinning of the Fermi level at the surface, and/or a variation of E_g at the surface because of reconstruction, strain, or adsorbed foreign atoms (Fahrenbruch and Aranovich, 1979). Each of the preceding types of measurement gives slightly different results for χ for the same pair of materials, and all are highly surface sensitive.

In whatever manner the χ values are measured for a HJ pair, it is likely that the band profile is modified in the formation of the junction according to the particular interface that is formed (depending on, for example, the chemical reactivity of the components). Thus published values of χ can serve only as a rough guide to the direction and magnitude of ΔE_c and ΔE_v.

An example of the possible effects of a conduction band spike on a solar cell is the Cu_2S/CdS junction (Chapter 10). Te Velde (1973) and Lindquist and Bube (1972) independently proposed that a conduction band spike controls photogenerated carrier transport from the Cu_2S to the CdS. Lindquist suggested that the thickness of the spike was the medium of control, as modulated by the depletion layer width in the CdS, whereas Te Velde thought that heat treatment varied ΔE_c per se, controlling V_{oc} and J_{sc}. Fahrenbruch and Bube (1974) and Rothwarf *et al.* (1975) later showed that a spike was not required to explain the observed variations, suggesting instead that the local electric field controlled the recombination at the interface. Böer (1978, 1979) resurrected the spike concept to explain the dependence of the J_{sc} and V_{oc} on y in the $Cu_x/Zn_yCd_{1-y}S$ junction.

5.3.3 The Question of Abruptness

Although Anderson's model is based on the assumption of a perfectly abrupt metallurgical junction, many junctions are fabricated at high enough temperatures so that at least some interdiffusion occurs, forming solid solutions of the semiconductor components in some cases. Oldam and Milnes (1963) and Cheung *et al.* (1975) have shown that junction grading over distances of as little as approximately 100 Å can smear out and lower the conduction band spike to the point where carrier flow is unimpeded. A calculated band profile is shown in Fig. 5.16. Photoexcitation and transport properties in such HJs with thin, graded layers are discussed by Hinkley and Rediker (1967) and Womac and Rediker (1972). Van Ruyven (1972) discusses the effects of perturbation of the electron wave functions by the changing periodic potential for band-gap energies graded over small distances (5–50 Å).

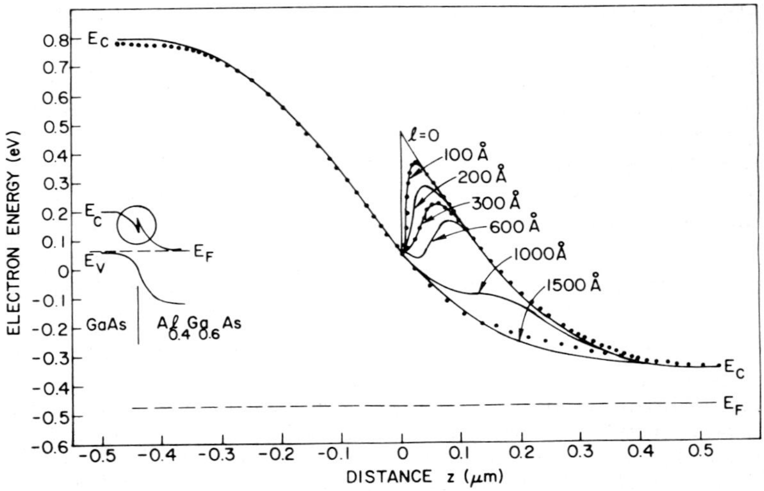

Fig. 5.16. The conduction band profiles of a p-GaAs/n-$Al_{0.4}Ga_{0.6}As$ heterojunction for different grading widths. The solid lines are obtained by a simplified model and the dots by a generalized graded-gap model. Note that W_d is larger than usual. (After D. Cheung, Ph.D. Thesis, Stanford Univ., California, Stanford, 1975.)

5.4 MODIFICATIONS TO THE SIMPLE HETEROJUNCTION MODEL

The modification of the Anderson band profiles to include the effects of electrically charged interface states and dipoles is illustrated in Fig. 5.17. These changes, plus the introduction of various tunneling mecha-

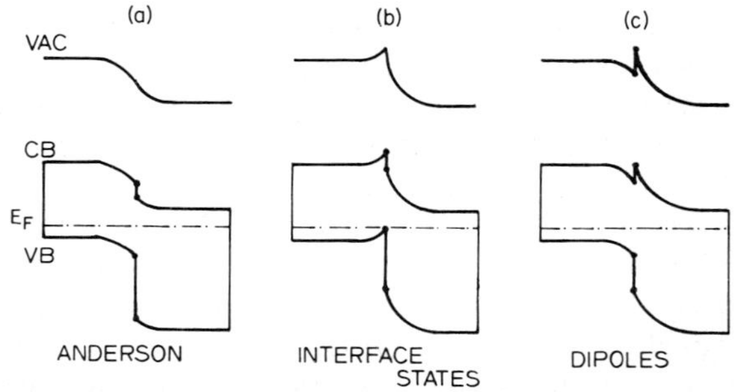

Fig. 5.17. Energy band profiles of a p/n heterojunction according to three models: (a) without interface states or dipoles (Anderson model), (b) including charged interface states, and (c) including dipoles. All cases have identical doping densities in the bulk materials. The opposite polarity of interface states or dipoles would give a second set of profiles.

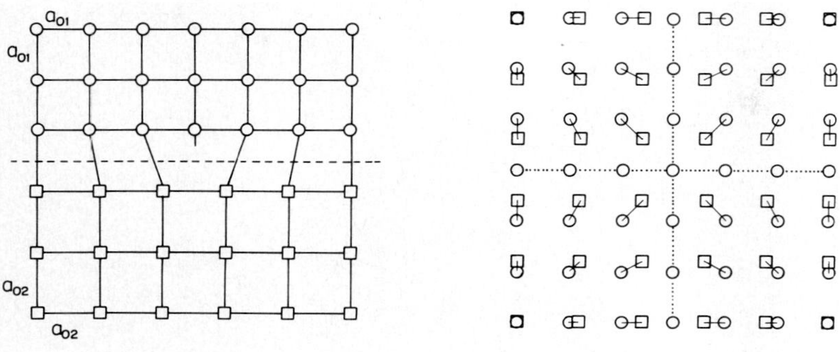

Fig. 5.18. Schematic diagrams of the interfacing of two single cubic crystals with lattice parameters a_{01} and a_{02} at a (100) plane. Complete abruptness and plastic relaxation of the misfit strain is assumed. In (a) a view parallel to the interface plane illustrates the formation of a pure edge dislocation. A view perpendicular to the interface plane is shown in (b), illustrating the corresponding set of orthogonal dislocation lines. The circles and squares represent the interface layers of atoms in the corresponding materials. (From J. Aranovich, Ph.D. Thesis, Stanford University, Stanford, California, 1980.)

nisms, bring the theory of HJs into better agreement with experimental observations.

5.4.1 The Physical Nature of Interface States

Extrinsic imperfection energy levels in the vicinity of the interface arise from the effects of lattice mismatch between the two components of the HJ and from impurities or defects introduced during fabrication. These interface states may or may not be electrically active.

The difference in lattice constant between the HJ components produces a periodic array of dangling bonds as shown in Fig. 5.18. These edge dislocations ideally form a regular net at the interface (e.g., Fig. 5.19) and each dislocation core is surrounded by a region of lattice strain. The area density of dangling bonds is given, in general, by

$$N_{db} = C_{12}[1/a_1^2 - 1/a_2^2], \tag{5.45}$$

where $C_{12} \sim 1$ is a constant depending on the two lattice structures and their mutual orientation and a_1 and a_2 are the relevant lattice constants. Such dangling bonds may be electrically active themselves and may serve as sites for impurity segregation. Experimentally it is found that a large amount of compensation takes place and so only a small fraction of the dangling bonds remains active and accounts for the observed electrical phenomena, as discussed by Tansley (1971). Since the amount of lattice

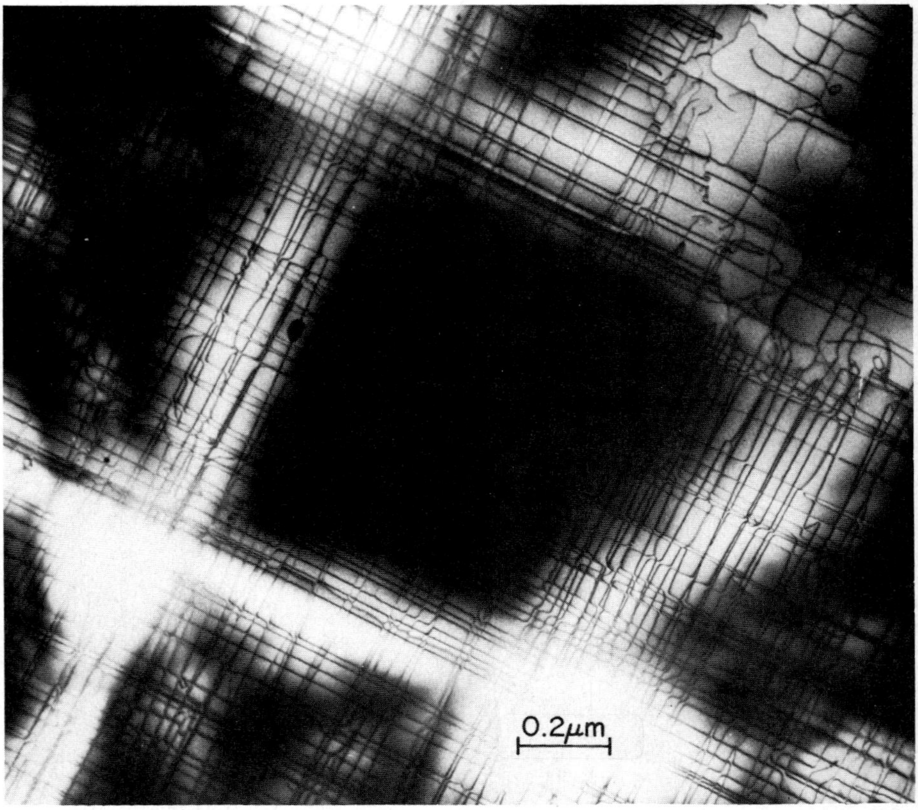

Fig. 5.19. TEM of the compositionally graded region of a VPE (100) $In_xGa_{1-x}P/GaP$ sample. The dislocations lie along [011] directions. The square black regions are thought to originate from surface undulations (cross-hatching) related to mismatch. [After G. H. Olsen, *J. Cryst. Growth* **31**, 223 (1975).]

mismatch to be accommodated is conserved across the junction, the total number of dangling bonds is ideally constant, whether they are concentrated at the interface or spread over a wider region containing the interface.

The lattice strain resulting from mismatch produces a local variation in the band-gap energy, which can also cause important electrical effects, such as enhanced conduction along the dislocation cores. These effects are discussed in Chapter 9 in connection with grain boundary effects in polycrystalline thin films. Above a critical strain, typically 1%, dense arrays of dislocations that propagate away from the interface along the

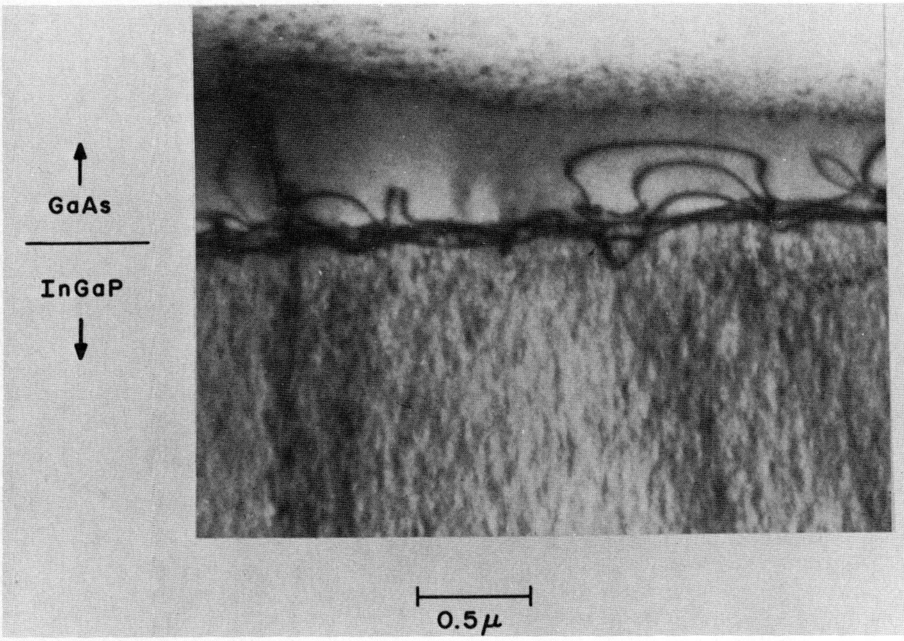

Fig. 5.20. Bright field {011} XTEM micrograph from a lattice mismatched GaAs/$In_xGa_{1-x}P$ interface. The [100] growth axis is vertical. A classic Frank–Read source in action can be seen. [After G. H. Olsen and M. Ettenberg, Growth effects in the heteroepitaxy of III–V compounds. *In* "Crystal Growth: Theory and Techniques" (C. H. L. Goodman, ed.), Vol. II, p. 1. Plenum, New York, 1978.]

growth direction are observed (Olsen and Ettenberg, 1978) (Fig. 5.20). Larger strains are taken up by cracking, particularly for thicker layers and when the layer is in tension rather than compression.

Olsen *et al.* (1978) fabricated $In_xGa_{1-x}As$/GaAs HJs by vapor phase epitaxy over a range of compositions corresponding to an interval in misfit strain around perfect fit. They were able to study the region of misfit where the strain is accommodated elastically and to determine its effect on lattice parameter and band-gap shift, as well as the conditions for the onset of plastic deformation via generation and multiplication of dislocations. Figure 5.21 illustrates their summary of critical strains and stresses for different mechanisms of accommodation and the relationship with mechanical and electrical properties in III–V compounds.

The lattice strain is of course affected by the difference in thermal expansion coefficients when HJs are fabricated at elevated temperatures.

Similar dislocations can be introduced into the interface region by

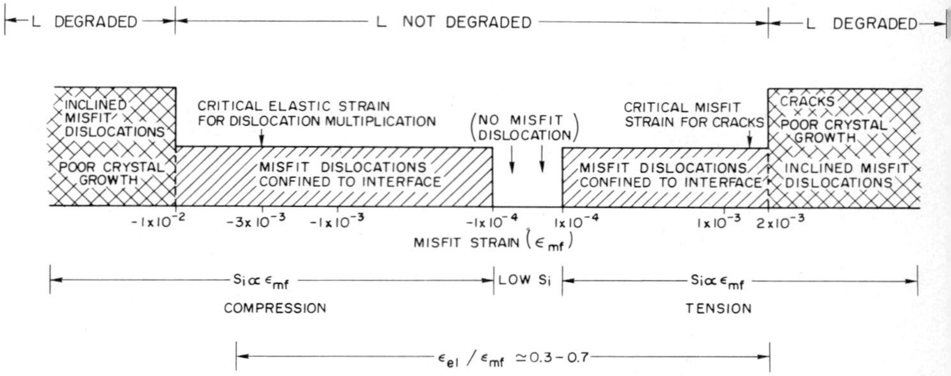

Fig. 5.21. Critical strains and stresses and their relationship to mechanical and electrical properties in III–V compound heterojunctions. L is the minority carrier diffusion length, ϵ_{el} is the elastic strain, and S_i is the interface recombination velocity. [After G. H. Olsen, C. J. Nuese, and R. T. Smith, *J. Vac. Sci. Technol.* **15,** 1410 (1978).]

sawing and/or polishing damage before the deposition of the second HJ component layer. A common "rule of thumb" is that such damage extends into the bulk for a distance on the order of the diameter of the polishing compound granules. Such damage must be removed completely by chem-mechanical polishing (with no abrasive), chemical etching, thermal etching, sputter etching, and/or annealing.

Extrinsic interface states also arise from impurities introduced during junction fabrication. Consider the situation in which a semiconductor surface is being prepared for deposition of a second semiconductor (or metal) component in the fabrication of a HJ or Schottky barrier. Even if surface contamination could be completely removed by chemical etching, thin oxide layers form almost instantaneously, and even adsorbed atoms can cling tenaciously to the surface. The chemical reactivity of the surface must be considered on a microscopic scale, since even 0.001 monolayer of electrically active impurity atoms is sufficient to alter completely the electrical properties of the junction. As an example, consider a p/p isotype junction in Si, both sides with a bulk doping density of $N_A = 10^{16}$ cm^{-3}. If a quantity of P, equivalent to 0.001 monolayer, is placed at the interface (and if the P atoms are all electrically active), the resulting charged interface density N_i is 6.8×10^{11} cm^{-2}. By charge conservation ($2qN_A W_d = qN_i$) this would produce a depletion layer extending ≈ 0.35 μm into the bulk p-Si on either side of the interface and, from Eq. (5.6), a potential barrier ≈ 0.76 eV high!

The effects of both intrinsic and extrinsic imperfections in HJ are discussed by Fahrenbruch and Aranovich (1979).

5.4.2 The Effect of Interface States on Electrical Properties of HJ

The presence of a large density of electrically active interface states provides two mechanisms in HJs:

(1) The charge stored in the states distorts the band profile, raising or lowering the conduction bands at the interface with respect to the equilibrium Fermi level. Ideally, the magnitude of this charge (Q_{ss} in semiconductor device parlance) can be measured by capacitance techniques.

(2) The states provide a large density of recombination centers needed to explain the high J_0 values observed. The recombination behavior of the interface states can be broadly quantified by an effective interfacial recombination velocity S_i. In simple situations the value of S_i can be derived using the Shockley–Read model (Many *et al.*, 1971).

Both Q_{ss} and S_i can be expected to vary with illumination and bias conditions.

In many cases the extremely high density of charged states at specific energy levels at the surface is sufficient to pin the surface (or interface) Fermi level at that energy. Fermi-level pinning has long been observed at dislocations and grain boundaries, sometimes to the extent of inversion as in the $p/n/p$ behavior of grain boundaries in Si observed by Weinreich *et al.* (1960). A considerable portion of the book by Materé (1971) is devoted to such effects in Ge bicrystals.

Some of the most convincing demonstrations of Fermi-level pinning in HJs are in the isotype heterojunctions studied by Van Opdorp and Vrakking (1967) (Ge/Si), by Van Ruyven *et al.* (1965; Van Ruyven, 1964) (Ge/GaP), and by Donnelly and Milnes (1966) (Ge/Si). By observation of several reversals in sign of the photogenerated current as wavelength was varied, together with capacitance and current–voltage measurements, Van Ruyven *et al.* (1965) deduced band profiles such as that shown for Ge/GaP in Fig. 5.22, where the pinning of E_F at the interface, raised the bands at the interface, adding substantially to the barrier seen by conduction band electrons.

Although pinning of E_F at surfaces and isotype HJs has had considerable study, much less is known about E_F pinning in anisotype HJs, where the effects are much more subtle. The properties would be expected to be considerably different, since the states responsible for pinning the Fermi level at a free surface are different, depending on the majority carrier type (acceptorlike for n-type and donorlike for p-type materials in general), and some neutralization might occur. Van Ruyven (1964) does not share this view, holding instead that the surface pinning of the Fermi level re-

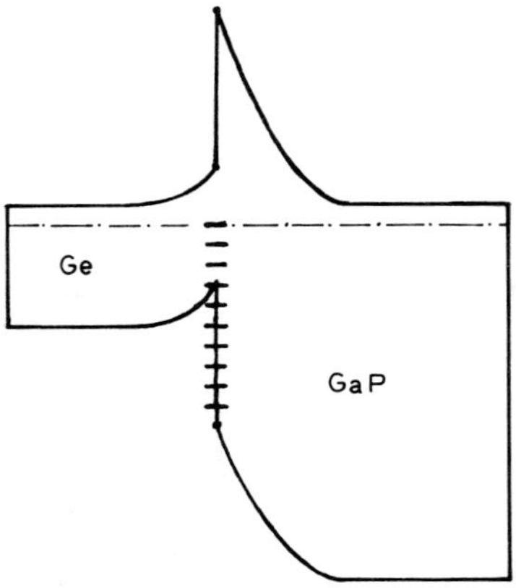

Fig. 5.22. Isotype Ge/GaP heterojunction showing the effects of charged interface states (lattice mismatch $\simeq 3.8\%$). (Redrawn from Van Ruyven *et al.*, 1965. Copyright 1965, Pergamon Press, Ltd.)

tains its identity for each of the two components when they are joined, creating an interfacial dipole to account for the difference.

The interface state distribution in space and energy at "free" surfaces has been extensively studied. Similarly, these properties of interface states at grain boundaries, dislocations, and to a lesser extent, isotype HJ have been appreciably researched. However, these quantities are more difficult to obtain in anisotype junctions and results tend to be very model dependent. The energy and spatial distribution of states in the depletion layer adjacent to the interface have been measured by the use of capacitance transient spectroscopy [deep-level transient spectroscopy (DLTS)] in connection with thermal excitation and photoexcitation. This topic is reviewed by Miller *et al.* (1977). Sah (1977) also compares the thermal and photostimulated capacitance methods. Unfortunately, the sensitivity of these techniques falls off as one probes closer to the interface so that distributions at or near the interface cannot be obtained unless an insulating layer is present.

Analysis of the capacitance–voltage data gives considerably more information about the energy levels, densities, and optical and thermal cross sections of the states at the interface. The intercept of the $1/C^2$ versus V curves as $1/C^2 \to 0$ gives the diffusion voltage modified by the charge stored at the interface Q_{ss} and the dipole voltage ϕ_d. Consider a HJ consisting of a p-type material of acceptor density N_{A1}, and n-type material of donor density N_{D2}. When an intervening layer of width W_i and donor density N_{Di} is introduced, the new, modified voltage intercept V_d' is given by

$$V_d' = V_d + \frac{q(N_{A1} + N_{Di})(N_{D2} - N_{Di})W_i^2}{2\epsilon_s(N_{A1} + N_{D2})} + \phi_d, \qquad (5.46)$$

where $\epsilon_{s1} = \epsilon_{s2} = \epsilon_{si} = \epsilon_s$ and where V_d is the diffusion voltage without the intervening layer. For a region of zero thickness, the charge stored at the interface Q_{ss} can be expressed as the limit of $W_i N_{Di}$ as $W_i \rightarrow 0$. Equation (5.46) then becomes

$$V_d' = V_d - [Q_{ss}^2 / 2q\epsilon_s (N_{A1} + N_{D2})] + \phi_d . \tag{5.47}$$

Thus, when the interface layer is thin, all the states near the interface behave as though they were *at* the interface, the spatial dependence is lost, and it is difficult (if not impossible) to discriminate between charged interface states and dipoles without additional information. Measurements as a function of frequency can be used to separate time-dependent interface state populations and ΔE_c (time-independent) contributions to the V_d offset, as well as the bias-voltage dependence of the occupation of interface states. These aspects are discussed by Donnelly and Milnes (1967). Interface state densities have been determined for the Au/*n*-GaAs Schottky diode by Borrego *et al.* (1977) using a combination of J versus V versus T and C versus V data.

With regard to the recombinative aspects of interface states it is again difficult to measure their spatial distribution. One would wish to do so with a resolution of 20–50 Å in order to establish whether tunneling is involved in the recombination pathway. In addition, one would wish to discriminate between recombination in the region *close* to the interface (within ≈ 100 Å, because of lattice mismatch dislocations, interdiffusion, and other "bulk" effects) and recombination in the thin (two to five atomic layers) region *at* the interface (because of electrically active dangling bonds, oxide layers, compound formation, and the like). In most cases these are both lumped into an effective surface recombination velocity S_i.

Several of the experimental methods of determining interface recombination values are summarized below:

(1) *Photogenerated Current Loss.* The relationship between J_{sc} and junction field can be used to obtain an effective S_i as described in Section 6.3.1 and Eq. (6.23), although the results can be quite model dependent.

(2) *Recombination in Double Heterostructures.* By studying recombination kinetics in the thin center layer of a *p*-AB/*p*-CD/*n*-CD or a *p*-AB/*n*-CD/*p*-AB structure, one can obtain approximate values for S_i. For the latter case, the recombination current in the *n*-type center layer is given approximately by $q(2S_i + UW)$, where the variable W is the intermediate layer thickness and U is the bulk recombination rate there. The lifetime in the slab can be measured by a variety of techniques, including electroluminescence time decay (Ettenberg and Olsen, 1977; Ettenberg and Kressel, 1976), photoluminescence time decay (Nelson, 1978), cur-

rent decay from pulsed laser excitation (for the first structure, where $E_{g,AB} > E_{g,CD}$) (Acket *et al.*, 1975), and pulsed reverse-bias recovery (Dean and Nuese, 1971).

(3) *$1/C^2$ versus V Analysis.* Using the C versus V analysis [Eq. (5.47)] with variation of T and/or illumination intensity and wavelength, the energies, densities, and thermal and optical cross sections of the states involved can be obtained, and in principle, used to calculate the recombination kinetics. A very thorough knowledge of the junction structure and the bulk properties is required. Such measurements have been done using these and the DLTS techniques on n- and p-type Si MOS structures by Schulz and Johnson (1977) and on InP MOS devices by Stannard (1979). Very few, if any, measurements on anisotype HJs appear to have been done.

(4) *Surface Photovoltage Spectroscopy.* The decay of photoexcited surface voltage of a semiconductor surface (Section 4.6.3) and V_{oc} decay in a HJ (Section 4.6.5) can be used to obtain information about recombination losses close to the surface of interest (in the narrow band-gap material for the HJ). The recombination current is schematically $\int U dx + S_i$, so that S_i can be obtained by careful analysis, particularly if one wishes to compare S_i for otherwise identical devices.

Measured or inferred values of $N_i = Q_{ss}/q$ range between $\sim 10^{12}/cm^2$ (with negligible effect on transport) and $> 10^{15}/cm^2$, when the equivalent of a whole lattice plane of atoms is electrically active and junction transport is dominated by interface losses.

Of course, separation of the charging and recombination effects is somewhat artificial, depending on how well these interface states communicate with the conduction and valence bands of the particular side of the junction of interest. A complete description requires simultaneous solution with respect to charging and recombination. This problem has been explored for grain boundary interfaces, as described in Section 9.4.2.

5.4.3 Interfacial Dipoles

Interfacial dipoles were introduced by Van Ruyven *et al.* (1965) to explain observed conduction band discontinuities in isotope HJs that differed from those predicted by the electron affinities. The presence of such dipoles follows from the pinning of the Fermi level at the surface of each semiconductor by a large density of surface states, forming an almost metallic layer at the surface. The joining of these metallic layers on each of the semiconductors on the creation of a HJ is accompanied by the formation of a dipole of atomic dimensions, much as occurs in the joining of two bulk metals with different work functions.

The HJ models of Van Ruyven (1972) and of Anderson (1960) represent two extremes. In the former case a dipole is established and the depletion layers are relatively unaltered by connection of the two component semiconductors, regardless of the doping levels in the bulk of the semiconductors. In the Anderson model the depletion layers are formed by the connection of the two semiconductors. Most HJs probably lie somewhere between these two extremes.

The physical nature of dipoles in semiconductors has been discussed by a number of authors. Such dipoles could arise from the mechanism discussed earlier, from charge on opposite sides of a thin insulating layer at the interface or from a spatial distribution of oppositely charged defects on either side of the interface in the bulk of the semiconductors. In the latter case the charge might be spread over several hundreds of angstroms, e.g., at dislocations propagating away from a lattice-misfit interface.

5.5 MODELS FOR HETEROJUNCTION TRANSPORT

A number of different models have been advanced for HJ transport in various types of junctions. The models differ chiefly in the dominant and/or limiting step in the transport route. In general, *parallel* transport routes (e.g., injection/diffusion into the QNR and recombination/generation in the depletion layer) do not interact appreciably and the resulting currents can be added, with each route dominating over a certain bias voltage or temperature range (see Fig. 5.13, for example). When the transport mechanisms are in *series* (e.g., thermal activation over a barrier to the interface, followed by interfacial recombination), the models describing transport usually focus on the limiting step. In only a few models, notably those of Tansley, and Padovani and Stratton, are more than one series-limiting step considered to interact appreciably. In HJs there are the options of emission (thermionic and possibly field-aided thermionic), diffusion, recombination in the bulk or at interface states, tunneling, and combinations of these to act as limiting steps.

5.5.1 Injection and Diffusion in the Quasi-Neutral Regions

In the Anderson model, discussed in Section 5.3.1, the limiting step is diffusion and recombination in the quasi-neutral regions, essentially the same as in the homojunction case. The model predicts J_0 values orders of magnitude lower than those observed experimentally in most cases.

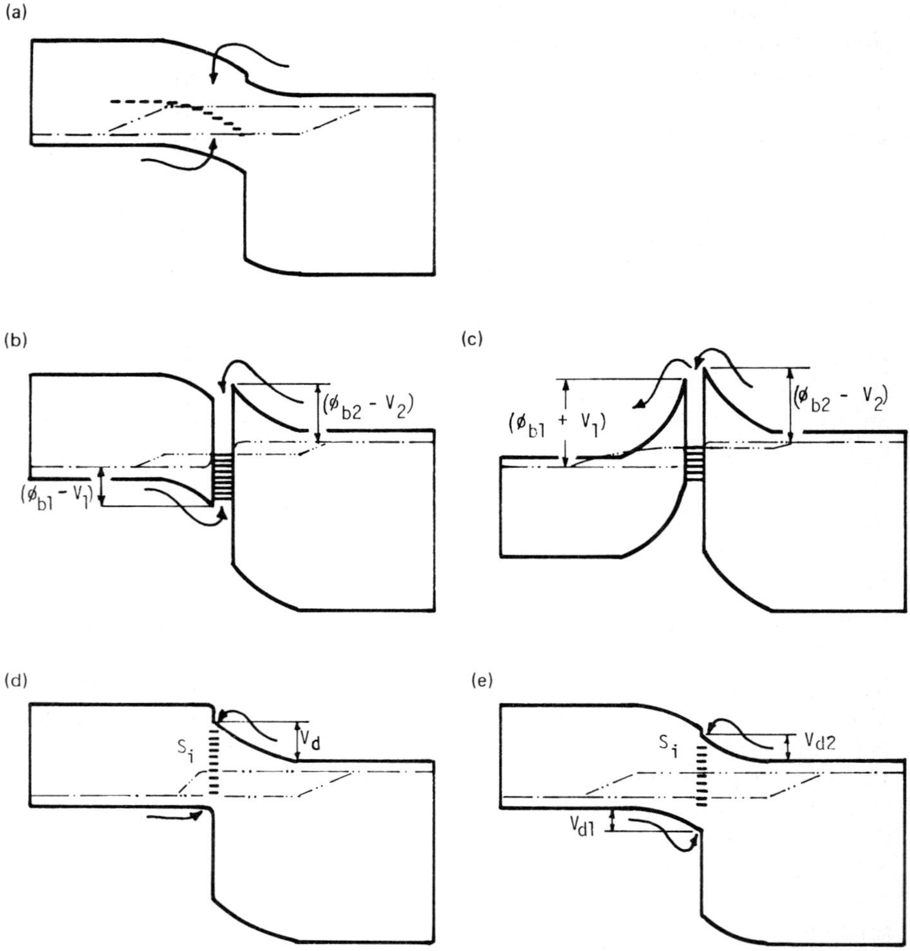

Fig. 5.23. (a) Recombination/generation in the depletion layer of a dark, forward-biased heterojunction. The Dolega model of junction transport for (b) anisotype and (c) isotype junctions. Interface recombination velocity controlled transport for (d) the Cu_xS/CdS case and (e) the general case.

5.5.2 Recombination/Generation in the Depletion Layer

In the typical HJ solar cell, recombination/generation is expected to be largely confined to the side of the depletion layer in the smaller band-gap component, because of the large barrier to minority carrier injection into the large band-gap component. For the example shown in Fig. 5.23a, the calculation of the $J-V$ relation is expected to be similar to that given in Section 5.2.4, provided that the position of maximum recombination lies well inside the depletion layer of material 1. Recombination currents in p/n heterojunctions are considered by Smith (1977).

5.5.3 Direct Recombination through Interface States (Schottky Barrier Control)

Direct recombination through states at the interface appears to be an important route of transport in many HJ but the degree of control exercised by this step varies from model to model. Dolega (1963) (p/n junctions) and Oldham and Milnes (1964) (isotype junctions) described the HJ as two Schottky diodes back-to-back on either side of an almost metallic layer of interface states where recombination was very strong, shown in Fig. 5.23b and c. This model predicts a $J-V$ relation given essentially by Eq. (5.42) with $1 < A < 2$, depending on the ratio $(\epsilon_{s1}N_{A1}/\epsilon_{s2}N_{D2})$. However, since the limiting step in these models is thermionic emission over the barriers, they cannot explain the temperature-independent slopes of the log J versus V curves seen for many HJ.

5.5.4 Transport Control by Interface Recombination

This model, used by Rothwarf (1975) to explain transport at moderately high injection levels in the Cu_xS/CdS thin-film HJ, is somewhat similar to Dolega's, except that the limiting step is interfacial recombination, characterized by a recombination velocity S_i, rather than the thermal velocity as in thermionic emission. In this case, shown in Fig. 5.23d, the Cu_xS is degenerate so that essentially all the diffusion potential appears in the CdS. The $J-V$ relation can then be written out of hand, giving

$$J(V) \simeq qS_i(n_n - n_{n0}) = qS_iN_D \exp(-qV_d/kT)[\exp(qV/kT) - 1]. \quad (5.48)$$

Note that an A factor of unity is predicted (and observed at moderate to high bias voltages in the illuminated case).

The general case is similar to the model for homojunction interfacial recombination shown in Fig. 5.11, which admits A factors larger than 1 as well as A factors that vary with bias voltage.†

5.5.5 Tunneling

In many HJs the slopes of the log $J-V$ curves are rather independent of temperature, suggesting that tunneling through a potential barrier dominates the transport. Quantum mechanical tunneling can become appre-

† Also see the correlation of $A \simeq 2$ currents in AlGaAs/GaAs HJs with peripheral surface recombination in Henry *et al.* (1978) and Henry and Logan (1978).

ciable for barriers thinner than 100–200 Å. Tunneling calculations can be handled by direct solutions of the Schrödinger equation, but reasonably accurate results for many barrier profiles can be obtained using the WKB approximation (Gundlach and Simmons, 1969). The calculations generally assume that the electron mean-free-path is much larger than the tunneling distance and they neglect electron–electron interactions. The tunneling transmission coefficient (WBK) is

$$\mathcal{T}(E) = \exp\left\{-\frac{2(2m_t^*)^{1/2}}{\hbar}\int_{x_2}^{x_1}[E - q\phi(x)]^{1/2}\,dx\right\}, \tag{5.49}$$

where x_1 and x_2 are the classical turning points, E is the electron energy, and $q\phi(x)$ is the potential barrier, both measured from the zero of kinetic energy for the electron. The value of the tunneling effective mass m_t^* is discussed by Barber (1967) and Crowell (1969).

For the band diagram shown in Fig. 5.24a, *intraband tunneling* through a conduction band spike is the limiting step for injection into the p-type material. The transmission coefficient for normal incidence tunneling through the bottom of a triangular barrier of height E_b is given by Nordheim (1928).

$$\mathcal{T}(\mathscr{E}) = \exp[-4(2m_t^*)^{1/2}\,E_b^{3/2}/3q\hbar\mathscr{E}]. \tag{5.50}$$

The bias-voltage dependence enters through the junction field $\mathscr{E}$ (which, together with E_b, defines the height and width of the barrier) and the result is a log J–V characteristic with slope independent of temperature. The model was apparently first used by Rediker *et al.* (1964) to explain transport through a conduction band spike for GaAs/Ge and GaAs/GaSb HJs. The intraband tunneling mechanism is discussed in a review by Tansley (1971).

Since the tunneling coefficient depends exponentially on the tunneling effective mass m_t^*, the junction current also depends strongly on m_t^*. This introduces a large dependence on the crystallographic orientation at the junction, as discussed by Anderson (1977).

A mechanism that includes *interband tunneling* through a thin portion of the barrier as the rate limiting step was used by Riben (1965) and extended by Riben and Feucht (1966b) to include tunneling into and subsequent recombination through states in the depletion layer. These theories follow from similar "excess-current" mechanisms for tunnel diodes, as, for example, discussed by Chynoweth *et al.* (1961). For the carrier transport route shown in Fig. 5.24b, Riben and Feucht's model gives

$$J_{t/r} = BN_t \exp\{[-4(2m_t^*q)^{1/2}(V_d - k_2V)]/3\hbar H_0\}, \tag{5.51}$$

where $H_0 = (2qN_A/\epsilon_{s2})^{1/2}$, $qV_d = E_{g1} + \Delta E_V - \delta_n - \delta_p$, and $k_2 = 1/[1 +$

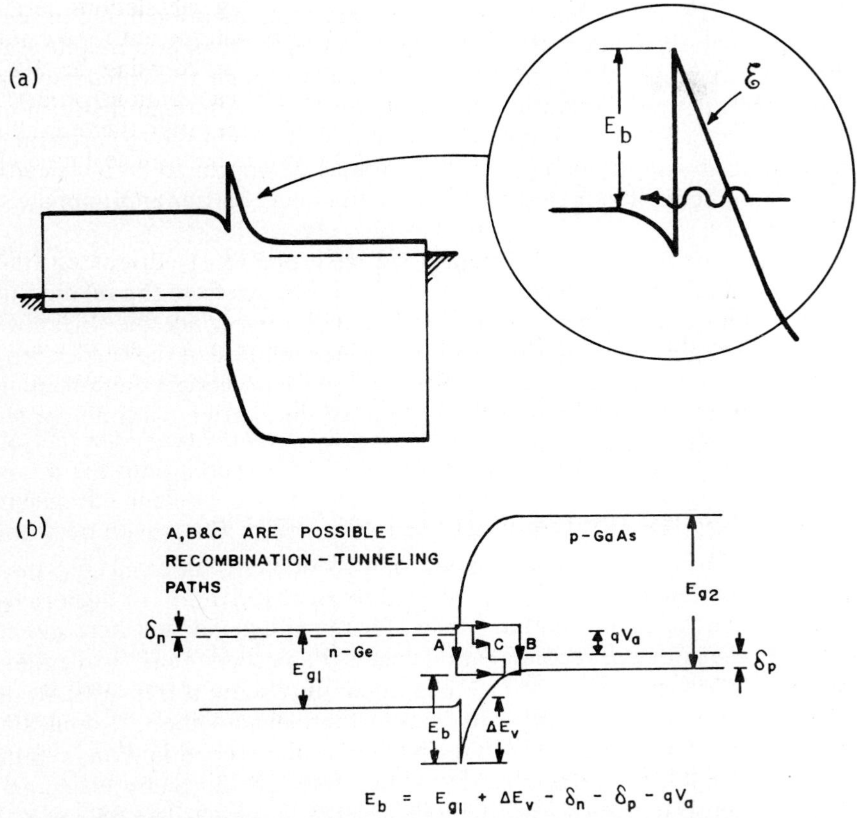

Fig. 5.24. (a) Intraband tunneling transport model. (b) Combined tunneling-recombination model of Riben and Feucht for forward bias. [(b) After A. R. Riben and D. L. Feucht, *Solid-State Electron.* **9,** 1055 (1966). Copyright 1966, Pergamon Press, Ltd.]

$\epsilon_{s1}N_A/\epsilon_{s2}N_D$] expresses the asymmetry of the junction. The density of tunneling/recombination centers is N_t, and B is a constant that contains N_c. Riben and Feucht found that Eq. (5.51) gave a good fit to the data for their n-Ge/p-GaAs diodes.

Equation (5.51) predicts a small temperature dependence of J_0 arising from the variation of the band-gap energy with temperature. This results in a simplified version of Eq. (5.51):

$$J_{t/r} = B' \exp[\beta T] \exp(\alpha V). \tag{5.52}$$

Stepwise tunneling and recombination through a staircase of closely spaced states in the interface region was used in a succeeding work by

Riben and Feucht (1966a). They introduced an additional factor depending on the number of tunneling steps and on the temperature into Eq. (5.51) and enabled the model to be applied to much thicker depletion layers than the model that included only one step. Although the form of this parameter is postulated by forcing the model to fit the experimental data, its form and the number of steps predicted appear to be physically reasonable. (e.g., about 20 steps across a tunneling barrier approximately 800 Å thick).

In contrast to any of the models we have previously discussed, the junction bias-voltage dependence for tunneling arises from the relation of the tunneling transmission to the junction field (and hence the thickness of the barrier), rather than from the voltage dependence of the carrier density at the top of the barrier. Note also that in these models the tunneling is assumed to take place near the bottom of the barrier, with no appreciable thermal activation involved.

5.5.6 Combined Thermal Activation and Tunneling

Many HJs exhibit a combination of two or more mechanisms: tunneling/recombination, as described, and thermal activation to higher energies where the barrier is thin enough for tunneling and/or where direct access to the interface recombination centers can occur. Thus, as the temperature is lowered, the log $J-V$ characteristics show a transition from slopes varying inversely with T, indicating thermal activation, to constant slopes, indicating tunneling. A large portion of the recombination in both cases occurs at the interface. Lindquist and Bube (1972) used such a combined model to explain forward-bias transport in the Cu_xS/CdS (single-crystal CdS) solar cell. Another example of this sort of mechanism is the n-CdS/p-CdTe HJ solar cell studied by Mitchell *et al.* (1977), who found tunneling-like transport of dark diode current at $T < 290°K$ following Eq. (5.52) with $\alpha = 22/V$ and $\beta = 0.015/°K$. For $T > 290°K$, a recombination-like mechanism prevailed, following Eq. (5.42), with $A = 1.89$ and an activation energy for J_0 of $\Delta E = 0.59$ eV. The zero-bias depletion layer width of this junction was 0.19 μm, predominantly in the p-CdTe.

There is considerable similarity between heterojunctions with large S_i (so that S_i is not the limiting step) and Schottky barriers, in which the surface recombination velocity is effectively infinite. Thermally assisted tunneling (also called thermionic-field emission, T-F) in Schottky barriers has been modeled extensively by Stratton (1962), Padovani and Stratton (1966) and others (Chang and Sze, 1970; Crowell and Rideout, 1969) and has been reviewed by Rhoderick (1974) and Padovani (1971). The current

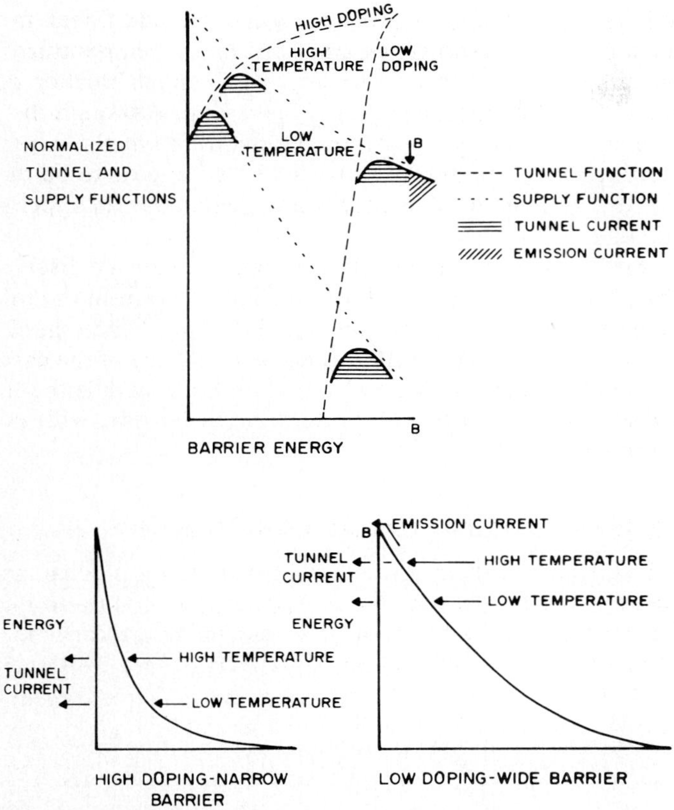

Fig. 5.25. Tunnel and supply functions for temperature and barrier width variations. [After S. J. T. Owen and T. L. Tansley, *J. Vac. Sci. Technol.* **13,** 954 (1976).]

through the barrier is the integrated product of the incident electron flux and the quantum mechanical transmission (or tunneling probability) of the barrier, both of which are energy dependent. The current flows in a narrow energy range centered around some energy E_m as suggested by Fig. 5.25. Stratton used a supply function†

$$J(E)\, dE = (4\pi q m^* kT/h^3)\, \{\exp[-(E + E_F)/kT]\}\, dE \qquad (5.53)$$

that represents a flux of electrons whose velocities are normal to the barrier and where E is the tunneling energy and E_F is the Fermi energy, both measured from the conduction band (or valence band) edge. Without

† Note that the dominant tunneling must occur at least several kT above the Fermi level for the formulation of Eq. (5.53) to be a good approximation.

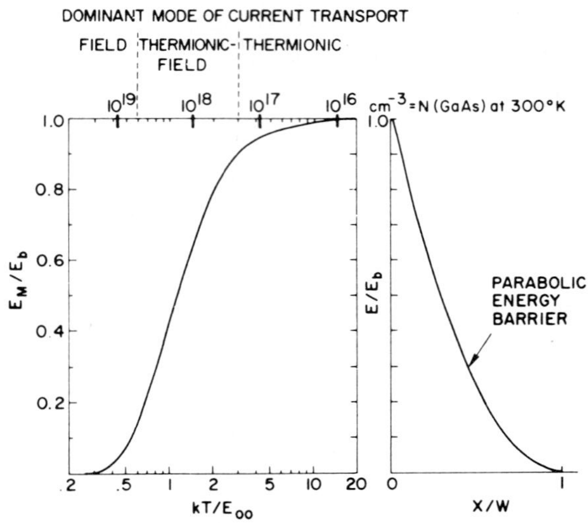

Fig. 5.26. Relative position of maximum transport through or over a Schottky barrier versus the parameter kT/E_{00}. Note the narrow range of doping in *n*-GaAs over which thermionic-field emission is an appropriate description of transport. The doping range scale is shifted considerably for Si; $N_D = 10^{19}$ cm^{-3} lies in the center of the T-F range for Si. [After V. L. Rideout, *Solid-State Electron.* **18,** 541 (1975). Copyright 1975, Pergamon Press, Ltd.]

approximations or simplifying assumptions the resulting integral is impossible to evaluate analytically. Padovani and Stratton assumed a Gaussian distribution of electron energies centered around the energy E_m for which the tunneling transport was a maximum. They were then able to expand the tunneling integrand in a Taylor's series about E_m and to obtain an "analytical" result for the *forward*-bias $J-V$ relation of the form

$$J(V) = J_{0f} \exp(qV/E_0) = J_{0f} \exp(qV/A_{T-F}kT), \tag{5.54}$$

where

$$E_0 = E_{00} \coth(E_{00}/kT) \qquad \text{and} \qquad E_{00} = (q\hbar/2)[N_D/\epsilon_s m_t^*]^{1/2} \tag{5.55}$$

for the Schottky barrier case.† For a given value of N_D, E_{00} is the height of a triangular potential barrier with a transmission through its base of about $1/e$ [Eq. (5.50)]. Because of the simplifying assumptions, the preceding expressions are valid only for the approximate range $0.25\,kT < E_{00} < 1.5kT$. The ranges in which the field, thermionic-field, and thermionic emission theories are applicable are shown in Fig. 5.26 in terms of the energy corre-

† In the calculation of E_{00}, if $\hbar$, N_D, ϵ_s, m_t^*, are in SI units, then E_{00} is in joules.

sponding to the maximum tunneling current E_m,† as a function of kT/E_{00}.‡ The value of J_0 for forward bias is given by

$$J_{0f} = \frac{A^* T^2 \pi^{1/2} E_{00}^{1/2} [q(V_d - V)]^{1/2}}{kT \cosh(E_{00}/kT)} \exp[-(\delta_n/kT) - (q\phi_b - \delta_n)/E_0], \quad (5.56)$$

where δ_n is negative when the conduction band lies below the Fermi level. At high temperatures and for thick barriers, J_{0f} is thermally activated with an energy $\simeq q\phi_b/A_{T-F}$, where ϕ_b is the barrier height and A_{T-F} is the thermionic-field emission ideality factor. At low temperatures and for thin barriers, J_{0f} becomes almost independent of temperature.

Tansley and Owen (1976) extended the preceding treatment to heterojunctions. They calculated the tunneling transmission using the WKB approximation and, for a parabolic potential barrier, the integral of Eq. (5.49) can be evaluated exactly, giving (Tansley, 1968)

$$\begin{aligned} \mathcal{T}(E) &= \exp\{-(2/q\hbar)[\epsilon_s m_t^*/N]^{1/2} (q\phi_n)[[1 - (E/q\phi_b)]^{1/2} \\ &\quad - (E/q\phi_b) \ln[(q\phi_b/E)^{1/2} + [(q\phi_b/E) - 1]^{1/2}]]\} \\ &= \exp[(-q\phi_b/E_{00})\Theta(E/q\phi_b)], \end{aligned} \quad (5.57)$$

where N is the effective donor or acceptor density, E is the electron energy, and ϕ_b is the total barrier height. The universal function $\Theta(E/q\phi_b)$ as well as $\mathcal{T}(E)$ and $\mathcal{T}(E)\exp(-E/kT)$ for a specific barrier are shown in Fig. 5.27. Note that the maximum current transport occurs around 0.15 eV in this case, despite the extremely small tunneling probability there ($<10^{-7}$). Tansley and Owen used Eq. (5.57) with the electron flux expression of Eq. (5.53) and integrated the product numerically to obtain the total junction current.§ The equivalent diode quality factor $A_{T-F} = E_0/kT$ of Eq. (5.54) must be multiplied by a voltage division factor for the HJ to account for the degree of junction asymmetry. For example, for a n^+/p structure with electron tunneling through the n-type depletion layer as the limiting step, Tansley and Owen give the equivalent HJ diode factor as

$$A_{HJ} = A_{T-F}(T, N_D)[\epsilon_{sn} N_D/(\epsilon_{sn} N_D + \epsilon_{sp} N_A)]. \quad (5.58)$$

† $E_m = (V_d - V)/\cosh^2(E_{00}/kT)$.

‡ Strictly speaking, "pure" field emission transport occurs only at $T = 0°K$, since tunneling transport above that temperature always involves some thermal activation and is thus really thermionic-field emission. Practically, the distinction is made between field and thermionic-field emission mechanisms when E_m is several kT above the bottom of the tunneling barrier. This distinction must not be confused with the limit of validity for the *mathematical* model presented here.

§ Carrier transport across a HJ interface has been considered in more detail by Wu and Yang (1979), who considered quantum mechanical reflection for carriers with energies greater than the potential barrier energy and the differences in effective masses in the HJ components.

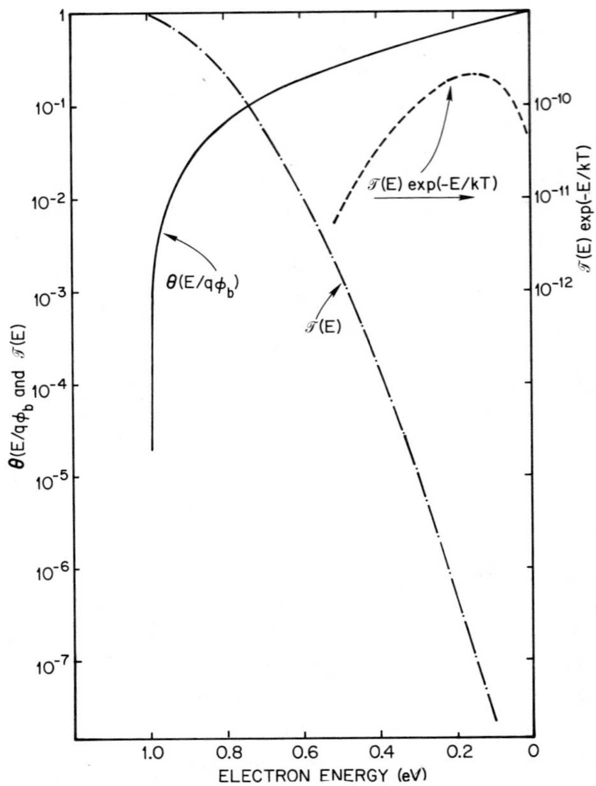

Fig. 5.27. The functions $\Theta(E/q\phi_b)$ and $\mathscr{T}(E)$ from Eq. (5.57) and the quantity $\mathscr{T}(E)$ $\exp(-E/kT)$ which is proportional to incremental junction current at an electron energy E. (Values assumed are $\phi_b = 1.0$ eV, $\epsilon_s/\epsilon_0 = 10$, $m_t^*/m_0 = 0.2$, and $N_D \simeq 10^{19}$ cm^{-1}, giving $E_{00} = 0.04$ eV.)

These authors found excellent agreement between the calculated variation of A_{HJ} with temperature and experimental values for several HJs between p-GaAs and n-Ge, n-$Ga_xIn_{1-x}As$, or n-$GaAs_yP_{1-y}$ as shown in Fig. 5.28.

Aranovich *et al.* (1980) have used an approach similar to that presented here to describe transport in the n-ZnO/p-CdTe HJ solar cell. In this case the n-ZnO is degenerate and the diffusion potential appears almost entirely in the CdTe. These authors followed the suggestion of Padovani (1971, see Section III.6 therein) who pointed out that the involvement of a large density of deep acceptors within the depletion layer could considerably reduce the width of a portion of the depletion layer to enable

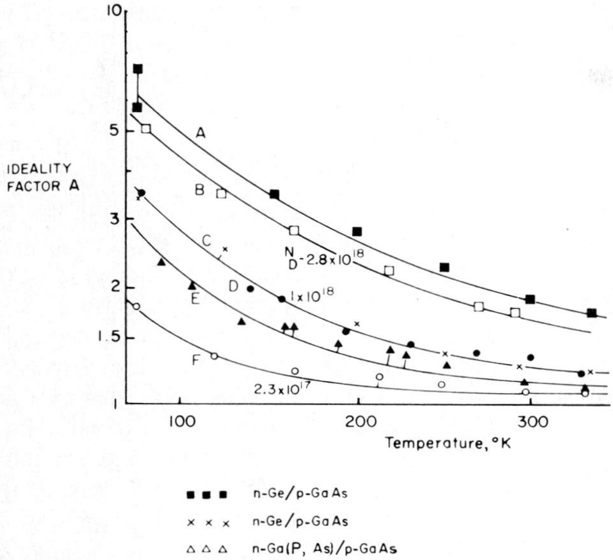

Fig. 5.28. Ideality factors versus temperature for six heterojunctions with donor and acceptor densities as summarized below. The correction for voltage division between the two semiconductor components F, as tabulated, has been included before plotting.

Diode	Component 1	N_D (cm^{-3})[a]	Component 2	N_A (cm^{-3})[a]	F[b]
A	n-Ge	1.2×10^{19}	p-GaAs	?	?
B	n-$Ga_{0.95}In_{0.05}As$	2×10^{18}	p-GaAs	7×10^{18}	0.78
C	n-Ge	3.5×10^{18}	p-GaAs	2×10^{19}	0.85
D	n-$Ga_{0.995}In_{0.005}As$	1.3×10^{18}	p-GaAs	4×10^{18}	0.76
E	n-$GaAs_{0.8}P_{0.2}$	1×10^{18}	p-GaAs	2×10^{18}	0.67
F	n-$Ga_{0.92}In_{0.08}As$	3×10^{17}	p-GaAs	4×10^{18}	0.93

[a] These are values used for best fit, which in most cases are very close to the values measured by other methods.

[b] F is the fraction of the applied voltage dropped across the depletion layer of the n-type material.

[Figure after S. J. T. Owen and T. L. Tansley, *J. Vac. Sci. Technol.* **13**, 954 (1976).]

tunneling to take place.† For Aranovich's diodes the involvement of these states gave rise to a J–V relation that was somewhat illumination and time dependent. By comparison of a number of diodes, a strong correlation

† The effective acceptor density used by Aranovich was considerably larger than that estimated from 300°K resistivity measurements ($\sim 10^{17}$ cm^{-3}) but was more consistent with that indicated by junction capacitance and reverse-bias breakdown voltages.

was found between the value of J_0 and the preparation-sensitive interface state density that controls the barrier height. Similarly, a correspondence was found between the diode factor E_0 and the effective acceptor density in the *p*-CdTe depletion layer, as suggested by Eq. (5.55).

5.5.7 Heteroface Structures

In a heteroface structure, like that shown in Fig. 1.4, the free surface of a homojunction is replaced with a large band-gap window material so that the original free surface recombination velocity ($S \simeq 10^6 - 10^7$ cm sec^{-1}) is replaced by an interface recombination velocity S_i, which is several orders of magnitude smaller. The value of S_i is controlled by careful interface preparation and lattice matching. In addition, an accumulation layer may exist because of differences in electron affinity that can produce electric fields that reduce the effective S_i still further. The $J-V$ characteristics of the heteroface diode can be obtained using the homojunction expressions for injection/diffusion [Eq. (5.15) with S replaced by an effective S_i] and for recombination/generation in the depletion layer [Eq. (5.27) or (5.30), for example].

As an example, the most common heteroface structure is the $Al_xGa_{1-x}As/GaAs$ solar cell in which (typically)

(a) recombination/generation dominates at V_{oc} values typical of unconcentrated sunlight with $A \simeq 2$ and $J_0 = 10^{-12} - 10^{-10}$ A cm^{-2};
(b) injection/diffusion dominates at V_{oc} values typical of highly concentrated illumination with $A \simeq 1$ and $J_0 = 10^{-19} - 10^{-18}$ A cm^{-2};
(c) $S_i \lessapprox 10^4$ cm sec^{-1}; and
(d) the indirect band-gap energy of the $Al_{0.9}Al_{0.1}As$ is about 2.1 eV.

A more detailed discussion of GaAs-based heretoface solar cells is found in Chapter 8.

5.6 SUMMARY OF HETEROJUNCTION TRANSPORT

The parameters A and J_0 are perhaps the most direct indicators of the dependence of solar efficiency on the electrical transport properties of the junction (see Fig. 6.5). The quality factor A for *homojunctions* with homogeneously distributed recombination centers arises from the type of transport involved: $A = 1$ for injection and diffusion into the quasi-neutral regions when $n_i^2/N_A = n_{p0}$ enters the J_0 expression (e.g., $J_0 = q(L_n/\tau_n)(n_i^2/N_A)$ for the p region), and $A \simeq 2$ for the mechanism in which recombination/generation in the depletion layer dominates the current transport and n_i enters the J_0 expression (e.g., $J_0 \simeq q(W_d/2\tau)n_i$, where τ is the recombination lifetime in the depletion layer).

Consider the case of injection and diffusion into a p-type QNR. The applied bias moves the electron quasi-Fermi level E_{Fn} directly with respect to the recombination center energy level E_r while E_{Fp} remains relatively constant in the highly doped p-type quasi-neutral region. In the recombination/generation case the applied bias moves *both* E_{Fn} and E_{Fp} with respect to the recombination center energy level in the depletion layer. Thus for a given change in applied bias, the change of $(E_{Fn} - E_r)/q$ and of $(E_r - E_{Fp})/q$ are about half as large, and viewed on a simplistic level, this introduces the factor of $A \simeq 2$. These concepts can be readily extended to heterojunctions. The total current is the sum of the diffusion and recombination/generation components, and for a heterojunction, the latter generally dominates the transport, especially at low to moderate bias, in the range of solar cell operation.

Sah *et al.* (1957) showed that the A factor for symmetrical homojunctions depends on the energy level of the effective recombination centers with respect to the intrinsic level E_i as well as on the magnitude of the applied bias, ranging from 1 for shallow recombination centers close to the band edges to almost 2 for centers with levels within a few kT of E_i. Choo extended this theory to asymmetrically doped homojunctions but retained the homogeneous distribution of recombination centers, providing the theoretical possibility for A factors greater than 2. For both homojunctions and heterojunctions with a nonuniform spatial distribution of recombination centers within the depletion layer, we have seen (Fig. 5.11) that the A factor depends on the junction asymmetry as well as on the energy levels of the centers involved. This leads to A factors varying from 1 to well above 2. In heterojunctions, the possibilities of tunneling, recombination, and thermal activation lead to wide possibilities for the A factor, including even $A < 1$.

Since the quantum efficiency in most heterojunction solar cells already lies in the 80–95% range, the largest potential for improvement in efficiency is from increases in A and decreases in J_0.† Such changes depend chiefly on the properties of the interface and depletion layer, rather than on the bulk collection or recombination properties. However, A and J_0 are loosely related as is indicated by a plot of log J_0 versus $1/A$ shown in Fig. 5.29 for a wide variety of HJs at room temperature. Some of these junctions are governed by expressions of the type $J = J_{00} \exp(-\alpha\phi_b) \exp(\alpha V)$, where α is relatively temperature independent. For these, an equivalent A factor is determined using the relation $\alpha = q/AkT$ with $T = 300°K$. The data of Fig. 5.29 have the form

$$J_0 = J_{00} \exp(-\Delta E/AkT) \tag{5.59}$$

† If J_0 and J_L are somehow held constant, then, for an ideal solar cell, the solar efficiency η_s is directly proportional to A, as is shown by Eq. (6.9).

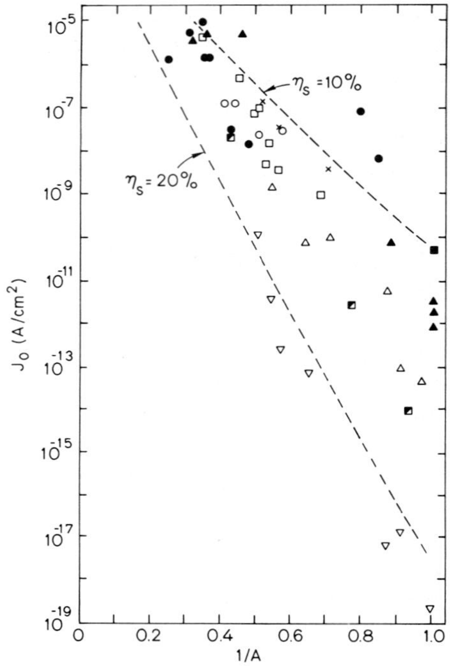

Fig. 5.29. The diode parameters J_0 and A for a variety of junctions at 300°K. Symbols: ▲ Si p/n, ▽ GaAs p/n (and heteroface), ◪ InP p/n, ● CdS/InP, ○ ITO/InP, □ CdS/CdTe, × ZnO/CdTe, ■ Cu_xS/CdS (thin-film polycrystal, under solar illumination), and △ GaAs MIS. All are single-crystal or are thin films on single-crystal substrates except for Cu_xS/CdS. Two theoretical constant solar efficiency curves are also shown for $J_L = 20$ mA cm^{-2}, $P_s = 83$ mW cm^{-2}, and $R_s = 0$. Note that the solar efficiency values are for one specific value of J_L, typical of cells with $E_g = 1.4-1.5$ eV.

and an energy ΔE of approximately $E_g/2$ can be obtained from the slope of the data for each of the junction types. This relationship is suggested by the form of the elementary homojunction expressions for J_0 for the injection/diffusion (I/D) and recombination/generation (R/G) cases [$J_0 \sim n_i^2 = N_c N_v \exp(-E_g/kT)$ for I/D, and $J_0 \sim n_i = (N_c N_v)^{1/2} \exp(-E_g/2kT)$ for R/G]. However, the theoretical basis for data for a continuum of A factors, ranging from 1 to 2 or greater, is not as obvious.

In the case of thermionic-field emission, J_0 and the equivalent A factor, $A_{T-F} = E_0/kT$, can be varied over a wide range by changing the doping density (at constant T). A similar relation between ln J_0 and $1/A_{T-F}$ holds with a slope of approximately $-\phi_b/kT$.

Using the general expression of Eq. (5.42), we can associate J_{00} for the junction of Fig. 5.15a, for example, with the density and cross section of the effective recombination centers. The A factor can then be associated with the location of the energy levels with respect to E_i, E_{Fn}, and E_{Fp} and with the asymmetry of the junction. For heterojunctions, the complexity of these relationships, coupled with the lack of specific knowledge about the centers involved in the depletion layer (particularly at the interface) have made A and J_0 largely into curve fitting parameters.

5.7 SCHOTTKY BARRIERS, MIS, AND SIS STRUCTURES

The utilization of metal/semiconductor (MS or Schottky diodes) and metal/insulator semiconductor (MIS) devices for solar cell applications is very attractive because of the apparent simplicity of these systems. Other advantages are the applicability to thin-film polycrystalline materials and the promise of relatively inexpensive processing technology. The simplest of these devices is the Schottky barrier in which the transport is ordinarily controlled by thermionic emission. The calculated and measured values of V_{oc} for such MS devices are considerably lower than for the corresponding p/n junctions, however. This is a consequence of the large values of J_0 that arise from the small values observed for the potential barrier height ϕ_b.

The barrier height ϕ_b in MS devices is controlled by interface phenomena. For the more covalent semiconductors, such as GaAs, Si, and CdTe, the Fermi level is pinned at the surface by metal-induced defect states and ϕ_b is relatively independent of the metal function ϕ_m. These materials include all those with band gaps that make them optimum absorber layers for solar cells ($E_g < 1.7$ eV). For ionic materials, which include the common window layer semiconductors, such as CdS, ZnO, and SnO_2, ϕ_b is much more strongly determined by ϕ_m (see Fig. 5.36).

It has been found that insertion of an insulating layer, 10–30 Å thick, between the metal and the semiconductor can have an advantageous effect, mainly by increasing the V_{oc} for the device, without appreciably decreasing J_{sc} or *ff*. For thicknesses of the insulating layer above about 30 Å, a strong reduction of J_{sc} and *ff* occurs. For properly controlled oxide layers increases of V_{oc} up to 50% have been observed. The study of such MIS structures is especially pertinent, since most Schottky barriers probably include very thin insulating layers formed during fabrication.

One of the MIS class of devices is the induced-junction structure (Fig. 11.3). The compensating charge layer usually associated with one of the depletion layers in a homojunction is instead contained in (or at the surface of) a thick insulating layer on the surface of a semiconductor. If the insulating layer is properly formed, the magnitude of this charge is sufficient to induce an inversion layer in the semiconductor that can be used both to separate the photogenerated carriers and for lateral carrier collection between grid lines.

Heterojunctions can also be improved by adding an insulating layer (resulting in an SIS structure) and high V_{oc} values are predicted for certain configurations, although very little experimental work has been done on such junctions intentionally.

Finally we close this section with a discussion of a relative of the Schottky barrier—the ohmic contact, which is essential to all solar cells.

5.7.1 Basic Schottky Barriers

Schottky barrier diodes have been under practical consideration since the early 1930s and the transport mechanisms that determine the $J-V$ relations have been the object of much theoretical research (Wagner, 1931; Schottky and Spenke, 1939; Sze, 1969), which has been reviewed by Rhoderick (1974) and Padovani (1971). The forward-bias transport mechanisms (for n-type semiconductors, for example) include

(a) thermionic emission of electrons from the semiconductor over the top of the barrier into the metal,
(b) recombination in the depletion region,
(c) quantum mechanical tunneling through the barrier, and
(d) hole (minority carrier) injection and diffusion.

All of these may operate simultaneously. Transport in Schottky barriers is particularly relevant, since it is the limiting behavior for a heterojunction when the interface recombination velocity becomes very large (cf. Sections 5.5.3 and 5.5.5).

The basic and dominant mechanism in Schottky barrier junctions is usually simple thermionic emission, which was theoretically treated by Bethe (1942). Bethe assumed that the barrier height $\phi_b \gg kT$ and that electron scattering within the depletion layer and image forces could be neglected. The procedure is to calculate the components of electron velocity v_x normal to the barrier assuming a Maxwellian distribution, and then integrate the product nv_x over all v_x such that

$$m^* v_x^2/2 \geq q(V_d - V), \tag{5.60}$$

where V_d, the diffusion voltage,† is defined in Fig. 5.30 and the carrier

† By convention, the diffusion voltage for Schottky and MIS devices is usually symbolized by V_{bi} (called the built-in voltage) rather than V_d. Similarly, the diode factor is symbolized by n, rather than A. We opt for consistency here.

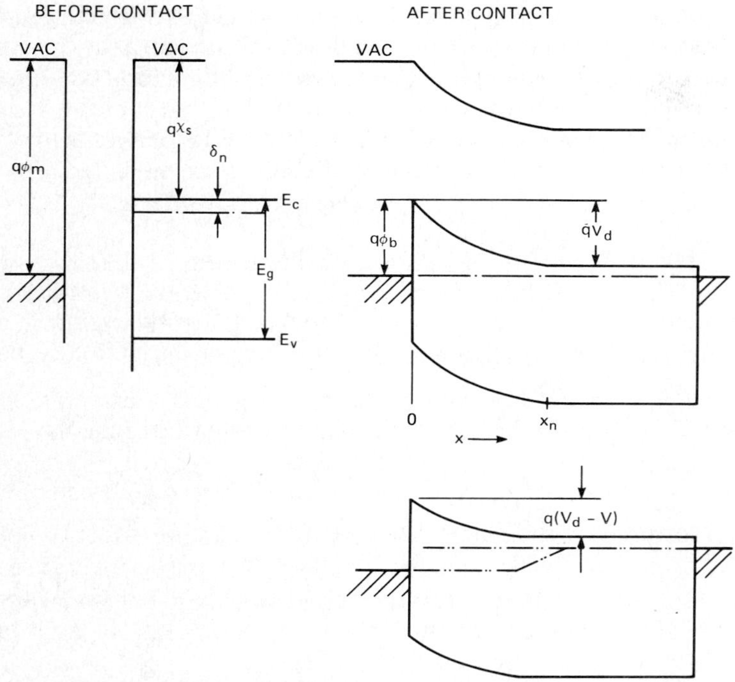

Fig. 5.30. Metal and semiconductor before contact and Schottky barrier at zero bias and forward bias V. In the Schottky limit $\phi_b = \phi_m - \chi_s = V_d + \delta_n/q$.

density at the top of the barrier n is given by

$$n = N_c \exp\{-[E_c(x = 0) - E_F]/kT\}$$
$$= 2[2\pi m^* kT/h^2]^{3/2} \exp(-\delta_n/kT) \exp[-q(V_d - V)/kT]. \quad (5.61)$$

The integration is also carried out over all values of v_y and $v_z \geq 0$ and the result is

$$J = A^* T^2 \exp(-q\phi_b/kT)[\exp(qV/kT) - 1] = J_0[\exp(qV/kT) - 1], \quad (5.62)$$

where $q\phi_b = qV_d + \delta_n$ and

$$A^* T^2 = 4\pi\, qm^* k^2 T^2/h^3 = qN_c[kT/2\pi m^*]^{1/2}. \quad (5.63)$$

Note that J_0 has the form $qnv_{th}/4$ where $n = N_c \exp(-q\phi_b/kT)$ and $v_{th} = (8kT/\pi m^*)^{1/2}$ is the mean thermal velocity of the electrons. For thermionic emission from a metal into a *vacuum*, $m^* \to m$ and $A^* \to A_R = 120\ A\ cm^{-2}\ {}^\circ K^{-2}$, the Richardson constant. The relationship of A^* to A_R

for various tensor effective masses encountered in multiple valley semiconductors is discussed by Crowell and Sze (1966a).

It is interesting to note that the forward-bias diffusion current for a p/n junction reduces (approximately) to the thermionic current for a Schottky barrier when the mean-free-time between phonon emission events for an electron in the metal is used for the lifetime:

$$\begin{aligned} J_{0,\text{diffusion}} &= qn_{p0}(D_n/\tau_n)^{1/2} = qn_{p0}(kT\mu_n/q\tau_n)^{1/2} \\ &= qn_{p0}[kT(\tau_r q/m^*)q\tau_n]^{1/2} \\ &= qn_{p0}(kT/m^*)^{1/2} \end{aligned}$$

Now if we identify the carrier density at the depletion layer edge in a p/n junction n_{p0} with the carrier density at the top of the Schottky barrier $N_c \exp(-q\phi_b/kT)$, then

$$qn_{p0}(kT/m^*)^{1/2} = q(kT/m^*)^{1/2}N_c \exp(-q\phi_b/kT) \simeq J_{0,\text{thermionic}}.$$

In brief, the lifetime in the metal is not zero but rather l/v_{th}, where l is a mean-free-path.

Field lowering of the potential barrier (the Schottky effect or field effect) can be appreciable for thin barriers (high N_D), particularly in reverse bias. Image forces exerted by the metal attract the electrons within the semiconductor adjacent to the metal, producing a potential given by

$$\phi_{FE}(x) = -q/16\pi\epsilon_{s,hf}x, \tag{5.64}$$

where $\epsilon_{s,hf}$ is the high-frequency permittivity (Crowell and Sze, 1966b). This potential is superimposed on the potential arising from the depletion layer. The total potential is then

$$\phi(x) = \phi_b - (qN_D/2\epsilon_s)(2xW_d - x^2) - q/16\pi\epsilon_{s,hf}x \tag{5.65}$$

for a parabolic potential barrier of original height ϕ_b. This is shown in Fig. 5.31 for several values of N_D. For all but the largest values of N_D, the barrier lowering $\Delta\phi$ is given by

$$\Delta\phi \simeq (q\mathscr{E}_{max}/4\pi\epsilon_{s,hf})^{1/2} \simeq (q/4\pi\epsilon_{s,hf})^{1/2}[2qN_D(V_d - V)/\epsilon_s]^{1/4}. \tag{5.66}$$

The barrier-lowering-modified quality factor is given by $A' = [1 - (\Delta\phi/4\phi_b)]^{-1}$. Surface state penetration of the barrier can also result in comparable magnitudes of barrier lowering, particularly for high N_D (Parker *et al.*, 1968).

The thermionic emission theory above assumes that the electron quasi-Fermi level is constant throughout the depletion layer. However, this is not strictly true if the electron mean-free-path is short compared with the depletion layer width. Transport is then limited by the diffusion velocity of the carriers, since transport into the metal is effectively trans-

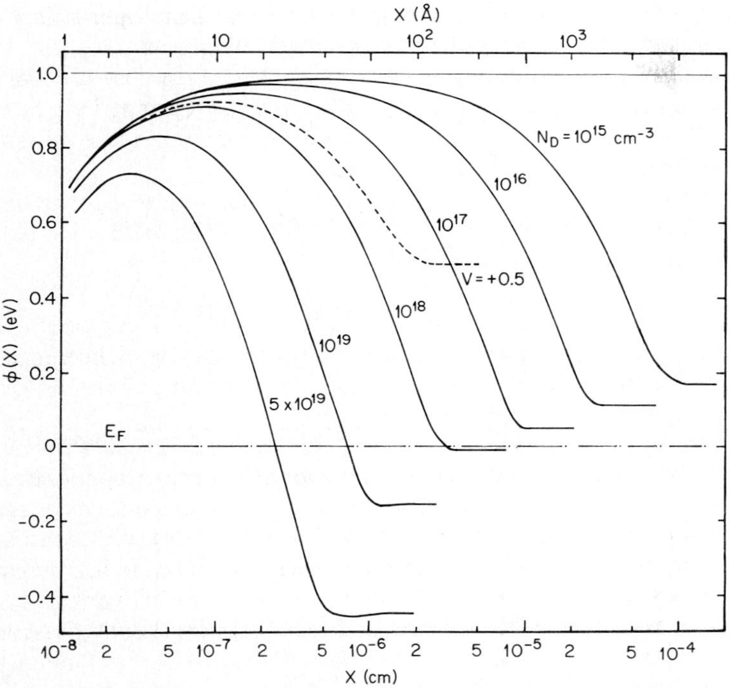

Fig. 5.31. Potential barrier profiles showing simple image force lowering for various doping densities. Parameters assumed are: $\phi_b = 1.0$ eV, $m^* = 0.11m_0$, and $\epsilon_{s,hf}/\epsilon_0 = 10$. Dashed curve is for $N_D = 10^{18}$ cm^{-3} at forward bias of 0.5 V.

port into a medium of exceedingly small minority carrier lifetime† (Schottky, 1938). Consequently, the quasi-Fermi level is lowered slightly near the interface (cf. Section 5.2.3). To deal with this situation, Schottky‡ started from the current density expression

$$J_n = q[n(x)\mu\mathscr{E} + D_n dn/dx] = qD_n\{-qn(x)(dV/dx)/kT + dn/dx\}, \quad (5.67)$$

which he integrated across the depletion layer, assuming the boundary conditions

$$n(0) = N_c \exp(-q\phi_b/kT) = N_c \exp\{-[E_c(0) - E_F]/kT\},$$
$$n(x_n) = N_c \exp(-\delta_n/kT)$$

† This lifetime is essentially the relaxation time for hot electrons in the metal, about 10^{-13} sec.

‡ Schottky used $\exp[-qV(x)/kT]$ as an integrating factor for Eq. (5.67). Details of both these theories are found in Henisch (1957).

to obtain a relation between J_n and the potential distribution across the depletion layer. For a parabolic potential barrier, this leads to a $J-V$ relation given approximately by

$$J_n \simeq q\mu_n[2q(V_d - V)N_D/\epsilon_s]^{1/2}N_c \exp(-q\phi_b/kT)[\exp(qV/kT) - 1]. \tag{5.68}$$

The quantity $[2q(V_d - V)N_D/\epsilon_s]$ is simply the maximum electric field for a parabolic barrier $\mathscr{E}_{max}$. In Eq. (5.67), J_n is the small difference between two large quantities and thus

$$\mathscr{E}_{max} = -dV/dx \simeq -[kT/qn(x)](dn/dx), \tag{5.69}$$

so that $\mu_n\mathscr{E}_{max}$ approximates the diffusion velocity near the interface. Since $\mathscr{E}_{max} \sim N_D^{1/2}$, the effect of diffusion is significant mainly for low values of N_D, where the field effect is small.

Whereas the thermionic emission result of Eq. (5.62) gives a $J-V$ relation that depends only on ϕ_b and is independent of the properties of the depletion layer, the diffusion theory of Eq. (5.68) introduces a specific dependence on μ_n, $V(x)$, and N_D in the depletion layer. The diffusion theory also introduces a slight dependence of J_0 on V but is somewhat less sensitive to temperature.

The thermionic model of Bethe and the diffusion theory of Schottky have been combined by Crowell and Sze (1966a, b) and extended to include electron–phonon interactions, quantum mechanical transmission over the barrier, and image force lowering of the barrier (the Schottky effect). A simplified form of their result is given by

$$J = \frac{qN_c}{(1/v_R) + (1/v_d)} \exp(-q\phi_b'/kT)[\exp(qV/kT) - 1], \tag{5.70}$$

where v_R is an effective interface recombination velocity, v_d is an effective diffusion velocity, and $\phi_b' = \phi_b - \Delta\phi$ expresses the field effect. For a mobility that is independent of electric field, $v_d \simeq \mu\mathscr{E}_{max}$. For higher values of N_D, $\mathscr{E}_{max}$ is large and $v_R < v_d$ with $v_R \simeq \bar{v}_{th}/4 = (kT/2m^*)^{1/2}$ and the thermionic process dominates. For low N_D and hence low $\mathscr{E}_{max}$, $v_d < v_R$ and the Schottky diffusion process dominates. Their results show that, except for very thin barriers with substantial tunneling current, the predominant transport mode is thermionic emission and Eq. (5.62) describes the $J-V$ relation quite adequately, given small adjustments to A^* and ϕ_b.

The Schottky diode is essentially a majority carrier device and thus the J_0 value is independent of middle values of N_D, to first order. However, for large ϕ_b and/or for semiconductors with low N_D and at moderately high forward-bias voltages, minority carrier injection into the valence band may add to the total current (Scharfetter, 1965). This is handled in a way similar to that for the homojunction [Eq. (5.12)]. The diffusion effects expressed by Eq. (5.68) are also felt at low N_D, tending to lower J_0. For high values of $N_D(\gtrsim 10^{18}-10^{19}\ cm^{-3})$, Schottky barrier lowering becomes significant and the barrier becomes thin enough for appreciable tunneling to occur.

Thermally assisted tunneling in thin Schottky barriers, discussed in Section 5.5.6, causes a temperature-dependent lowering of the effective potential barrier and a complex variation of the diode quality factor A. The relative position in energy for maximum transport through such a Schottky barrier is shown in Fig. 5.26. Tunneling through states within the depletion layer has been discussed by Sarrabayrouse *et al.* (1977) and Parker (1969a, b). They calculated the tunneling probability for extended states (conduction electrons) to localized states (trapping centers) in the depletion layer and vice versa, using a Dirac delta function to express the wave function of the localized states. The results are compared with those of Padovani for thermally assisted reverse-bias tunneling, showing that considerably greater tunneling current can flow with the aid of a moderate density of traps.

The various modes of Schottky diode carrier transport are summarized in Fig. 5.32. The domains of applicability of the various theories of Schottky barrier transport are illustrated in Fig. 5.33, as a function of doping density.

Large edge effect currents can easily mask or alter the Schottky diode currents, especially for reverse bias and/or in small area test devices, so recourse is generally made to guard-ring structures when the phenomenology of transport is studied (Padovani, 1968). Other nonideal transport mechanisms such as recombination in the depletion layer (Low, 1955) have been observed, but we shall not consider them here. These mechanisms have similar effects on the values of J_0 and the diode quality factor A as observed for heterojunction devices.

5.7.2 Consideration of the Barrier Height

A most important result of the consideration of Schottky barrier transport is the dependence of J_0 on the barrier height ϕ_b. The magnitude of the barrier height determines the achievable values for the open-circuit voltage for a Schottky barrier solar cell. If the interface did *not* play a role in the determination of ϕ_b, its value would be given directly by

$$\phi_b = \phi_m - \chi_s, \tag{5.71}$$

where ϕ_m is the metal work function and χ_s is the semiconductor electron

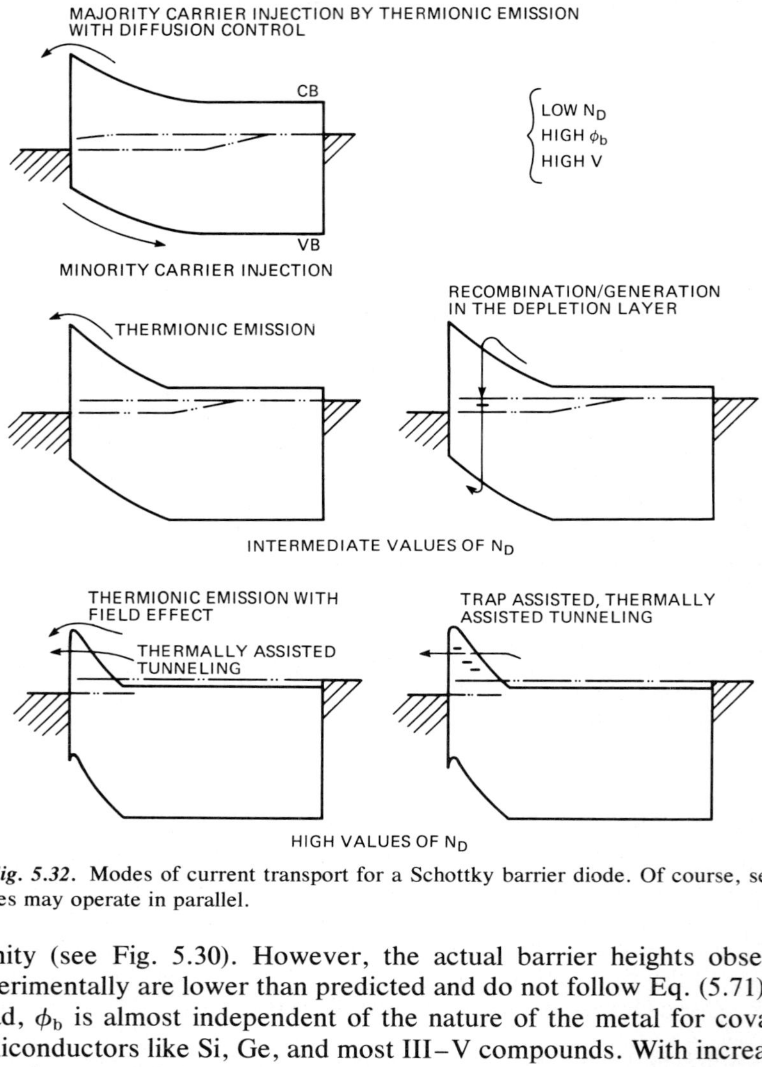

Fig. 5.32. Modes of current transport for a Schottky barrier diode. Of course, several modes may operate in parallel.

affinity (see Fig. 5.30). However, the actual barrier heights observed experimentally are lower than predicted and do not follow Eq. (5.71). Instead, ϕ_b is almost independent of the nature of the metal for covalent semiconductors like Si, Ge, and most III–V compounds. With increasing ionic character of the semiconductor there is an increasing dependence of ϕ_b on ϕ_m, but not usually as strong, or even linear, as suggested by Eq. (5.71). This independence of the experimentally determined values of ϕ_b on the nature of the metal expresses the lack of reaction of the Fermi level to the deposition of metal layers under practical conditions, as if E_F were pinned by interface states.†

† Extensive listings of ϕ_b values is given by Mead (1966), Milnes (1980), and Kurtin *et al.* (1969).

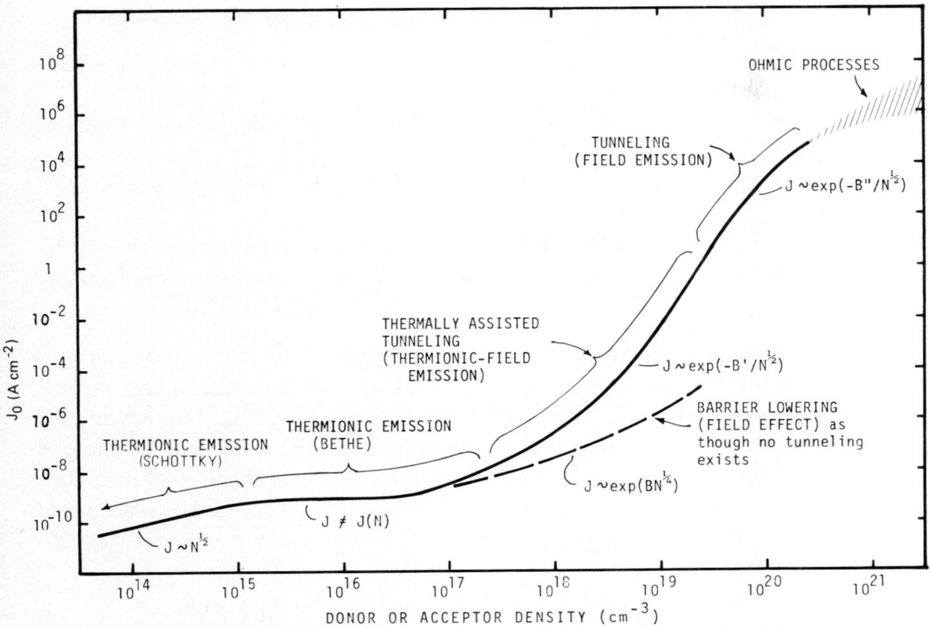

Fig. 5.33. Qualitative plot of zero-bias log J versus V intercept current J_0 versus doping density N for the Schottky diode. B, B', and B'' are constants and $T = 300°\text{K}$. The boundaries of the regions and the values of J_0 depend on the detailed properties of the semiconductor. A barrier height of 1.0 eV is assumed.

The fact that barrier heights measured in metal/covalent–semiconductor structures do not fit neatly into the model of Eq. (5.71) was found early in MS structure research.† Bardeen (1947) postulated that intrinsic states at the semiconductor surface were responsible for the Fermi-level pinning and that the further contact to the metal produced only second-order effects. Heine (1965) and others argued that, strictly speaking, localized semiconductor states cannot exist at an MS junction, but virtual resonance (metallike) surface states could exist and account for the observed behavior. The study of the MS interaction has been the object of an immense amount of experimental and theoretical research and a substantial part of it (at least until a few years ago) related to intrinsic surface states. Lindau *et al.* (1978) give a brief but complete review.

† A thorough introduction to semiconductor surfaces can be found in Many *et al.* (1971).

The barrier height can be measured by any of several methods:

(a) from the photon energy threshold of internal photoemission current from the metal into the semiconductor,

(b) by measurement of J versus $1/T$, assuming thermionic emission or diffusion-limited current [Eq. (5.62) or (5.68)],

(c) by junction capacitance measurements, extrapolating $1/C^2$ versus V data to $1/C^2 \to 0$, and

(d) use of far-ultraviolet photoemission spectroscopy, where the band bending is measured with respect to deep valence band-energy levels.

The processes determining ϕ_b can be represented macroscopically in terms of the effect of net interface charges and/or dipoles on the "true" metal and semiconductor work functions and the resulting barrier height. This has been the traditional way of attacking the problem by device physicists. One simple model assumes a constant distribution of interface states at the semiconductor side of the interface, separated from the metal by a very thin insulating layer, leading Crowell and Sze (1966a), for example, to derive the expression

$$\phi_b = C_2(\phi_m - \chi_{sc}) + (1 - C_2)[(E_g/q) - \phi_0^*] + C_0, \qquad (5.72)$$

where $C_2 = \epsilon_i/(\epsilon_i + q^2\delta N_{ss})$, with ϵ_i and δ as the permittivity and thickness of the interfacial layer, respectively, and N_{ss} (states cm^{-2} eV^{-1}) as the density of interface states. The position of the Fermi level if the metal were absent is ϕ_0^*, measured from the valence band, C_0 is a small term that for practical considerations can be disregarded; if

$$N_{ss} \to \infty, \quad C_2 \to 0, \quad \text{and} \quad \phi_b = [(E_g/q) - \phi_0^*], \qquad (5.73)$$

for metal/n-type barriers, for example. In this case the Fermi level is pinned by interface states and the barrier height is independent of the metal and entirely determined by the interface properties. This is known as the "Bardeen limit." If

$$N_{ss} \to 0, \quad \text{then} \quad C_2 \to 1 \quad \text{and} \quad \phi_b \simeq (\phi_m - \chi_s), \qquad (5.74)$$

as in Eq. (5.71). This is called the "Schottky limit."†

It has been customary to fit experimental data with an expression such as

$$\phi_b = a\phi_m + b \qquad (5.75)$$

† The reasonableness of the Bardeen and Schottky limits is discussed by Cohen (1979).

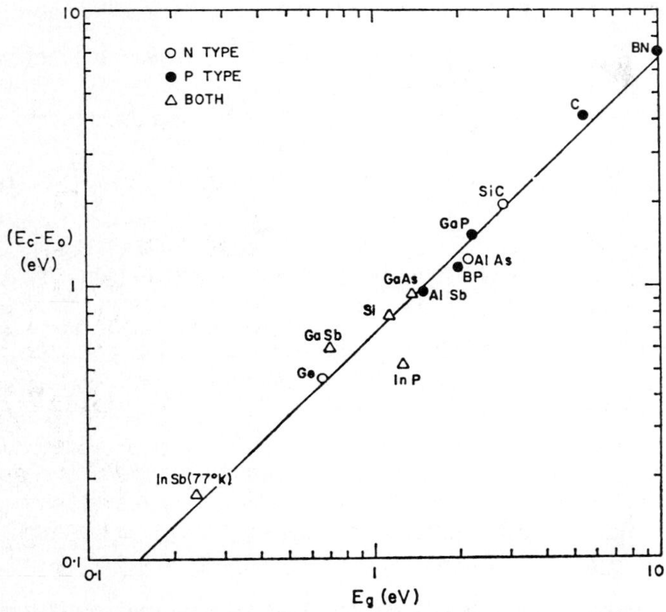

Fig. 5.34. Location of interface Fermi level relative to conduction band edge for Au contacts on various surface-state-controlled materials. The line is $(E_c - E_0) = 2E_g/3$ where $E_0 \equiv \phi_0^*$ here. Symbols: (○) n-type, (●) p-type, and (△) both types. [After C. A. Mead, *Solid-State Electron.* **9,** 1023 (1966). Copyright 1966, Pergamon Press, Ltd.]

for a given semiconductor. From the experimental values of the constants a and b, the magnitudes of N_{ss} and ϕ_0^* could then be determined. For a variety of covalent semiconductors data such as those shown in Fig. 5.34 indicate an effective pinning of the Fermi level at $q\phi_0^* \sim \frac{2}{3}E_g$ below the conduction band, the well-known "$\frac{2}{3}$ rule."†

The dependence of the actual barrier height on the metal and semiconductor work functions has also been expressed by a linear relation of the form

$$\phi_b = S'(\phi_m - \chi_s) + b', \tag{5.76}$$

where S' is called the index of interface behavior. Then $S' = 1$ expresses the Schottky limit whereas $S' = 0$ is the complete stabilization of E_F with independence of the metal work function. Since ϕ_m follows a linear rela-

† InP is a notable exception, with E_F pinned 0.2 eV to 0.4 eV below the conduction band edge.

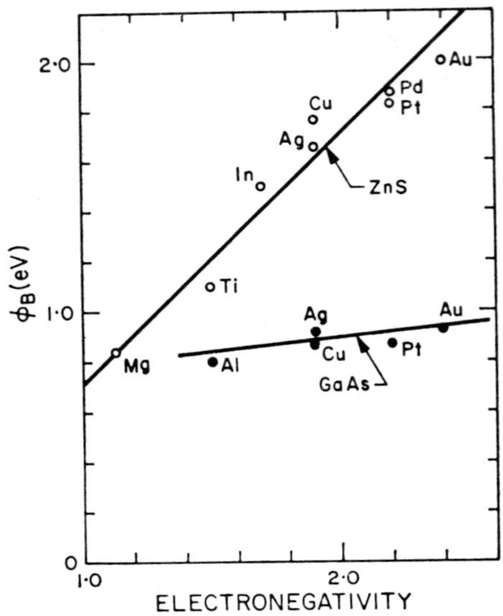

Fig. 5.35. Barrier heights versus electronegativity for various metals on ZnS and GaAs. [After C. A. Mead, *Solid-State Electron.* **9,** 1023 (1966). Copyright 1966, Pergamon Press, Ltd.]

tionship with the electronegativity X_m,† the above equation can be written

$$\phi_b = S(X_m - \chi_s) + b, \tag{5.77}$$

which has the advantage of using the more basic electronegativity instead of the metal work function, which experimentally depends strongly on the surface conditions.

In a landmark paper Kurtin *et al.* (1969) presented an extensive compilation of experimental values of ϕ_b. The index of interface behavior S, defined by fitting the data with an expression like Eq. (5.77), as shown in Fig. 5.35, was found to follow a definite trend, for $S \simeq 0$ (complete pinning of the Fermi level for covalent semiconductors) to $S \simeq 1$ for high-polarity tetrahedral semiconductors and ionic compounds. When S was plotted as a function of the electronegativity difference $\Delta X = X_A - X_B$ for semiconductors,‡ a sharp transition was found at $\Delta X \simeq 0.8$. We show in Fig. 5.36 a plot of S as a function of the semiconductor heat of formation ΔH_F. This parameter is related to ΔX but provides a better basis for discussion in terms of interface chemistry.

† The relation $\phi_m = 2.3X_m + 0.34$ is given by Louie *et al.* (1977).
‡ The heat of formation is also proportional to ΔX.

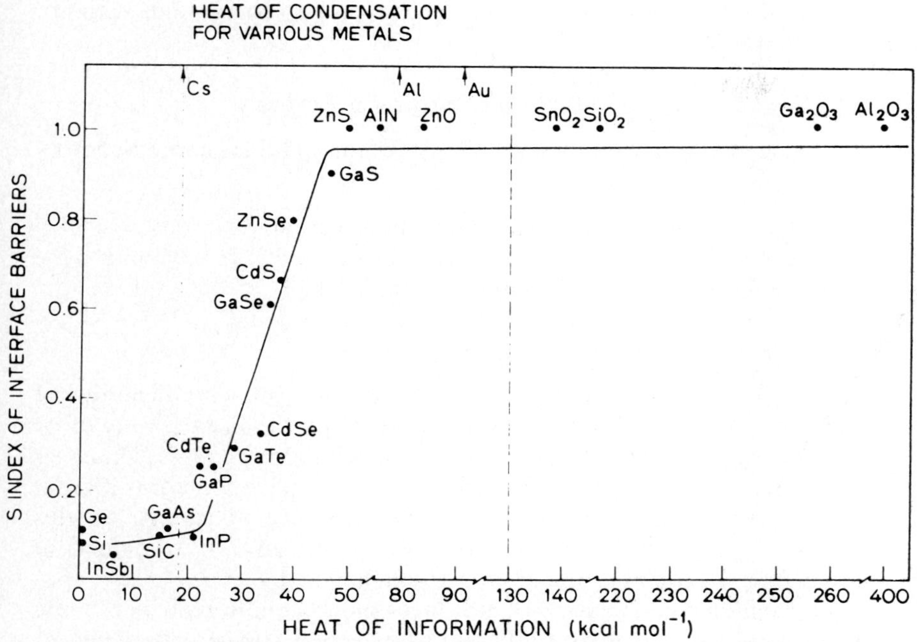

Fig. 5.36. The index of interface behavior S plotted against the chemical heat of formation. The values for S are those reported in S. Kurtin, T. C. McGill, and C. A. Mead, *Phys. Rev. Lett.* **22,** 1433 (1969). The range for the heat of condensation for various metals is also shown. Note that almost the same curve could be drawn with E_g as the abcissa (except for SiC for which E_g = 2.9 eV). [After Lindau *et al., J. Vac. Sci. Technol.* **15,** 1332 (1978).]

Two other observations about ϕ_b merit attention. First, when measurements are made on materials with both carrier types, yielding both ϕ_{bn} and ϕ_{bp}, it is found that $\phi_{bn} + \phi_{bp}$ approximately equals E_g/q for a wide variety of semiconductor and metal combinations (unless one of the contacts is ohmic, in which case its equivalent ϕ_b could not be measured). This is consistent with *both* Eq. (5.73) and Eq. (5.74). Thus the barrier height is approximately independent of doping.

Second, it is found that for Au barriers on various III–V and II–VI semiconductors, the energy of the valence band maximum at the interface depends only on the anion (McCaldin *et al.,* 1976). Au is chosen because of its small chemical reactivity. For example, the barrier heights ϕ_{bn} for Au/*n*-CdS and Au/*n*-ZnS differ by the same energy as the band gaps of CdS and ZnS. The band bending in each case is such as to bring the valence band connection to the Au to about the same energy for each cation.

These results hold for a variety of Ga, In, Cd, and Zn compounds with Sb, As, P, Te, Se, and S.

5.7.3 The Origin of Surface and Interface States

In the last section we discussed a set of empirical rules about Schottky barrier formation:

(a) the "$\frac{2}{3}$ rule" for covalent semiconductor barriers,
(b) the characteristic relation of the index of interface behavior to the ionicity of the semiconductor shown in Fig. 5.36,
(c) the anion rule for nonreactive Au contacts, and
(d) the summation rule: $\phi_{bn} + \phi_{bp} \simeq E_g/q$.

At this point it is important to note that these are but simplifications of more complex underlying processes. Indeed even the linear fitting of ϕ_b by Eq. (5.76) or (5.77) has been questioned by Schluter (1978). Other effects such as the reactivity of the metal with the semiconductor (different for each crystallographic orientation), the presence of foreign species (such as oxygen), and the possibility of self-compensation of defects at the interface make the problem very formidable.

The detailed consideration of interfaces on an atomic scale is but one view of the problem. Taking a different vantage point, the macroscopic effects of interface states can be discussed in terms of interface state densities and dipole layers and their associated charge and recombination properties. A vastly oversimplified view is that such orchestrations at the interface merely serve to "dope" it much as one would dope the volume of a bulk crystal. Then one is led to ask, "What is the nature of this doping?"

Refined experiments in the past few years (see, e.g., Lindau *et al.*, 1978) have shown that at bare, hyperclean surfaces† with a high degree of perfection there are no intrinsic states, contrary to previous theory. This is the case for a large variety of compounds and surfaces examined. These experiments indicate that metal-induced phenomena are responsible for the pinning of E_F. These phenomena are truly microscopic and the properties of the MS interface are determined by a very small coverage of metallic atoms, less than a monolayer. The difficulty of the problem can be easily understood—ideal surfaces are unachievable. Observed surface states are to be considered intrinsic if a high degree of perfection in surface preparation allows the assumption that all extrinsic causes have been eliminated. As pointed out in the thorough paper by Spicer *et al.* (1976), earlier agreement that the Fermi level at the (110) GaAs surface was in-

† Such surfaces are produced by careful cleaving of single crystals under ultrahigh vacuum (10^{-10} Torr) conditions.

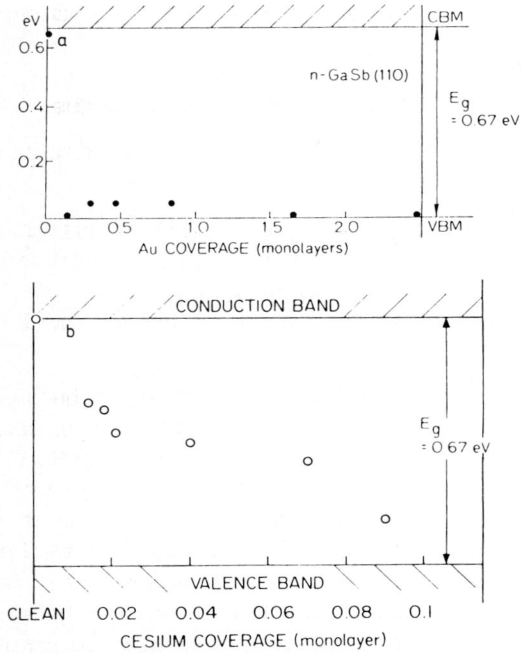

Fig. 5.37. (a) The Fermi level position for Au metal overlayers. (b) Cs metal overlayers on the (100) *n*-GaSb surface shown as a function of coverage. [After Lindau *et al.*, *J. Vac. Sci. Technol.* **15,** 1332 (1978).]

trinsically pinned near midgap, was later shown to be not necessarily the case. Instead, pinning should be attributed to external causes rather than intrinsic properties. This result is doubly important. First, it discloses the basic properties of semiconductor interfaces. Second, it shows that unless close-to-ideal surfaces are prepared, Fermi level pinning is unavoidable and arises from accidental processes and, paradoxically, at apparently the same position.

Lindau *et al.* (1978) examined clean (110) surfaces of the III–V semiconductors GaAs, GaSb, and InP prepared by cleaving under UHV conditions ($p \sim 1 \times 10^{-10}$ Torr), on which overlayers of Au, Al, and In were evaporated. Photoemission measurements were then carried out using synchrotron radiation. The measured Fermi-level position for (110) GaSb as a function of Au coverage is shown in Fig. 5.37a. Within less than 0.1 of a monolayer, band bending over the entire band gap takes place followed by a much smaller variation between 0.1 and 1 monolayer, after which complete stabilization is achieved. Thus the Fermi-level stabilization

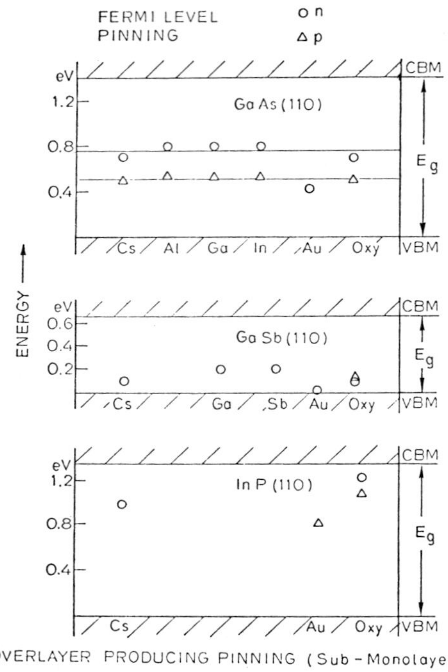

Fig. 5.38. The final pinning positions for various adatoms on GaAs, GaSb, and InP. Coverage is approximately 10^{15} cm^{-2}. (After Spicer *et al.*, 1980.)

starts taking place long before a bulklike metal overlayer is deposited. Similar results are observed for Cs coverage, Fig. 5.37b, and for Au on (110) GaAs and (110) InP. By studying the photoemission spectrum of the latter system it was determined that dispersed Au atoms rather than bulklike Au clusters were present. Another important fact is the apparent independence of the Fermi level position on the nature of the overlayer. In Fig. 5.38 a schematic diagram of the Fermi level position for the three III–V compounds for different adatoms is shown. In all the cases the same early Fermi-level pinning is observed and occurs approximately at the same energy, even though the metal adatoms are metals with very different electronegativities (e.g., Cs $\simeq$ 0.7, Al $\simeq$ 1.5, and Au $\simeq$ 2.4, Pauling scale), and one, O, is a nonmetal.

In Fig. 5.39 we show the far ultraviolet photoemission energy distribution curves for (110) n-GaSb with increasing Au coverages. The photoemission curves were obtained at a radiation energy of 120 eV such that most of the signal comes from the last two molecular layers. We see a decrease of the Ga 3d signal until complete disappearance with increasing

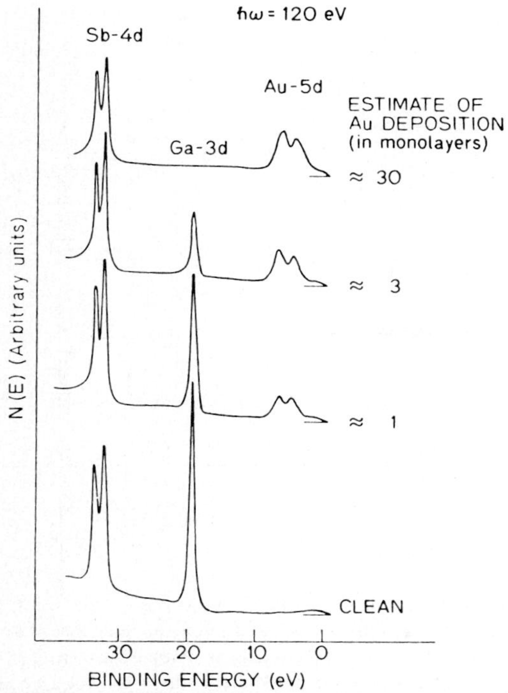

Fig. 5.39. Photoemission spectra taken at a photon energy of 120 eV for (110) GaSb exposed to different amounts of Au. (After Lindau *et al.*, 1978.)

Au coverage while the Sb 4d emission is constant, after an initial drop, even for coverages 15 times larger than the sampling depth. The results show that the surface layer of GaSb decomposes as a consequence of interface formation. Sb moves to the metal layer surface (as shown by Auger profiling) and a nonstoichiometric interface results. However, for InP the cation and anion signals decrease in a similar fashion, still giving sizable emission for coverages 20 times the sampling depth. The decrease of both signals together in InP does not preclude the assumption of nonstoichiometry at the interface, since such a slight difference in concentration is beyond the detection capability of the technique.

In these systems at least, we are dealing with strong interactions at the interface that modify the semiconductor surface chemically. This occurs very early and to a first approximation independently of the nature of the adatom. This fact should make Schottky barrier heights independent (within certain limits) of cleanliness and degree of perfection, for instance, which is apparently the case. Lindau *et al.* (1978) hypothesize that

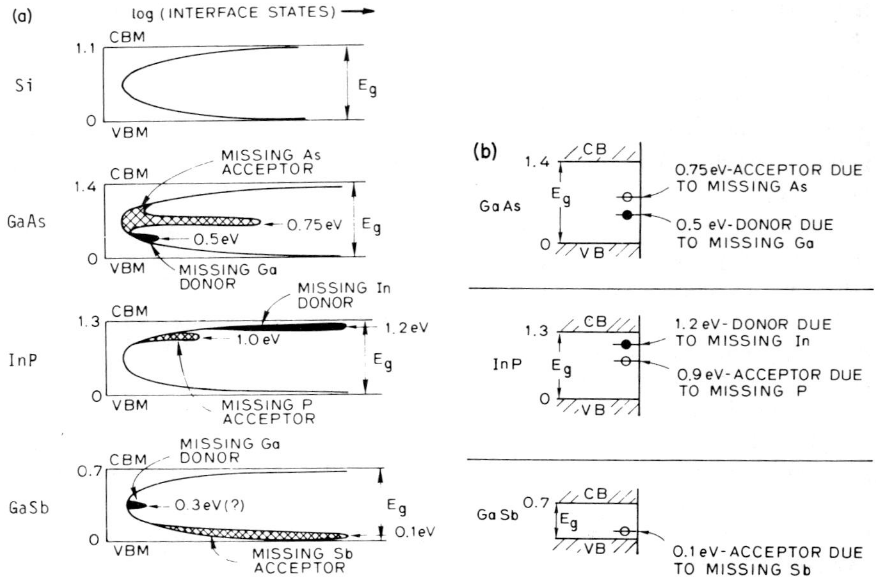

Fig. 5.40. The defect energy levels induced by overlayers on GaAs, InP, and GaSb. (a) Defect density versus energy, against the continuum of interface states usually found at semiconductor/oxide interfaces. Si is included for comparison. (b) The position in energy and the donor or acceptor nature of the defects are shown as well as a suggestion as to the missing atom responsible for the defect. [After W. E. Spicer, I. Lindau, P. Skeath, and C. Y. Yu, *J. Vac. Sci. Technol.* **17**(5), 1019 (1980).]

the strong interaction produces defect states at the semiconductor surface and that these defects are responsible for the Fermi level pinning.

Spicer *et al.* (1980) suggest a unified model of the formation of interface states and Schottky barriers for III–V compounds which gives the surface state assignments for GaAs, InP, and GaSb shown in Fig. 5.40. It is rather ironic, in view of the controversies involved historically, that these states are intrinsic in the sense that they are native defects (vacancies) but they are not present until induced by a metal or oxide overlayer.

For materials with no pinning it is probable that self-compensation phenomena at the interface neutralize the charge of the defect states. This view is supported by the predominance of self-compensation in bulk crystals of the large band-gap, ionic semiconductors.

The results of these and other experiments than *suggest* a second, "deeper" set of empirical rules for III–V compound semiconductors:

(1) No intrinsic states are present within the band gap at hyperclean surfaces of covalent semiconductors such as GaAs, InP, and GaSb. De-

fect states (antistructure or vacancies), induced by the presence of metal or oxygen adatoms, appear to be the cause of the Fermi-level pinning in these materials.†

(2) For covalent semiconductors, the properties of the surface, in particular the band bending, are determined by metal or oxygen coverages of less than 0.1 monolayer and change relatively little with increased coverage.

(3) Decomposition, involving monolayers of the semiconductor surface, takes place at least in some semiconductor-metal systems (e.g., Au/InP, Au/GaSb). In some cases quantities of one of the semiconductor components (on the order of monolayers) are ''carried along'' on the growing metal surface (e.g., Sb on Au from Au/*n*-GaSb). In other cases the semiconductor components are distributed through many monolayers of the metal overlayer.

(4) The band bending and decomposition appear to be, to first order, independent of the nature of the adatom for these III–V compound semiconductors. This conclusion makes ϕ_b independent (within limits) of cleanliness and degree of imperfection for all but the most nearly perfect, hyperclean surfaces. This conclusion seems to be justified by the barrier heights reported in the literature.

Because of limitations of time and scope, the results discussed in this section are drawn largely from one group. Of course a much larger body of research exists, some of it with alternative points of view, as reported, for example, in the yearly *Physics of Compound Semiconductor Interfaces* conferences.‡ Notable bodies of research of particular relevance to photovoltaics are, for example, the consideration of chemical reactivity and charge redistribution at metal/semiconductor interfaces by Brillson (1978, 1979) and the work on the anion effect for nonreactive metal Schottky barriers by McCaldin *et al.* (1976).

These results suggest that solar cell materials may be divided into two classes:

(1) Ionic materials for which $S \simeq 1$ and interface modification can readily be accomplished. Unfortunately, this class contains no common semiconductors with suitable band gaps for absorber layers.

(2) Covalent, low-polarity materials for which $S \simeq 0$ and interface modification is difficult, chancy, or impossible. This class appears to include all the useful absorber materials.

† Hyperclean surfaces of column IV semiconductors *do* show intrinsic midgap surface states however.

‡ The proceedings of the *PCSI* conferences are printed in the *Journal of Vacuum Science and Technology*.

The results further suggest that in most of the practical means of fabrication of solar cell devices, the surfaces of covalent components have been "fixed" long before metal (or other) layers are applied. Therefore deliberate modification of interface behavior† must either use particular interfacial dopants that are extremely reactive toward the surface to be modified (enabling them to displace existing species) or use active dopants that are applied *in situ* as the film or surface is grown. Further, the desired defect configuration that is established must be very stable. An example of such interfacial doping for covalent materials is the use of atomic H at Si grain boundaries (Seager and Ginley, 1980). Another such example is the use of Ru to passivate surface recombination (without increase in barrier height) for *n*-GaAs-based Schottky barrier solar cells (Nelson *et al.*, 1980).

5.7.4 MIS Devices

It is probable that most Schottky barrier devices include at least several atomic layers of some kind of oxide between the metal and semiconductor layers, introduced inadvertently during or before fabrication.‡ The deliberate incorporation of a carefully grown insulating layer (usually an oxide) at the interface has been found to produce substantial increases in V_{oc}. The study of MIS structures is thus doubly important.

The presence of an insulating layer has the following effects: (a) the layer acts as a passive dielectric separation between the metal and the semiconductor, lowering ϕ_b; (b) the flow of carriers may be limited by tunneling through the insulating layer, reducing the current for a given bias voltage; (c) the applied bias is sustained partially by the insulating layer and thus the barrier height is not constant, resulting in values of the A factor greater than unity; and (d) trapped charge within the insulator layer or at the insulator/semiconductor interface acts either to increase or to decrease the effective ϕ_b, depending on the polarity of the charge.

For a Schottky barrier *without* an insulating layer, the dark diode current is dominated by the thermionic emission current corresponding to majority carrier transport over the barrier. Smaller effects of tunneling through the top of the barrier and/or recombination of majority carriers in

† Aside from the deliberate introduction of an intervening insulator layer, discussed in the following section.

‡ Native oxides formed at *room temperature* on various semiconductor surfaces: *InP*, immediately after chemical etching: 10–20 Å (Wager and Wilmsen, 1980). *Si*, after etch in 20% HF, quenched in isopropanol: rapid growth up to 10 Å, to 12 Å within 1 hour at room temperature, and a terminal thickness of $\simeq 20$ Å after 1 week (Card and Rhoderick, 1971). *GaAs*, after methanol-Br etch, 18–20 Å (Stirn and Yeh, 1977).

the depletion layer are observed but can be neglected for most solar cells. Minority carrier injection from the metal into the semiconductor is also present but is usually very small compared to the majority carrier transport. The expression for V_{oc} for a simple Schottky barrier with transport by thermionic emission is

$$V_{oc} \simeq \phi_b + (AkT/q)\ln[J_{sc}/A^*T^2], \tag{5.78}$$

where A is the diode factor ($A = 1$ for simple theory), ϕ_b is the barrier height, and A^* is the effective-mass modified Richardson constant.

For the MIS diode the dark current is the sum of four components: $J_D = J_{th} + J_{rg} + J_{id} + J_s$, i.e., the contributions of thermionic flow, recombination/generation in the depletion layer, injection and diffusion into the quasi-neutral bulk, and recombination at the semiconductor/insulator interface, respectively, as shown in Fig. 5.41. The addition of an intervening insulating layer can reduce the majority carrier transport toward a level comparable to that of the minority carrier components J_{rg} and J_{id}, resulting in substantial increases in V_{oc}. Since the photogenerated minor-

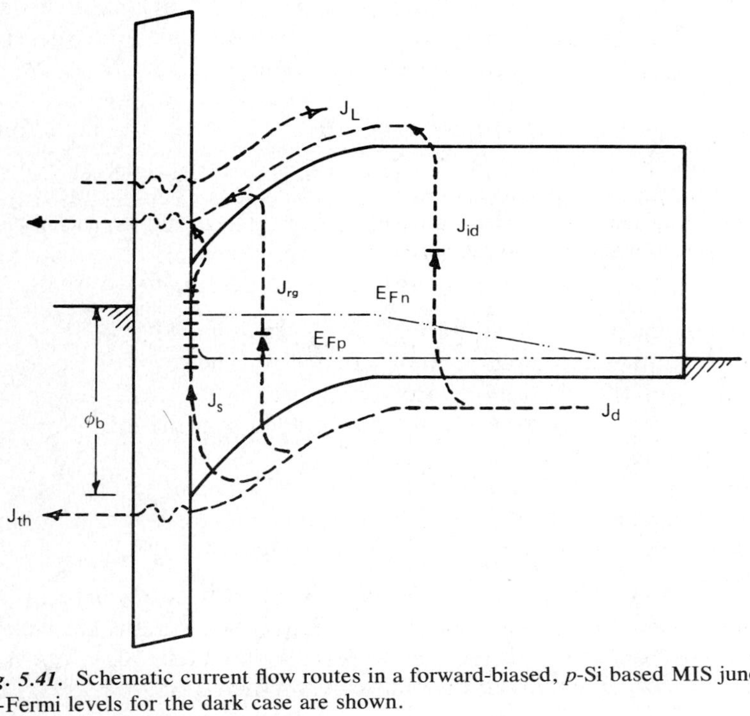

Fig. 5.41. Schematic current flow routes in a forward-biased, *p*-Si based MIS junction. Quasi-Fermi levels for the dark case are shown.

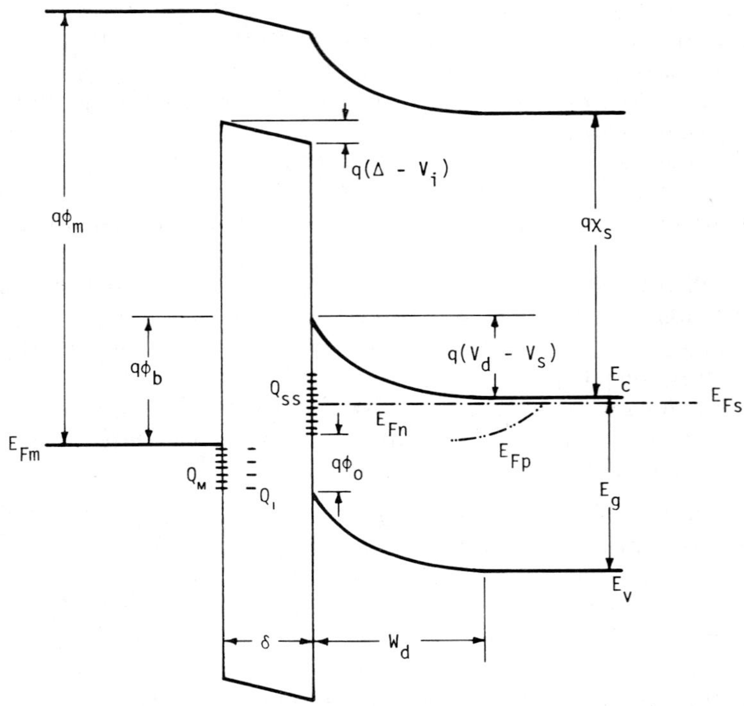

Fig. 5.42. Schematic of an MIS solar cell under illumination. The applied voltage is partially supported by the interfacial layer (V_i) and partially by the semiconductor (V_s). The potential drop across the insulator is Δ for the unbiased device. The net charge at the metal surface is Q_M and the trapped charge within the insulating layer is Q_I. The potential ϕ_0 characterizes the distribution of interface states.

ity carriers must also tunnel through this layer, it must not be thicker than 20–30 Å to maintain high quantum efficiency. The form of Eq. (5.78) is still valid for the majority carrier transport for the MIS device, given changes in the value of ϕ_b and A and provided that the insulator layer is thin enough to permit free tunneling.

Equation (5.78) shows that the value of V_{oc} can be increased if ϕ_b and/or A are increased by the effects of the insulator layer. A number of theoretical models have been offered (see Fig. 5.42):

(1) Card and Rhoderick (1971) calculated a value of A in terms of the insulator layer thickness δ, the depletion layer width, and the charge on surface states Q_{ss}. In this model, for δ between 10 and 30 Å, the surface states equilibrate with the semiconductor, predicting an increase in A with increasing δ and/or increasing Q_{ss}. In this model the value of A can be in-

creased with only small variation in ϕ_b (and hence J_0) for certain values of the parameters.† This leads to the possibility of maximizing V_{oc} by proper choice of δ and Q_{ss} (i.e., a horizontal movement on the J_0 versus $1/A$ plot of Fig. 5.29). The model was in agreement with their experimental results.

(2) An increase in ϕ_b has been predicted by Pulfrey (1976) as the result of charge located within the insulator layer Q_I.

(3) An increase in the diode factor A has been predicted by Lillington and Townsend (1976) as the result of an increased insulator layer thickness.

(4) By the introduction of another term in Eq. (5.78), corresponding to a tunneling transmission coefficient, Card and Yang (1976) predict an increased V_{oc} up to oxide layer thicknesses of about 30 Å, above which J_L is limited by tunneling.

(5) A theoretical treatment by Fonash (1975a, 1976) predicts a decrease in the effective work function of the metal due to the fixed insulator charge, and an increased A value.

Other theoretical treatments of MIS junctions have been given by Landsberg and Klimpke (1977, 1978), McQuat and Pulfrey (1976), Kar (1977), Viktorovitch and Kamarinos (1977), and Green *et al.* (1974). A review of MIS solar cell theory and practice has been given by Pulfrey (1978). Ghosh *et al.* (1978) provide an interesting comparison of HJ, Schottky barrier, and MIS theories as applied to the Sn_2O/n-Si solar cell.

Fonash (1975a, b, 1976, 1977; see also Fonash *et al.*, 1978) has proposed a number of basic mechanisms that can lead to V_{oc} enhancement. The schematic MIS energy band diagram shown in Fig. 5.42 illustrates the case. The junction is under illumination, developing a voltage V, a portion of which V_s is supported by the semiconductor, and the remainder V_i by the insulating layer. Assuming that the insulating layer interfaces an otherwise ideal MS junction, the barrier height is given by $\phi_b = \phi_m - \chi_s - \Delta$, and the total current density for the device under illumination can be expressed as

$$J = \mathcal{T}_e A^{**} T^2 \exp(-q\phi_b/kT)[\exp(qV_s/kT) - \exp(-qV_i/kT)] + (\mathcal{T}_h p_{n0} D_n/L_n)[\exp(qV/kT) - 1] - \mathcal{T}_h J_L. \tag{5.79}$$

The first term represents the majority carrier thermionic emission transport mechanism, modulated by a transport coefficient $\mathcal{T}_e$, which expresses tunneling processes through the insulating layer. The second term, the hole (minority carrier) ideal diffusion curent, has been included,

† Such an experimental result has been reported by St. Pierre *et al.* (1976), in which ϕ_b varied from $\simeq$0.93 to 0.8 eV for variation of A from $\simeq$1.2 to 1.7, using various oxide preparation methods on Si.

since reduction of the first term (e.g., via $\mathcal{T}_e$) leads eventually to current levels where minority carrier transport becomes important. The factor $\mathcal{T}_h$ represents hole tunneling through the insulating layer. The third term is the light-generated current, which for simplicity is assumed to be independent of bias and interface characteristics (except for tunneling). Fine details have been omitted for simplicity. The portion of the applied bias dropped across the interfacial layer $V_i = V - V_s$ depends on the characteristics of the interfacial layer,† trapped charge within the insulating layer Q_I, interface charges and dipoles, and the electrical characteristics of the semiconductor. By electrostatic analysis, a rather complex expression for $V_i(V)$ can be deduced. The voltage drop on the semiconductor side V_s can then be written $V_s = V - V_i = V/A$, giving $A = \{1 - [V_i(V)/V]\}^{-1}$ and Eq. (5.79) takes the form

$$J \simeq J_{oe}[\exp(qV/AkT) - 1] + J_{oh}[\exp(qV/kT) - 1] - \mathcal{T}_h J_L. \quad (5.80)$$

This simple equation shows, within the limitations of the assumptions, the parameters that can be varied to increase V_{oc}: (a) most important, the barrier height that determines the majority carrier saturation current J_{oe}, (b) the diode quality factor A that, in this formulation, expresses the ratio (V/V_s) (what Fonash calls "field shaping"), and (c) the magnitudes of J_{oe} and J_{oh} that are controlled by $\mathcal{T}_e$ and $\mathcal{T}_h$, the tunneling transport coefficients. These variations, all of which essentially reduce the thermionic component of the diode current, are explored in the following paragraphs.

Increasing the Effective ϕ_b The Fermi level is normally pinned at the surface of Si and GaAs by surface states. The relevant interface is now between the insulator and the semiconductor, however, and the deliberate addition of the proper interfacial oxide might satisfy dangling bonds at the semiconductor surface, reducing the density of the interface states. Such a surface passivation might relieve the Fermi-level pinning so that ϕ_b would depend more strongly on the metal work function.‡

In addition, the insulating layer is expected to be electrically active per se and trapped charge there can modify the effective barrier height. Fixed charge within the insulating layer produces a change in ϕ_b proportional to

† The electric field that such an insulator can sustain is limited by its dielectric strength, of the order of 10^6 V cm^{-1} or less. For the usual insulator thickness $\simeq 20$ Å, this gives a limit on $V_i \leq 0.2$ V, not including the voltage drop across dipole layers that might be present at the metal/insulator and insulator/semiconductor interfaces.

‡ Experimental observation of Schottky diodes using Al, Cu, Ag, Pd, and Au on GaAs and Si (Fonash *et al.*, 1978) has shown higher barrier heights for the MIS configurations. However, the values of ϕ_b reported (Fonash *et al.*, 1978; Childs *et al.*, 1978) appear to have the same lack of correlation with the metal work function, and so the situation is unresolved.

$-Q_\mathrm{I}$ (over a limited range) and, for an *n*-type semiconductor, the use of an oxide containing negative charge would increase ϕ_b. In this case the diode factor A should remain practically equal to unity, but the J_0 value should change exponentially with the amount of trapped charge (as long as the value of the trapped charge density is relatively independent of bias).

If the fixed charge is great enough, an inversion layer can be created at the semiconductor/insulator interface, perhaps aided by the presence of a low work function metal for a *p*-type substrate (Salter and Thomas, 1976; Schewchun *et al.*, 1974). This approach has been highly successful for MIS devices on *n*-Si giving $V_\mathrm{oc} \simeq 0.655$ V with $A \lessapprox 1.2$ at solar current densities (Godfrey and Green, 1979). (Also see Section 11.1.1.)

Decreasing Majority Carrier Tunneling Although the minority carrier tunneling coefficient must remain close to unity, the tunneling coefficients for majority and minority carriers need not be equal. By increasing the insulator thickness (to perhaps 100 Å) majority carrier transport could be attenuated by many orders of magnitude. The light-generated current would be maintained in this case by incorporating a distribution of traps within the insulator layer at the proper energy level to assist minority carrier tunneling (Kipperman *et al.*, 1978).

The effect of an increasing insulator thickness on the J–V characteristics of a practical MIS structure can be seen, for example, in the article by Rhoderick (1974).

Advantageous changes in the respective tunneling energies for electrons and holes might also be obtained by proper choice of the electron affinity of the insulator layer. Thus for a *p*-type base, the tunneling barrier height for holes might be made considerably smaller than that for electrons (Childs *et al.*, 1976). For sufficient tunneling control, the lower bound for the forward-bias current transport is then imposed by the minority carrier diffusion current. This mechanism reduces the forward saturation current, leaving the diode factor equal to unity.

Field Shaping The insulating layer partially sustains the total bias voltage V across the device. The V_s/V ratio and, therefore, the diode factor A depend on the characteristics of the insulating layer, and especially on the interfacial energy states at the insulator–semiconductor interface. This mechanism, called "field shaping" by Fonash (1975b, 1976), permits the possibility of higher V_oc values, even higher than the diffusion voltage V_d, since a portion of V_oc is sustained across the insulator. The V_s/V ratio depends on the actual distribution of interface states and the relative capture cross section for electrons (which should be high for an *n*-type base for the process to be effective) as compared to that for holes. This mechanism has a small effect on the saturation current but leads to A

factors that are greater than unity and bias dependent. Several possible interface level distributions, with their effects on the electrical characteristics of MIS devices, have been examined by Fonash *et al.* (1978; Fonash, 1975a, b, 1976; Childs *et al.*, 1976, 1978). In this case, the improvement in V_{oc} follows from the increase in the A factor.

The three basic mechanisms discussed here have been compared with experimental evidence by Fonash *et al.* (1978) using forward and reverse $C-V$, dark and light $J-V$, and IR photoemission measurements. They find that an increase in ϕ_b and field shaping are the processes that can describe most of the results.

The role of tunneling across the insulating layer has been stressed in an alternative model proposed by Shewchun *et al.* (1974; St. Pierre *et al.*, 1976) that compares successfully with experimental evidence on *p*-Si MIS structures.

We briefly consider the electrochemical stability of the insulating layer and its interface with the semiconductor. Diffusion of O_2 through the metal into the insulating layer appears to produce aging effects in some junctions, for example, showing a change with time of the dark $J-V$ characteristics and the photovoltaic response.

Ponpon and Siffert (1978) studied such aging effects for Au/*n*-Si Schottky diodes, finding that the electrical properties of these devices are directly correlated to oxygen diffusion through the metal. Before exposure to air their diodes showed no rectification; on exposure to air, increasing rectification ratios were observed, finally reaching the "normal" values seen by other researchers. A reduction of the J_0 values by several orders of magnitude could be correlated with exposure time. The diode quality factor remained unchanged. These effects were associated with an increase of ϕ_b, which was also consistent with V_{oc} and IR photoemission measurements. Ponpon and Siffert suggest that the initial low values of ϕ_b are to be associated with a fixed positive charge Q_I trapped in the insulator layer. Oxygen diffusion through the metal neutralizes the charge, restoring ϕ_b to its normal value. Metals with a low oxide heat of formation (e.g., Au) allow O_2 to diffuse easily and aging effects appear. Conversely, metals with a high oxide heat of formation (e.g., Al) prevent O_2 diffusion and small aging effects are observed.

The same authors also studied Schottky barriers of a variety of metals on *n*-CdTe (Ponpon and Siffert, 1977). Contrary to the case of Si, the Au barrier on CdTe is formed immediately after metal evaporation. An increase in V_{oc} is still observed when the diode is exposed to air, however. By studying the exposure time dependence of the forward $J-V$ characteristics it was found that the main change in this case arises from the A factor, whereas values of J_0 remained practically constant. This result suggests that field shaping, because of interface states, is the actual improvement mechanism for the Au/*n*-CdTe barrier.

5.7.5 The SIS Structure

Practical heterojunctions often show large values of J_0 and severe limitations on V_{oc}, apparently of the same kind as for Schottky barrier devices. This consideration is supported by the view that heterojunctions are a large class of devices of which Schottky barriers are extreme examples (cf. Section 5.5.2). Thus one is led to consider the introduction of a thin insulator layer at the heterojunction interface. The photovoltaic behavior of the SIS structure was analyzed theoretically by De Visschere and Pauwels (1978). Because of the complexity of the problem, restrictive assumptions had to be made, but different cases could still be distinguished. They found that when collection of photogenerated carriers takes place in the more heavily doped semiconductor, the insulating layer is disadvantageous. However, when collection takes place in the weakly doped semiconductor, the insulating layer should have a favorable effect if the band discontinuity has the appropriate sign (in the other case, the layer has no influence). These conclusions indicate that an insulating layer would have detrimental effects on solar cells of the Cu_xS/CdS type but might be good for conductive-metal-oxide/absorbing-semiconductor systems, depending on the character of the interface.

Many practical heterojunctions are in fact SIS structures, since native insulating oxides are introduced during fabrication. Shewchun *et al.* (1978) have proposed a SIS model to describe the ITO/Si system, where experiments indicate the presence of a thin interfacial insulating layer. This was done by extending the analysis of MIS structures in an earlier work (Green *et al.*, 1974) to the SIS structure where one of the semiconductors is a degenerate wide-band-gap oxide. Similar considerations were made for the SnO_2/*n*-Si solar cell by Ghosh *et al.* (1978). The SIS junction is also discussed in Section 11.1.3.

5.8 OHMIC CONTACTS

Ohmic contacts are an important but somewhat neglected area of solar cell research. Despite recent theoretical advances, their fabrication still involves considerable art and personal recipe. Indeed, in 1971 Gossick commented, "the history of ohmic contacts is still rather like that of a primitive society as it consists mostly of unrecorded reports passed around informally among experimenters. . . ." (Gossick, 1971). In Section 6.2.2 it is shown that a reasonable allocation for contact-resistance power loss in a solar cell is about 1.0% of its output, corresponding to a 6-mV drop across each contact. For operation at 1 sun this is equivalent to a contact resistivity of ρ_c of 0.2 ohm cm^2 for a back, full coverage contact and about 0.01 ohm cm^2 for a front, grid contact. The ρ_c allocation is inversely proportional to the concentration ratio, to a first approximation, so that the contact technology for high-intensity cells is very demanding.

Two definitions of "ohmic contact" are common. The first, for which we reserve the term *ohmic*, has these characteristics:

(a) a linear $J-V$ characteristic,
(b) negligible resistance,
(c) negligible photovoltaic effect,
(d) absence of $1/f$ noise.

Rose (1955) defines the ohmic contact as "providing a reservoir of carriers ready to enter the insulator . . . (or semiconductor) . . . as needed." Our second definition is "that the voltage drop across the contact be small with respect to the device voltage drop, but not necessarily linear with current." We refer to this nonohmic, low-resistance contact as a *quasi-ohmic* contact.

The specific contact resistivity is defined at zero applied bias by

$$\rho_{c0} = [dV/dJ]_{V=0} \tag{5.81}$$

and for small voltage drops (<20 mV) this is a good approximation to the effective contact resistivity for a cell in operation (described in Section 6.2.3). In general, ρ_{c0} decreases strongly with increasing carrier density in the semiconductor and increases as temperature is decreased.

The theory and practice of ohmic contacts are reviewed by Rideout (1975) and in a symposium volume edited by Schwartz (1969).

The formation of an ohmic contact between a metal and a semiconductor in which ϕ_m and χ_s determine the potential barrier height ϕ_b (the Schottky limit) is illustrated in Fig. 5.43. For the n-type case, if ϕ_m is smaller than χ_s, then the contact should be strictly ohmic with extremely small values of ρ_{c0}. The value of ρ_{c0} for such contacts is dependent on the carrier density in the semiconductor, scattering phenomena in the metal and semiconductor regions adjacent to the barrier, and quantum mechanical reflection at the potential variations at the interface. The situation is considered in several theoretical papers by Gossick (1969a, 1969b, 1970, 1971a, 1971b).

In many cases, however, the production of the ideal ohmic contact described previously is not possible. For covalent semiconductors, the Fermi level is at least partially pinned by surface states, fixing ϕ_b relatively independent of ϕ_m (approaching the Bardeen limit). For some ionic

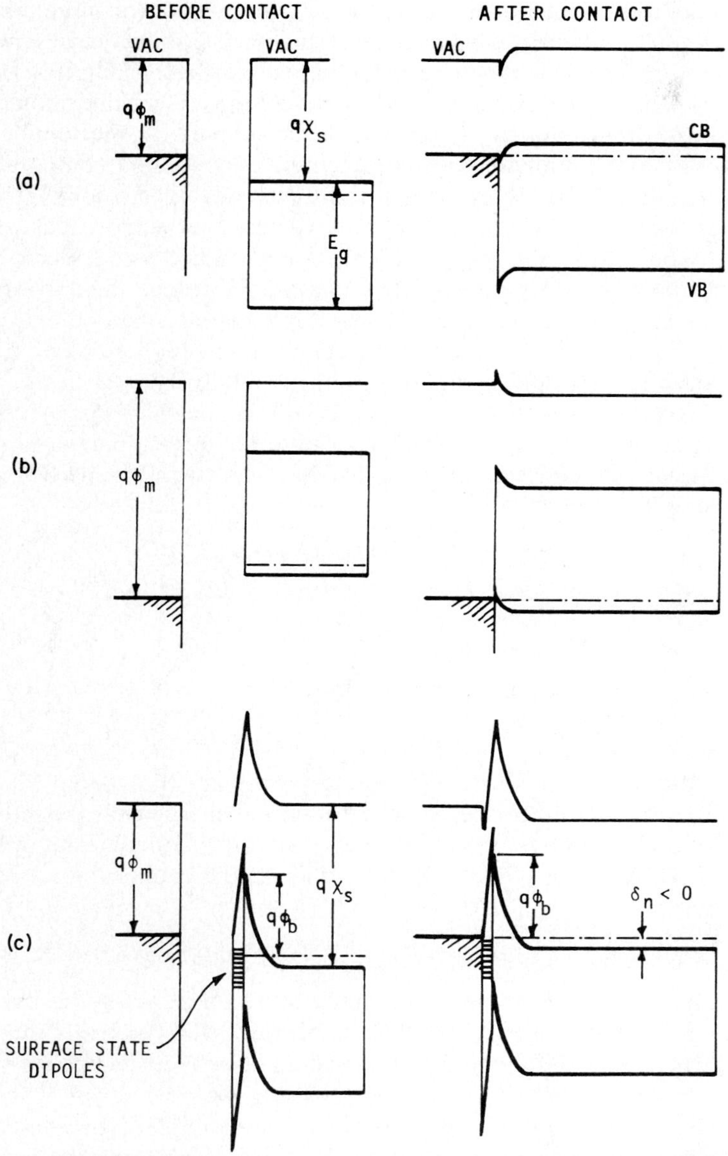

Fig. 5.43. Ideal ohmic contacts on (a) *n*- and (b) *p*-type semiconductors. (c) Contact on surface-state-dominated material (Bardeen barrier). This latter contact might equally well be expressed in terms of an interface charge density rather than an interface dipole as shown here. On the left is the equilibrium band structure before contact.

semiconductors, a metal cannot be found with a small (or large) enough ϕ_m to produce an ideal ohmic contact to n-type (or p-type) semiconductors. In these cases recourse must be made to leaky Schottky diodes in which the barrier is either small enough to permit large thermionic currents and/or thin enough to permit easy transport by thermionic-field emission or direct tunneling processes, i.e., quasi-ohmic contacts. The usual procedure is to highly dope a thin layer of the semiconductor adjacent to the contact metal and thereby reduce the barrier thickness to $\lesssim 100$ Å. This thin region must be highly doped without introducing compensating impurities or imperfections that would reduce the carrier density. Thus the properties of these quasi-ohmic contacts can be extremely dependent on the details of the preparation techniques involved.

At room temperature and above, and for relatively light to moderate doping levels ($N_{A,D} < 10^{18}$ cm^{-3}), the dominant mechanism for current transport across a metal/semiconductor interface with a barrier to transport is thermionic emission over the barrier. The current density is given by the simple diode equation

$$J = A^*T^2 \exp[-q(\phi_b - \Delta\phi_b/kT][\exp(qV/kT) - 1],$$

or

$$J = J_0[\exp(qV/kT) - 1], \tag{5.82}$$

where $\Delta\phi_b$ is the image force barrier lowering:

$$\Delta\phi_b = [q^3\phi_b N_D/8\pi\epsilon_{s,hf}^2\epsilon_s]^{1/4} + \alpha\mathscr{E}_m. \tag{5.83}$$

The quantity α is an empirical parameter necessary to account for additional lowering caused by the presence of bound charge near the interface (Andrews and Lepselter, 1968) the detailed nature of which is not fully understood (typically, $\alpha \simeq 2 \times 10^{-7}$ cm). The specific contact resistivity is then given by

$$\rho_{c0} = [dV/dJ]_{V=0} = kT/qJ_0 = [kT/qA^*T^2]\exp[q(\phi_b - \Delta\phi_b)/kT]. \tag{5.84}$$

To obtain $\rho_{c0} \lesssim 0.1$ Ω cm^2 at room temperature for Schottky barrier contacts to Si, a value of $\phi_b \leq 0.45$ eV is required for electrons (or $\phi_b \leq 0.42$ eV for holes), assuming $\Delta\phi_b = 0$ and an effective Richardson's constant $A^* = 112$ A cm^{-2} °K^{-2} for electrons and $A^* = 32$ A cm^{-2} °K^{-2} for holes. This can be easily achieved for p-Si, for example, by using PtSi or RhSi contacts ($\phi_b = 0.25$ eV and 0.33 eV, respectively), but the resulting contact resistivities have large temperature coefficients and are not useful at liquid N_2 temperatures. A contact of Ag/p-Si with $\phi_b \simeq 0.43$ eV without any tunneling or barrier lowering would be marginally useful.

The majority of semiconductor applications make use of tunneling through a thin potential barrier on a highly doped semiconductor substrate layer. The simple theory is a rather straightforward extension of thermionic-field theory. The barrier is both lowered by image forces ($\Delta\phi_b$) and thinned by the large carrier density, but the effects of thinning proceeds more rapidly as $N_{D,A}$ increases, so that the decrease in ρ_{c0} is predominantly due to tunneling.

Chang *et al.* (1971) used a generalized majority carrier transport model of Chang and Sze (1970) [similar to the models of Padovani and Stratton (1966) and Crowell and Rideout (1969)] to calculate ρ_{c0}, considering quantum mechanical transmission both above and through the barrier, the Schottky effect, and the variation of m_t^* with doping density. Calculations were made for contacts to degenerate *n*- and *p*-type Si and GaAs. Although the results are obtained by numerical analysis, the authors give aproximate analytic results for two limiting cases. For thermionic emission, tunneling is neglected and the result of Eq. (5.81) is obtained. If transport over the barrier is neglected compared to tunneling through the bottom of the barrier, then

$$\rho_{c0} \simeq [kT\,E_{00}^{1/2}/qA^*T^2(qV_d)^{1/2}]\exp(qV_d/E_{00})\exp(\delta_n/kT), \tag{5.85}$$

with

$$E_{00} = (q\hbar/2)[N/m_t^*\epsilon_s]^{1/2}. \tag{5.86}$$

where m_t^* is the tunneling effective mass (Barber, 1967) and N the appropriate doping density, N_D or N_A.† Thus ρ_{c0} decreases exponentially with $[N/m_t^*]^{1/2}$. The value of m_t^* introduces a dependence of the tunneling current on crystallographic orientation. The values of ρ_{c0} at 300°K for *n*- and *p*-Si as a function of doping density and barrier height are shown in Fig. 5.44a and for *n*- and *p*-GaAs in Fig. 5.44b. Values of ρ_{c0} as a function of temperature are shown in Fig. 5.45. These results show that for $N < 10^{17}$ cm^{-3}, the value of ρ_{c0} depends very weakly on N. For $N > 10^{19}$ cm^{-3}, transport is almost entirely by tunneling and ρ_{c0} is no longer dependent on temperature and in some cases may even have a positive temperature coefficient. For large doping densities ($N > 10^{20}$ cm^{-3}), the barrier height becomes less important and all the curves converge.

Pellegrini and Salardi (1975) have done an analysis that is similar except that it starts from a somewhat different generalized barrier transport expression and neglects the Schottky effect. Their results for ρ_{c0} versus T and N are qualitatively similar to those described above.

The heavily doped layer is usually quite limited in thickness and Popovic (1978) has considered the calculation of ρ_{c0} for M/n^+ quasi-ohmic

† Note that the value of δ_n is negative when E_F lies above the conduction band edge.

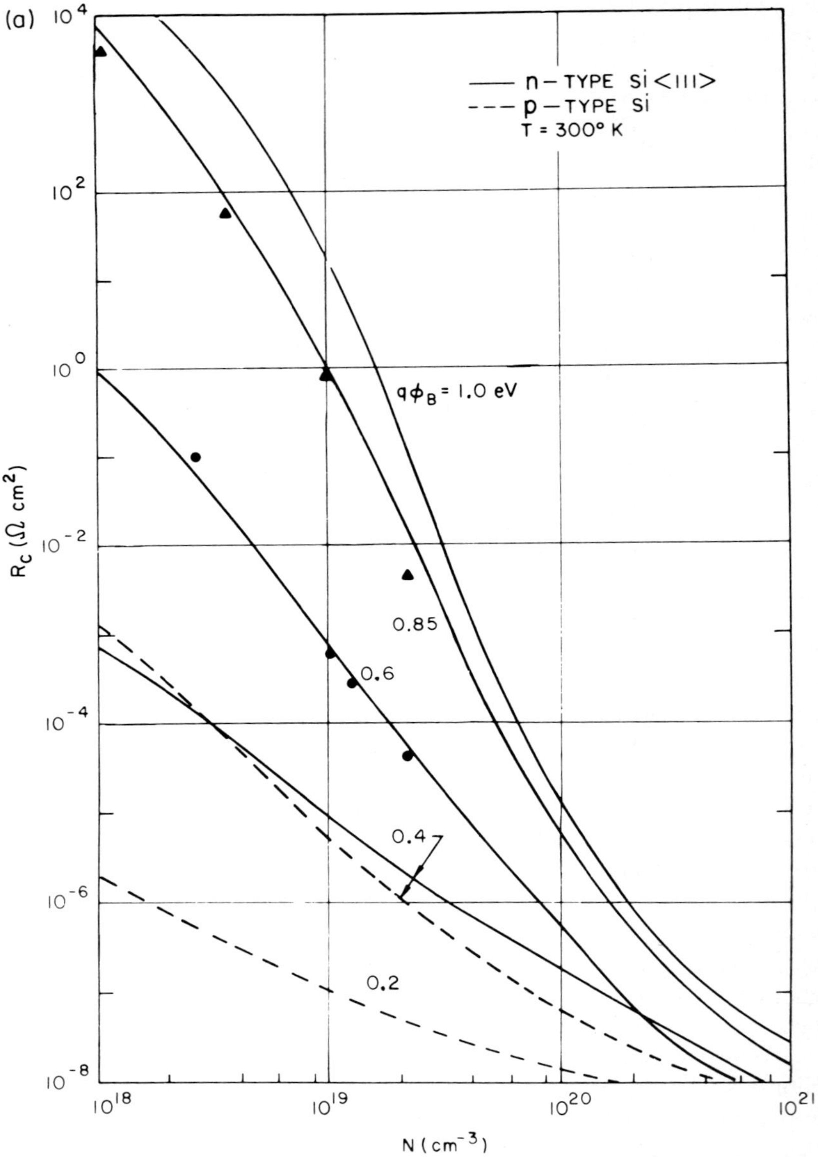

Fig. 5.44a. Theoretical specific contact resistance at 300°K for (111) oriented *n*-Si (solid lines) and *p*-Si (dotted lines). The filled circles are experimental results for Al/*n*-Si and Mo/*n*-Si barriers ($q\phi_b$ = 0.6 eV), and triangles for PtSi/*n*-Si barriers ($q\phi_b$ = 0.85 eV).

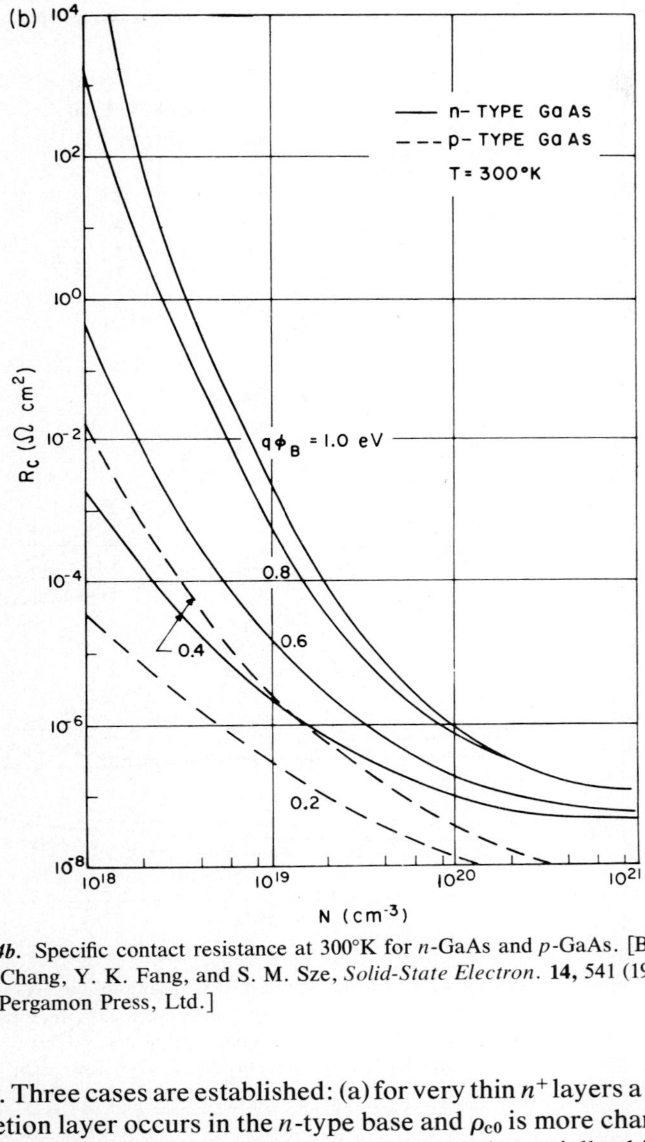

Fig. 5.44b. Specific contact resistance at 300°K for n-GaAs and p-GaAs. [Both figures after C. Y. Chang, Y. K. Fang, and S. M. Sze, *Solid-State Electron.* **14**, 541 (1971). Copyright 1971, Pergamon Press, Ltd.]

contacts. Three cases are established: (a) for very thin n^+ layers a portion of the depletion layer occurs in the n-type base and ρ_{c0} is more characteristic of an equivalent M/n contact, (b) for n^+ layers substantially thicker than the surface depletion layer, ρ_{c0} is essentially that of an M/n^+ contact, and (c) for the intermediate case, where $x_{n^+} > W_d$, carriers pass over the n^+ "valley" without scattering so only those with $E \geq E_{CB}$ in the n-base are available for tunneling. This introduces a dependence on the base doping

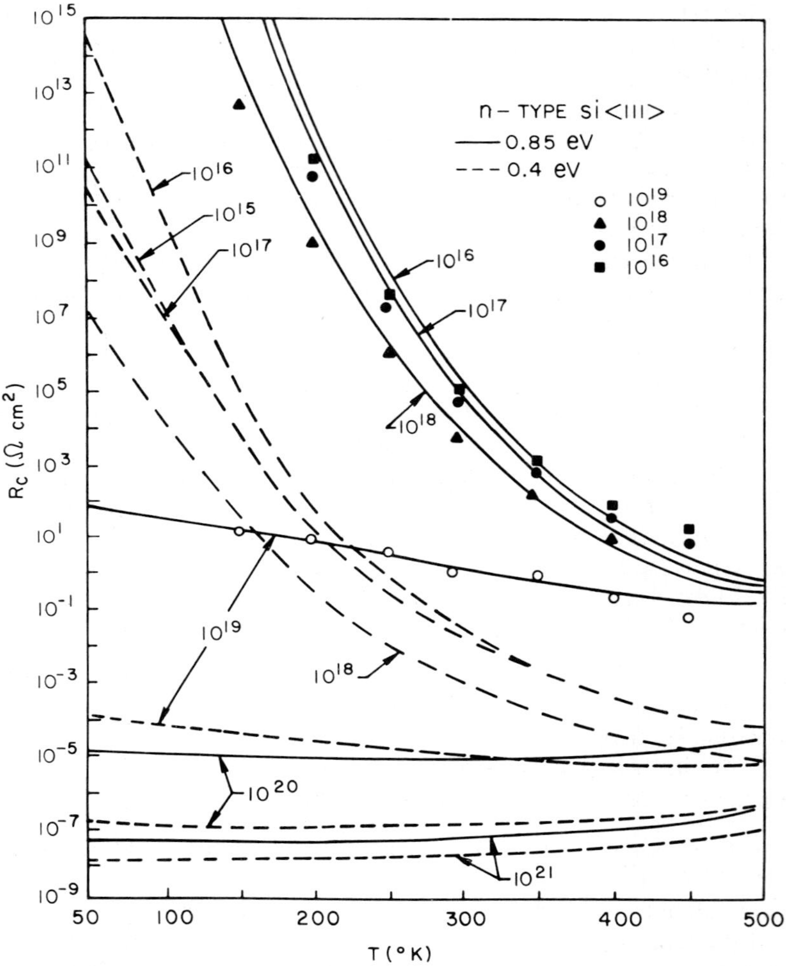

Fig. 5.45. Specific contact resistance for (111) oriented n-Si samples versus temperature. Barrier height and doping density are parameters. Also shown are experimental results of PtSi/n-Si barriers made on silicon substrates of various doping densities. [After C. Y. Chang, Y. K. Fang, and S. M. Sze, *Solid-State Electron.* **14,** 541 (1971). Copyright 1971, Pergamon Press, Ltd.]

density $N_{D,n}$ given approximately by

$$\rho_{c0} \sim \rho_{c00}[N_c/N_{D,n}]^m, \tag{5.87}$$

where $m \leq 1$. For high N_{D,n^+} and/or low temperatures, $m \to 1$.

It is clear that defect states within the tunneling barrier can enhance the tunneling probability and decrease ρ_{c0} under favorable circumstances.

These defects have been introduced by mechanical or electrical damage (capacitor discharge through the contact), and more recently by ion implantation with a neutral species. However, defects also offer the deleterious possibility of compensation of the high doping density at the surface, particularly in wide band-gap semiconductors. In addition, during subsequent heat treatment, such defects can act as gettering sites for nearby impurities (Eltoukhy and Greene, 1979), which could also reduce the doping efficiency. Finally, damage markedly increases the atomic diffusion rate, which can tend to broaden the doping profile and decrease its peak value. Thus compromises must be sought in determining the effectiveness of induced damage techniques.

Ideally, the value of ρ_c can be calculated theoretically if the effective ϕ_b, $N(x)$, ϵ_s, and m_t^* values are known. In practice, however, ρ_c can be strongly affected by interface layers (e.g., native oxides), surface damage, minority carrier injection, and energetically deep-lying impurity levels. "Because of these complications, the correlation of theory and experiment is often difficult and, consequently most studies of ohmic contacts usually consist of little more than an examination of processing techniques which lead to low impedance . . . contacts" (Rideout, 1975).

To summarize, the practical fabrication of quasi-ohmic contacts utilizes several fundamental principles:

(1) a highly doped surface layer that enhances tunneling transport,
(2) an auspicious combination of ϕ_m and χ_s to reduce or eliminate ϕ_b, if possible, and
(3) the introduction of defect states within the tunneling barrier by mechanical or electrical damage or ion implantation.

The range of values for ρ_c is covered by a patchwork of theories, as suggested by Table 5.II.

Quasi-ohmic contacts can be broadly classified into those that are ohmic by virtue of having a small barrier height and the much larger class in which transport is by tunneling. Tunneling contacts can be further separated into two classes: (a) *passive* metallizations that form a tunneling contact to an already present n^+ or p^+ layer, and (b) *active* metallizations in which the metal (or some component of alloy) is subsequently incorporated into the semiconductor by diffusion or alloy-regrowth to form the n^+ or p^+ layer.

The metallization is commonly done by vacuum evaporation, sputtering, electroplating, chemplating, or application of thin metal sheets that have been previously cut to size (preforms). Each method has its special virtues. For example, the ability of sputtering to "cut through" or remove surface contamination or native oxide layers can be beneficial in some cases. In other cases, the damage produced by sputtering appears to com-

Table 5.II

Dependences of ρ_{c0} on Doping Density and Temperature

Type of transport	N range (cm^{-3})[a]	$\rho_{c0}(N)$[b]	$\rho_{c0}(T)$
Thermionic	10^{15}	Negligible variation with N	$(1/T)\exp(q\phi_b/kT)$
Thermionic with barrier lowering	$10^{15}-10^{19}$	$\exp(-BN^{1/4})$	$(1/T)\exp(q\phi_b/kT)$
Thermionic-field emission	$10^{17}-10^{19}$	$\exp(B'/N^{1/2})$	Complex (decreasing with increasing with N)
Field emission[c] (tunneling)	10^{19}	$\exp(B''/N^{1/2})$	$(1/T)\exp(\delta_n/kT)$
Ohmic	Any	Negligible variation with N	Negligible variation with T.

[a] Note that the doping density N is that characteristic of the material adjacent to the contact, which, in most cases, is different from that of the bulk semiconductor. It is assumed that all doping impurities are ionized.

[b] B, B', and B'' are approximately constant with respect to N.

[c] The value of δ_n is negative when E_F lies above the conduction band.

promise contact performance and frequently has major effects on the nature of the contact region.

To obtain the required high doping density in the surface layer a number of techniques are available:

(1) high temperature diffusion from solid film or gaseous sources,

(2) doping by diffusion from a contact metal or alloy (such as diffusion of Zn from Au/Zn/Au into p-InP),

(3) ion implantation of a dopant with subsequent heat treatment to anneal out damage and activate the dopant,

(4) epitaxial growth of n^+ or p^+ layers (especially easy when the active layers of the cell are made by CVD or LPE), and

(5) alloy-regrowth (called alloyed contacts).

The alloy-regrowth technique is somewhat like LPE. A metal contact alloy is chosen that will dissolve a portion of the semiconductor when the alloy is melted. A dopant is included in the alloy; on cooling the dissolved semiconductor is forced out of solution and the newly grown layer is highly doped. Since the surface is dissolved, less stringent surface cleanliness is required. In some cases the regrowth process takes place below the melting point of the alloy [e.g., Al–Si and Al–Ge (Ottaviani *et al.*, 1974)].

To achieve the maximum electrical activity of the dopant, a favorable balance between the competing mechanisms of substitutional doping and compensation resulting from lattice disorder must be obtained. For example, in producing n^+-GaAs layers, using Ge as a dopant, a layer of Au is utilized as a getter for Ga to ensure that the Ge goes into the GaAs lattice substitutionally (Yoder, 1980). However, conditions must be such that excess gettering does not occur, for then Ga vacancies might compensate the Ge donors and lower the electrical activity of the dopant. In situations where the dopant source is limited, care must also be taken to produce the high density over a distance in the semiconductor somewhat larger than the resulting depletion layer width. Further diffusion, however, would broaden the distribution of dopant and lower the maximum. Thus fast, hot diffusions would be preferable to take advantage of increased solubility at higher temperatures.

Some of the most extreme values of ϕ_b for Si are obtained with the platinum silicide contact with $\phi_b = 0.25$ eV on p-Si and 0.85 eV on n-Si. The PtSi layer may be formed by heat treatment after application of a thin Pt film, or by application of PtSi by vacuum evaporation or sputtering. In the first case, PtSi is formed by a solid–solid reaction at $T \simeq 500°C$ (the melting point of PtSi is 980°C). This contact system also has the advantage of growing beyond surface contamination or damage and is thus highly reproducible. The contact to p-Si is quite conducting without surface doping ($\rho_{c0} \simeq 10^{-4}\ \Omega\ cm^2$ at 300°K), but the contact to n-Si must be preceded by an $n+$ layer. The PtSi contact is also chemically highly stable. The characteristics of PtSi, RhSi, and $ZrSi_2$ contacts to p- and n-type Si are explored by Lepselter and Andrews (1968).

A brief listing of ohmic and NOLR contacts is given in Table 5.III. The measurement of contact resistivity is somewhat difficult, particularly for the low values required for solar cell grid contacts. Some of the methods are outlined here and shown schematically in Fig. 5.46:

(1) *Comparison of 2-point and 4-point measurements.* By comparison of 2- and 4-point geometry measurements of bulk resistivity, the value of ρ_c can be separated out ($2R_c = R_4 - R_2$). However, the accuracy of measurement of ρ_{bulk} severely limits the sensitivity of the method. If the contacts are semirectifying, only the "reverse" branches of the J–V curves for each of the contacts can be obtained.

(2) *Three-point method.* By this method, outlined in Fig. 6.18, a full J–V curve of the contact can easily be obtained. In this case the sensitivity is limited by geometric factors and possible conductive surface films between the contacts.

(3) *Disk-to-plane.* In this measurement, explored by Cox and

Table 5.III

Ohmic and Quasi-ohmic Contacts

Semi- conductor	ρ or n, p[a]	Contact[b]	Type[c]	ρ_{c0} (Ω cm^2)	Reference, comment
p-Si	10^{19} cm^{-3}	Al	Active	10^{-6}	Berger (1972)
p-Si	10^{15} cm^{-3}	Al	Active	5×10^{-4}	Ting and Chen (1971)
p-Si	0.5 Ω cm	Al	Active	10^{-3}	Hooper (1965)[d]
p-Si	0.001 Ω cm	Al	Active	10^{-6}	Sello (1968)
p-Si	$\simeq$2 Ω cm	Ni	Passive	Low	Kukulka *et al.* (1978) (uses p^+ layer)
p-Si	Any	Pt/PtSi	Thermionic	10^{-4}	Lepselter and Andrews (1968) (low Schottky barrier)
n-Si	0.005 Ω cm	Al	Passive	4×10^{-3}	Hooper (1965)
n-Si	10^{19} cm^{-3}	Ag/Pd/Ti	Passive	—	Bickler *et al.* (1978)
n-Si	2×10^{20} cm^{-3}	Al	Passive	8×10^{-7}	Ting and Chen (1971)
n-Si	—	Ni/Pd_2Si	?	—	Coleman *et al.* (1978)
p-GaAs	0.1 Ω cm	Ag–In–Zn	Active	10^{-4}	Cox and Strack (1967)
n-GaAs	2×10^{16} cm^{-3}	Ni/AuGe/Ni	Active	8×10^{-5}	Heime *et al.* (1974)
n-GaAs	2×10^{18} cm^{-3}	Ni/AuGe/Ni	Active	10^{-6}	Heime *et al.* (1974)
n-GaAs	2.6 Ω cm	Ag–In–Ge	Active	$<10^{-3}$	Cox and Strack (1967)
p-InP	1–10 Ω cm	Au/Zn/Au	Active	10^{-3}	Theil *et al.* (1977)
n-InP	10^{15} cm^{-3}	Ag/Sn/In	Active	$<10^{-4}$	Mills and Hartnagel (1975)
p-CdTe	1–10 Ω cm	Au–Cu	?	0.2	Anthony *et al.* (1981)
n-CdTe	$>10^{16}$ cm^{-3}	In	Active or ohmic	—	
n-CdS	Any	In	Ohmic	—	Thompson and Cornwall (1972)

[a] Resistivity or carrier density in bulk semiconductor.
[b] Material next to semiconductor cited last.
[c] Type of contact is described in text.
[d] Hooper (1965) also gives ρ_{c0} values for Al, V, Mo, Ni, Co, and Chromel on both n- and p-type Si of various resistivities.

Strack (1967), contact is made to one side of a thin semiconductor slab with several small disks of various diameters d and a full-coverage contact on the reverse side. In the limit of small disk contact size to large slab size the total resistance is given by

$$R = V/I = (\rho/\pi d) \tan^{-1}(4t/d) + 4\rho_c/\pi d^2 + R_0, \tag{5.88}$$

where ρ is the bulk resistivity, t is the thickness of the slab, and R_0 includes other residual resistances not dependent on d. For small d the

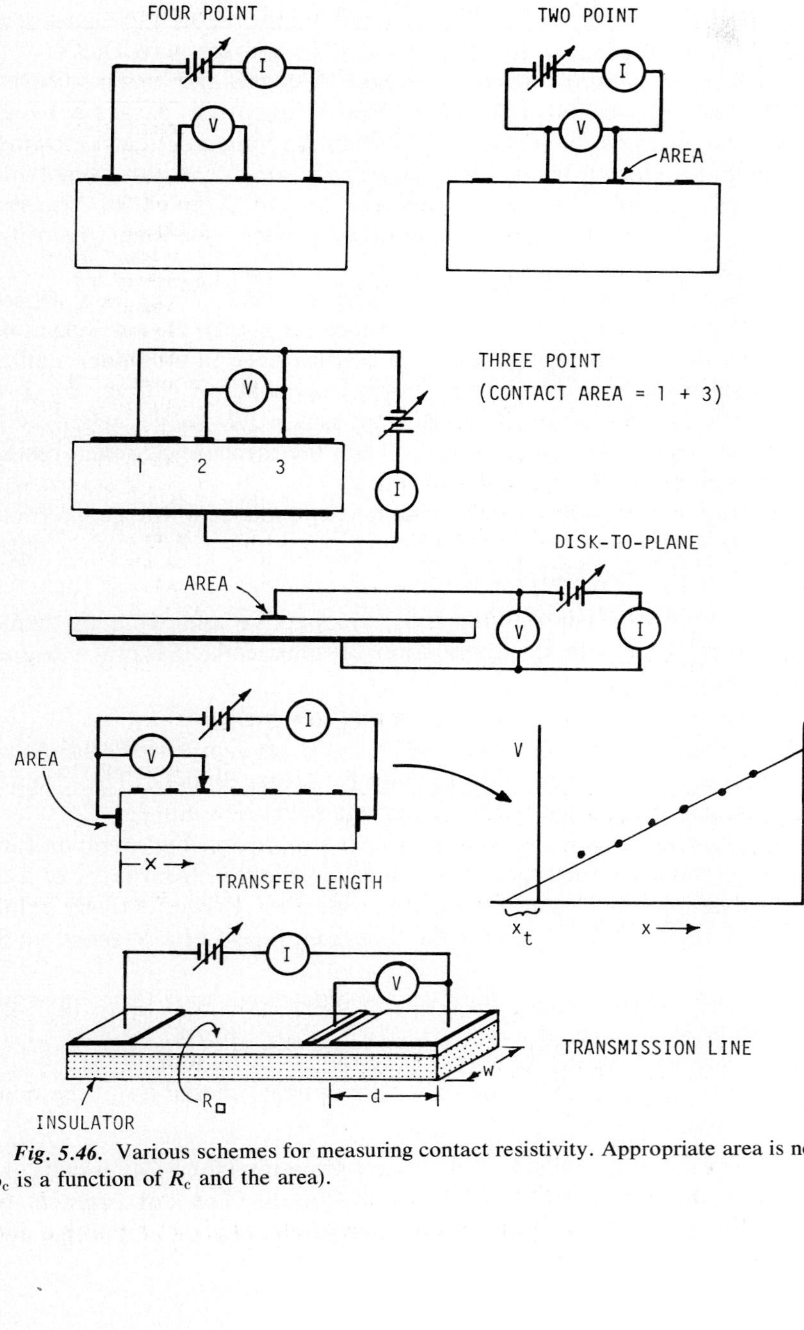

Fig. 5.46. Various schemes for measuring contact resistivity. Appropriate area is noted (ρ_c is a function of R_c and the area).

value of R is dominated by ρ_c and by comparing measurements with various values of d even very small values of ρ_c can be determined ($\sim 10^{-4}\ \Omega\ cm^2$). Heime *et al.* (1974) have refined the method to measure ρ_c values down to $10^{-6}\ \Omega$ cm for (Ni/Au–Ge/Ni) contacts to n-GaAs.

(4) *Transfer length.* In this method (Mengali and Seiler, 1962) a number of equally spaced contacts are used, with current passed between the two end contacts, as in Fig. 5.46. When the voltage at each intermediate contact is plotted against the contact spacing, the resulting line (with slope = $I\rho_{bulk}$/Area) can be extrapolated to $V = 0$, giving a "transfer length" x_t for the end contact. The contact resistivity is then given by $\rho_c = \rho_{bulk}x_t$.

(5) *Transmission line analysis* (Berger, 1972; Ting and Chen, 1971). For contacts on thin films of sheet resistivity $R_\square$ on insulating substrates, the current distribution can be described in a manner similar to a transmission line. The characteristic impedance Z is given by $Z = [R_\square/\rho_c]^{1/2}/w$ and the attenuation constant is $\alpha = [R_\square/\rho_c]^{1/2}$, where w is the contact width and d is the length. Then the measured contact resistance R_c is given by $R_c = Z \coth(\alpha d)$.

(6) *Magnetoresistance.* The angular dependence of the geometrical magnetoresistance can be used to obtain ρ_c to within $\sim 0.5\%$ of the total resistance (Gutai and Mojzes, 1975).

At this point we wish to broaden our perspective somewhat and make further correlation with the fabrication of solar cells. Several requirements are evident.

(1) The contact must have sufficiently low resistivity.

(2) To reduce the resistance of the grid contact strips, the usual practice is to thicken them substantially by solder dipping. The contact metallization should tolerate this treatment with good graces.

(3) The contact must be stable against subsequent high-temperature processing steps. In particular, the contact to the thin front layer of a Si cell, for example, must not degrade its properties. This effectively eliminates alloy-regrowth contacts for the thin front layers of UV-sensitive Si cells.

(4) With encapsulation, the contact must have long-term environmental stability.

(5) The use of costly or scarce materials is to be minimized.

(6) For back-wall reflection cells, the optical reflectivity of the contact must be considered.

Probably the most widely used contact structures for Si solar cells are vacuum-evaporated Al/p-Si and Ag/Pd/Ti/n^+-Si. The development of the latter contact illustrates the strong interaction between metallurgy and solid-state electronics that is required for contact technology. Although Ag makes a good tunneling contact to n^+-Si ($\phi_b = 0.6$–0.7 eV), its adhesion is unsatisfactory. The introduction of a thin, intervening layer of Ti provides good adhesion while retaining the electrical properties. Early contacts of this type used 2–3 μm of Ag on 1 μm of Ti with a thin layer of native oxide on the Si.† However, in solar cell field tests, corrosion problems were encountered; a solid–solid reaction between Ti and Si was apparently enhanced by the presence of small amounts of moisture and the contact separated.† The presence of Ti also reduced the allowable temperature of subsequent fabrication operations somewhat (Faith, 1978). By the introduction of a thin interlayer of Pd or Mo, which prevented the reaction, a contact with good electrical properties *and* excellent stability was obtained.‡ For the back, p-type contact, an active metallization of Al is vacuum evaporated on and subsequently diffused at 600 to 800°C, producing both a p^+ layer and a tunneling contact ($\phi_b \simeq 0.6$ eV for Al/p-Si).

Although the Ag/Pd/Ti contact discussed here has good stability and electrical properties, it is too expensive for large-area, terrestrial cells. Not only are two of the metals involved costly, but the deposition procedure per se is expensive and wastes over 90% of the metals. Thus other methods are being developed for n^+-Si such as electroless plating of Au then Ni (Bickler *et al.*, 1978). Electroless plating of Ni over Pd yields a Ni/Pd_2Si/n^+-Si contact on mild heat treatment (Coleman *et al.*, 1978). Another alternative is screen-printed pastes of Ag–Ti (for n^+-Si) or Ag–Al (for p-Si) (Salles *et al.*, 1978). Both plating and screen printing have the advantage of putting metal only where it is needed.

Application of contacts on AlGaAs and GaAs solar cells presents slightly different problems. Because their intended use is usually for concentrator systems, cost is not so much of a factor, but ρ_c must be reduced to values $< 10^{-4}\ \Omega\ cm^2$ for the front-grid contact. In order to make low ρ_c contacts for Gunn effect oscillators, Cox and Strack used Ag–In–Ge alloyed contacts for n-GaAs and an Ag–In–Zn alloy for p-GaAs, obtaining $\rho_c < 10^{-4}\ \Omega\ cm^2$ (Cox and Strack, 1967). However, the n-type contact was subject to aging effects (Cox and Hasty, 1968), a problem that was resolved by using epitaxial n^+ layers between the contacts and the bulk material. Heime *et al.* (1974) obtained $\rho_c < 10^{-5}\ \Omega\ cm^2$ using an alloyed contact of Ni/Au–Ge/Ni/n-GaAs. Some contacts used for concentrator AlGaAs solar cells are listed in Table 5.IV.

† See Springgate (1968). It was later found that the Ti thickness could be reduced by at least a factor of 10^2 without affecting the properties of the contact significantly.

‡ The Ag/Pd/Ti contact has also been used as a tunneling contact to the back p-type layer of Si solar cells. In this case a p^+ layer must have been added by other means.

Table 5.IV

Contacts Used for GaAs-Based Concentrator Solar Cells[a]

Reference	Front layer	Base
Van der Plas *et al.*[b]	Metal/p^+-GaAs/p-AlGaAs	?
Ewan *et al.*[c]	Ag/Au-Zn/p-AlGaAs	Au-Ge-Ni/n-GaAs
Sahai *et al.*[d]	Ag-Zn/p-GaAs[e]	?
Fan *et al.*[f]	Au/n^+-GaAs	Au/p^+-GaAs

[a] The first three cells are designed for high-concentration systems, whereas the last is an n^+/p GaAs homojunction designed for one sun.

[b] H. A. Van der Plas, L. W. James, R. L. Moon, and N. J. Nelson, *Proc. 13th IEEE Photovoltaic Specialists Conf.*, p. 934 (1978).

[c] J. Ewan, R. C. Knechtli, R. Loo, and G. S. Kamath, *Proc. 13th IEEE Photovoltaic Specialists Conf.*, p. 941 (1978).

[d] R. Sahai, D. D. Edwall, and J. S. Harris, Jr., *Proc. 13th IEEE Photovoltaic Specialists Conf.*, p. 946 (1978).

[e] Contact was made *through* openings in the AlGaAs layer, directly to the p-GaAs layer in this cell.

[f] J. C.-C. Fan and C. O. Bozler, *Proc. 13th IEEE Photovoltaic Specialists Conf.*, p. 953 (1978).

Perhaps the only truly ohmic contact ($\phi_m < \chi_s$) that is widely used for solar cells is that between n-CdS and In or Zn used in the p-Cu_xS/n-CdS cell.

5.9 SUMMARY

In this chapter we have examined the determining factors for homojunction, heterojunction, Schottky, and MIS device characteristics. The theory for homojunction transport is well established (except for the V_{oc} discrepancy discussed in Section 7.5.3). For heterojunctions, the Anderson model was extended to include interface states and dipoles in order to explain experimental results. It appears that heterojunction and heteroface devices can be divided into two broad classes, depending on degree of lattice mismatch.

First, there are those with near zero lattice mismatch and extremely careful preparation, giving negligible interface state densities. The carrier transport is determined by bulk properties. Despite the expense of their preparation, heteroface structures show great promise for concentrating systems with already realized solar efficiencies in the 20–30% range. However, zero lattice mismatch and careful preparation may not always be sufficient to ensure the absence of interface states except for fortuitous choices of heterojunction components and fabrication methods.

Table 5.V

Transport Parameters and Solar Efficiency of Selected Junctions

Junction	ϵ_{mf} (%)[a]	J_0 ($A\ cm^{-2}$)	A	η_s (%)	Transport mechanism
n-CdS/p-InP[b]	0.3	3.0×10^{-8}	2.38	12.8	Recombination
n-CdS/p-InP	0.3	1.5×10^{-6}	2.93	12.5	Recombination
n-CdS/p-InP	0.3	1.3×10^{-8}	2.14	14.4	Recombination
n-CdS/p-InP	0.3	1×10^{-5}	2.99	12	Recombination/tunneling
n-ITO/p-InP	Large	1.2×10^{-7}	2.35	12.4	Tunneling
	Large	2.5×10^{-8}	2.02	8.3	Tunneling
n/p InP	0	2×10^{-8}	2.36[c]	?	Recombination/diffusion
		$\approx 1 \times 10^{-14}$	1.08[c]		
n-CdS/p-CdTe	9.7	1.7×10^{-8}	1.89	8.0	Recombination
n-ZnO/p-CdTe	29.1	1.2×10^{-7}	1.98	9	Recombination
p/n CdTe	0	1.5×10^{-10}	≈ 2	10.7	Recombination
p-Cu_2S/n-CdS	4.6	5.5×10^{-11}	1.0	≈ 8	Recombination at[d] interface
n-ZnSe/p-GaAs	0.3	2×10^{-10}	1.4	Low?	Recombination?
n-Ge/p-GaAs	0.07	3.4×10^{-8}	1.68	?	Tunneling
n-Ge/p-GaAs	0.07	8.1×10^{-9}	1.17	?	Tunneling
p/n GaAs[e]	0	1.1×10^{-10}	2.0	14.8	Recombination
p/n GaAs[e]	0	3.8×10^{-12}	1.85[c]	12	Recombination/diffusion
		7.4×10^{-14}	1.54[c]		
	0	2.7×10^{-13}	1.75[c]	12	Recombination/diffusion
		6.6×10^{-18}	1.15[c]		
Au/I/n-GaAs	Large	1×10^{-10}	1.40	~15	Schottky
n-ITO/p-Si	Large	1×10^{-8}	1.56	12.1	MIS
n-SnO_2/n-Si	Large	10^{-6}	≥ 2[c]	—	Schottky
		10^{-11}	1.0[c]	8.8	Diffusion
n/p Si	0	5.5×10^{-6}	2.86[c]	11.6	
		3.3×10^{-12}	1.0[c]		Recombination/diffusion

Note: All data are at $\approx$300°K, η_s is the solar efficiency. An equivalent diode quality factor A is given even though tunneling dominates in some cases. The tunneling exponent is then $\alpha = q/Ak(300°K)$.

[a] ϵ_{mf} ~ lattice mismatch.

[b] First named component is deposited on the second except for homojunctions.

[c] Indicates two branches of curve for single device.

[d] Calculated from data given for illuminated case.

[e] With AlGaAs window layer.

In the second class are the heterojunctions with small to large mismatch in which the transport is almost completely dominated by the presence of interface states, both by distortion of the bands and by recombination and/or tunneling. Solar cells made with these materials show promising efficiencies as well (although perhaps not as high as in the zero-lattice-mismatch junctions).

Most junctions exhibit a combination of two or more modes of carrier transport, depending on temperature and bias voltage. In Table 5.V are presented data on several of the more common devices with values of A, J_0, and solar efficiency taken from the literature. The implications of these data are the following.

(1) Near zero lattice mismatch is beneficial but by no means required for small J_0 and large solar efficiency.

(2) The values of J_0 are in almost all cases many orders of magnitude larger than predicted by simple theory.

The foregoing discussion on heterojunctions, heteroface structures, Schottky barriers, and MIS devices has led us to focus on the interface properties as determining factors for J_0 and A. Only in fortuitous cases does carrier transport depend principally on the bulk properties of the components. Hence control of interface properties is a major pathway toward increased solar efficiency. In the section discussing MIS and SIS junctions we saw that there are ways to increase V_{oc} and solar efficiency by manipulation of the interface in some cases, particularly by the introduction of a thin insulating layer with the proper charge states.

REFERENCES

Acket, G. A. *et al.* (1975). GaAs and related compounds. *Conf. Ser.—Inst. Phys.* No. 24, p. 181.

Anderson, R. L. (1960). *IBM J. Res. Dev.* **4,** 283.

Anderson, W. W. (1977). *Infrared Phys.* **17,** 147.

Andrews, J. M., Jr., and Lepselter, M. P. (1968). *Tech. Dig.—IEEE Int. Electron Devices Meet., Washington, D.C.* p. 78.

Anthony, T., Fahrenbruch, A. L., and Bube, R. H. (1982). *J. Electronic Matl.* **11,** 89.

Aranovich, J., Golmayo, D., Fahrenbruch, A. L., and Bube, R. H. (1980). *J. Appl. Phys.* **51,** 4260.

Aven, M., and Garwacki, W. (1967). *J. Electrochem. Soc.* **114,** 1063.

Aven, M., and Swank, R. K. (1968). *In* "Ohmic Contacts to Semiconductors" (B. Schwartz, ed.), p. 69. Electrochem. Soc., Pennington, New Jersey.

Barber, H. D. (1967). *Solid-State Electron.* **10,** 1039.

Bardeen, J. (1947). *Phys. Rev.* **71,** 717.

Berger, H. H. (1972). *J. Electrochem. Soc.* **119,** 507.

Bethe, H. A. (1942). "Theory of the Boundary Layer of Crystal Rectifiers," MIT Radiat. Lab. Rep. 43-12. Mass. Inst. Technol., Cambridge, Massachusetts.

Bickler, D. B., Gallagher, B. D., and Sanchez, L. E. (1978). *Proc. 13th IEEE Photovoltaic Specialists Conf.,* p. 241.

Böer, K. W. (1978). *Phys. Status Solidi A* **49,** 455.

Böer, K. W. (1979). *J. Appl. Phys.* **50,** 5356.

Borrego, J. M., Gutmann, R. J., and Ashok, S. (1977). *Solid-State Electron.* **20,** 125.

Brillson, L. J. (1978). *J. Vac. Sci. Technol.* **15,** 1378.

Brillson, L. J. (1979). *J. Vac. Sci. Technol.* **16,** 1137.

Card, H. C., and Rhoderick, E. H. (1971). *J. Phys. D* **4,** 1589.
Card, H. C., and Yang, E. S. (1976). *Appl. Phys. Lett.* **29,** 51.
Chang, C. Y., and Sze, S. M. (1970). *Solid-State Electron.* **13,** 727.
Chang, C. Y., Fang, Y. K., and Sze, S. M. (1971). *Solid-State Electron.* **14,** 541.
Cheung, D., Chiang, S. Y., and Pearson, G. L. (1975). *Solid-State Electron* **18,** 263.
Childs, R., Fortuna, J., Geneczko, J., and Fonash, S. J. (1976). *Proc. 12th IEEE Photovoltaics Specialists Conf.,* p. 862.
Childs, R., Ruths, J. M., Sullivan, T. E., and Fonash, S. J. (1978). *J. Vac. Sci. Technol.* **15,** 1397.
Choo, S. C. (1968). *Solid-State Electron.* **11,** 1069.
Chynoweth, A. G., Feldmann, W. L., and Logan, R. A. (1961). *Phys. Rev.* **121,** 684.
Cohen, M. L. (1979). *J. Vac. Sci. Technol.* **16,** 1135.
Coleman, M. G., Pryor, R. A., and Sparks, T. G. (1978). *Proc. 13th IEEE Photovoltaic Specialists Conf.,* p. 597.
Cox, R. H., and Hasty, T. E. (1968). *In* "Ohmic Contacts to Semiconductors" (B. Schwartz, ed.), p. 88. Electrochem. Soc., Pennington, New Jersey.
Cox, R. H., and Strack, H. (1967). *Solid-State Electron.* **10,** 1213.
Crowell, C. R. (1969). *Solid-State Electron.* **12,** 55.
Crowell, C. R., and Rideout, V. L. (1969). *Solid-State Electron.* **12,** 89.
Crowell, C. R., and Sze, S. M. (1966a). *Solid-State Electron.* **9,** 1035.
Crowell, C. R., and Sze, S. M. (1966b). *J. Appl. Phys.* **37,** 2683.
Dean, R. H., and Nuese, C. J. (1971). *IEEE Trans. Electron Devices* **ED-18,** 151.
De Mari, A. (1968). *Solid-State Electron.* **11,** 33.
DeVisschere, P., and Pauwels, H. (1978). *Proc. CEC Photovoltaic Sol. Energy Convers. Conf., 1st, 1977,* p. 330. Reidel Publ., Dordrecht, Netherlands.
Dolega, U. (1963). *Z. Naturforsch., Teil A* **18,** 653. (In German.)
Donnelly, J. P., and Milnes, A. G. (1966). *Solid-State Electron.* **9,** 174.
Donnelly, J. P., and Milnes, A. G. (1967). *IEEE Trans. Electron Devices* **ED-14,** 63.
Ellis, B., and Moss, T. S. (1970). *Solid-State Electron.* **13,** 1.
Eltoukhy, A. H., and Greene, J. E. (1979). *J. Appl. Phys.* **50,** 505.
Ettenberg, M., and Kressel, H. (1976). *J. Appl. Phys.* **47,** 1538.
Ettenberg, M., and Olsen, G. H. (1977). *J. Appl. Phys.* **48,** 4275.
Fahrenbruch, A. L., and Aranovich, J. (1979). Heterojunction phenomena and interfacial defects in photovoltaic converters. *In* "Solar Energy Conversion" (B. O. Seraphin, ed.), Topics in Applied Physics, Vol. 31, p. 257. Springer-Verlag, Berlin and New York.
Fahrenbruch, A. L., and Bube, R. H. (1974). *J. Appl. Phys.* **45,** 1264.
Faith, T. J. (1978). *Proc. 13th IEEE Photovoltaics Specialists Conf.,* p. 321.
Fonash, S. J. (1975a). *Proc. 11th IEEE Photovoltaic Specialists Conf.,* p. 376.
Fonash, S. J. (1975b). *J. Appl. Phys.* **46,** 1286.
Fonash, S. J. (1976). *J. Appl. Phys.* **47,** 3597.
Fonash, S. J. (1977). *J. Appl. Phys.* **48,** 3953.
Fonash, S. J., Sullivan, T. I., Childs, R., and Ruths, J. (1978). *Proc. 13th IEEE Photovoltaic Specialists Conf.,* p. 645.
Ghosh, A. K., Fishman, C., and Feng, T., (1978). *J. Appl. Phys.* **49,** 3490.
Godfrey, R. B., and Green, M. A. (1979). *Appl. Phys. Lett.* **34,** 790.
Goodman, A. M. (1963). *J. Appl. Phys.* **34,** 329.
Gossick, B. R. (1964). "Potential Barriers in Semiconductors." Academic Press, New York.
Gossick, B. R. (1969a). *Phys. Stat. Solidi* **35,** 997.
Gossick, B. R. (1969b). *Surf. Sci.* **18,** 181.

Gossick, B. R. (1970). *Surf. Sci.* **21,** 123.
Gossick, B. R. (1971a). *Surf. Sci.* **25,** 465.
Gossick, B. R. (1971b). *Surf. Sci.* **28,** 469.
Green, M. A., King, F. D., and Shewchun, J. (1974). *Solid-State Electron.* **17,** 551.
Grove, A. S. (1967). "Physics and Technology of Semiconductor Devices." Wiley, New York.
Gückel, H., Thomas, D. C., Iyengar, S. V., and Demirkol, A. (1977). *Solid-State Electron.* **20,** 647.
Gundlach, K. H., and Simmons, J. G. (1969). *Thin Solid Films* **4,** 61.
Gutai, L., and Mojzes, I. (1975). *Appl. Phys. Lett.* **26,** 325.
Heime, K., König, U., Kohn, E., and Wortmann, A. (1974). *Solid-State Electron.* **17,** 835.
Heine, V. (1965). *Phys. Rev. A* **138,** 1689.
Henisch, H. K. (1957). "Rectifying Semiconductor Contacts." Oxford Univ. Press (Claredon), London and New York.
Henry, C. H., and Logan, R. A. (1978). *J. Vac. Sci. Technol.* **15,** 1471.
Henry, C. H., Logan, R. A., and Merrit, F. R. (1978). *J. Appl. Phys.* **49,** 3530.
Hinkley, E. D., and Rediker, R. H. (1967). *Solid-State Electron.* **10,** 671.
Hooper, R. C., Cunningham, J. A., and Harper, J. G. (1965). *Solid-State Electron.* **8,** 831.
Hovel, H. J. (1973). *Proc. 10th IEEE Photovoltaic Specialists Conf.,* p. 34.
Kar, S. (1977). *Tech. Dig.—IEEE Electron Devices Meet., Washington, D.C.* p. 56A.
Kipperman, A. H. M., Backerra, S. C. M., Maaskamp, H. J., and van Zolingen, R. J. C. (1978). *Proc. CEC Photovoltaic Sol. Energy Convers. Conf., 1st, 1977,* p. 961. Reidel Publ., Dordrecht, Netherlands.
Kroemer, H. (1975). Problems in the theory of heterojunction discontinuities. *CRC Crit. Rev. Solid State Sci.* **5,** 555.
Kukulka, J., Shewchun, J., Kazandjian, A., and Burk, D. (1978). *Proc. 13th IEEE Photovoltaic Specialists Conf.,* p. 268.
Kurtin, S., McGill, T. C., and Mead, C. A. (1969). *Phys. Rev. Lett.* **22,** 1433.
Landsberg, P. T., and Klimpke, C. (1977). *Proc. R. Soc. London, Ser. A* **354,** 101.
Landsberg, P. T., and Klimpke, C. (1978). *Proc. 13th IEEE Photovoltaic Specialists Conf. Proc.,* p. 665.
Lawrence, H., and Warner, R. M. (1960). *Bell Syst. Tech. J.* **39,** 389.
Lepselter, M. P., and Andrews, J. M. (1968). *In* "Ohmic Contacts to Semiconductors" (B. Schwartz, ed.), p. 159. Electrochem. Soc., Pennington, New Jersey.
Lillington, D. R., and Townsend, W. G. (1976). *Appl. Phys. Lett.* **28,** 98.
Lindau, I., Chye, P. W., Garner, C. M., Pianetta, P., Su, C. Y., and Spicer, W. E. (1978). *J. Vac. Sci. Technol.* **15,** 1332.
Lindholm, F. A., and Fossum, J. G. (1977). *IEEE Trans. Electron Devices* **ED-24,** 325.
Lindquist, P. F., and Bube, R. H. (1972). *J. Appl. Phys.* **43,** 2839.
Louie, S. G., Chelikowsky, J. R., and Cohen, M. L. (1977). *Phys. Rev. B* **15,** 2154.
Low, G. G. E. (1955). *Proc. Phys. Soc., London, Sect. B* **68,** 310.
McCaldin, J. O., McGill, T. C., and Mead, C. A. (1976). *J. Vac. Sci. Technol.* **13,** 802.
McOuat, R. F., and Pulfrey, D. L. (1976). *J. Appl. Phys.* **47,** 2113.
Many, A., Goldstein, Y., and Grover, N. B. (1971). "Semiconductor Surfaces," p. 194. North-Holland Publ., Amsterdam.
Mataré, H. F. (1971). "Defect Electronics in Semiconductors." Wiley (Interscience), New York.
Mead, C. A. (1966). *Solid-State Electron.* **9,** 1023.
Mengali, O. J., and Seiler, M. R. (1962). *Adv. Energy Convers.* **2,** 59.

Middlebrook, R. D. (1957). "An Introduction to Junction Transistor Theory," p. 106. Wiley, New York.
Miller, G. L., Lang, D. V., and Kimmerling, L. C. (1977). Capacitance transient spectroscopy. *Annu. Rev. Mater. Sci.* **7,** 377.
Mills, H. T., and Hartnagel, H. L. (1975). *Electron. Lett.* **10**(25–26), 621.
Milnes, A. G. (1980). "Semiconductor Devices and Integrated Electronics." Van Nostrand Reinhold, New York.
Milnes, A. G., and Feucht, D. L. (1972). "Heterojunctions and Metal-Semiconductor Junctions." Academic Press, New York.
Mitchell, K. W., Fahrenbruch, A. L., and Bube, R. H. (1977). *J. Appl. Phys.* **48,** 4365.
Nelson, R. J. (1978). *J. Vac. Sci. Technol.* **15,** 1475.
Nelson, R. J., Williams, J. S., Leamy, H. J., Miller, B., Casey, H. C., Jr., Parkinson, B. A., and Heller, A. (1980). *Appl. Phys. Lett.* **36,** 76.
Nordheim, L. (1928). *Proc. R. Soc. London, Ser. A* **121,** 626.
Nussbaum, A. (1962). "Semiconductor Device Physics," p. 129. Prentice-Hall, Englewood Cliffs, New Jersey.
Oldham, W. G., and Milnes, A. G. (1963). *Solid-State Electron.* **6,** 121.
Oldham, W. G., and Milnes, A. G. (1964). *Solid-State Electron.* **7,** 153.
Olsen, G. H., and Ettenberg, M. (1978). Growth effects in the heteroepitaxy of III-V compounds. *In* "Crystal Growth: Theory and Techniques" (C.H.L. Goodman, ed.), Vol. 2, p. 1. Plenum, New York.
Olsen, G. H., Nuese, C. J., and Smith, R. T. (1978). *J. Vac. Sci. Technol.* **15,** 1410. *J. Appl. Phys.* **49,** 5523.
Ottaviani, G., Sigurd, D., Marrello, V., Mayer, J. W., and McCaldin, J. O. (1974). *J. Appl. Phys.* **45,** 1730.
Padovani, F. A. (1968). *Solid-State Electron.* **11,** 193.
Padovani, F. A. (1971). The voltage–current characteristics of metal-semiconductor contacts. *In* "Semiconductors and Semimetals" (R. K. Willardson and A. C. Beer, eds.), Vol. 7A, p. 75. Academic Press, New York.
Padovani, F. A., and Stratton, R. (1966). *Solid-State Electron.* **9,** 695.
Parker, G. H., and Mead, C. A. (1969a). *Phys. Rev.* **184,** 780.
Parker, G. H., and Mead, C. A. (1969b). *Appl. Phys. Lett.* **14,** 21.
Parker, G. H., McGill, T. C., Mead, C. A., and Hoffman, D. (1968). *Solid-State Electron.* **11,** 201.
Pellegrini, B., and Salardi, G. (1975). *Solid-State Electron.* **18,** 791.
Ponpon, J. P., and Siffert, P. (1977). *Rev. Phys. Appl.* **12,** 427.
Ponpon, J. P., and Siffert, P. (1978). *Proc. 13th IEEE Photovoltaic Specialists. Conf.,* p. 639.
Popovic, R. S. (1978). *Solid-State Electron.* **21,** 1133.
Pulfrey, D. L. (1976). *IEEE Trans. Electron Devices* **ED-23,** 587.
Pulfrey, D. L. (1978). *IEEE Trans. Electron Devices* **ED-25,** 1308.
Rediker, R. H., Stopek, S., and Ward, J. H. R. (1964). *Solid-State Electron.* **7,** 621.
Rhoderick, E. H. (1974). *Conf. Ser.—Inst. Phys.* No. 22, p. 3.
Riben, A. R. (1965). Thesis, Carnegie Inst. Technol., Pittsburgh, Pennsylvania.
Riben, A. R., and Feucht, D. L. (1966a). *Int. J. Electron.* **20,** 583.
Riben, A. R., and Fuecht, D. L. (1966b). *Solid-State Electron.* **9,** 1055.
Rideout, V. L. (1975). *Solid-State Electron.* **18,** 541.
Rose, A. (1954). *Onde Electr.* **34,** 645. [Also see Johnson, E. D., and Rose, A., *Proc. IRE* **47,** 407 (1959).]

Rose, A. (1955). *Phys. Rev.* **97,** 1538.
Rose, A. (1965). *J. Appl. Phys.* **35,** 2664.
Rothwarf, A. (1975). *Int. Workshop CdS Sol. Cells Other Abrupt Heterojunctions, Univ. Del.* NSF-RANN AER 75-15858, p. 9.
Rothwarf, A., Burton, L. C., Hadley, H. C., and Stork, G. M. (1975). *Proc. 11th IEEE Photovoltaic Specialists Conf.,* p. 476.
Sah, C.-T. (1962). *IRE Trans. Electron Devices* **ED-9,** 94.
Sah, C.-T. (1977). *IEEE Trans. Electron Devices* **ED-24,** 410.
Sah, C.-T., Noyce, R. N., and Shockley, W. (1957). *Proc. IRE* **45,** 1228.
St. Pierre, J. A., Singh, R., Shewchun, J., and Loferski, J. J. (1976). *Proc. 12th IEEE Photovoltaic Specialists Conf.,* p. 847.
Salles, Y., Diguet, D., and Lauvray, H. (1978). *Proc. 13th IEEE Photovoltaic Specialists Conf.,* p. 257.
Salter, G. C., and Thomas, R. E. (1976). *Proc. 11th IEEE Photovoltaic Specialists Conf.,* p. 862.
Sarrabayrouse, G., Buxo, J., Myszkowski, A., and Esteve, E. (1977). *Rev. Phys. Appl.* **12,** 433.
Scharfetter, D. L. (1965). *Solid-State Electron.* **8,** 299.
Scaife, D. E. (1980). *Solar Energy* **25**(1), 41.
Schluter, M. (1978). *J. Vac. Sci. Technol.* **15,** 1374.
Schottky, W. (1938). *Naturwissenschaften* **26,** 843.
Schottky, W., and Spenke, E. (1939). *Wiss. Veroeff. Siemens-Werken* **18,** 225.
Schulz, M., and Johnson, N. M. (1977). *Appl. Phys. Lett.* **31,** 622.
Schwartz, B., ed. (1969). "Ohmic Contacts to Semiconductors." Electrochem. Soc., Pennington, New Jersey.
Schwuttke, G. H. (1974). *Proc. Symp. Mater. Sci. Aspects Thin Film Syst. Sol. Energy Convers.*
(NSF—RANN), *Tucson, Ariz.* p. 49.
Seager, C. H., and Ginley, D. S. Barrier heights and passivation of grain boundaries in polycrystalline Si. (1980). *Sol. Cells* **1,** 293.
Sello, H. (1968). Ohmic contacts and integrated circuits. *In* "Ohmic Contacts to Semiconductors" (B. Schwartz, ed.) p. 277. Electrochem. Soc., Pennington, New Jersey.
Shewchun, J., Green, M. A., and King, F. D. (1974). *Solid-State Electron.* **17,** 551, 563.
Shewchun, J., Dubow, J., Myszkowski, A., and Singh, R. (1978). *J. Appl. Phys.* **49,** 855.
Shockley, W., and Queisser, H. J. (1961). *J. Appl. Phys.* **32,** 510.
Smith, D. L. (1977). *Phys. Stat. Sol.* (a) **44,** 381.
Spicer, W. E., Lindau, I., Gregory, P. E., Garner, C. M., Pianetta, P., and Chye, P. W. (1976). *J. Vac. Sci. Technol.* **13,** 780.
Spicer, W. E., Lindau, I., Skeath, P., and Su, C. Y. (1980). *J. Vac. Sci. Technol.* **17**(5), 1019.
Springgate, W. F. (1968). *In* "Ohmic Contacts to Semiconductors" (B. Schwartz, ed.), p. 253. Electrochem. Soc. Pennington, New Jersey.
Stannard, J. (1979). *J. Vac. Sci. Technol.* **16,** 1462.
Stirn, R. J., and Yeh, Y. C. M. (1977). *IEEE Trans. Electron Devices* **ED-24,** 476.
Stratton, R. (1962). *Phys. Rev.* **125,** 67.
Stringfellow, G. B. (1976). *J. Vac. Sci. Technol.* **13,** 908.
Sze, S. M. (1969). "Physics of Semiconductor Devices," Chap. 8. Wiley, New York.
Tansley, T. L. (1968). *Proc. Int. Symp. GaAs, 2nd, Dallas, Tex.* p. 222. IPPS, London.
Tansley, T. L. (1971). Heterojunction properties. *In* "Semiconductors and Semimetals" (R. K. Willardson and A. C. Beer, eds.), Vol. 7A, p. 294. Academic Press, New York.
Tansley, T. L., and Owen, S. J. T. (1976). *IEEE Trans. Electron Devices* **ED-23,** 1123.

TeVelde, T. S. (1973). *Solid-State Electron.* **16,** 1305.

Theil, F. A., Bacon, D. D., Buehler, E., and Bachmann, K. J. (1977). *J. Electrochem. Soc.* **124,** 317.

Thompson, M. J. and Cornwall, M. G. (1972). *Solid-State Electron.* **15,** 861.

Ting, C. Y., and Chen, C. Y. (1971). *Solid-State Electron.* **14,** 433.

Van Opdorp, C., and Vrakking, J. (1967). *Solid-State Electron.* **10,** 955.

Van Ruyven, L. J. (1964). *Phys. Status Solidi* **3,** K109.

Van Ruyven, L. J. (1972). Phenomena at heterojunctions. *Annu. Rev. Mater. Sci.* **2,** 501.

Van Ruyven, L. J., Papenhuijzen, J. M. P., and Verhoeven, A. C. J. (1965). *Solid-State Electron.* **8,** 631.

Van Vliet, K. M., (1966). *Solid-State Electron.* **9,** 185.

Viktorovitch, P., and Kamarinos, G. (1977). *J. Appl. Phys.* **48,** 3060.

Wager, J. F., and Wilmsen, C. W. (1980). *J. Appl. Phys.* **51,** 812.

Wagner, C. (1931). *Phys. Z.* **32,** 641.

Weinreich, O., Mataré, H. F., and Reed, B. (1960). "Solid State Physics in Electronics and Telecommunications," Vol. 1, p. 97. Academic Press, New York.

Womac, J. F., and Rediker, R. H. (1972). *J. Appl. Phys.* **43,** 4129.

Wu, C. M., and Yang, E. S. (1979). *Solid-State Electron.* **22,** 241.

Yoder, M. N. (1980). *Solid-State Electron.* **23,** 117.

CHAPTER 6

THE CALCULATION OF SOLAR EFFICIENCY

One of the goals of the past few chapters has been the calculation of the solar efficiency η_s. The current–voltage $J-V$ characteristics of an ideal cell with a single dominant current transport mechanism can be represented by an expression such as

$$J = J_0[\exp(qV/AkT) - 1] - J_L, \tag{6.1}$$

at least over the range of J values that determine the solar efficiency. The dark $J-V$ curve is translated downward by the magnitude of the light-generated current J_L without change in shape, as shown in Fig. 6.1. This ideal characteristic is the result of superposition and in this case the short-circuit current $J_{sc} = -|J_L|$.

In most cases the following general steps are involved in the calculation of solar efficiency.

1. Calculation of the total light-generated current J_L from the incident spectral photon flux, the bulk semiconductor parameters [the absorption constant $\alpha(\lambda)$, the minority carrier diffusion length L_d, and the surface recombination velocity S], and the geometry of the device.†
2. Calculation of the maximum power P_m using the $J-V$ relationship of the diode and the J_L value.

† Considered in detail, this calculation would also involve grid coverage and reflection losses.

3. Calculation of the solar efficiency

$$\eta_s = |P_m|/P_s \tag{6.2}$$

using the total solar input power P_s.

These steps are not always independent, as is discussed toward the end of this chapter.

Two additional quantities are important. The open-circuit voltage is given by

$$V_{oc} = (kT/q)\ln[(J_L/J_0) + 1].$$

The fill factor ff, given by

$$\text{ff} \equiv P_m/J_{sc}V_{oc}, \tag{6.3}$$

is a measure of the "squareness" of the $J-V$ curve. In terms of V_{oc} and ff the solar efficiency is given by

$$\eta_s = V_{oc}|J_{sc}|\text{ff}/P_s.$$

Although the *fundamental parameters* of the solar cell are J_L, J_0, A, and R_s, the usual description is in terms of V_{oc}, J_{sc}, and ff. These parameters are useful, intuitive, and easily measured. For nearly ideal cells J_{sc}, V_{oc}, and ff have been usually treated almost as independent parameters in the literature. This independence is even more of an approximation for less-than-ideal cells, such as those under high-intensity illumination or those with transport dominated by interface recombination. Properly, the nonideal cells must be treated as a whole. There are three methods of attacking the solar efficiency problem.

1. By superposition, assuming that the light and dark currents are linearly additive, an exact solution can be obtained for the ideal cell, including the presence of simple, lumped values of series and parallel resistance.
2. The nonideal cell can be treated by analysis in terms of perturbations on the ideal cell parameters.
3. By the computer-numerical solution of the transport equation simultaneously with the other relevant equations, nonideal and three dimensional cells may be treated by a holistic approach.

The implications of superposition are discussed by Lindholm *et al.* (1976, 1979) for a general case, by Rothwarf (1978) for the Cu_2S/CdS cell, and by Tarr and Pulfrey (1979) for Si and GaAs cells at $C = 1$.

In the first section of this chapter, we discuss the calculation of η_s for

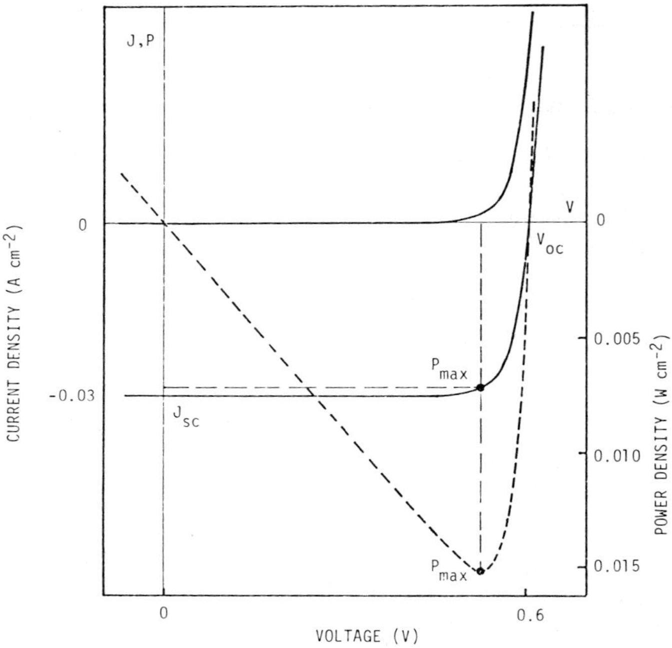

Fig. 6.1. Light and dark *J–V* curves for an ideal cell. The power output is shown by the dashed curve.

an ideal cell and derive the relation of η_s to J_0 and A. Next we discuss the dependence of η_s on series and parallel resistance (R_s and R_p) and, finally, we consider four related topics: the calculation of solar efficiency by perturbation analysis (Section 6.3.1) and by global computer numerical solution, the dependence of J_0 and A on illumination, the variation of η_s with temperature and illumination level, and, finally, an energy loss analysis.

6.1 THE IDEAL CELL UNDER ILLUMINATION

In this section we calculate the solar efficiency of an ideal cell in which the law of superposition holds. The implications of this model are as follows.

1. For a range of illumination intensities below and including that for intended use, J_0 and the slope of the log J versus V curve are not functions of illumination intensity or wavelength, and so the log J_{sc} versus V_{oc} curve is identical with the dark log J versus V curve.
2. J_L is not a function of bias voltage.

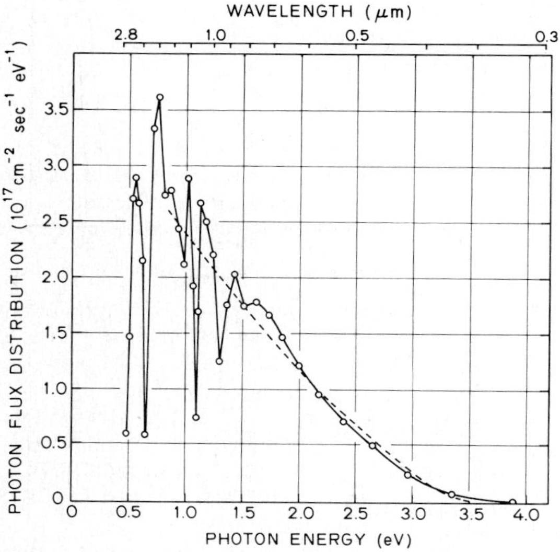

Fig. 6.2. Solar photon flux distribution $d\Gamma/dE$ at AM1.5 versus photon energy. The dashed curve is the derivative of the polynomial approximation of Fig. 6.3. [Drawn from data in "Terrestrial Photovoltaic Measurement Procedures," NASA (prepared for ERDA), ERDA/NASA/1022.77/16, NASA TM 73702 (1977).]

3. $R_s \to 0$ and $R_p \to \infty$, i.e., there are no resistance losses.

The total light-generated current is given by the integration of the product of $\eta_Q(\lambda)$ [or $\eta_Q(E)$] and the solar photon flux $d\Gamma/d\lambda$ (or $d\Gamma/dE$):

$$J_L = q \int_0^\infty \eta_Q(\lambda)(d\Gamma/d\lambda)\, d\lambda = q \int_0^\infty \eta_Q(E)(d\Gamma/dE)\, dE \qquad (6.4)$$

Note that the quantum efficiency may be a function of intensity as well. The AM1.5 solar photon flux is shown in Fig. 6.2.† The integral of the solar photon flux, shown in Figs. 6.3 and 6.4, is useful for estimation purposes; if we assume $\eta_Q = 1$, then $J_L \simeq q \int_E^\infty (d\Gamma/dE)\, dE \simeq -J_{sc}$.

The power produced by the cell is $P = J(V)V$ as indicated in Fig. 6.1 and the maximum power point P_m is found by maximization of $J(V)V$. The value of P_m (or equivalently ff) cannot be obtained explicitly. However, by iteration P_m can be obtained to any desired accuracy; such a scheme is described here.‡

† These data are tabulated in Appendix 1.

‡ This treatment has been expanded to include R_s, R_p, and a bias-voltage-dependent quantum eficiency by Mitchell *et al.* (1976).

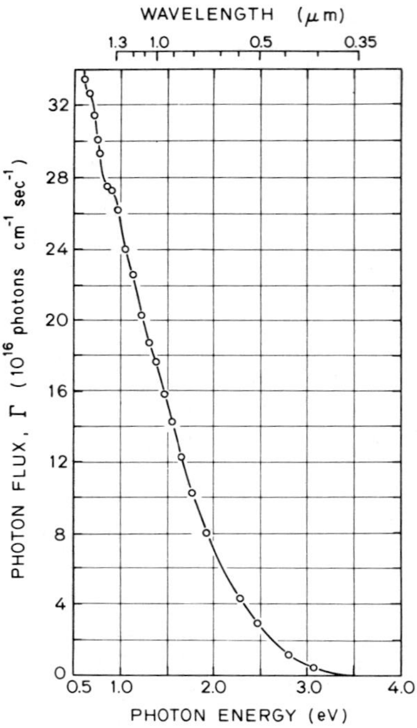

Fig. 6.3. AM1.5 solar photon flux distribution integrated with respect to photon energy versus lower integration limit $[\Gamma(E) = \int_E^\infty (d\Gamma/dE)\,dE$ versus $E = hc/\lambda]$. Total irradiance $P_s = 83.2$ mW cm^{-2}. Curve may be fitted approximately for $E > 0.9$ eV by $\Gamma(E) = A(B - E)^m$, where $A = 2.99 \times 10^{16}$ photons cm^{-2} sec^{-1} eV$^{-2.35}$, $B = 3.485$ eV, and $m = 2.35$. [Data from "Terrestrial Photovoltaic Measurement Procedures," NASA (prepared for ERDA) ERDA/NASA/1022-77/16, NASA TM 73702 (1977).]

Since the active, light-absorbing area of the cell $\mathscr{A}_A$ is likely to be slightly smaller than the total area seen by the diode $\mathscr{A}_D$, it is convenient to work in terms of currents:

$$I_L = J_L \mathscr{A}_A, \qquad I_0 = J_0 \mathscr{A}_D. \tag{6.5}$$

At the maximum power point $P_m = V_m I_m$, the derivative $dP/dV = 0$. A parameter β is defined as

$$\beta = (I_L' + I_m)/I_L', \tag{6.6}$$

where $I_L' = I_L + I_0 \simeq I_L$. This is the ratio of the current passing in the forward direction through the diode biased at V_m in the dark to I_L. By convention, $I_L > 0$ and $I_m < 0$. From the equation obtained by setting $dP/dV = 0$ for Eq. (6.1) it can be shown that

$$\beta = (I_0/I_L') \exp(qV_m/AkT) = 1/[(qV_m/AkT) + 1]. \tag{6.7}$$

Equation (6.7) can be rearranged to eliminate (qV_m/AkT) and obtain

$$\beta = [1 + \ln(\beta I_L'/I_0)]^{-1}. \tag{6.8}$$

Only a few iterations are required to obtain β to four significant figures. For good cells, β is a slowly varying constant such that $0.04 < \beta < 0.10$.

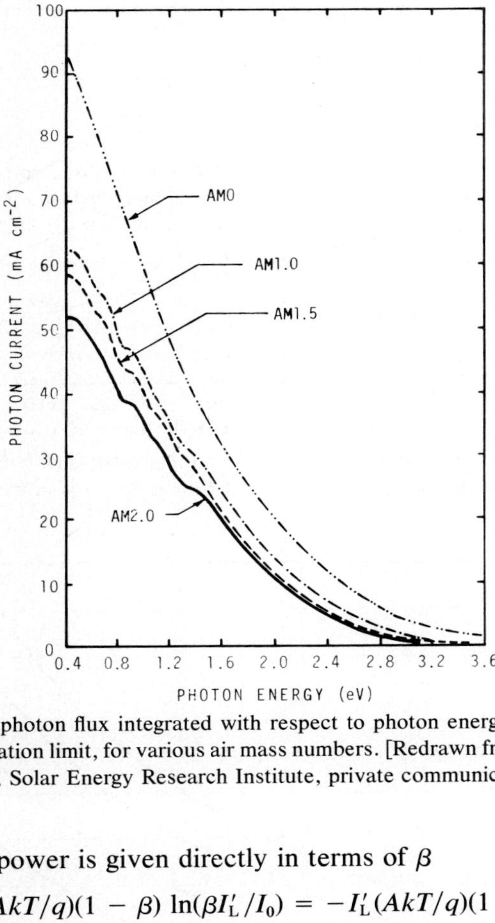

Fig. 6.4. Solar photon flux integrated with respect to photon energy (multiplied by q) versus lower integration limit, for various air mass numbers. [Redrawn from curves supplied by K. W. Mitchell, Solar Energy Research Institute, private communication (1979).]

The maximum power is given directly in terms of β

$$P_m = -I'_L(AkT/q)(1 - \beta)\ln(\beta I'_L/I_0) = -I'_L(AkT/q)(1 - \beta)^2/\beta, \quad (6.9)$$

and the fill factor can be obtained by dividing by $I_{sc}V_{oc}$ (since for $R_s = 0$, $-I_{sc} = I_L \simeq I'_L$):

$$\text{ff} = P_m/V_{oc}I_{sc} = (1 - \beta)\ln(\beta I'_L/I_0)/\ln(I'_L/I_0). \quad (6.10)$$

From this result, it is clear that, for an ideal cell without resistance losses, the fill factor is *not* a function of the diode factor†; the diode factor A is just a scaling factor for the voltage.‡ On the other hand, P_m and hence η_s

† Series or parallel resistance does introduce a small A dependence on ff. The extent of this dependence for nonzero values of R_s is shown by Pulfrey (1978).

‡ Ghosh *et al.* (1980) give an alternative approximate expression for ff for $R_s = 0$:

$$\text{ff} \simeq \left\{1 - \frac{1}{\ln(I_{sc}/I_0)}\right\}\left\{1 - \frac{\ln[\ln(I_{sc}/I_0)]}{\ln(I_{sc}/I_0)}\right\}.$$

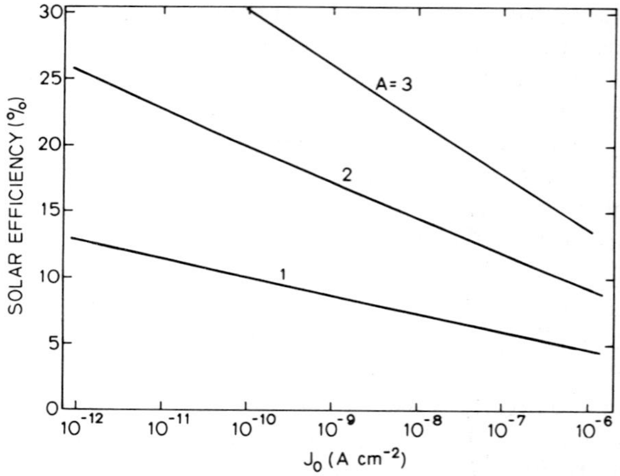

Fig. 6.5. Ideal solar efficiency versus J_0 for various diode quality factors assuming constant J_L. Calculated for $J_L = 25$ mA cm^{-2}, solar input $P_s = 100$ mW cm^{-2}, and $R_s = 0$.

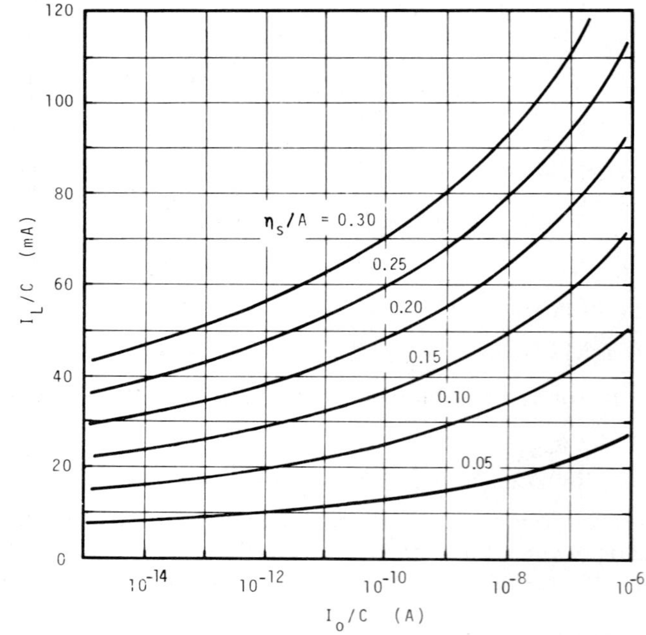

Fig. 6.6. Solar conversion efficiency η_s versus I_L, I_0, and A for an ideal cell with $R_s = 0$. Temperature is 300°K and the incident solar power density is $P_s = 100$ mW per unit sun (i.e., for a concentration ratio $C = 1$). The running parameter is η_s/A, where A is the diode factor. [Redrawn from K. W. Mitchell, Solar Energy Research Institute, private communication (1979).]

are directly proportional to A. The η_s for such an ideal diode is shown in Fig. 6.5. The relationship of the photovoltaic parameters is shown in a more general way in Fig. 6.6, where a solar concentration ratio can also be accounted for.

From these results, it might seem that the best way to increase η_s would be to raise A or decrease J_0. However, Fig. 5.29 shows that J_0 increases almost exponentially with increasing A. Two constant η_s curves, calculated using Eq. (6.9) and shown in Fig. 5.29, demonstrate this relationship and show that maximizing η_s is more subtle than merely decreasing J_0 or increasing A.

No treatment of photovoltaics is complete without at least some comment on maximum theoretical efficiency. The relation between theoretical efficiency and semiconductor band gap for homojunction cells was established early by a number of researchers including Prince (1955), Loferski (1956), Wolf (1960), and others, using rather ideal material properties and

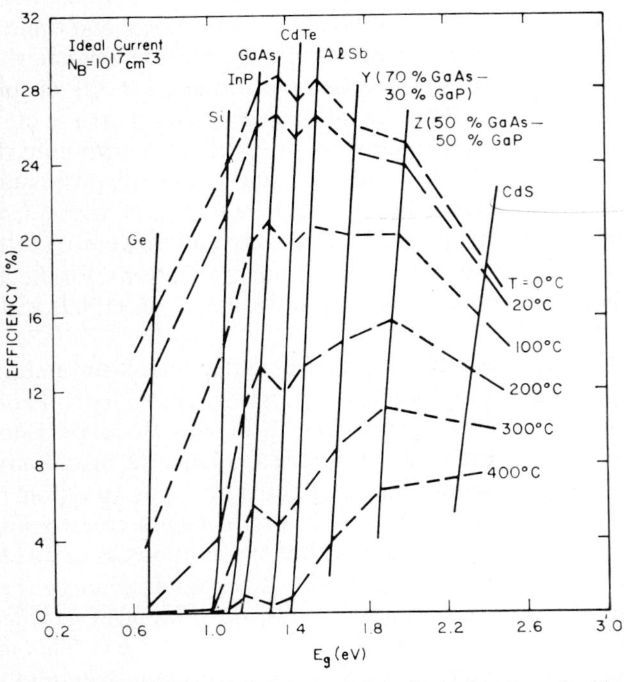

Fig. 6.7. Theoretical solar efficiency versus semiconductor band gap for ideal homojunction cells ($A = 1$) with no surface recombination losses. The variation with temperature is also shown. For AM0 solar input. [Figure from S. M. Sze, "Physics of Semiconductor Devices," p. 644, Wiley, New York (1969) (redrawn by Sze after J. J. Wysocki and P. Rappaport, *J. Appl. Phys.* **31,** 571 (1960).]

idealized junction models. Although the magnitudes of the predicted η_s are quite dependent on the parameters chosen, the maxima of the curves lie close to 1.4–1.5 eV (Fig. 6.7). For higher temperatures the maxima move to higher band gaps. These curves represent the compromise between the number of solar photons usefully absorbed by the cell (which decreases as E_g increases) and the thermally activated forward-bias diode current (which decreases as E_g increases, thereby increasing V_{oc}).

Wolf (1971) shows a set of theoretical efficiency calculations for the Si cell for various assumptions about the materials and device properties. For example, he obtains $\eta_s = 22\%$ and $V_{oc} = 0.788$ at AM0 (for $\tau_n = 10$ μsec and $J_0 = 2.4 \times 10^{-15}$ A cm^{-2}). The effects of variation of minority carrier lifetime on η_Q, ff, V_{oc}, and η_s are given by Graff and Fischer (1979) for Si-based cells.

Theoretical efficiency predictions for heterojunction solar cells are difficult to make because of lack of knowledge about junction transport. In many heterojunctions the dominant mechanism is recombination/generation in the depletion layer, but the transport may be modified so much by interfacial recombination, band-edge discontinuities, and tunneling that J_0 and A cannot be predicted. However, if we naively assume that junction transport is dominated by injection and diffusion in the absorber quasi-neutral region ($A = 1$), then the theoretical solar efficiency curves shown in Fig. 6.8 can be readily calculated. No useful absorption in the window layer and reasonably high values of τ and μ have been assumed. These considerations again favor an absorber band gap of about 1.4 eV, since both absorption and the principal diode current are determined by the absorber in this example. More elaborate predictions for heterojunction solar efficiency have been made by Sreedhar *et al.* (1969) and Sahai and Milnes (1970).

Since the predicted efficiencies are so dependent on material and device properties, it is of interest to determine a *maximum* theoretical efficiency based on broader principals. This was done by Shockley and Queisser (1961)† on the basis of detailed balance of blackbody radiation fluxes between the sun and the solar cell and the assumption, "that each photon (entering the cell) with $h\nu > E_g$ produces one electronic charge q at a voltage of $V = E_g/q$." For the further assumptions of unit solar concentration ratio and that the only mode of recombination in the semiconductor was radiative and band-to-band, they found a *detailed balance limit* of 30% for an optimal band-gap energy of 1.1 eV. This is curiously close to the band-gap energy of Si—the most abundant solid element.

† The discussion section of this paper points out several of the problems facing photovoltaic science (such as discrepancies between observed and predicted A factors, J_0 values, and V_{oc} values) that are still unresolved.

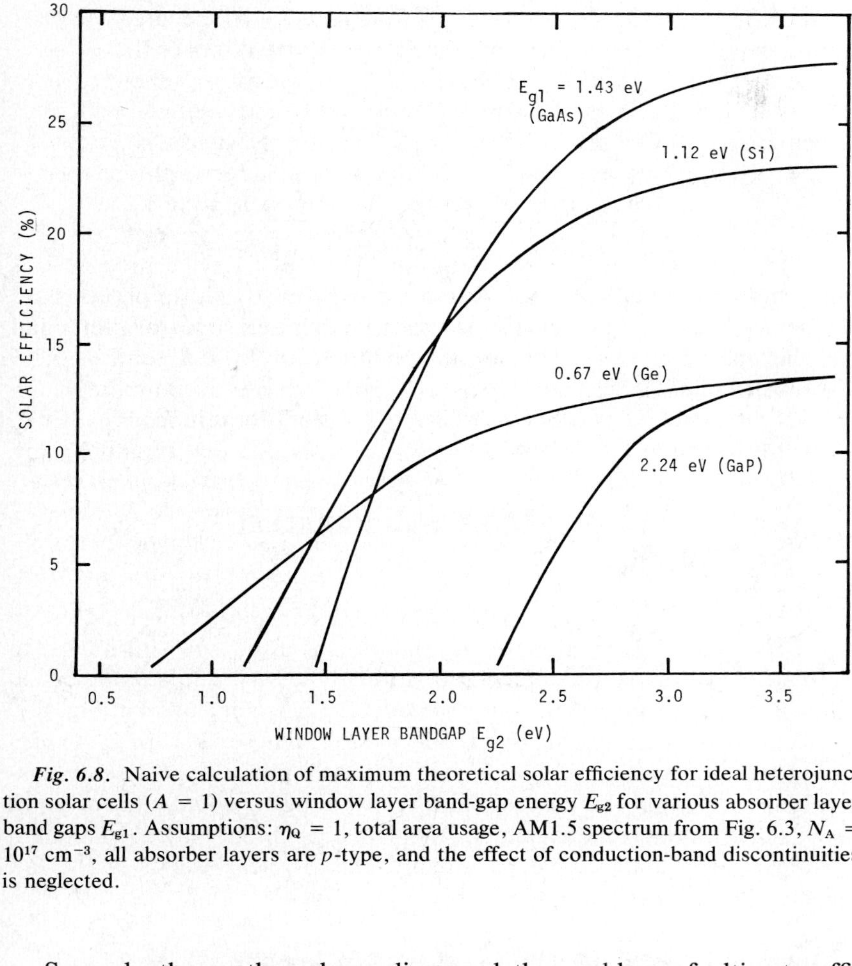

Fig. 6.8. Naive calculation of maximum theoretical solar efficiency for ideal heterojunction solar cells ($A = 1$) versus window layer band-gap energy E_{g2} for various absorber layer band gaps E_{g1}. Assumptions: $\eta_Q = 1$, total area usage, AM1.5 spectrum from Fig. 6.3, $N_A = 10^{17}$ cm^{-3}, all absorber layers are p-type, and the effect of conduction-band discontinuities is neglected.

Several other authors have discussed the problem of ultimate efficiency, including Mathers (1977), Landsberg (1977), and De Vos (1980). De Vos, using arguments similar to those of Shockley and Queisser, found a detailed balance limit of $\eta_s = 40\%$ for a single cell under the theoretical maximum solar concentration ratio $C = 4.6 \times 10^4$ (see Section 12.1.1). For an infinite number of stacked, tandem cells (Section 12.2.4), they found a limit of $\eta_s = 68\%$ for $C = 1$ and $\eta_s = 87\%$ for $C = 4.6 \times 10^4$.

Using Carnot efficiency arguments, Rose (1960) found an ultimate limit for the open-circuit voltage of $V_{oc} = (E_g/q)[1 - (T_c/T_R)]$, where the cell temperature is T_c and the temperature of the blackbody radiator illuminating the cell is T_R. This gives $V_{oc} = 1.07$ V for Si. The implicit assump-

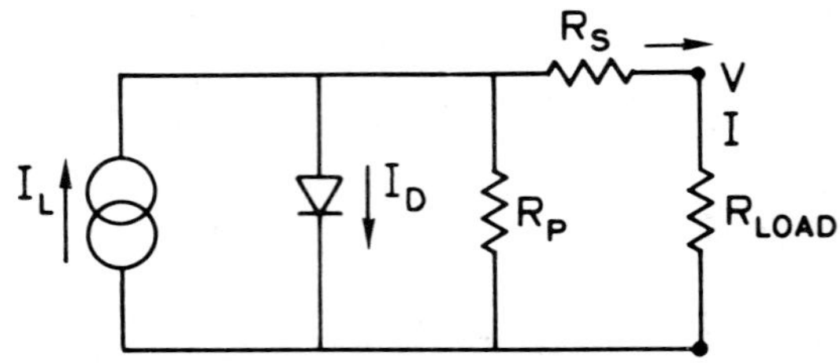

Fig. 6.9. Simplified equivalent circuit for a solar cell.

tions are that the cell–radiator system is reversible, that no photons are lost from the cell–radiator system (implying maximum concentration), and that the exchange is nearly monochromatic with $E_g < h\nu < E_g + kT_c$. The results of Shockley and Queisser (1961) indicate an ultimate V_{oc} = 0.83 V (they used E_g = 1.09 eV for Si; if this is adjusted to E_g = 1.11 eV, V_{oc} = 0.85 V).

6.2 THE EFFECTS OF SERIES AND PARALLEL RESISTANCE

In any real cell $R_s > 0$ and $R_p < \infty$, resulting in some power losses. In this section the contributions to R_s and R_p are discussed and means to evaluate their effects are considered. In many devices, it is sufficient to lump R_s and R_p contributions into the equivalent circuit shown in Fig. 6.9, i.e., a current source shunted by a diode and R_p, all in series with R_s, to produce a terminal voltage V and current I. To obtain more precise evaluation, particularly in the case of thin, resistive films in the current path, various distributed resistance models must be used that may be evaluated by numerical or analytic means. Finally, the physical origins of R_s, R_p, and the contact resistance are briefly considered.

6.2.1 First-Order Evaluation of R_s and R_p Losses

The effect of simple R_s on a cell can be illustrated by the graphical addition, at constant current, of the $I-V$ characteristic of the cell and a resistance characteristic with slope $1/R_s$ in Fig. 6.10. It is seen that V_{oc} is unchanged by simple R_s and that I_{sc} is changed very little unless R_s is quite large, in which case the cell characteristic approaches $1/R_s$ and ff approaches 0.25. Similarly, the characteristic for simple R_p can be obtained by addition, at constant voltage, of the zero resistance $I-V$ curve and $1/R_p$. It is seen in this case that I_{sc} is unchanged whereas V_{oc} may be changed slightly.

An approximate limit on R_s for small power loss can be obtained by

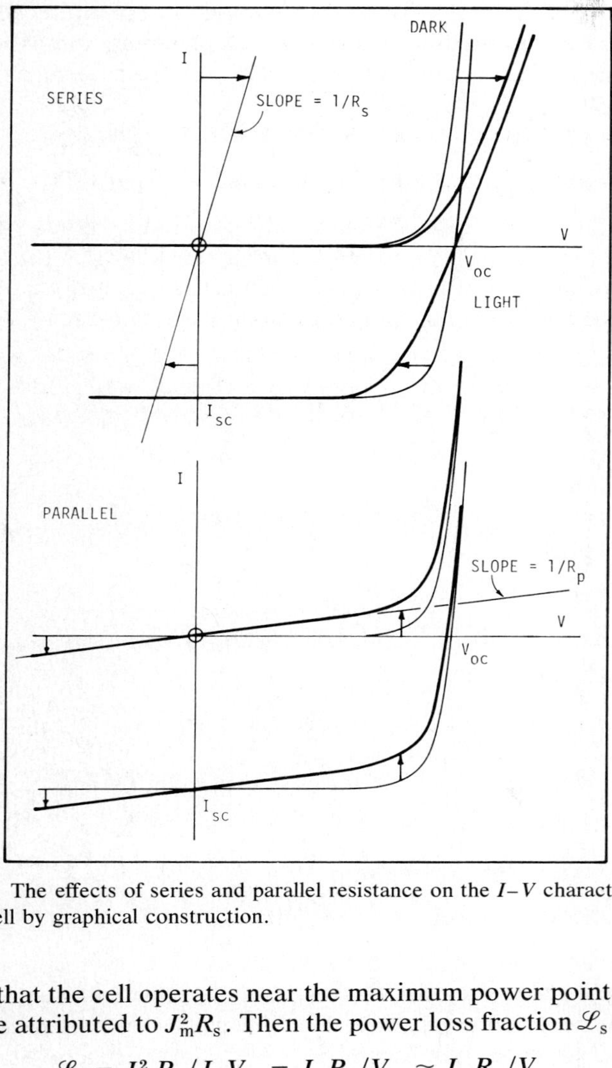

Fig. 6.10. The effects of series and parallel resistance on the I–V characteristics of an ideal solar cell by graphical construction.

assuming that the cell operates near the maximum power point and all the loss can be attributed to $J_m^2 R_s$. Then the power loss fraction $\mathscr{L}_s$ is given by

$$\mathscr{L}_s = J_m^2 R_s / J_m V_m = J_m R_s / V_m \simeq J_{sc} R_s / V_{oc}. \tag{6.11}$$

For a 3% loss at $J_{sc} = 40$ mA cm^{-2} and $V_{oc} = 0.6$ V, R_s must be less than 0.5 Ω for each square centimeter of cell area.

Similarly, the power loss fraction due to R_p is given by

$$\mathscr{L}_p = (V_m^2 / R_p) / J_m V_m = V_m / J_m R_p \simeq V_{oc} / J_{sc} R_p \tag{6.12}$$

and for a 3% loss due to R_p, R_p must be greater than 500 Ω for each square centimeter of cell area. These approximations are quite accurate for $\mathscr{L}_{s,p} < 5\%$. For the small-loss case V_{oc} and J_{sc} remain almost unchanged and the major effect is the reduction of ff, approximately by the factor $(1 - J_{sc}R_s/V_{oc} - V_{oc}/J_{sc}R_p)$.

The cell characteristic including R_s and R_p is given by

$$I = I_0\{\exp[q(V - IR_s)/AkT] - 1\} + (V - IR_s)/R_p - I_L. \quad (6.13)$$

With the addition of $R_s > 0$, superposition no longer holds and the light $J-V$ curve cannot be described by a simple translation of the dark curve.

The β formulation approach of Section 6.1 can be used to obtain accurate η_s and ff values for the simple lumped element circuit of Fig. 6.9. For the case involving only R_s,†

$$\beta = \left[1 + \frac{1}{1 + 2R_sI'_L\beta(q/AkT)} \ln(\beta I'_L/I_0)\right]^{-1} \quad (6.14)$$

with P_m given by

$$P_m = -(AkT/q)I'_L(1 - \beta)\ln(\beta I'_L/I_0) + R_sI'^2_L(1 - \beta)^2 \quad (6.15)$$

and ff by

$$\text{ff} = P_m/[I_{sc}(AkT/q)\ln(I'_L/I_0)]. \quad (6.16)$$

The term I_{sc} must remain in the denominator of Eq. (6.16) since for large R_s, I_L may be different from I_{sc} (which may be quickly found by iteration).

The variation of ff with J_L (or, equivalently, the light intensity), shown in Fig. 6.11 with the effects of series and parallel resistance, provides a valuable diagnostic tool for preliminary investigation of experimental cells.

6.2.2 Distributed Resistance Models

Most solar cells contain a thin front layer in which current is collected laterally by a grid structure. The resistance loss is distributed within the film and more accurate models must be used. An example of such a gridded cell structure is shown in Fig. 6.12. The series resistance of the device includes the following contributions:

- R_g Resistance of the front grid structure
- R_{c1}, R_{c2} Contact resistances (inversely proportional to contact area)

† Equation (6.6) still holds, but Eq. (6.7) is not applicable here.

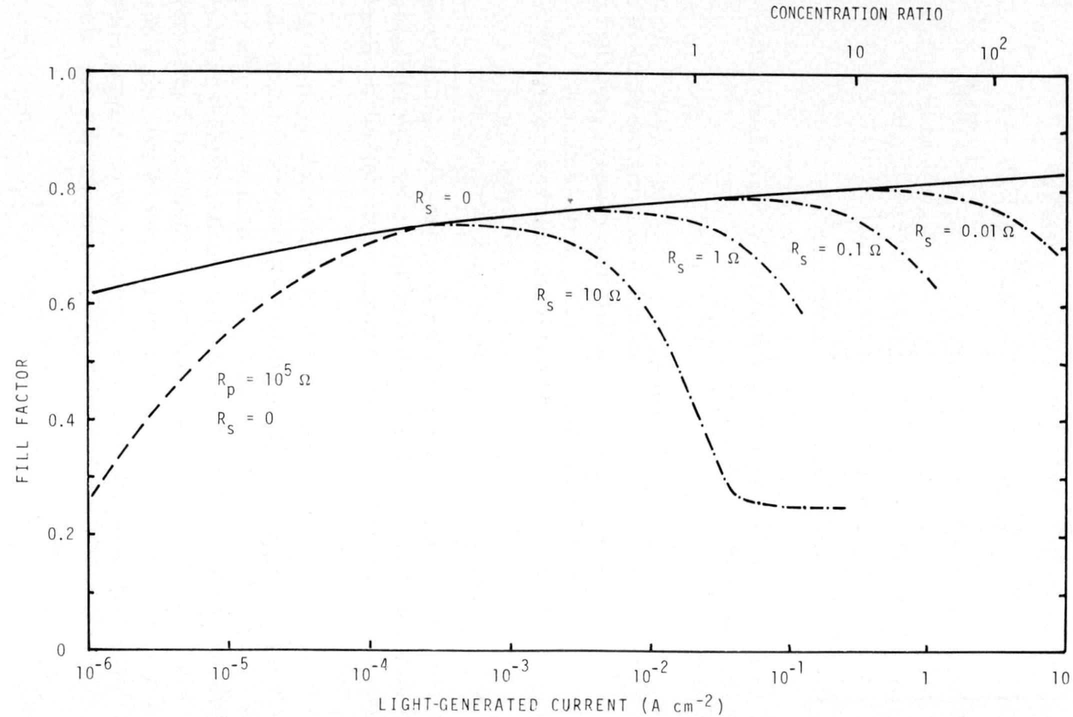

Fig. 6.11. Fill factor as a function of I_L showing the effect of R_s and R_p. Assumed: $I_0 = 10^{-9}$ A cm^{-2}, $A = 1$, total area active $= 1$ cm^2, and $R_p \to \infty$ for all cases except for the dashed curve, where $R_p = 10^5\ \Omega$.

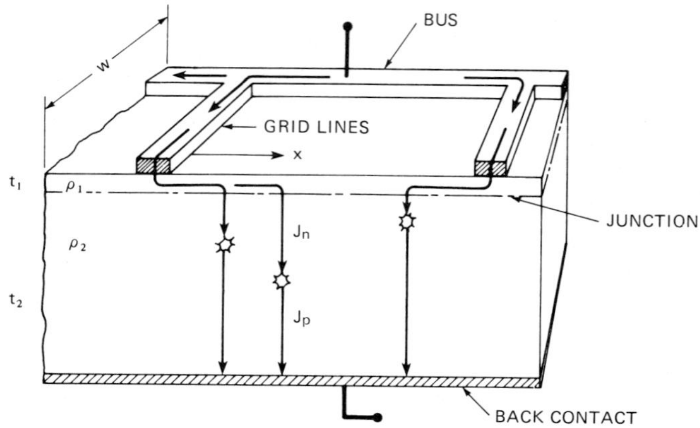

Fig. 6.12. Current flow paths in a gridded cell structure with the thickness of the front layer t_1 much less than that of the base t_2.

$R_1(x) = \rho_1 x / w t_1$ Lateral resistance to current flow along the plane of the film depending on the distance x (where ρ_1 is the bulk resistivity of the film and t_1 is its thickness)

$R_2 = \rho_2 t_2 / \mathscr{A}_D$ Transverse resistance through the base layer (of bulk resistivity ρ_2, thickness t_2, and area $\mathscr{A}_D$)

R_{s2} Spreading resistance of the back contact sheet.

Given the total allowed R_s, the cell designer can then allocate the contributions according to the material constraints of the device. The application of these ideas to the design of grid structures is discussed by Riemer (1978) and Surreze (1978).

The problem of distributed resistance may be handled approximately by various lumped element equivalent circuits, by analytic solutions using simplifying assumptions, or, more exactly, by computer-numerical solutions for finite element models. Wolf and Rauschenbach (1963) examined models in which R_s and R_p were lumped into second- and higher-order versions of the equivalent circuit shown in Fig. 6.9. Fang and Hauser (1978) and Heizer and Chu (1976) discuss solutions for two-dimensional structures, and Spaderna and Navon (1978) give results for three-dimensional structures at high concentration ratios.

The analytic approach can yield useful results for some simple geometries such as the one-dimensional example that follows. Current flow in the front layer is assumed to be in the plane of the layer as shown in Fig. 6.13 whereas the current flow in the base and at the junction is perpendicular to the cell area. Consider an elemental volume of the front layer

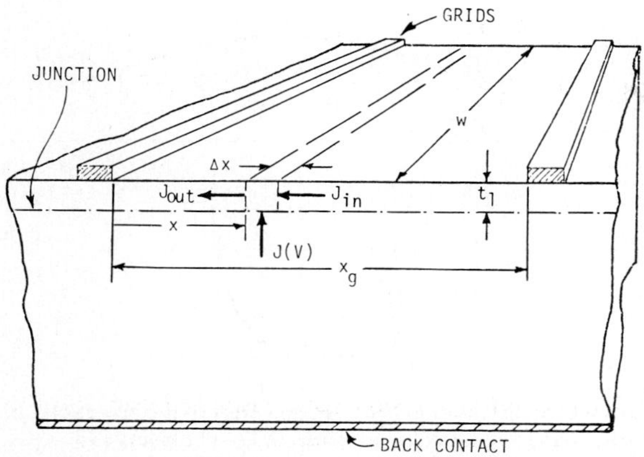

Fig. 6.13. Cross section of front gridded solar cell for analytical distributed resistance analysis.

$\Delta x t_1 w$ bounded by x and $x + \Delta x$. The lateral current density at the two boundaries is $J_{in} = (dV/dx)|_x/\rho$ and $J_{out} = (dV/dx)|_{x+\Delta x}/\rho$. The difference $J_{out} - J_{in}$ is balanced by the current flow through the junction plane $J(V)$ at the bias voltage considered:

$$J_{out} t_1 w - J_{in} t_1 w = J(V)\,\Delta x w.$$

The quantity $(dV/dx)|_{x+\Delta x}$ can be expanded in a Taylor series about x to obtain

$$d^2V/dx^2 = J(V)\rho/t_1. \qquad (6.17)$$

A solution of Eq. (6.17) can be obtained easily by assuming that $J(V) = J_m$ is a constant (the current at the maximum power point) giving a parabolic relation for $V(x)$, shown in Fig. 6.14.† If the resistance loss is not too great, the approximation is quite good. The distributed resistance power loss per unit area may be obtained directly in terms of the grid spacing x_g:

$$P_{loss} = J_m^2\, \rho x_g^2/12t_1 \qquad (6.18)$$

(or an "equivalent" series resistance of $R_s^{equiv} = \rho x_g^2/12t_1$). Such analytic results have been obtained by Handy (1975) and Wyeth (1977) for two-dimensional geometries.

By using finite element models for the treatment of the distributed re-

† A solution for Eq. (6.17) for the case of $J = J(V)$ is given by Wysocki (1961).

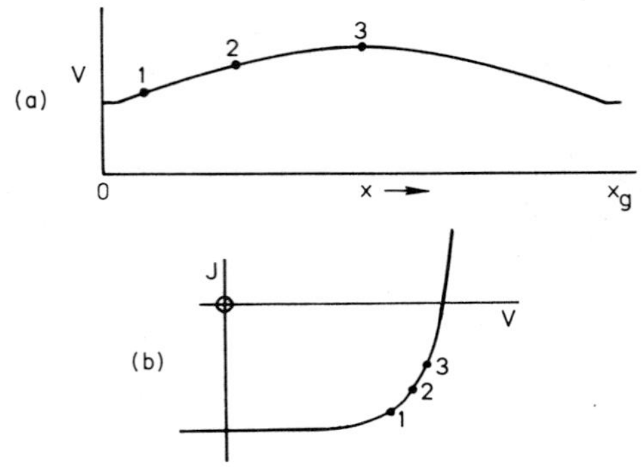

Fig. 6.14. (a) Voltage distribution between grid lines of the cell of Fig. 6.13, for operation near P_{max} and (b) corresponding voltage values on $J-V$ curve.

sistance problem, accurate results can be obtained for more complex diode relations and geometries, and for both distributed series and parallel resistances. The essense of this method is shown in Fig. 6.15, where the cell is first transformed into a long, half-grid-spacing-wide section, which is then divided into finite elements of width Δx. Since $x_g/2$ is a symmetry point, no current flows to the right side of the element labeled zero. By *assuming* a trial voltage $V_{(0)}$ for this element, the current through the element $I_{(0)}$ can easily be calculated, and then $V_{(1)}$ and $I_{(1)}$, and the succeeding values up to the terminals of the cell. By varying the trial parameter $V_{(0)}$, the terminal $I-V$ curve for the device can be generated for even the most complex diode characteristic. The model is readily extended to two dimensions as has been done by Mitchell (1977).

Calculations are done here for a hypothetical heterojunction cell with the following parameters:

$x_g = 0.10$ cm	$J_0 = 10^{-9}$ A cm^{-2}
$\mathcal{A}_D = 1$ cm^2	$A = 2$
$w = 1$ cm	$J_L = 0.03$ A cm^{-2}

all other R_s contributions $= 0.1\ \Omega$ cm^2

$t_1 = 2 \times 10^{-4}$ cm grid coverage $= 5\%$.

These calculations indicate a limiting bulk resistivity of about 0.2 Ω cm for a negligible power loss in the front layer, as shown in Fig. 6.16.

A result of this analysis is that V_{oc} can be reduced from its $R_s = 0$ value by appreciable distributed series resistance because of shunting by the portion of the cell diode structure that is shadowed by the grid. This is especially important for experimental cells when R_s may be quite large, and the grid shadowing area can be a large fraction of the active area of the cell.

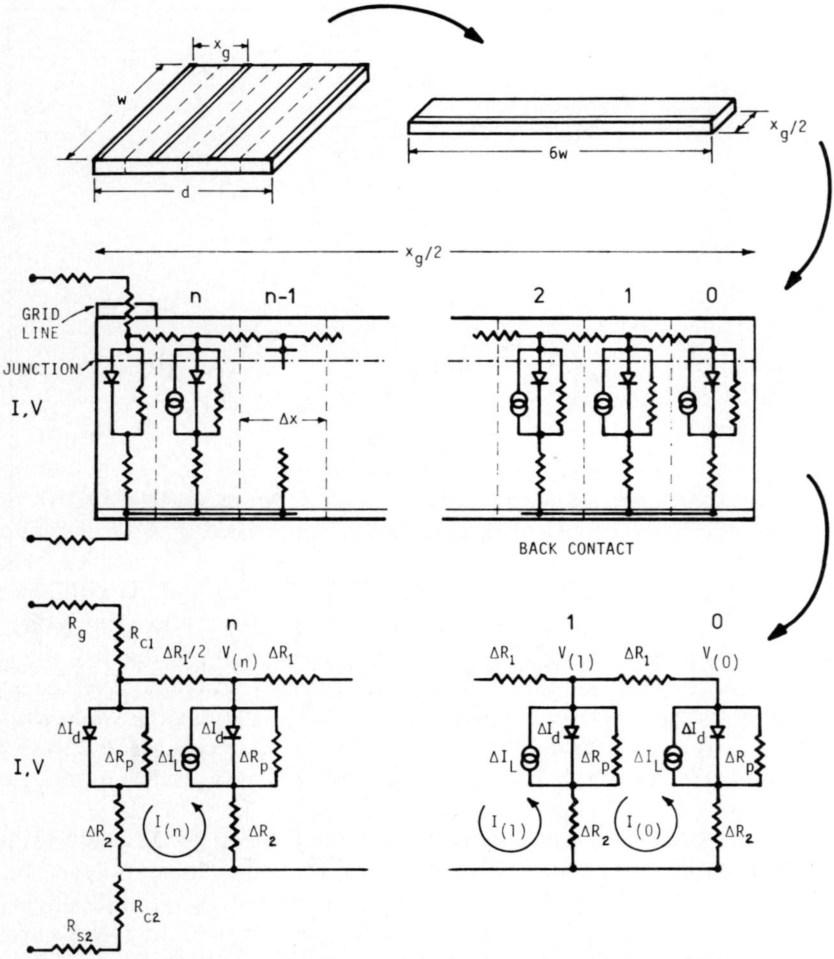

Fig. 6.15. One-dimensional, finite element, distributed parameter model.

6.2.3 Physical Origins of Series and Parallel Resistances

In this section we briefly comment on the physical basis for the contributions to series and parallel resistance listed in Section 6.2.2.

The bulk resistivity contributions of the emitter and base regions to R_s are quite different because of the geometrical factors involved. Thus, for negligible loss in the thin front layers, where current flow is along the layer, resistivities of $\rho_1 < 0.001-1\ \Omega$ cm are required, whereas for the

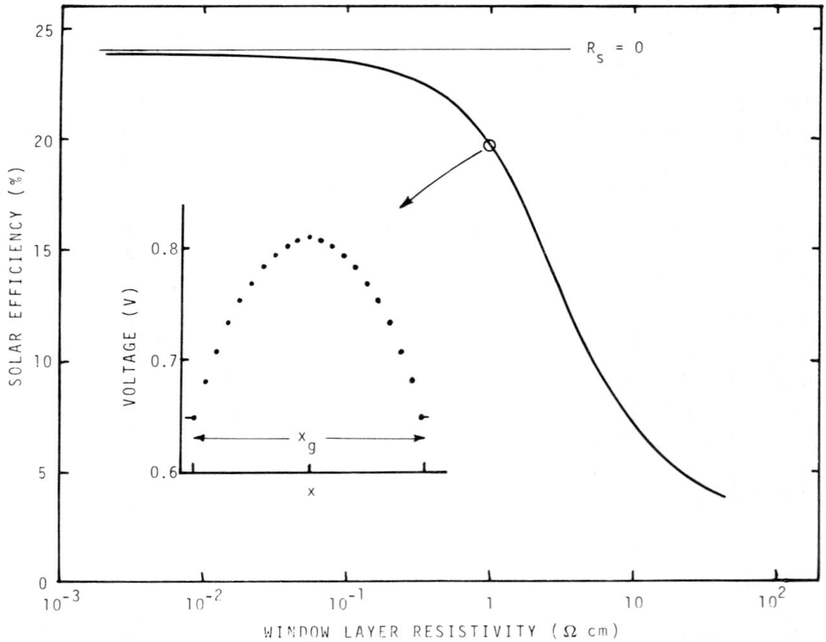

Fig. 6.16. Calculated solar efficiency versus window layer resistivity. Cell parameters given in text. Efficiency for total $R_s = 0$ is also shown. Inset shows voltage distribution between two grid lines for 1 Ω cm resistivity when cell is operating at the maximum power point.

base region, where current flow is through the layer, $\rho_2 < 25$–1000 Ω cm is sufficient. The thin front layers of heterojunction cells are polycrystalline in some cases so that the mobility is limited by grain-boundary barriers and the resistivity may be highly anisotropic. In these cases either very high carrier densities are required or a transparent conducting window layer must be applied. This is discussed in more detail in Chapter 9.

The bulk contribution from insulating layers within the cell (such as depletion layers or the insulating layer in a $p/i/n$ structure) is more complex. If the layer is thinner than the carrier diffusion length as modified by the local electric field (Section 4.5.1), then the resistance is essentially zero. For i-layers much thicker than L_n and/or L_p, the current transport is modified by space-charge-limited flow and, for high injected currents, a $J \propto V^2$ dependence occurs rather than an ohmic resistance. These cases are discussed by Rose (1964).

For high solar concentration ratios, the photogenerated carrier density can modulate the conductivity as *both* minority and majority carrier density are increased by photogeneration ($n \simeq p$). Under these conditions the larger the current generated, the greater the majority carrier density; the voltage drop across the cell may be limited to a few kT, independent of J_L.

If the grid lines on the front layer of a cell can be spaced less than a majority carrier diffusion length apart, then the transport is controlled by diffusion rather than drift and the voltage drop between grid lines is of the order of only kT.

The contact resistance R_c is perhaps the most overlooked (and frustrating) contribution to R_s in experimental cells. Indeed many "ohmic" contacts are, in reality, leaky Schottky diodes with nonlinear characteristics. Figure 6.17 shows the contribution of such a "quasi-ohmic" contact to the dark and light cell characteristics. In this case the "forward" characteristic of the contact is operative in the illuminated, power-generating mode. The theory of ohmic contacts is outlined in Section 5.8.

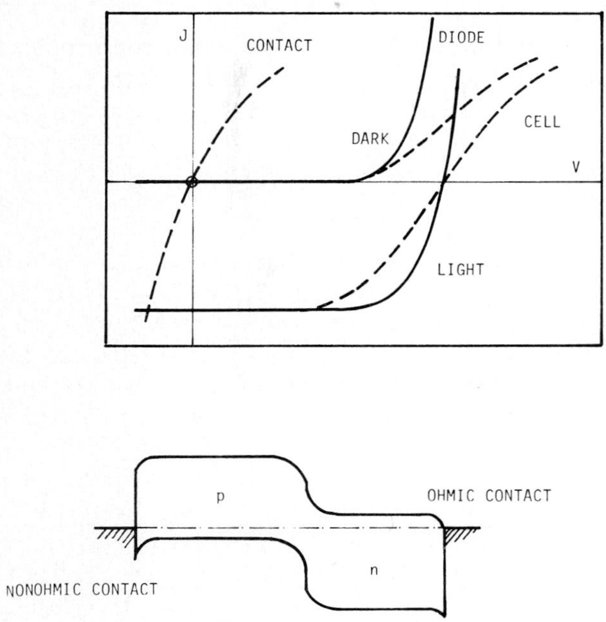

Fig. 6.17. The effect of a nonohmic contact on the $J-V$ characteristics of a solar cell. The dashed lines are the $J-V$ characteristic of the nonohmic contact alone and the resulting cell terminal characteristics.

Contact resistance can be measured by various schemes as discussed by Berger, (1972) and Cox and Strack (1967). The essence of one of these measurements is illustrated by the three-point technique shown in Fig. 6.18. Contact 3, connected to a high-impedance voltmeter, draws negligible current and thus the voltage drop across this contact is also negligible; contact 3 senses the potential just inside the semiconductor. Variation of the current through contacts 2 (and 2′) and 1 yields the $I-V$ characteristic of contacts 2 and 2′ and the contact resistance R_c. Such data must be interpreted with care since some contacting operations can leave a skin of highly conducting material on the surface of the semiconductor that essentially shorts contacts 2 and 3 and gives exaggeratedly small values of R_c. The specific contact resistivity is defined by

$$\rho_{c0} = dV/dJ|_{V=0} = R_{c0}\mathscr{A}_c \quad (\Omega\ \text{cm}^2), \tag{6.19}$$

where $\mathscr{A}_c$ is the contact area. Most contacts are only semi-ohmic so that $\rho_c = \rho_c(V)$. Since we are interested in the solar cell contact voltage drop ΔV during operation, it is convenient to use an effective contact resistivity given by

$$\rho_c = \Delta V/J_m,$$

where J_m is the current density at the maximum power point for the desired solar concentration ratio, as shown in Fig. 6.18.

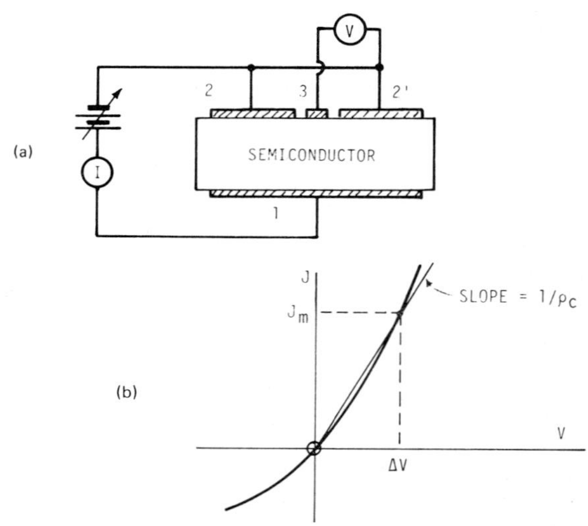

Fig. 6.18. (a) Three-point measurement of contact resistivity (of contacts 2 and 2′) with (b) resulting $J-V$ curve. As a working value for ρ_c, it is convenient to give the reciprocal of the slope of the line through the $J-V$ value at $J = J_m$(AM1) for a cell operating at AM1.

For negligible I^2R_s loss, the contact resistivity beneath grid lines usually must be of the order of less than 0.01 Ω cm² (for AM1 conditions and 7.5% grid coverage) whereas the large area contact at the rear of the cell can have $\rho_c \leq 0.2$ Ω cm².

The contributions ot R_p have been discussed in Section 5.2.5.

6.3 OTHER TREATMENTS OF THE CALCULATION OF SOLAR EFFICIENCY

6.3.1 Collection Function Analysis of the Light Current

In the foregoing examples the quantum efficiency η_Q was assumed to be independent of the applied bias voltage. However in many real cells, especially non-lattice-matched heterojunctions, η_Q is decreased by forward bias, resulting in reduced ff and V_{oc}. These cases are difficult to analyze because of the nonlinear equations involved and the lack of concrete information about the properties of the material in the depletion layer. Consideration of the variation of η_Q with bias as a perturbation on the ideal cell characteristics is a means of gaining perspective on the processes of photogenerated carrier transport without resorting to the solution of a nonequilibrium transport equation in the depletion layer. This approach assumes a linear superposition of J_{dark} and J_L in the depletion layer and is therefore good only to first order. In many cases this approach simply involves including the bias- and/or wavelength-caused perturbation in each of the components and factors of the quantum efficiency expression. The result can be assembled into a factor called a collection function $H(\lambda,V)$ that multiplies the ideal cell J_L, giving a general expression for the $J-V$ characteristic:

$$J = J_0 \,[\exp(qV/AkT) - 1] - H(\lambda,V)J_{L0} \qquad (6.20)$$

where J_{L0} is the $H = 1$ value of the light current. J_{L0} can sometimes be obtained by measurement at large reverse bias.

The general effect of an $H < 1$ on the $J-V$ characteristics is shown in Fig. 6.19, where the value of the light-generated current collected is reduced with respect to that generated in the absorber, especially for forward bias. As a result V_{oc} is reduced slightly and there is a considerable reduction in ff. The reduction in J_{sc} is usually quite small in efficient cells. The collection function H can be measured by comparison of light and dark $J-V$ curves or by ac methods, but the experimenter must first verify that the diode parameters A and J_0 are relatively invariant with illumination, or at least account for such variations.

An example of the collection function analysis is that given by Mitchell *et al.* (1977) for a CdS/CdTe cell where a considerable portion of the

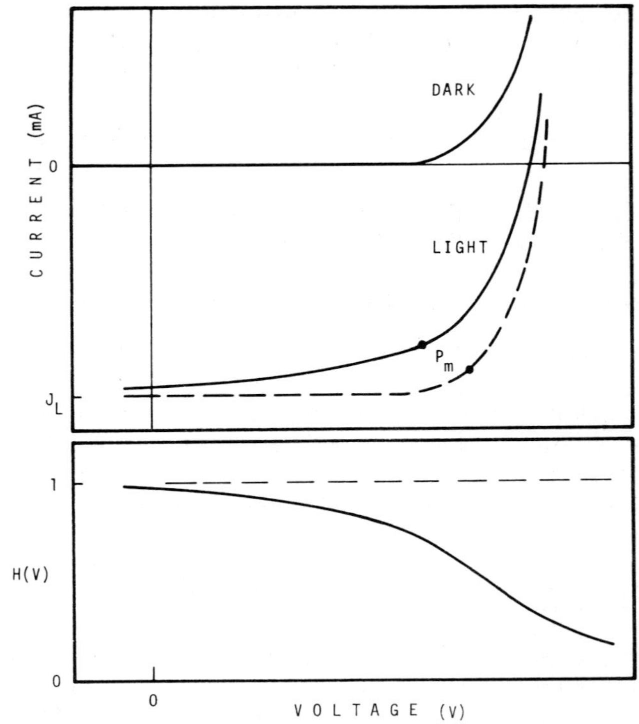

Fig. 6.19. Schematic $J-V$ curves showing the effect of a collection function. Dashed curve is the dark $J-V$ curve displaced by the large reverse bias value of J_L.

useful absorption occurs in the CdTe side of the depletion layer. No appreciable useful absorption takes place in the CdS window layer, and J_0 and A are almost independent of light intensity. The collection function H is a product of two factors; the first, $g(\lambda, V)$, represents the bulk processes of absorption and recombination, and the second, $h(V)$, represents the bias dependence of interfacial recombination loss. An example of a linear formulation for the collection of photogenerated carriers from the thick, quasi-neutral region is (see Fig. 5.1 for coordinates):

$$g_1(\lambda, V) = \exp[-\alpha(\lambda)W_d(V)]/[1 + 1/\alpha(\lambda)L_n] \qquad \text{for } x_p < x < \infty. \tag{6.21}$$

A second term for collection in the depletion layer of width $W_d(V)$, where the carriers are assumed to be swept out by the junction field without suffering recombination loss, is

$$g_2(\lambda, V) = \{1 - \exp[-\alpha(\lambda)W_d(V)]\} \qquad \text{for } 0 < x < x_p, \tag{6.22}$$

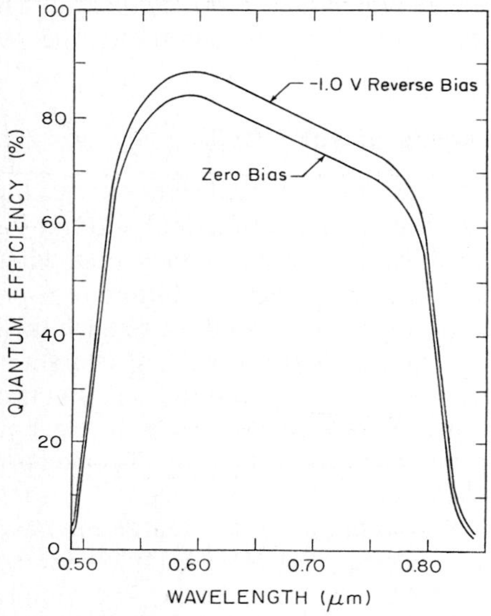

Fig. 6.20. Measured (and smoothed) spectral response for a CdS/CdTe heterojunction prepared by vacuum evaporation of CdS onto single-crystal *p*-CdTe, at applied biases of zero and −1 V, reverse. Data are corrected for 13% reflection loss. A value of $S_i = 2 \times 10^6$ cm sec^{-1} was deduced from these data. [From K. W. Mitchell, A. L. Fahrenbruch, and R. H. Bube, *J. Appl. Phys.* **48,** 4365 (1977).]

where $x = 0$ is the interface position and $g = g_1 + g_2$. The depletion layer width $W_d = |x_p - x_n|$ is almost all in the CdTe for a heavily doped window layer and $W_d(V) \simeq [2\epsilon_s(V_d - V)/qN_A]^{1/2} \simeq x_p$.

In this case, the recombination of photogenerated carriers at centers close to the disordered interface of the CdS/CdTe heterojunction was not a function of wavelength, and Mitchell *et al.* (1977) made use of an approximate interfacial collection function $h(V)$ due to Rothwarf (1976).

$$h(V) = 1/[1 + (S_i/\mu\mathscr{E})], \tag{6.23}$$

where the electric field at the interface is $\mathscr{E} \simeq 2(V_d - V)/W_d(V)$, S_i is the interfacial recombination velocity, and μ is the local mobility. The spectral response of Mitchell's cell in Fig. 6.20 shows no appreciable variation of the ratio $H(V = 0)/H(V = -1)$ over most of the wavelength range, indicating that in this range $h(V)$ is constant with wavelength. The shape of the curves depends on wavelength and is almost independent of bias voltage in this range, so $g = g(\lambda)$ and $g \neq g(V)$. Thus this analysis argues that the observed variation in H with voltage over most of the wavelength range is due only to bias-dependent interface recombination losses occurring in regions that are thin compared to the absorption length $1/\alpha \simeq 0.2\ \mu m$.

6.3.2 Holistic Analysis of Solar Cells

For more complex systems, a global treatment of the entire solar efficiency calculation with computer-numerical solutions may be required. An example of such a system is a high-intensity cell where a high injection condition exists and the minority carrier lifetime is not constant with carrier density or position. Forward bias may influence the lifetime in the absorber quasi-neutral region and alter the $J-V$ characteristics as well. In this case, one must return to a solution of the transport equations (perhaps simultaneous solution of the electron and hole equations with Poisson's equation) using a variable lifetime. The cell is divided into finite elements, and a relaxation solution is sought for each bias voltage in order to generate an $I-V$ relation. Such treatments have been applied by Fossum (1976) and Spaderna and Navon (1978) for concentrator cells, Dunbar and Hauser (1976a, b) for Si cells, and Hutchby and Fudurich (1976; see also Sutherland and Hauser, 1976) for graded-gap AlGaAs/GaAs structures.

6.4 THE EFFECT OF TEMPERATURE AND ILLUMINATION ON CELL EFFICIENCY

The ranges of intensity and temperature encountered for space use of photovoltaics are $0.03 < P_{s0} < 2$ (where P_{s0} is the solar intensity at the earth's radius) and -125 to $+140$°C, corresponding to the orbits of Jupiter and Mercury. For terrestrial concentrator systems, the intensity may exceed $10^3 P_{s0}$, and the temperature is controllable to within 5–10°C above the heat sink temperature (which may be held at $T > 100$°C if extraction of thermal energy is desired).

6.4.1 Dependence of J_0 and A on Illumination Level

Some of the heterojunctions, MIS cells, and the Cu_xS/CdS cell in particular show a variation of the diode J_0 and/or A factor with illumination level and/or wavelength. This is particularly true when trapping centers at or near the junction are not in good thermal communication with the conduction or valence bands (i.e., "slow states") and can have their occupancy, and hence charge, changed by illumination. This effect has also been observed in Si cells to a more limited extent. The change in J_0 and A

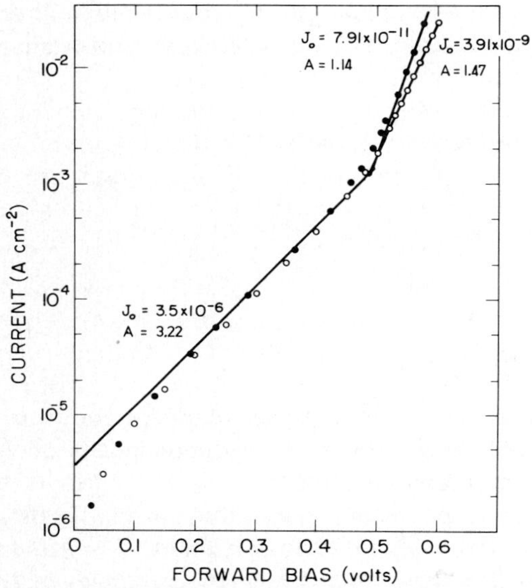

Fig. 6.21. Comparison of dark log J versus V data (open circles) with log J_{sc} versus V_{oc} data (filled circles) for a p/n junction Si cell. This cell (a NASA terrestrial reference standard) has a solar efficiency of 13.6% at AM1. Apparent J_0 and A values are shown.

can be due to optical absorption by states at the interface, in the bulk material near the junction, or in the insulating layer of MIS cells, any of which can produce changes in the junction profile. In the case of Cu_xS/CdS, the effect is caused by a change in ionized donor or acceptor density upon illumination, which in turn modifies the depletion layer width, the shape of the junction barrier, and, finally, the junction transport. Although the same impurity centers are responsible for photoconductivity in homogeneous CdS:Cu (CdS doped with Cu), with similar spectral and temperature response, the junction effects are not a result of simple photoconductivity in the Cu_xS/CdS case (Fahrenbruch and Bube, 1974).

Differences between dark log J versus V characteristics and a log J_{sc} versus V_{oc} plot, obtained by varying the light intensity, reveal a dependence of J_0 and/or A on illumination. Data for a Si homojunction cell are shown in Fig. 6.21. However, since the illumination level varies so widely in these measurements, it is usually impossible to discriminate between changes in J_0 or A or to quantify these effects. This method also cannot discriminate between changes in J_0 and/or A and a bias-voltage-dependent quantum efficiency $H(V)$. The log J_{sc} versus V_{oc} plot does remove the effect of small to moderate R_s from the characteristic however. In the extreme case of the Cu_xS/CdS cell, J_0 and/or A change sufficiently that the light and dark J versus V curves cross.†

More quantitative information about the dependence of J_0 and A can sometimes be obtained by comparison of the dark J versus V curve with the J versus V curve taken at constant illumination level (by adding J_L to the latter curve). The range of data for this measurement is limited to higher currents by the accuracy of the measurement of J_L however. The effect of $H(V)$ must be either negligible or accounted for in some way as well.

6.4.2 Thermal Behavior

The temperature dependence of the solar efficiency can be interpreted in terms of the individual temperature dependences of J_{sc}, V_{oc}, and ff (e.g., Wysocki and Rappaport, 1960).

The value of J_{sc} depends on temperature primarily through the terms involved in the minority carrier diffusion length: $L = (\mu kT\tau/q)^{1/2}$. If a cell has high quantum efficiency η_Q initially, then changes of L with temperature do not substantially affect η_Q [see Eq. (4.19)]; only in cells with low initial η_Q can changes in L appreciably affect J_L. With this caveat in mind, the effects of temperature on L and J_L are discussed.

The variation of minority carrier lifetime τ with temperature is complex, depending on the relative location of the energy levels of the recombination centers and the quasi-Fermi levels as well as the temperature dependence of the recombination cross sections of the centers themselves. As a simple example, consider a donorlike recombination center with energy level at E_i and a coulomb-attractive cross section σ_n that decreases with increasing temperature as T^{-2}. Thus, $\tau_{n0} = 1/N_r\sigma_n v_{th}$ varies as $T^{3/2}$. Using the Shockley–Read formulation of Eq. (3.56), calculation shows that the electron lifetime τ_n near 300°K is approximately τ_{n0} and increases slowly with increasing temperature until n_i approaches the value $N_A\tau_{n0}/\tau_{p0}$. For values of n_i exceeding $N_A\tau_{n0}/\tau_{p0}$, the strong temperature dependence of n_i causes τ to increase almost exponentially with T. The experimentally observed variation of τ_n for Si is a rather gradual increase with increasing T (approximately as $T^{+1.5}$ when T is close to 300°K and for low injection conditions (Othmer and Chen, 1978). Mathur *et al.* (1981) give additional experimental results.

The dependence of the mobility on temperature is determined by the dominant scattering mechanisms. For example, if acoustic scattering is

† By itself, $H(V)$ cannot cause crossing of the light and dark J–V curves since even if $H(V)$ approached zero, J could be no higher than the dark value.

dominant, then the mobility varies with T^{-m}, where $m = 1.5$ in the simplest case (Section 3.3). For Si, the observed variation of μ is as approximately T^{-2} for moderately doped material, and thus L varies quite slowly with temperature. In the case of GaAs, the increase in L with increasing T is somewhat stronger due to a larger variation of τ with T.

The small shift of the optical absorption edge energy with temperature produces changes in J_{sc} when convoluted with the solar spectrum. These changes are relatively minor [e.g., for Si, $(\Delta J_{sc}/J_{sc})/\Delta T \simeq 0.03\%/°C$].† A small increase in J_{sc} with increasing temperature also results from the change of the absorption coefficient itself with temperature (Shumka, 1970).

Experimentally, J_{sc} for most good Si cells is nearly constant, increasing slowly with temperature. An example is shown in Fig. 6.22.

The decreases in V_{oc} and ff with increasing temperature arise mainly from changes in n_i; the value of J_0 increases exponentially with decreasing $1/T$ causing V_{oc} to decrease almost linearly with increasing T. The variations shown by the following equations, written for one side of a homojunction with $A = 1$, serve to describe the general case:

$$V_{oc} = (kT/q)\ln(J_{sc}/J_0) \tag{6.24}$$

with

$$J_0 = (D/\tau)^{1/2} n_i^2/N_A = BT^3(D/\tau)^{1/2}\exp[-E_g/kT], \tag{6.25}$$

where the terms that are relatively temperature independent have been collected in the constant B. Then

$$V_{oc} = (E_g/q) - (kT/q)\ln[(D/\tau)^{1/2}T^3B/J_{sc}], \tag{6.26}$$

and the logarithmic term is larger than zero and varies slowly with temperature. An example is shown in Fig. 6.22.

The temperature variation of V_{oc} is often used to estimate the effective barrier height for experimental photovoltaic devices. For example, from Eq. (6.26) for one half of a Shockley diode, extrapolation of experimental V_{oc} versus T data to $T = 0°K$ should yield an intercept of the $T = 0°K$ value of E_g/q. Similarly, extrapolation of V_{oc} data for a heterojunction (e.g., with small band gap E_{g1}) or a Schottky diode (with barrier height ϕ_b), in which thermal activation dominates the current transport, yields approximately E_{g1}/q or ϕ_b, respectively. This is true even if the diode factor for the homojunction or heterojunction is greater than one, so long

† For Si, $dE_g/dT = -4 \times 10^{-4}$ eV/°C and, at $\simeq 1.1$ eV, the AM1.5 spectrum is such that $d\Gamma/dE \simeq 1.5 \times 10^{17}$ photons cm^{-2} sec^{-1} eV^{-1}. Spaderna and Navon (1978) discuss the influence of temperature variation on optical absorption.

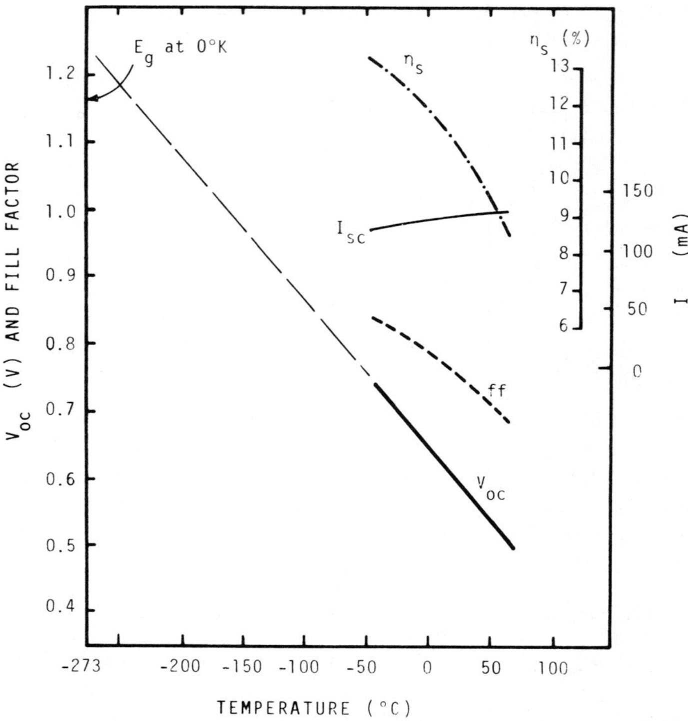

Fig. 6.22. Photovoltaic parameters for an experimental, 4-cm² Si cell versus temperature (for P_s = 140 mW cm^{-2}). The V_{oc} curve has been extrapolated to T = 0°K. [Data from R. K. Yasui and L. W. Schmidt, *Proc. 8th IEEE Photovoltaic Specialists Conf.* (1970), p. 110.]

as tunneling processes are not involved. The extrapolation to 0°K does *not* yield V_d in any case. For most simple transport models that do not involve tunneling, the 0°K extrapolation from data in the 77°K to 300°K range overestimates E_g/q or ϕ_b by at least 5%, because of the small temperature variation of the logarithmic term in Eq. (6.26).

In general, ff decreases quite strongly with temperature as shown in Figs. 6.22 and 6.23. The variation can be calculated using Eqs. (6.8) and (6.10), but the mathematics are too involved to be particularly enlightening.

For Si-based homojunction cells, the solar efficiency $\eta_s = V_{oc}J_{sc}\,\text{ff}/P_s$ reaches a maximum at $-150°$ to -100°C; near 25°C, the efficiency decreases at the rate of $\Delta\eta_s/\Delta T \simeq -0.05$ percentage points per Celsius degree, as shown in Fig. (6.22). A good comparison of the variation of theoretical and experimental photovoltaic parameters with temperature for Si cells is given by Arora and Mathur (1981).

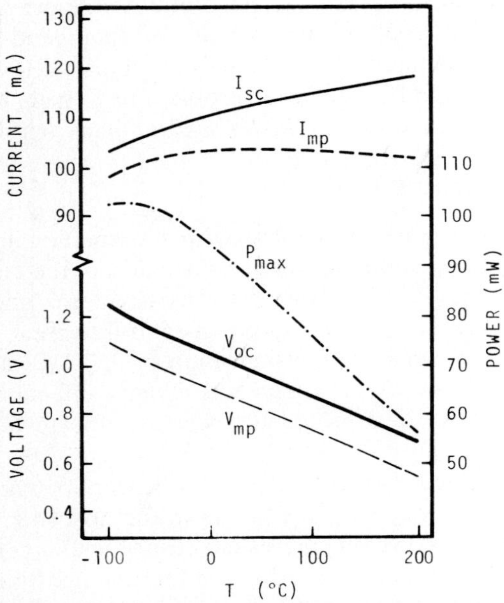

Fig. 6.23. Photovoltaic parameters for an AlGaAs/GaAs solar cell versus temperature ast AM0. At 25°C, η_s = 16.4%. [Data from "GaAs Concentrator Photovoltaic Power System Feasibility Investigation," AFAPL-TR-76-C-2142. Hughes Aircraft Co., Torrance, California, 1977. Figure redrawn from C. Stuerke, *Proc. 13th IEEE Photovoltaic Specialists Conf.* (1978), p. 551, © 1978 IEEE.]

For GaAs-based cells, the peak in η_s occurs at a higher temperature, $-100°$ to $-50°$C, because of the relatively stronger increase in L with temperature commonly seen in GaAs. The rate of decrease of η_s, in percentage points, is $\Delta\eta_s/\Delta T \simeq -0.033$ per Celsius degree near 25°C as shown in Fig. 6.23.

The temperature dependences of other devices where the current transport is mainly by thermal injection, such as heterojunctions and Schottky barrier cells, are qualitatively similar to the preceding example (Fischer-Colbrie *et al.*, 1976). For heterojunctions, η_s at elevated temperatures generally increases quite strongly with increasing band-gap energy of the smaller band-gap component as suggested by Fig. 6.5. Thus higher band gaps are more suited to concentrator and other high temperature systems.

In cases where the current transport is controlled at least partially by tunneling, such as in Cu_xS/CdS and MIS cells, the V_{oc} and ff might be expected to be somewhat more independent of temperature. The thermal variation of the photovoltaic parameters of MIS cells appears to be relatively unexplored at this time.

Low-temperature operation poses special problems: L, and hence J_{sc}, can decrease considerably if the mobility is dominated by charged impurity scattering. Contacts may become nonohmic at low temperatures, causing large losses in ff. The latter problem in Si cells has been largely overcome by the technology of Al-diffused p^+ back surface layers.

6.4.3 High-Intensity Effects

The variation of solar efficiency with illumination intensity is best treated holistically by computer analysis because of the complexity of the equations involved. This has been done by Spaderna and Navon (1978) for example. However, it is physically insightful to treat the variation of η_s with intensity by separate consideration of J_{sc}, V_{oc}, and ff, as in the previous section. It is found that series resistance effects on ff are usually more crucial than the rise in cell temperature due to high illumination levels.†

The light-generated current ($J_L \simeq -J_{sc}$) is proportional to photon flux Γ as long as the minority carrier lifetime in the absorber is constant. At higher photon fluxes however, increased carrier traffic begins to saturate the recombination centers, increasing the lifetime and thus producing an increase in quantum efficiency (Vasil'ev *et al.*, 1975). For the example of donorlike recombination centers with energy level $E_r = E_i$ in a p-type absorber, the lifetime is approximately constant ($\tau_n \simeq \tau_{n0}$) with increasing photogenerated carrier denisty until n_p approaches $N_A(\tau_{n0}/\tau_{p0})$, as indicated by Eqs. (3.45) and (3.56).‡ At electron densities exceeding this, the lifetime increases substantially, as shown in Fig. 3.15, producing a superlinear relation between J_L and Γ for cells with initially low η_Q. For cells with initially high η_Q, the change in η_Q with illumination level is small, e.g., because of the form of Eq. (4.19) and the superlinearity is negligible. Ho *et al.* (1977) and Ho and Mathias (1978) observed such a superlinearity of the J_L versus Γ relation for Si EFG ribbon cells. Their data correlated well with separate measurements showing an increase of the minority carrier diffusion length with increasing photon flux.

The open-circuit voltage increases with intensity as $\ln[J_L(\Gamma)/J_0]$ until, at high injection levels, more complex effects come into play. These effects include the voltage drop across the depletion layer at the junction, which becomes appreciable at high V_{oc}, and a change from $A \simeq 1$ transport to high injection, $A = 2$ transport as outlined in Section 5.2.5.

An interesting example of high-intensity effects in a $p/i/n$ junction is

† The thermal and electronic aspects of high-intensity operation are discussed in Chapter 12.

‡ For a coulomb-attractive, donorlike recombination center $\tau_{p0} \simeq 10^3\tau_{n0}$, for example.

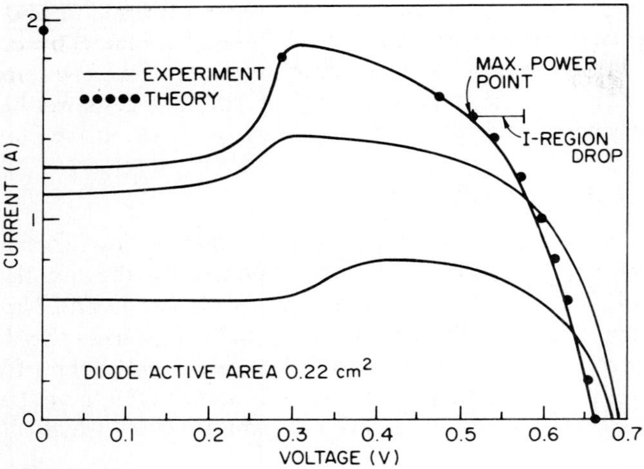

Fig. 6.24. Current–voltage characteristics for a Si *p/i/n* junction solar cell under high illumination. [Redrawn from R. Swanson, "Silicon Photovoltaic Cells in TPV Conversion," EPRI ER-633, Proj. 790-1, Interim Rep. (1978).]

shown in Fig. 6.24, where the lifetime in the front part of the absorber layer is thought to increase when the density of minority carriers there is increased by forward-bias injection. Thus, at forward bias, J_L actually appears to increase with respect to its zero bias value.†

For zero R_s, the fill factor ff increases slowly with intensity, as shown by Fig. 6.11. The effect of nonzero R_s on ff becomes crucial at high intensities. In order for the R_s loss to be negligible we must have $R_s < V_{oc}\mathscr{L}/J_{sc}$, where $\mathscr{L}$ is allowed fractional I^2R_s power loss given by Eq. (6.11). For AM1 conditions a rule-of-thumb value is 0.5 Ω cm² so, at higher intensities, $R_s \lesssim [0.5/C]\,\Omega\,\text{cm}^2$. For a $C = 1000$ system, the necessary $R_s < 0.0005\,\Omega\,\text{cm}^2$ is difficult to achieve and requires closely spaced grid lines, low contact resistivities, and/or special grid geometries.‡ The variation of ff with concentration ratio for several values of R_s is also shown in Fig. 6.11, assuming no thermal effects.

Given good heat sinking, η_s increases almost logarithmically with increasing intensity until high injection and series resistance effects cause

† R. Swanson, personal communication, Dep. Electr. Eng., Stanford Univ., Stanford, California (1979).

‡ Such special geometries include, for example, the stacked vertical multijunction cell, Section 7.6.1; and the V-groove cell, Section 12.2.3.

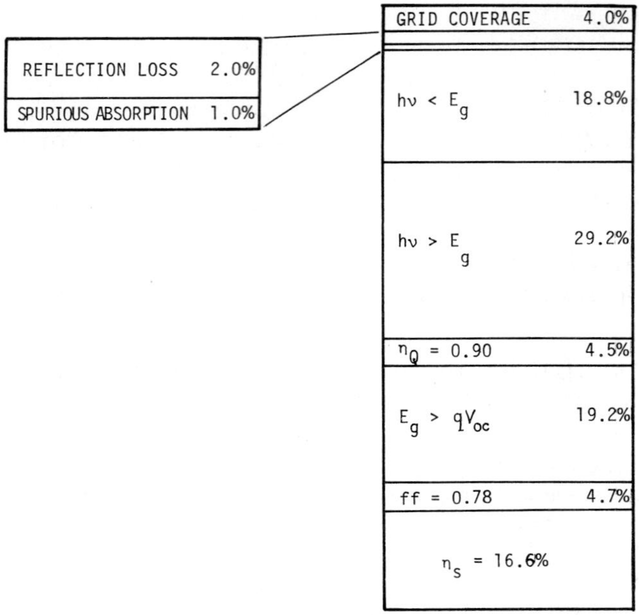

Fig. 6.25. Power loss chart for Si cell operating at AM1.5 (Fig. 6.3). Percentage of total input power P_s lost to each of the loss mechanisms is noted. The R_s loss is included in ff. Values of $J_0 = 5 \times 10^{-12}$ A cm^{-2} and $A = 1$ for the diode are assumed.

it to saturate or even decrease with increasing C (see Fig. 8.11, for example).

6.5 LOSS ANALYSIS

Energy loss analysis is a powerful tool for cell optimization; each loss mechanism can be put into proper perspective and attacked separately to optimize the efficiency of the device. Such an analysis, similar to that done by Wolf (1971), is shown in Fig. 6.25. The major contributions are listed here:

(1) $h\nu > E_g$ thermalization of hot carriers to near the band edge,

(2) $h\nu < E_g$ low-energy photons pass through the cell without pair generation,

(3) η_Q collection efficiency for photogenerated carriers,

(4) $qV_{oc} < E_g$ determined by diode parameters, principally thermalization of carriers over potential barrier,

(5) Fill factor (ff) loss in the diode forward-biased to V_m,

(6) R_s, R_p I^2R losses in series and parallel resistance (usually included in ff),

(7) Reflection loss reflection of photons at light-incident surfaces,

(8) Area factor grid coverage,

(9) Spurious absorption absorption in antireflection coating, at defects, etc.

Of these, the first two are unavoidable except by the use of graded gap structures, spectral splitting cells, or spectral conversion (see Chapter 12). Item 3, $qV_{oc} < E_g$, is a consequence of both the thermodynamic efficiency of the diode for separation of charge and of the diode parameters themselves.† Items 3 through 9 remain available for optimization.

REFERENCES

Arora, J. D., and Mathur, P. C. (1981). *J. Appl. Phys.* **52,** 3646.

Berger, H. H. (1972). *J. Electrochem. Soc.* **19,** 507.

Cox, R. H., and Strack, H. (1967). *Solid-State Electron.* **10,** 1213.

DeVos, A. (1980). *J. Phys. D.* **13,** 839.

Dunbar, P. M., and Hauser, J. R. (1976a). *Proc. 13th IEEE Photovoltaic Specialists Conf.*, p. 23.

Dunbar, P. M., and Hauser, J. R. (1976b). *Solid-State Electron.* **19,** 95.

Fahrenbruch, A. L., and Bube, R. H. (1974). *J. Appl. Phys.* **45,** 1264.

Fang, C. R., and Hauser, J. R. (1978). *Proc. 12th IEEE Photovoltaic Specialists Conf.*, p. 1306.

Fischer-Colbrie, E., Wichner, R., and Charlson, E. J. (1976). *Proc. 12th Photovoltaic Specialists Conf.*, p. 40.

Fossum, J. G. (1976). *Solid-State Electron* **19,** 269.

Ghosh, A. K., Fishman, C., and Teng, T. (1980). *J. Appl. Phys.* **51,** 446.

Graff, K., and Fischer, H. (1979). Carrier lifetime in silicon and its impact on solar cell characteristics. *In* "Solar Energy Conversion—Solid State Physics Aspects" (B. O. Seraphin, ed.), Topics in Applied Physics, Vol. 31, p. 173. Springer-Verlag, Berlin and New York.

Handy, R. J. (1967). *Solid-State Electron.* **10,** 765.

Heizer, K. W., and Chu, T. L. (1976). *Solid-State Electron.* **19,** 471.

Ho, C. T., and Mathias, J. D. (1978). *IEEE Trans. Electron Devices* **ED-25,** 1332.

Ho, C. T., Bell, R. O., and Wald, F. V. (1977). *Appl. Phys. Lett.* **31,** 463.

Hutchby, J. A., and Fudurich, R. L. (1976). *J. Appl. Phys.* **47,** 3140.

† For a Shockley diode, $qV_{oc} = E_g - kT \ln\{qN_cN_v[(D_n/\tau_n)^{1/2}/N_A + (D_p/\tau_p)^{1/2}/N_D]/J_L\}$, for example.

Landsberg, P. (1977). *Photochem. Photobiol.* **26,** 313. [Also see P. Landberg and J. Mallinson, *Proc. CNES-CNRS Int. Conf. Solar Electricity,* Paris: CNRS, 1976, p. 27.]

Lindholm, F. A., Fossum, J. G., and Burgess, E. L. (1976). *Proc. 12th IEEE Photovoltaic Specialists Conf.;* p. 33.

Lindholm, F. A., Fossum, J. G., and Burgess, E. L. (1979). *IEEE Trans. Electron Devices* **ED-26,** 165.

Loferski, J. J. (1956). *J. Appl. Phys.* **27,** 777. [Reprinted *in* "Solar Cells" (C. E. Backus, ed.), p. 96. IEEE Press, New York, 1976.]

Mathers, C. D. (1977). *J. Appl. Phys.* **48,** 3181.

Mathur, P. C., Sharma, R. P., Saxena, P., and Arora, J. D. (1981). *J. Appl. Phys.* **52,** 3651.

Mitchell, K. W. (1977). *Tech. Dig. IEEE Int. Electron Devices Meeting, Washington, D.C.,* p. 229.

Mitchell, K. W., Fahrenbruch, A. L., and Bube, R. H. (1976). *Solid-State Electron.* **20,** 559.

Mitchell, K. W., Fahrenbruch, A. L., and Bube, R. H. (1977). *J. Appl. Phys.* **48,** 4365.

Othmer, S., and Chen, S. C. (1978). *Proc. 13th IEEE Photovoltaic Specialists Conf.,* p. 1238.

Prince, M. B. (1955). *J. Appl. Phys.* **26,** 534. [Reprinted *in* "Solar Cells" (C. E. Backus, ed.), p. 89. IEEE Press, New York, 1976.]

Pulfrey, D. L. (1978). *Solid-State Electron.* **21,** 519.

Riemer, D. E. (1978). *Proc. 13th IEEE Photovoltaic Specialists Conf.,* p. 603.

Rose, A. (1960). *J. Appl. Phys.* **31,** 1640.

Rose, A. (1964). *J. Appl. Phys.* **35,** 2664.

Rothwarf, A. (1976). "Theoretical Prospects of the CdS/Cu_2S Solar Cell," Tech. Rep. NSF/RANN/AER 72-03478 A04/TR 76/1. Inst. Energy Convers., Newark, Delaware.

Rothwarf, A. (1978). *Proc. 13th IEEE Photovoltaic Specialists Conf.,* p. 1312.

Sahai, R., and Milnes, A. G. (1970). *Solid-State Electron.* **13,** 1289.

Shockley, W., and Queisser, H. J. (1961). *J. Appl. Phys.* **32,** 510.

Shumka, A. (1970). *Proc. 8th IEEE Photovoltaic Specialists Conf.,* p. 96.

Spaderna, D. W., and Navon, D. H. (1978). *IEEE Trans. Electron Devices* **ED-25,** 1290.

Sreedhar, A. K., Sharma, B. L., and Purohit, R. K. (1969). *IEEE Trans. Electron Devices* **ED-16,** 309.

Surreze, H. B. (1978). *Proc. 13th IEEE Photovoltaic Specialists Conf.,* p. 609.

Sutherland, J. E., and Hauser, J. R. (1976). *Proc. 12th IEEE Photovoltaic Specialists Conf.,* p. 939.

Tarr, N. G., and Pulfrey, D. L. (1979). *Solid-State Electron.* **22,** 265.

Vasil'ev, A. M., Evdokimov, V. M., Landsman, A. P., and Milovanov, A. F. (1975). *Geliotekhnika* **11,** 18. [Engl. transl.: *Appl. Sol. Energy* **11,** (1-2), 72 (1975).]

Wolf, M. (1960). *Proc. IRE* **48,** 1246. [Reprinted *in* "Solar Cells" (C. E. Backus, ed.), p. 118. IEEE Press, New York, 1976.]

Wolf, M. (1971). *Energy Convers.* **11,** 63. [Reprinted *in* "Solar Cells" (C. E. Backus, ed.), p. 191. IEEE Press, New York, 1976.]

Wolf, M., and Rauschenbach, H. (1963). *Adv. Energy Convers.* **3,** 455. [Reprinted *in* "Solar Cells" (C. E. Backus, ed.), p. 146. IEEE Press, New York, 1976.]

Wyeth, N. C. (1977). *Solid-State Electron.* **20,** 629.

Wysocki, J. J. (1961). *RCA Rev.* **22,** 57.

Wysocki, J. J., and Rappaport, P. (1960). *J. Appl. Phys.* **31,** 571.

CHAPTER 7

SILICON SOLAR CELLS

Among all the semiconductors, the properties of silicon are the most widely known, and conceptually, the silicon homojunction solar cell is perhaps the simplest of all photovoltaic devices. In 1981 the Si p/n junction cell was the only solar cell widely available commercially. The cost of Si cells was \$10–\$15 per peak watt in 1983 for terrestrial arrays, in spite of relatively inexpensive raw materials and a fairly well-developed technology.

This chapter begins with a short historic perspective on the development of the Si cell. The purification and crystal growth of Si are treated in some detail to illustrate the complexities and the wealth of options involved in obtaining a semiconductor material of high purity and low defect density. The electrical properties of several representative device types and a sample processing sequence are included to demonstrate the interaction between electrical parameters and processing. The chapter concludes with a section outlining the optimization of cell properties both by fine-tuning the parameters of "standard" cell designs and by innovative approaches of a more basic nature.

7.1 HISTORY

There was already an appreciation for future need when the Si solar converter was invented at Bell laboratories in 1953. At that time D. M. Chapin was investigating power sources for remote communication systems while C. S. Fuller was exploring solid-state diffusion as a means of producing large-area *p/n* junctions in Si. Then G. L. Pearson, in his investigations of power rectifiers made by Fuller's techniques, noticed the extreme sensitivity to light of these devices and matched solution to need. When Pearson's diode (a Li-diffused *n*-on-*p* structure) was tested in sunlight, a solar conversion efficiency of $\simeq 4\%$ was obtained—some five times more efficient than previously known Se cell converters. Because of the high atomic diffusivity of Li the devices were somewhat unstable, however, even at room temperature. Fuller developed a boron diffusion technology, and, in 1954, Chapin *et al.* (1954) published a description of a 6% efficient, stable Si solar cell. The first practical trial was a 9-W array powering a telephone repeater installed in Americus, Georgia, in 1955. During 6 months of use it was a technical success, but it proved to be an uneconomical power source for the intended use.

The Sputnik era provided the impetus to continue Si cell research; the first use of solar photovoltaic converters in space occurred on March 17, 1958, when Vanguard I was put into orbit. Its solar-powered radio transmitter operated for 8 years before radiation damage to the cells stopped its signals.

Since that time the cost of Si cells has rapidly decreased as the technology has developed (Fig. 7.1) and the efficiencies have risen to the 15–17% level for space cells (AM0). Until 1977, when GaAs cells were first used on a few Russian satellites, the only serious competitor for Si cells had been the Cu_xS/CdS all-thin-film cell, which promised increased radiation resistance but failed to reach high enough efficiency levels to penetrate the space market. With the advent of terrestrial photovoltaics the picture has again become more complex, and many devices, including GaAs concentrator systems and various thin-film cells, are being investigated as possible economical alternatives to the single-crystal Si cell.

One trend for commercial Si cells in the past few years has been to "fine-tune" orthodox devices, principally by decreasing short-wavelength losses in the front *n*-type layer, increasing the obtainable V_{oc} by manipulation of doping levels, and increasing resistance to radiation damage.

The detailed history of technological improvement of the Si cell is beyond the scope of this book; the interested reader is referred to articles by Wolf (1972) and Smits (1976). However, a few of the notable improvement milestones have been the "standard" *n/p* cell (η_s = 10.4% at AM0), the "violet cell" (14%), and the Comsat nonreflecting (CNR) cell (16%). A fourth major milestone was the use of a "back-surface electric field" (BSF) to decrease carrier loss at the back surface of the cell, an idea conceived in 1960 (Dale and Rudenberg, 1960) but practically demonstrated by Mandelkorn and Lamneck (1972).

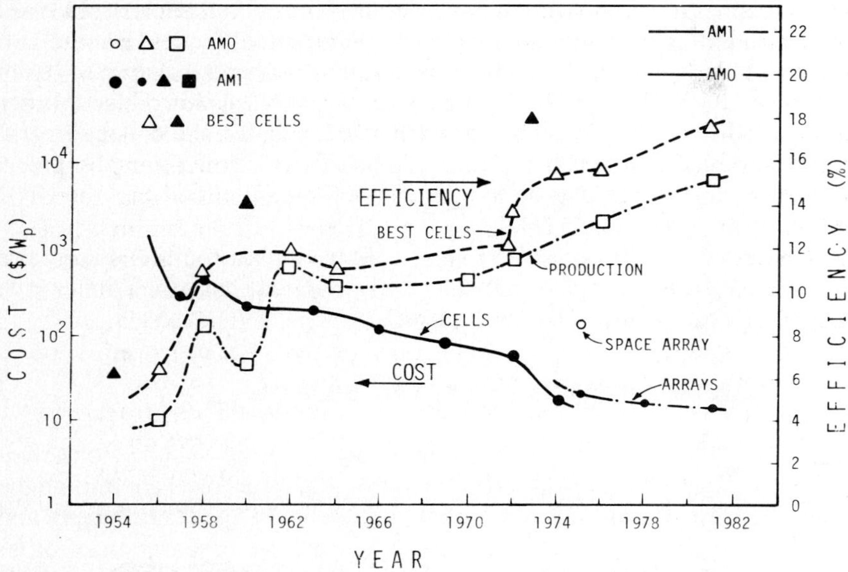

Fig. 7.1. Single-crystal silicon cell solar efficiency and cost per peak watt for cells and arrays over the years. $T \simeq 300°K$. Costs are given for the year in which they are quoted; inflation is not accounted for. Cost data are for terrestrial cells and arrays except for a 1975 datum for a space array. [Much of the early data is redrawn from Wolf (1972). The remainder is drawn from a variety of sources. The AM0 and AM1 efficiency lines at the top of the figure are practical maximum design goals by M. Wolf, *IEEE Trans. Electron Devices* **ED-27,** 781 (1981).]

The violet cell, developed at Comsat by Lindmayer and Allison (1973), serves to illustrate the direction of these improvements. The response to blue-violet light in this cell is substantially enhanced relative to its predecessors by decreasing the thickness of the diffused *n*-type front layer to 0.1–0.2 μm and by increasing the minority carrier lifetime there. In previous cells the diffusion of P to depths of $\simeq$0.5 μm was accompanied by the formation of a "dead layer" $\simeq$0.1 μm thick, where the P concentration approached the solubility limit in Si and the hole lifetime was about 10^{-10} sec. Dislocations propagating from the dead layer were thought to penetrate the depletion region, forming additional recombination centers there.† Although these latter centers did not appreciably affect J_L, they did increase the space-charge-region recombination current, increasing

† This view has been questioned and the "dead layer" is now thought instead to be due to phosphorus-vacancy pairs, various precipitates, and Auger recombination.

the diode loss at the maximum power point. Thus elimination of the dead layer also led indirectly to a considerable increase in ff and V_{oc} in the violet cell. To accommodate the increased range of spectral response (from 0.45 to 0.35 μm) an antireflection coating of Ta_2O_5 was introduced rather than the SiO_x or TiO_x coatings formerly used. Finally, the collection grid was redesigned to offset the reduced conductivity of the thinned *n*-layer. The next principal escalation in efficiency of the violet cell was the CNR cell described in Section 7.4.4.

Important reviews of silicon solar cell principles and technology are given by Brandhorst (1977) and Hall (1981). Calculations of the limits of Si cell efficiency are given by Wolf (1980).

7.2 GROWTH OF SINGLE-CRYSTAL SILICON

In this section we follow Si from the raw material—sand—to the finished crystal, ready for solar cell fabrication. Our purpose is to outline the procedure for a material in a rather mature stage of development and therefore to gain some perspective on the possible preparation of other less-well-researched materials.

7.2.1 Sand to Silicon

The word *silicon* comes from the Latin *silex,* meaning "flint." Silicon is the second-most abundant element, comprising about 20% of the earth's crust, mainly in the form of SiO_2-based minerals (quartzite, agate, jasper, opal, and flint) and silicates (micas, feldspars, zeolites, garnets, and clays). Silicates are the major constituents of glass. Although Lavoisier suggested that SiO_2 was the oxide of a then unknown element (ca. 1787), it was 1823 before Si was first isolated by Berzelius.

The present-day uses of elemental Si are principally in steelmaking, in the production of SiC abrasives, and in silicones. The use of elemental Si for optical components for infrared transmission served as an impetus for single-crystal growth technology until methods were discovered to make polycrystalline optical material with low infrared loss. Its first use as an electronic component appears to be as a point contact rectifier in 1906. Despite the rapid expansion of the semiconductor industry, relatively little tonnage is used there: 45 tons in 1964, 500 tons in 1972, and $\approx$1500 tons in 1978.

Silicon is a light, brittle element that is ductile only when heated relatively close to its melting point of 1410°C. The element is very resistant to chemical attack. A layer of SiO_2 10–20 Å thick forms almost immediately on exposure to air, then more and more gradually increases in thickness to 50–60 Å. Silicon is not attacked by common acids except a mixture of HF and HNO_3, but it is quite soluble in KOH or NaOH, liberating H_2. The material takes a high optical polish rather easily, but the high index of refraction makes antireflection coatings necessary on optical elements. The physical and electronic properties of Si are listed in Table 7.I and the optical absorption constant is shown in Fig. 3.8.

The transformation of raw sand to high-purity Si involves the following general steps:

(1) reduction of SiO_2 to impure Si with C in an electric arc furnace,
(2) transformation to a chemical intermediate such as trichlorosilane,
(3) purification by distillation and other means,
(4) reduction of the intermediate to Si under high-purity conditions,
(5) casting into a form suitable for crystal growth, and
(6) crystal growth, providing an additional purification step by segregation of certain of the impurities.

These purification steps bring impurity levels from between 1 and 10% down to the parts per billion range. Standard analytical techniques such as emission spectroscopy are much too crude to measure impurity concentration at this level and sophisticated techniques such as mass spectroscopy and neutron activation analysis must be used. Often the electrical properties of the finished bulk crystal or the device are used to monitor the ultimate impurity levels. Particular attention is paid to elements from columns III-A and V-A, which act as dopants, and to strong recombination centers ("lifetime killers") such as Au, Cu, and Fe.

The reduction of SiO_2 to Si with C in an electric arc furnace is a large-volume industrial process (200,000 tons per year in the United States in 1973) yielding up to 98–99% pure Si at prices of $\approx$\$1 kg^{-1}. There have been promising attempts at preliminary purification for solar cell destined material by unidirectional solidification of the arc furnace pour-off (Hunt *et al.*, 1976). Many methods exist by which the metallurgical-grade Si is reacted to form compounds that can be more easily purified. Most include Si/halogen compounds since these are gaseous or liquid at low temperatures and a large degree of purification can be effected by simple distillation. The purified compound is then reduced with H_2, an active metal, or pyrolytically. Representative of such reactions are

$$SiCl_4 + Zn, Al, \text{ or } H_2 \rightarrow Si + ZnCl_2, Al_2Cl_3, \text{ or } HCl$$

$$SiHCl_3 + H \rightarrow Si + HCl$$

$$SiH_4 \text{ (pyrolysis)} \rightarrow Si + H_2$$

$$SiI_4 \text{ or } SiBr_4 + H_2 \text{ (or pyrolysis)} \rightarrow Si + HI \text{ or } HBr \text{ (or } I_2 \text{ or } Br_2)$$

$$SiCl_4 + LiAlH_4 \rightarrow SiH_4 + LiCl + AlCl_3$$

$$\hookrightarrow \text{pyrolysis} \rightarrow Si + H_2$$

Table 7.I

Physical and Electronic Properties of Si[a]

Physical and optical properties			
Crystal form	Diamond cubic		
Lattice constant	5.431 Å		
Melting point	1412°C		
Boiling point	3145°C		
Thermal expansion coefficient	2.33×10^{-6} °C^{-1}		
Thermal conductivity	1.41 W cm^{-1} °C^{-1}		
Density	2.328 g cm^{-3}		
Index of refraction	≈ 3.5 ($\lambda > hc/E_g$)		
Expansion on freezing	9% volume		
Electric parameters		Practical[b]	Typical maximum observed values
E_g, direct	2.6 eV		
E_g, indirect	1.12 eV		
dE_g/dT	-2.8×10^{-4} eV °C^{-1}		
μ_e	—	500–700	1360 cm^2 V^{-1} sec^{-1}
μ_h	—	200–350	495 cm^2 V^{-1} sec^{-1}
ϵ_r (dc)	11.7	—	
τ_{n0}[d]	—	5–80	800 μsec
τ_{p0}[d]	—	5[c]	15 μsec
χ	4.01 eV		
n_i[e]	1.4×10^{10} cm^{-3}		
N_c	2.86×10^{19} cm^{-3}		
N_v	1.02×10^{19} cm^{-3}		
Effective masses			
Electrons			
Longitudinal	0.98		
Transverse	0.19		
Conductivity	0.26		
Density of states	0.33		
Holes			
m^*_{heavy}	0.49		
m^*_{light}	0.16		
$m^*_{spin-orbit}$	0.24		
Density of states	0.55		

[a] All properties at 300°K unless inapplicable.

[b] In fabricated solar cells.

[c] In a diffused layer, hole lifetime is reduced to 0.001 to 0.1 μsec.

[d] See Section 3.5.2.

[e] The intrinsic carrier density is given by $n_i = 3.87 \times 10^{16}\, T^{3/2} \exp(-0.605/kT)$. [R. A. Smith, *Semiconductors*, Cambridge Univ. Press. (1959), p. 359.]

Several of these reactions have serious drawbacks in that the reactants are expensive (SiI_4 or $SiBr_4$), yields are low, or the products are dangerous. Runyan (1965) and Wolf *et al.* (1976) provide more information about these and other Si processes.

A widely used commercial method is based on the simplified reaction

$$\left.\begin{array}{c} SiCl_4 \\ \text{or } 2SiHCl_3 \end{array}\right\} + 2H_2 \xrightarrow[\text{Heat}]{} \left\{\begin{array}{c} Si + 4HCl \\ \text{or } 2Si + 6HCl. \end{array}\right.$$

The Si is chlorinated in a fluidized bed forming $SiCl_4$, which is then distilled at $\approx 58°C$ (with some possible intervening purification steps) and finally deposited on hot quartz, Ta, or, more often, an rf induction-heated Si rod in the presence of H_2 at $\approx 950°C$. The latter procedure has the advantage of eliminating the casting step discussed later.

In some cases the purified Si must be formed into a shape favorable to crystal growth by casting. This step can introduce considerable difficulty. Hot molds are not good since molten Si dissolves all metals to some degree and there is even slow dissolution of SiO_2 crucibles, which adds impurities present in the SiO_2 to the melt. Si has a 9% increase in volume on freezing so that SiO_2 molds are destroyed on cooling. Si_3N_4 and SiC have been used with some success for hot molds. Some casting is done using cold molds so that impurities are confined to the surface layers (Runyan, 1965).

Figure 7.2 shows the cost of Si versus impurity content. Research has been carried out to specify a "solar grade" of Si in which the primary consideration is lifetime rather than the general high-purity, low-defect concentration requirements of integrated circuit fabrication (Wakefield *et al.*, 1975). It is the purpose of these research efforts to identify which impurities and what impurity concentrations are deleterious to solar cell efficiency (Hill *et al.*, 1976).

The growth of single crystals of Si can be accomplished by a wide variety of methods involving growth from gas, liquid, or solution (see, e.g., Runyan, 1965, p. 29). The two most widely used techniques for Si, Czochralski (CZ) and float zone (FZ), are discussed briefly here. Several of the more unusual techniques, aimed at formation of thin layers without slicing, are mentioned as well.

In the process of crystal growth it is desirable for solidification to occur as close to thermal equilibrium as possible. Crystal growth can be divided into three broad categories:

1. *Single-Component Systems.* Material is solidified from its own liquid or vapor.
2. *Multicomponent Systems.* Material is frozen out of a supersaturated solution, e.g., Si from In.

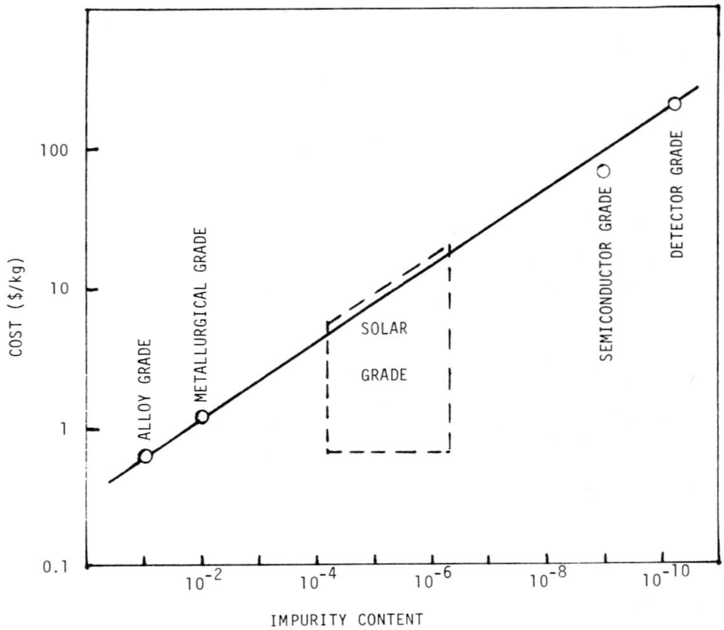

Fig. 7.2. Cost of Si versus purity, used in the identification of a "solar-grade" material (before crystal growth). [Redrawn from G. F. Wakefield, P. D. Maycock, and T. L. Chu, *Proc. 11th IEEE Photovoltaic Specialists Conf.* (1975), p. 49. © 1975 IEEE.]

3. *Multicomponent Reactive Systems.* Material is formed by a reaction in a vapor or liquid phase close to or at the growth site, e.g., $SiH_4 \rightarrow Si + 2H_2$ by pyrolysis.

The basic example of the first category is the Bridgman technique in which solidification occurs as a closed tube containing the molten liquid is passed slowly through a temperature gradient. This technique cannot be used for Si, however, because of dissolution of the tube, sticking to the walls, and expansion on freezing.

To avoid these problems much of the Si is grown today by the Teale–Little modification of the Czochralski method in which the effect of the container is partially eliminated by pulling the freezing crystal from the melt, Fig. 7.3. The container is eliminated altogether in the float-zone process, where surface tension is used to contain a thin moving zone of molten Si in a cylinder of the material.

In crystal growth from the liquid material there are two primary factors controlling growth rate. The first, latent heat removal, is the major

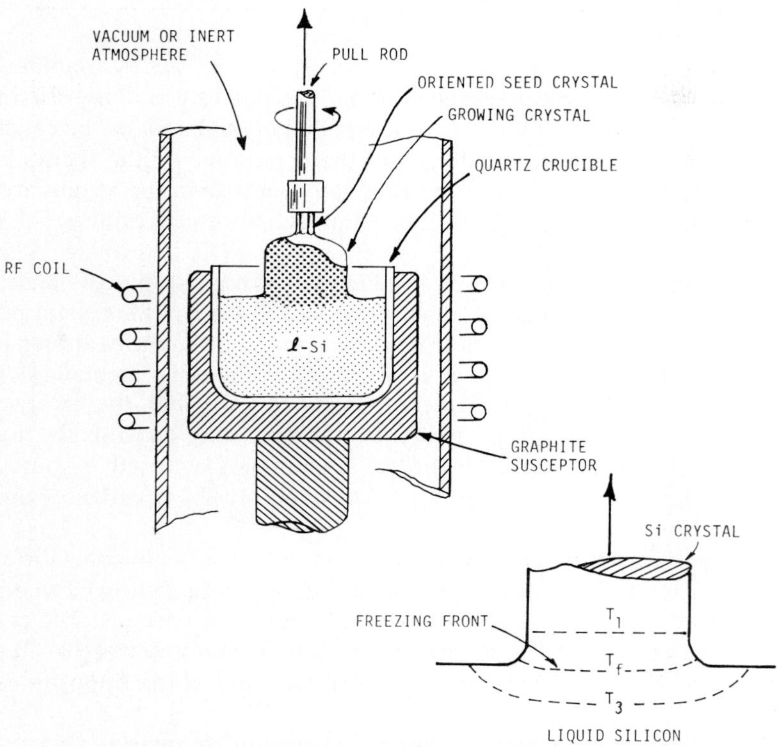

Fig. 7.3. Schematic of Czochralski crystal-growing apparatus.

rate-limiting step. The second, the time necessary for atoms in the liquid to diffuse to an appropriate crystal site, is usually minor in systems in which the liquid is not supercooled. The diffusion time is different for different crystal faces; this generally produces artifacts on the boule and may influence the impurity distribution. Many other factors, including the shape and magnitude of the temperature gradient in the liquid and the solid, the impurity concentration, crystalline line and surface defects, and the shape of the growth interface, influence the growth more or less strongly.

As a greater portion of the liquid is supercooled, the influence of these factors can become dominant. For example, dendritic growth is controlled by sets of twin planes that facilitate the selection of available nucleation sites in a supercooled melt where the growth rate is extremely fast. Similarly, whisker growth from the vapor is seen in which a single screw dislocation provides an axis for rapid nucleation.

7.2.2 The Czochralski (CZ) Process

The CZ process is shown schematically in Fig. 7.3. Variables controlled are the temperature of the crucible, the pull rate, and the stirring of the melt due to the rotation rate of the pulled crystal and/or the crucible. Inherent in the construction details of the apparatus are the thermal gradient shape near the growth interface, the heat loss in the region around the emerging crystal, and the heat loss through the chuck holding the seed end of the crystal.

To begin growth an oriented seed crystal is dipped into the melt, the temperature is allowed to decrease slowly, and finally the crystal pull is initiated. In most CZ systems the primary mode of latent heat loss is by radiation from the bulk of the crystal (proportional to its area and dependent on the temperature of the environment) so that the interaction between the pull rate and the gradients of temperature controls the diameter of the crystal. The effect of surface tension on the growth is somewhat smaller unless the growth interface is considerably above or below the liquid level (Fig. 7.3).

Typically, the liquid Si is held in a highly pure SiO_2 crucible (softening point $\simeq 1600°C$). Dopants are dissolved in the melt and allowed to equilibrate before growth. The environment is vacuum or inert gas. The crystal is pulled at a rate of 10^{-4}–10^{-2} cm sec^{-1} while being rotated at 10–40 rpm. Frequently, the crucible is simultaneously rotated in the opposite direction.

Perhaps the best crucible material at present is quartz. Quartz dissolves slowly in the liquid Si with the reaction

$$2SiO_2 \rightarrow 2SiO + O_2$$

until the melt is saturated with oxygen (2–4×10^{18} cm^{-3}) (Runyon, 1965, p. 18) and then only SiO is formed. The SiO vaporizes while impurities from the dissolved quartz are incorporated into the melt. The use of as small a surface-to-volume ratio as possible and of very pure quartz is desirable. Carbon has been used as a crucible material, but is somewhat soluble in Si, producing SiC inclusions. Impurities may also be introduced because of the difficulty in purifying the carbon.

Boules up to 12 inches in diameter and several meters long have been grown by the CZ process.

7.2.3 Impurity Segregation

Impurities from the melt are segregated linearly across the freezing interface so that

$$N_S = k_0 N_L \tag{7.1}$$

Table 7.II

Impurity Segregation Coefficients, Solubility Limits, and Concentration Tolerances in Silicon

Impurity	Segregation coefficient	Solubility limits	Tolerance[e]
Dopants			
O	0.5–1.0[b]	2×10^{18}[b]	
C	0.07[d]	9×10^{17}[d]	
B	0.8[a]	—	
Al	0.002[a]	2×10^{19}[d]	
Ga	0.008[a]	—	
In	0.0004[a]	—	
P	0.35[a]	—	
As	0.30[a]	—	
Sb	0.023[a]	10^{19}[b]	
Li	0.01[b]	—	
Lifetime killers			
Au	—	—	7×10^{14}
Cr	10^{-5}	5×10^{15}	5×10^{15}
Cu	4×10^{-4}	10^{17}	1×10^{15}
Fe	8×10^{-6}	3×10^{16}	1×10^{15}
Mg	2×10^{-3}	5×10^{18}	$>2 \times 10^{15}$
Mn	10^{-5}	4×10^{16}	2×10^{16}
Na	2×10^{-3}	9×10^{18}	2×10^{14}
Ni	10^{-4}	10^{17}	$>4 \times 10^{16}$
Ti	10^{-5}	10^{15}	4×10^{13}
V	10^{-5}	5×10^{15}	1×10^{14}
Zr	—	—	4×10^{13}
C	—	—	$\simeq 1 \times 10^{17}$

[a] H. J. Hovel, "Solar Cells," Semiconductors and Semimetals, Vol. 11. Academic Press, New York, 1975.

[b] W. R. Runyan, "Silicon Semiconductor Technology," pp. 18, 108. McGraw-Hill, New York, 1965.

[c] G. F. Wakefield, P. D. Maycock, and T. L. Chu, *Proc. 11th IEEE Photovoltaic Specialists Conf.* (1975), p. 49.

[d] D. E. Hill, H. W. Gutsche, M. S. Wang, K. P. Gupta, W. F. Tucker, J. D. Dowdy, and R. J. Crepin, *Proc. 12th IEEE Photovoltaic Specialists Conf.* (1976), p. 112. More complete data on solubility vs. temperature in Si will be found in F. A. Trumbore, *Bell Syst. Tech. J.* **39,** 205 (1960).

[e] Based on permissible impurity concentration such that a $\eta_s \simeq 10\%$ efficient solar cell can be fabricated from B-doped 0.5-Ω cm material. From D. E. Hill *et al., Proc. 12th IEEE Photovoltaic Specialists Conf.* (1976), p. 112.

at the interface, where N_S and N_L are the impurity concentrations in the solid and the liquid and k_0 is the equilibrium segregation coefficient, examples of which are shown in Table 7.II. In the CZ case the remaining impurities are concentrated in the melt, and, for the simple situation of

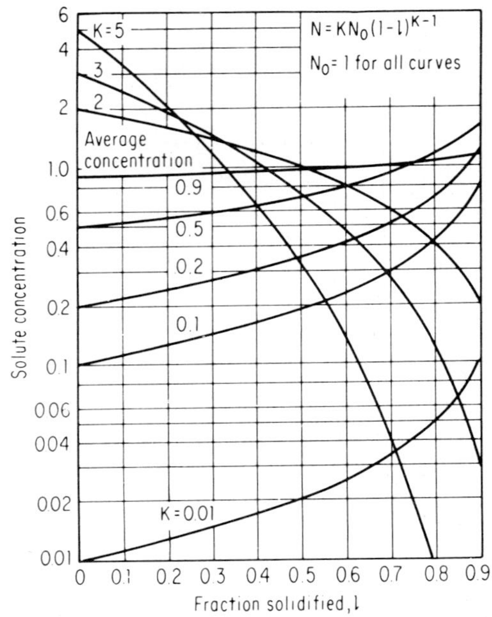

Fig. 7.4. Impurity concentration versus the fraction of the melt-solidified z (e.g., the CZ process). [After W. G. Pfann, "Zone Melting," 2nd ed. Wiley, New York, 1966.]

one-dimensional freezing,

$$N_S(z) = N_0 k_0 (1 - z)^{(k_0 - 1)}, \tag{7.2}$$

where N_0 is the impurity concentration in the melt before growth and z is the fraction of the melt frozen. This relation is plotted in Fig. 7.4. Note that a low k_0 is required to remove impurities from the melt, whereas a high k_0 is required for uniform doping. Too high a concentration of a doping impurity may inhibit single-crystal growth, however; for example, $\lesssim 2\%$ Sb will prevent Si from growing as a single crystal. For this reason B is preferred over Al as a dopant for CZ growth.

In the float-zone (FZ) process to be discussed next, impurities accumulate in the molten zone since they are being added at a rate $N_0\,dx$ at the melting front of the zone while at the freezing end they are being incorporated at a rate $k_0 N_L\,dx$. The relation for the case of impurity removal is

$$N_S(x) = N_0\{1 - (1 - k_0)\exp[-k_0 x/x']\}, \tag{7.3}$$

where x' is the length of the molten zone and x is the distance along the rod. This relation is plotted in Fig. 7.5. A single FZ pass is not as effective

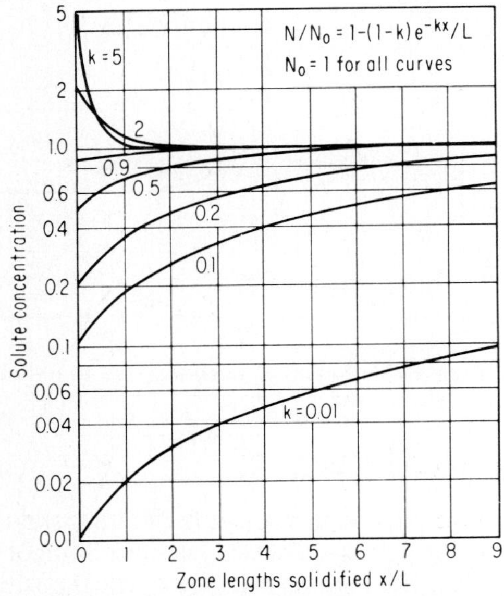

Fig. 7.5. Impurity concentration versus distance along a zone-refined rod (normalized by zone length L'). [After W. G. Pfann, "Zone Melting," 2nd ed. Wiley, New York, 1966.]

as CZ in removing impurities but in the FZ case it is much easier to make repeated passes to purify the crystal.

When it is desired to dope the FZ crystal intentionally the relation is

$$N_S = k_0 N_d \exp(-k_0 x/x'), \tag{7.4}$$

where N_d is the initial doping concentration in the zone. For N_S to be nearly constant with distance, the concentration in the zone must be almost constant and a low segregation coefficient is required in contrast to the CZ case. One method of circumventing this difficulty is to distribute dopant along the rod in a series of sawed notches before zoning so that the dopant is replenished as growth proceeds. Other methods are by addition of a dopant gas (e.g., PH_3 or B_2H_6) to the growth ambient or by neutron transmutation doping (NTD).

During growth the impurity is constantly being rejected at the freezing front. It must be transported back into the bulk of the liquid by diffusion so that a concentration gradient develops near the growth interface. This effectively increases k by a factor depending on the growth rate and stirring speed (see Fig. 7.6). The ratio of k_{eff}/k_0 can range from 1 to 4 for commonly used growth and stirring rates (Runyan, 1965).

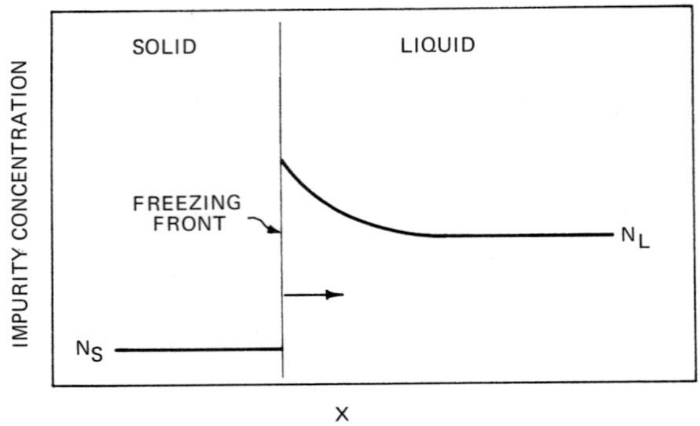

Fig. 7.6. Enhanced effective distribution coefficient due to diffusion near advancing freezing front.

7.2.4 The Float-Zone (FZ) Process

In the FZ processes, a modification of the Pfann zone-refining method, a narrow molten region is passed slowly along a Si ingot in a vacuum or inert gas environment. The ingot is held in a vertical position and the heat is supplied by rf induction. The molten Si is held in place both by surface tension and by the levitation effect of the rf field. Similar considerations with respect to control of lateral and radial temperature gradients and growth rate hold as for CZ. FZ crystals have a somewhat higher purity than CZ material due to the absence of contamination from a crucible; in particular the oxygen concentration can be reduced by a factor of 20–100. Minority carrier lifetime and mobility also tend to be higher than for CZ material. On the other hand, the dislocation density of the FZ material is higher because steeper temperature gradients are involved and because compromises must be made with regard to the temperature profile due to rf heating considerations. These differences are particularly important with respect to radiation resistance effects (Section 7.5.1).

The FZ process starts with a polycrystalline ingot of about the same size as the crystal to be grown. Both the end of the ingot and the end of an oriented seed crystal are locally heated and then skillfully joined, much like the comparable part of the CZ growth. In FZ growth, the heated zone usually travels in the upward direction and is about 2 cm long. The taper from the small seed to the large portion of the crystal is established by the difference in drive rates between the upper chuck, holding the polycrystalline ingot, and the lower chuck, holding the seed. Growth of crystals 50 to 100 cm long and 7.5 cm in diameter is routine, and crystals up to 10 cm in diameter have been grown. The processing is somewhat faster than with CZ growth. A general comparison of CZ and FZ Si growth is given by Matlock (1979).

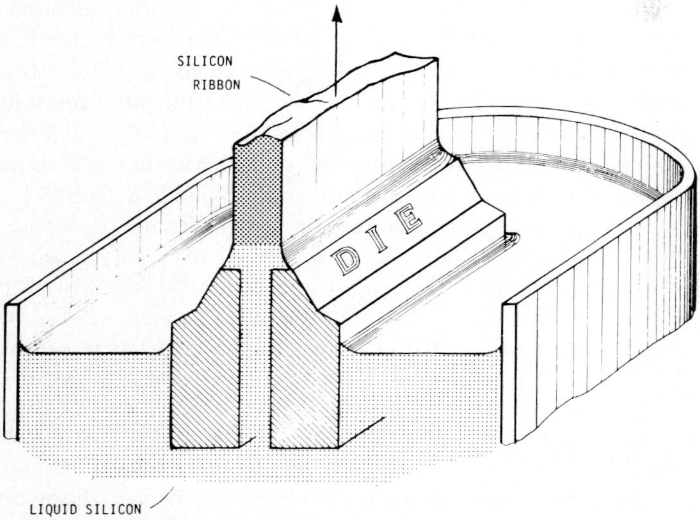

Fig. 7.7. Simplified schematic diagram of edge-fed growth of silicon ribbon.

7.2.5 Other Growth Methods

The expense of slicing and polishing wafers has led to development of methods for the growth of Si directly in the form of thin ribbons by "edge-defined, film-fed growth (EFG)" or by dendritic web growth.

EFG, apparently first applied to Si by Ciszek (1972; Ciszek and Schwuttke, 1975),† has been used as a commercial method for growing sapphire tubes, ribbons, and other cylindrical shapes for a number of years. Si growth by this technique has recently undergone intense development by LaBelle and Mlavsky (1971), and Ravi *et al.* (1975). In this process a carbon die with a slot-shaped aperture is immersed part way in a vessel of molten Si. The liquid Si wets the die and, in operation, flows through the slot to feed the solid zone above as in Fig. 7.7. The shape of the ribbon is controlled by the shape of the top surface of the die, by surface tension, temperature gradients, and the pull rate.

Rates reported in 1980 were up to 5 cm min^{-1} for ribbons 0.05 cm thick and up to 5 cm wide (Kalejs, 1980). Ribbons are flexible enough to be wound on large drums and ribbons 20 m long have been grown in a single run. No polishing is required. Using multiple dies and continuous processing, Mlavsky estimates the potential cost of ribbon at less than \$0.002 cm^{-2}.

† EFG is similar to the Stepanov technique for ribbon growth except that in the latter case the die is not wetted by the Si.

EFG ribbon has rather high dislocation and twinning densities. Since C is slightly soluble in liquid Si, erosion of the die material and SiC precipitates appear to be the largest problem areas. Some success has been achieved with SiO_2 dies by Jewett (1978). Diffused junction cells made from ribbon exhibited 10.6% AMl efficiencies in 1977 and higher efficiencies were expected with improved cell-processing techniques. An interesting development has been the growth of hollow, tubular cells by the EFG process, which should prove advantageous to concentrator systems since coolant can flow through the cell (Mlavsky *et al.*, 1976).

Dendritic web growth was brought to the point of pilot plant production of solar cells ($\eta_s \simeq 10\%$) by Westinghouse in 1966–1967, but there was an insufficient market at that time and the effort was discontinued (Riel, 1973). Terrestrial conversion has revived interest in dendritic web growth techniques and the Westinghouse effort was restarted in 1977–1978 (Seidensticker *et al.*, 1978). In this technique two parallel dendrites form the boundaries of a web or ribbon pulled from a supercooled melt. One or more twin planes parallel to the plane of the ribbon and at its center stabilize the growth (Dermatis *et al.*, 1965) and no dies are needed. Growth rates are $\simeq 10$ cm min^{-1} and widths of 4 cm have been achieved, with area output rates of 27 cm^2 sec^{-1} (Duncan *et al.*, 1980). Temperature control must be very precise. The quality of the ribbon is equal to good CZ material, and the twin planes appear not to be detrimental to the cell efficiency (Seidensticker *et al.*, 1975; Davis *et al.*, 1976). Cells with $\approx 15.5\%$ efficiency (AM1) have been produced (Duncan *et al.*, 1980).

A technique for exceptionally high-speed growth of Si ribbon has been introduced by Tsuya *et al.* (1980). Using gas pressure, molten Si is squirted from a slot in the bottom of a holding crucible onto a set of cold, rotating rollers, producing ribbons at a rate of 10 to 40 m sec^{-1}! Polycrystalline ribbons 20 to 200 μm thick and 0.1 to 5 cm wide have been grown. The grain size is large (10 to 100 μm) and cells with $\eta_s = 5\%$ (without an antireflection coating) were fabricated from the roller-quenched ribbon in the early stages of research.

Silicon can be grown from solution at somewhat lower temperatures than its melting point using Ga or Sn as a solvent, but rates are low and the technique has not received much attention for solar cells (Wolf, 1975).

When Si films are grown by vacuum evaporation, high source temperature (> 1800°C) and good vacuum ($\lesssim 10^{-7}$ Torr) are required, the latter to prevent formation of SiO. Solar cells grown by this method have reached $\eta_s \simeq 3\%$ (Feldman *et al.*, 1980). A high substrate temperature (> 1000°C) is required as well to achieve epitaxy or large-grain polycrystal growth. Higher temperatures also tend to reduce stacking fault and dislocation densities. The temperature can be reduced by fluxing with thin films of Pt or other metals (several monolayers), which remain on top of the growing film and increase atomic surface mobility sufficiently to promote crystal growth (Siekhaus, 1976).

Chemical vapor deposition (CVD) has been a more widely used method. CVD utilizes the decomposition of $SiCl_4$, $SiHCl_3$ (Chu *et al.*, 1967), or organosilicon compounds on a hot substrate. Growth is reaction-rate-limited at low temperatures and diffusion-limited at high temperatures, depending strongly on gas flow and pressure. At 1100–1300°C growth rates of 6 to 14 μm min^{-1} can be achieved, although rates of $\simeq 1$ μm min^{-1} are more commonly used to achieve better crystalline quality (Runyan, 1965, p. 74). Among the advantages of CVD are easy control of doping by introduction of gaseous dopants such as diborane, phosphine, or arsine into the gas stream and the possibility of etching substrates *in situ*. Layers that are *p*- and *n*-type can be easily grown sequentially merely by changing dopant flow. The quality of the grown layers is quite good for CVD epitaxial growth (Kressel *et al.*, 1976).

Other means of forming Si into sheets include

(a) dipping aluminum silicate or alumina ceramic substrates into molten Si (Heaps *et al.*, 1980),
(b) rolling Si into thin sheets at temperatures of $\simeq 1380$°C (Noel *et al.*, 1976),
(c) casting with directional solidification followed by slicing (so that all grain boundaries are perpendicular to the plane of the slice) (Fischer and Pschunder, 1976; Fischer, 1978),
(d) liquid phase epitaxy from a Sn solution of Si (Wolf, 1975, p. 306).

The directional solidification casting method (c) has received considerable attention for low-cost Si. In one modification, a heat exchanger is positioned at one end of the reservoir of molten Si to establish an almost one dimensional temperature gradient. The castings produced by the "Heat Exchanger Method" (HEM) are of excellent quality with grain sizes of several millimeters or more (Khattak and Schmid, 1980).

The quality of sheet Si can be improved by FZ processing in sheet form using scanning electron beam or laser energy sources (Lesk *et al.*, 1976).

Schwuttke (1979) and Ciszek (1982) give good comparisons of the CZ, FZ, HEM, EFG, and dendritic web growth methods.

7.3 IMPERFECTIONS, DOPING, AND LIFETIME

The efficiency of solar cells depends critically on the imperfection concentration because of its effect on the minority carrier lifetime τ in the active layers. Not only is J_{sc} increased with increased τ but an increase in

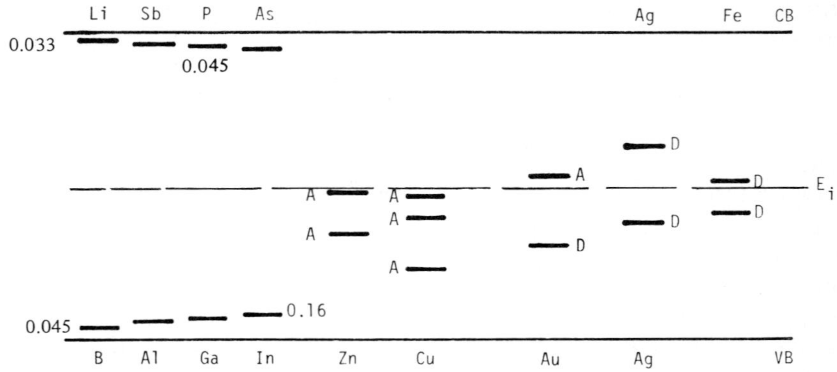

Fig. 7.8. Measured ionization energies for various impurities in Si. Donors (D) and acceptors (A) are indicated and energies are measured from the CB (VB) for centers above (below) midgap.

V_{oc} can be expected because of the decrease in J_0. Resistivity control is the second-most important consideration.

Imperfections may be separated into two broad categories:

(1) structural defects such as dislocations, interstitials, vacancies, twin and low angle grain boundaries in the lattice, and

(2) impurities: foreign species present either interstitially or substitutionally.

Often complexes are formed between foreign impurities and structural defects. Imperfections may or may not be electrically active.

The effectiveness of electrically active imperfections as dopants or as recombination centers is partially determined by their energy level within the forbidden energy gap and the position of the quasi-Fermi levels (Section 3.5.2). In general, levels within a few kT of the conduction or valence bands are effective as donors or acceptors but ineffective as recombination centers. The reverse is true for levels close to the center of the gap—especially deep donorlike (acceptorlike) states in p-type (n-type) materials which are strong recombination centers or "lifetime killers." Figure 7.8 shows the energy levels of various impurities in Si. A comprehensive review is given by Chen and Milnes, 1980.

7.3.1 Dopants

The criteria for selection of a dopant for a semiconductor are that it have a suitable energy level, solubility, and an acceptable diffusion constant D. Too large a D allows the dopant atoms to be mobile at ambient

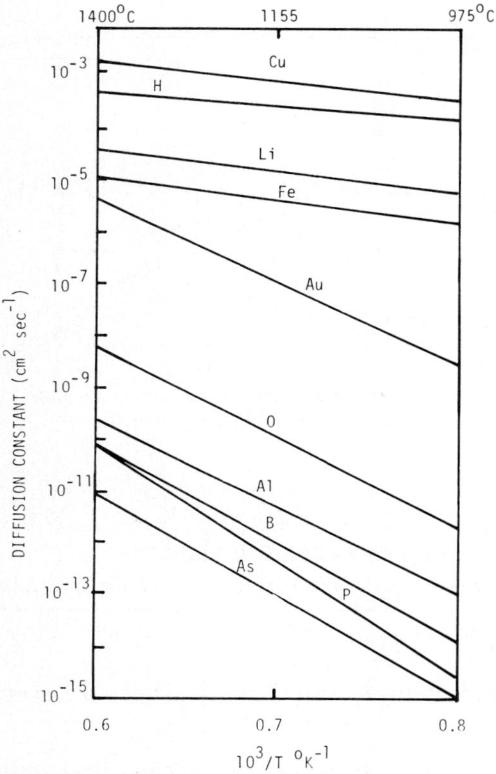

Fig. 7.9. Diffusion constants of various elements in Si as a function of inverse temperature. [Redrawn from S. M. Sze, "Physics of Semiconductor Devices," p. 31. Wiley, New York, 1969.]

temperature and leads to instability (in a 20-year time frame). The diffusion constants of several elements in Si are given in Fig. 7.9. A further consideration is that the impurity have a suitable segregation coefficient for the type of growth system used. P and B are probably the most widely used donor and acceptor dopants for Si.

The electronic mobility in the semiconductor decreases as the donor or acceptor concentration (N_D and N_A) increases because of charged impurity scattering as shown in Fig. 7.10. The onset of this decrease is N_D or $N_A \simeq 10^{16}$ cm^{-3} in both *p*- and *n*-Si. In addition, the presence of a large dopant concentration can decrease the minority carrier lifetime because of lattice distortion and complexing.

In Si cells the mechanism of Auger recombination (Section 3.5.1) is

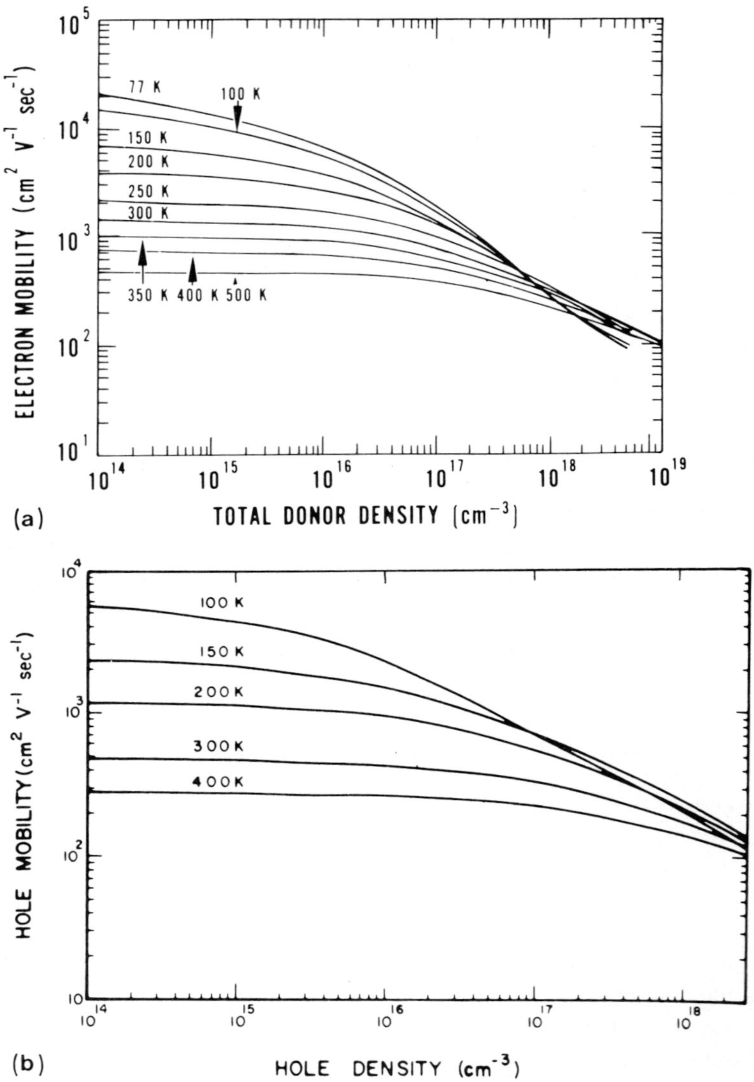

Fig. 7.10. (a) Mobility of electrons in n-Si (phosphorus doped) versus donor density and temperature. [After S. S. Li and W. R. Thurber, *Solid-State Electron*. **20,** 609 (1977). Copyright 1977, Pergamon Press, Ltd.] (b) Mobility of holes in p-Si (boron doped) versus acceptor density and temperature. [S. S. Li, *Solid-State Electron*. **21,** 1109 (1978). Copyright 1978, Pergamon Press, Ltd.]

especially important. The effective minority carrier lifetime is given by

$$\tau^{-1} = \tau_A^{-1} + \tau_{SRH}^{-1},$$

where τ_{SRH} is the Shockley–Read–Hall lifetime given by $U = (n_p - n_{p0})/\tau_{SRH}$, with U from Eq. (3.56), for the example of p-type material. The Auger lifetime is given by $\tau_{An} = 1/C_n n^2 \simeq 1/C_n N_A^2$ for n-Si and by $\tau_{Ap} = 1/C_p p^2 \simeq 1/C_p N_A^2$ for p-Si. The Auger constants are $C_n = 2.8 \times 10^{-31}$ cm^6 sec^{-1} and $C_p = 1 \times 10^{-31}$ cm^6 sec^{-1} (Hall, 1981). For good material, with $\tau_{SRH} \simeq 50$ μsec, Auger recombination dominates the effective lifetime for N_A or $N_D \gtrsim 5 \times 10^{17}$ cm^{-3} (Fischer and Pschunder, 1975).

These effects of doping on mobility and lifetime mark the limits for doping concentration in the active absorbing layers of solar cells.

7.3.2 Lifetime Killers

A partial list of midgap levels acting as lifetime killers is shown in Table 7.II with the permissible concentrations for solar cell material ($\eta_s \gtrsim$ (10%) (Wakefield *et al.*, 1975; Hill *et al.*, 1976). The lifetime τ is directly related to the concentration of such impurities as shown in Fig. 7.11. The concentration of Au, for example, must be held at the low ppb level. Particularly detrimental to Si are Na, Cu, and Fe since they also have high

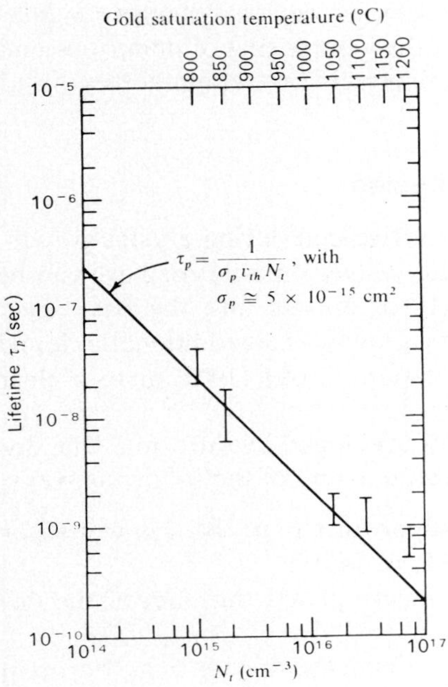

Fig. 7.11. Dependence of minority carrier lifetime in n-Si on Au impurity concentration N_t. The data indicate a hole capture cross section of $\sigma_p = 5 \times 10^{-15}$ cm^2. [From A. S. Grove, "Physics and Technology of Semiconductor Devices," p. 142. Wiley, New York, 1967.]

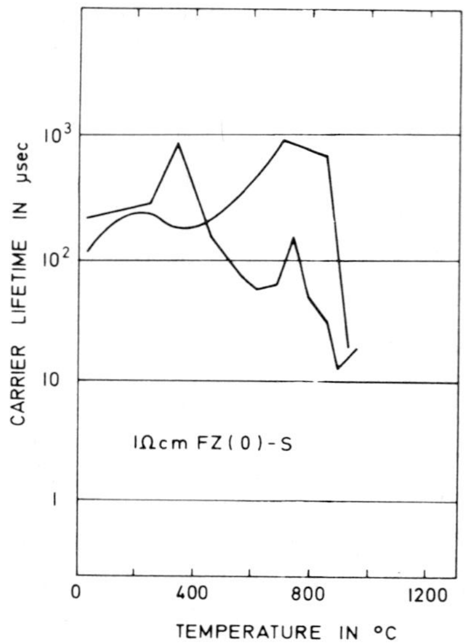

Fig. 7.12. Carrier lifetime versus isochronal annealing temperature for 1 Ω cm FZ Si. [From K. Graff, H. Pieper, and G. Goldbach, "Semiconductor Silicon" (H. R. Huff and R. R. Burgess, eds.), p. 170. Electrochem. Soc., Pennington, New Jersey, 1973.]

diffusivities. Defects or impurities introduced during crystal growth can have a major influence, producing variations in lifetime over many orders of magnitude. However, subsequent processing can also change τ significantly as shown in Fig. 7.12 because of the complexing of impurities and other defects during heat treatments at relatively low temperatures (Graff and Fischer, 1979).

7.3.3 Introduction of Dopants—Diffusion

The base layer dopant is generally introduced during crystal growth. Then, for example, the production of an *n*-layer on a *p*-type base can be accomplished by diffusion of a density of donors into the base layer exceeding the acceptor density there, by growth of an additional *n*-doped layer (e.g., by CVD), or by ion implantation. Grove (1967) gives a clear discussion of diffusion doping in Si.

At this time virtually all Si solar cells are doped by diffusion. The dopant can be supplied to the diffusion region in any of the following ways:

(1) by introduction in gaseous or vapor form (e.g., BCl_3 in wet H_2 or N_2, or P_2O_5 in a dry gas) (Runyan, 1965, p. 146);

(2) from a vacuum evaporated or electroplated thin film of the dopant;

(3) from an adjacent inert solid containing the dopant (e.g., boron in glassy layers of potassium silicates and borates);

(4) from a doped SiO_2 powder layer applied by painting, spinning, spraying, or screen printing (Shah and Fuller, 1976; Chandler *et al.*, 1976); and

(5) from a doped layer of SiO_2 applied by CVD.

A typical diffusion of P in Si may be carried out at 950°C for 10 min giving a diffused layer thickness on the order of 0.25 μm.

The diffusion ideally follows Fick's laws with a concentration profile given by

$$N(x, t) = N_0[1 - \mathrm{erf}(x/2\sqrt{Dt})] \tag{7.5}$$

for a constant source concentration. In most cases, however, the profile is more complex, involving a diffusion constant that depends on concentration, the number of vacancies present, and on temperature in a more complex way than the simple Ahrenius relationship (Shewmon, 1963).

As the diffusion proceeds, a concentration plateau forms at the solubility limit for the dopant—as is the case for the P diffusion shown in Fig. 7.13. As the Si cools, the high concentration of dopant damages the lat-

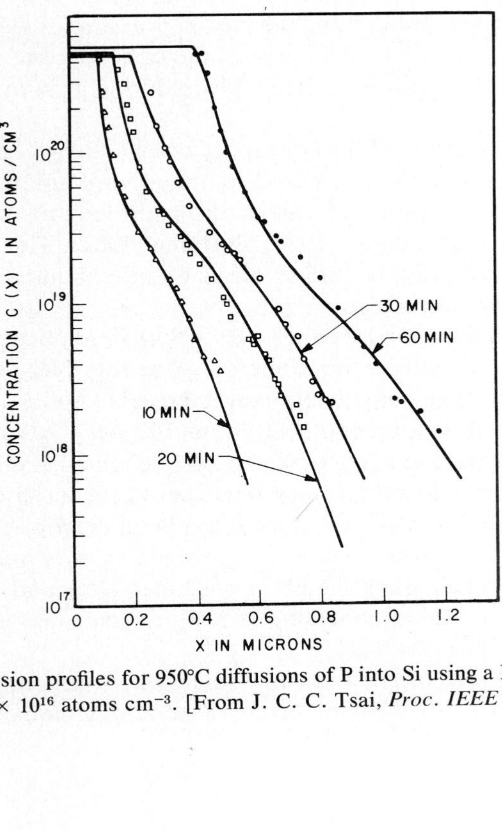

Fig. 7.13. Diffusion profiles for 950°C diffusions of P into Si using a $POCl_3$ source. Substrate doping = 2×10^{16} atoms cm^{-3}. [From J. C. C. Tsai, *Proc. IEEE* **57**, 1499 (1969). © 1969 IEEE.]

tice, forming dislocations and other imperfections with deep levels that greatly reduce the lifetime there (to 10^{-10} sec, for example; Lindmayer and Allison, 1973). This gives rise to the so-called dead layer at the surface that limited the high-energy response of many early versions of Si cells (ca. 1970–1975). The dead layer can be eliminated by

(1) reducing the dopant concentration at the surface during diffusion, perhaps by using a "glassy" source, which holds the surface concentration at some value below the solubility limit;
(2) diffusion from a limited source (e.g., a vacuum evaporated thin film);
(3) diffusing for a short time (called "predeposition"), removing the dopant source, and then continuing the diffusion (called "drive-in"), in effect producing a limited source, and
(4) removing the damaged layer after diffusion.

7.3.4 Other Means of Introducing Dopants

An important method of producing a doped layer is simply to grow it on the base using chemical vapor deposition. By CVD it is relatively easy to introduce dopants by varying the partial pressure of a dopant gas in the flow system. The sticking coefficient of the dopant atoms must be considered and some cross diffusion takes place during growth (Runyan, 1965, p. 127).

Silicon can be converted to phosphorus by a flux of neutrons; neutron transmutation doping has been used to obtain very uniform impurity concentrations and long minority carrier diffusion lengths in polycrystalline Si (Westbrook and Polgreen, 1976; Hoffman, 1981). This result is important because rapid grain boundary diffusion may dominate bulk diffusion in polycrystalline materials.

Implantation using high-energy (60–300 keV) ion beams of the dopant has been used to produce thin doped regions for solar cells (Varma and Swerdling, 1976). The beam energy can be used to control the depth (e.g., 60 keV yields an average depth of about 500 Å). The implantation produces considerable damage so that the sample must be annealed during or after implantation. Despite a somewhat low V_{oc} because of residual damage, J_{sc} is high and η_s = 10.4% (AM0) has been demonstrated by Kirkpatrick *et al.* (1976). Pulsed electron beam heating was used to anneal the implantation damage in these cells as well as to alloy and sinter both front and rear contact. Laser annealing is also a promising technique for removing implantation damage.

A detail that is of interest for MIS solar cells concerns the exchange of impurities with a growing oxide layer. If the segregation coefficient is less in the oxide than in the Si (as is frequently the case), a pile-up of impurity atoms at the Si/SiO_2 interface can occur (Runyan, 1965, p. 133).

7.4 THE FABRICATION AND PARAMETERS FOR A TYPICAL Si CELL

In this section the configuration of a typical cell, the fabrication steps, and the operating characteristics are outlined. This is done both to describe Si cell technology and because there is considerable similarity, at least in principle, to photovoltaic cells made from other materials.

7.4.1 Geometrical Cross Section and Band Diagram

The cross section and plan view of a Si *n/p* homojunction solar cell are shown in Fig. 7.14. The cell consists of a slice of single-crystal Si

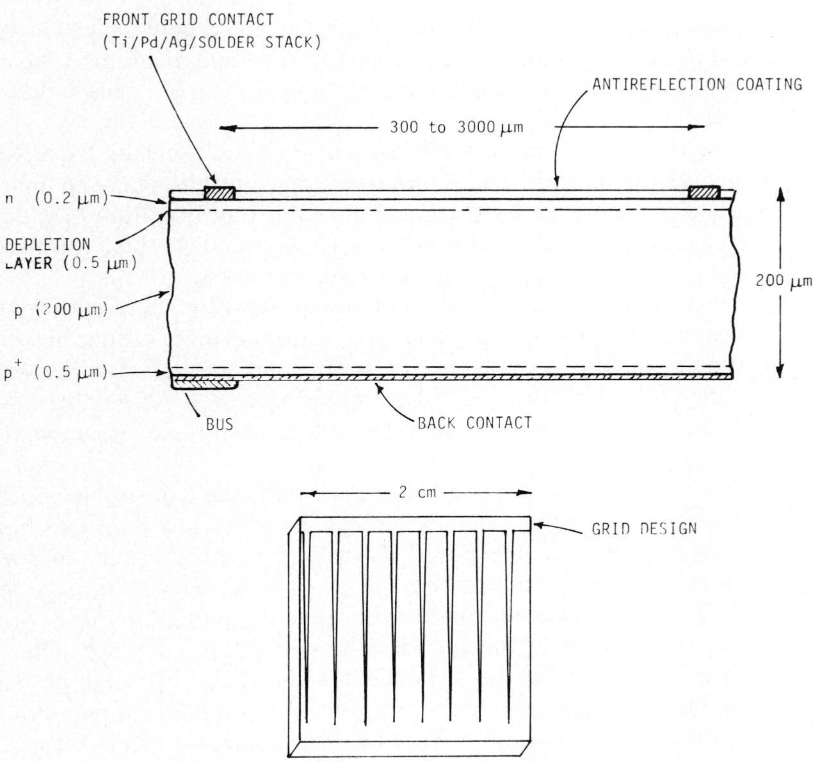

Fig. 7.14. Typical Si cell geometry.

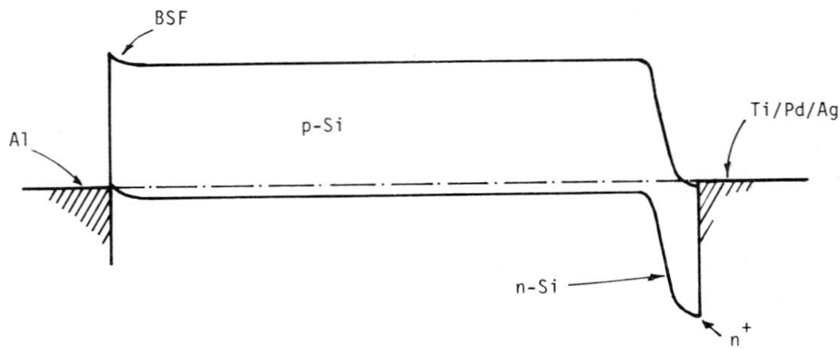

Fig. 7.15. Energy band diagram for typical Si cell with *n*-layer thickness exaggerated.

200–500 μm thick. The thickness is most often chosen by structural criteria rather than by capacity to absorb light. Indeed, cells 50 μm thick with η_s = 11.8% have been made (Storti *et al.*, 1980; Chiang *et al.*, 1978b).

An *n*-type layer 0.1–0.5 μm thick is diffused and finally the cell is contacted and the antireflection coating applied. The band diagram of the cell is shown in Fig. 7.15 and parameters for typical $n^+/p/p^+$ cells follow in Table 7.III.

The junction is produced by diffusion of P or As, typically from PH_3, $POCl_3$, or AsH_3 in a carrier gas. The surface concentration may be limited as described in the previous section to prohibit the formation of a dead layer. Oxygen is sometimes introduced to accomplish this, forming a SiO_2-based "diffusion glass," which is later removed.

The back contact is usually aluminum deposited by vacuum evaporation. A heat treatment (e.g., 500–800°C, 15 min) is used to diffuse in a portion of the Al, making a p^+-layer that lowers the contact resistivity and can, if sufficiently thick (0.2 μm thick, requiring 4 hr at 700–800°C), produce a back surface field to reduce the effect of surface recombinaton there (Section 7.5.2).

The front grid contact is complex. Ag would make a desirable contact but it has low adhesion to a clean Si surface. To produce a tenacious bond a thin Ti layer (400 Å) is interposed. However, in the presence of moisture an electrochemical reaction takes place between Ti and Ag, so a thin layer of Pd (200 Å) is often introduced between them to inhibit this corrosion. The front contact is annealed at 500–600°C for 5–30 min. The Ag may be thick enough to conduct the collected current (≃5 μm), or perhaps a layer of Pb–Sn solder is added by dipping to provide a large enough cross-sectional area to minimize R_s. The front contact is in the form of a grid and, depending on the resistivity and thickness of the *n*-layer, the grid

Table 7.III

Parameters for "Violet"-type Cells[a]

p^+-Layer	Dopant	Al	
	Thickness	0.5	μm
	τ_e	0.4	μsec
	L_n	1–10	μm
p-Type (base) layer	Dopant	B	
	Thickness	250	μm
	p	10^{16}	cm^{-3}
	μ_p	300	$cm^2\ V^{-1}\ sec^{-1}$
	μ_n	~1200	$cm^2\ V^{-1}\ sec^{-1}$
	ρ	2–3	Ω cm
	τ_e	6–50	μsec
	L_n	100–400	μm
	S^b	$0.3–1 \times 10^3$	cm sec^{-1}
	S^c	10^6	cm sec^{-1}
n^+-Type (diffused) layer	Dopant	P	
	Thickness	0.1–0.2	μm
	n, average	$1–30 \times 10^{19}$	cm^{-3}
	μ_p	~50	$cm^2\ V^{-1}\ sec^{-1}$
	D, average	1	$cm^2\ sec^{-1}$
	ρ, average	$10^{-3}–10^{-1}$	Ω cm
	ρ, sheet	25–500	$\Omega\ square^{-1}$
	τ_h	$10^{-4}–10^{-1}$	μsec
	S^d	$\sim 10^4$	cm sec^{-1}
Cell	Total area	4	cm^2
	W_d	~0.4	μm
	V_d	~0.94	V
	R_s	0.05–0.3	$\Omega\ cm^2$
	Grid spacing	5–30	cm^{-1}
	Grid coverage	5–10	%

[a] Although many of the data are for the "violet" cell, they have been gleaned from other sources as well and so do not represent a single specific cell. The violet cell (ca. 1973) had a p^+ back layer but no deliberate BSF.

[b] Effective S of base layer with back-surface field effect.

[c] Without back-surface field.

[d] Oxide passivated surface.

lines are 0.03 (in the violet cell) to 0.3 cm apart, running to a central bus. The grid and bus cover from 5 to 10% of the total area. The Ti/Pd/Ag contact is sometimes used on the back of the cell as well, as a tunneling contact to the Al-doped p^+-layer.

Since bare Si has a reflection coefficient that ranges between $\simeq 33\%$ and $\simeq 54\%$ over the desired spectral range (0.35 μm–1.1 μm) (Runyan,

1965), antireflection coatings (ARC) must be used. An optimal single-layer coating can reduce the average reflection over this interval to $\simeq 10\%$ and two layers can reduce it to $\simeq 3\%$. Layers of SiO, SiO_2, Si_3N_4, Al_2O_3, TiO_2, and Ta_2O_5 have been used, the last being particularly good for extended-response cells because of its high UV transmission. The use of surface texturing to reduce reflection is discussed in Section 7.4.4. The ARC coating completes the cell and also acts to passivate the surface and protect it from atmospheric corrosion. If the cell is intended for space use, a cover glass must be added to protect against low-energy proton damage.

7.4.2 Fabrication Steps

A typical fabrication sequence is outlined below:

(1) Si boule quality check (boules may be 6 inches in diameter or larger and several feet long),
(2) wafer slicing (slices are typically 0.5 mm thick, saw kerf loss is 0.2 mm; sawing introduces defects that must be removed by polishing and/or etching the surface),
(3) degreasing and cleaning,
(4) mechanical polishing (this step has been largely superseded by deep chemical polishing),
(5) etching (e.g., 40% NaOH + H_2O at 110°C gives a chemical polish, after which a texturizing etch might be used),
(6) final cleaning (HCl/H_2O, rinse, HF, rinse, etc.),
(7) P diffusion (PH_3 + Ar + O_2 at 865°C for 15 min; both sides of wafer are diffused),
(8) removal of diffusion glass with HF etch,
(9) Al deposition on back surface by vacuum evaporation (5000 Å),
(10) Al diffusion (to produce p^+-layer), 800°C for 15 min (through the back n-layer),
(11) Masking for grid (by photolithography or shadow mask),
(12) Ti, Pd, and Ag depositions, front surface and rear surface,
(13) mask removal,
(14) final contact anneal at $\simeq 550$°C for $\simeq 10$ min,
(15) solder dip,
(16) ARC application (e.g., Ta_2O_5 and sinter at 450°C for 1 min),
(17) dicing and edge treatment (diffused edge would short the cell if not removed),
(18) quality check and send to array fabrication.

The preceding schedule has been greatly simplified so many of the cleaning steps and quality checks that are necessary in practice have been

Table 7.IV

Electrical Properties of Several Si Solar Cells

Cell type	V_{oc} (V)	J_{sc} (mA cm^{-2})^a	ff	η_s (AM0)^a	J_0 (A cm^{-2})	η_s (AM1)	J_{sc} (mA cm^{-2}) AM1
Violet[b]	0.595	40	0.78	14%	10^{-12}	17%	—
CNR[c]	0.591	46	0.77	15.3%	6×10^{-12}	≃18%	34
Practical[d] maximum	0.69	47	0.84	18.5–19.0%	—	—	—
Theoretical[e] maximum	0.79	48	0.86	22.0%	2.4×10^{-15}	≃25%	—
Theoretical[f] extremum	0.85	54	0.87	29.0%	2×10^{-16}	—	—

[a] Total area basis. The *A* factors of these cells are ≃1.0 in the high bias region near V_{oc}. AM0 except as noted.

[b] J. Lindmayer and J. F. Allison, *COMSAT Tech. Rev.* **3,** 1 (1973). Reprinted *in* "Solar Cells" (C. E. Backus, ed.), p. 209. IEEE Press, New York, 1976.

[c] R. A. Arndt, *Proc. 11th IEEE Photovoltaic Specialists Conf.* (1975), p. 40.

[d] J. R. Hauser and P. M. Dunbar, *IEEE Trans. Electron Devices* **ED-24,** 305 (1977). Also see M. Wolf, *IEEE Trans. Electron Devices* **ED-27,** 751 (1980).

[e] M. Wolf, *Energy Convers.* **11,** 63 (1971). Reprinted *in* "Solar Cells" (C. E. Backus, ed.), p. 191. IEEE Press, New York, 1976. Zero area loss due to gridding, assuming an overall η_Q of 0.88, 3% reflection loss, and 3% R_s loss.

[f] Based on Wolf's calculations with $\eta_Q = 1$, zero area, reflection, and R_s loss.

omitted. In some cases the gridding steps have been replaced for terrestrial cells by screen print application of a Ag paste grid with subsequent heat treatment.†

7.4.3 Electrical Parameters

Electrical parameters for Si cells at several technology stages are given in Table 7.1V. It is evident that considerable potential for improvement in J_{sc} exists for the "violet cell." The situation is elucidated by Fig. 7.16, which shows the calculated spectral quantum efficiency contributions for the *p*-, *n*-, and depletion layers of a cell that has negligible front-surface loss. Compare this with the spectral response curves for practical cells shown in Fig. 7.17. A loss in high-energy response takes place in the *n*-layer because of excessive recombination there in the bulk as well as at the surface. By improving the properties of the *n*-layer, the

† A low-cost, large-volume fabrication sequence for terrestrial cells is outlined by Bickler *et al.* (1978).

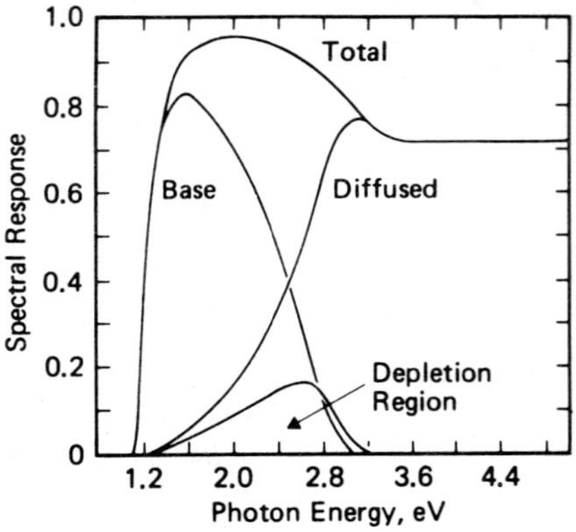

Fig. 7.16. Computed internal response of a Si n/p cell showing the individual contributions from each of the three regions. The front surface recombination velocity is $\approx 10^4$ cm sec^{-1} in this example. [From H. Hovel "Solar Cells," Semiconductors and Semimetals, Vol. 11, p. 27. Academic Press, New York, 1975.]

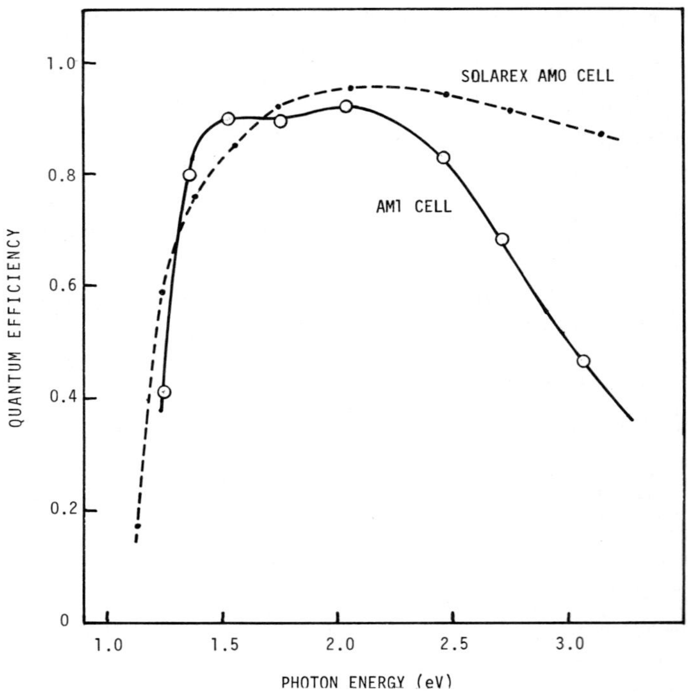

Fig. 7.17. Measured quantum efficiency versus photon energy for typical AM1 Si p/n junction solar cell. (AM1 efficiency for this cell is 13.7%.) Also shown is the response for a textured AM0 Si cell for space use. Reflectance and shadowing losses have not been subtracted from the response curve of the AM0 cell and J_{sc} = 42 mA cm^{-2}. [Data for AM0 cell courtesy of George Storti, Solarex Corp., Rockville, Maryland, 1979.]

quantum efficiency has been extended to >85% at 3.0 eV. Of course, reflection at the front surface also contributes to the J_{sc} loss. The potential for improvement in V_{oc} is discussed in Section 7.5.3.

7.4.4 The CNR Cell

The so-called Comsat nonreflective (CNR) cell is described here as an example of a highly developed device. Its structure, photoelectronic properties, and fabrication technique serve as a technological milestone with respect to cells designed for space use. The band diagram of the CNR cell is essentially that shown in Fig. 7.15. The properties of the quasi-neutral regions are similar to those in Table 7.III and the other cell parameters are listed in Table 7.IV.

This cell is novel in that the front surface is texturized—formed into a three-dimensional surface composed of tiny pyramids 1–2 μm high, as shown in Fig. 7.18. Certain basic etches such as 2–3% NaOH in MeOH are anisotropic, etching (100) oriented Si in such a manner as to expose the (111) planes forming four-sided pyramids of included angle 70.5° at the surface. The junction is formed by P diffusion after texturizing, after which the front grid contact of Ti/Pd/Ag is deposited.

The texturizing serves two purposes: (1) it reduces reflection losses by multiple incidence, and (2) it allows the light to travel in paths that are not perpendicular to the junction interface. Thus photogeneration takes place closer to the junction on the average, increasing η_Q, especially for low-energy photons. Figure 7.19 illustrates the antireflection properties of the texturized surface with respect to planar Si with and without Ta_2O_5 antireflection coatings. The final single layer ARC of Ta_2O_5 on the CNR cell further reduces the reflectivity to ≃2% over most of the useful range. The effect of nonperpendicular light incidence is to increase the "effective"

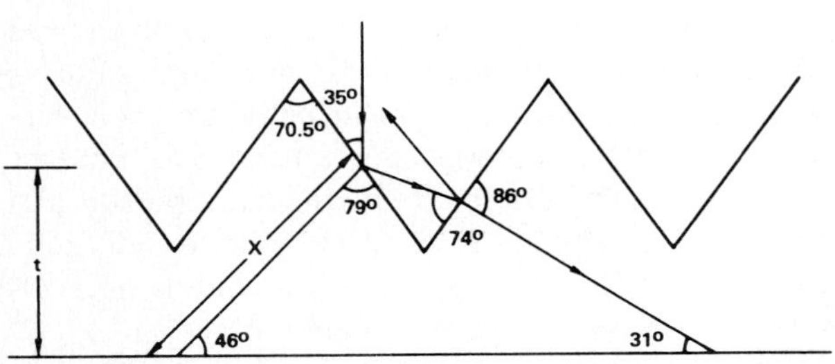

Fig. 7.18. Light paths at an idealized texturized cell surface for a surrounding medium of index 1.0 and a Si index of 3.8. [From R. A. Arndt, J. F. Allison, J. G. Haynos, and A. Meulenberg, Jr., *Proc. 11th IEEE Photovoltaic Specialists Conf.* (1975), p. 40. © 1975 IEEE.]

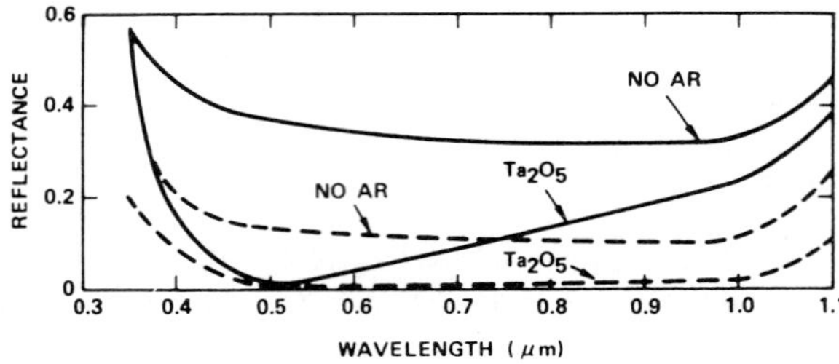

Fig. 7.19. Reflectance as a function of wavelength for normally incident light. (Upper and lower solid curves) Flat, polished Si surface without and with quarter-wave AR coating (Ta_5O_3), respectively. (Upper and lower dashed curves) CNR texturized Si surface without and with quarter-wave AR coating (Ta_5O_3), respectively. [From R. A. Arndt, J. F. Allison, J. G. Haynos, and A. Meulenberg, Jr., *Proc. 11th IEEE Photovoltaic Specialists Conf.* (1975), p. 40. © 1975. IEEE.]

absorption coefficient (e.g., at $\lambda = 0.9\ \mu m$, α_{eff} is increased from 300 to 420 cm^{-1} by texturizing). This cell has $J_{sc} \simeq 46$ mA cm^{-2} at AMO—some 10–15% higher than the best "violet" cell.

7.5 FINE TUNING

In this section considerations applying to conventional cell designs are discussed such as the effect of high-energy radiation on space cells, increases in efficiency brought about by using electric fields to collect carriers or reduce surface recombination, and the optimization of solar efficiency by proper choice of base and diffused layer resistivity.

7.5.1 Radiation Effects

The degradation of solar efficiency by high-energy radiation is crucial in space applications; an unshielded cell would become useless in a matter of days in the high fluxes of hostile particles present in outer space. Since almost every satellite is powered by solar cells, this is a very important concern. The primary effect of radiation damage is to increase the defect concentration in the semiconductor, reducing the minority carrier lifetime and ultimately the cell's efficiency. The tolerance to radiation damage depends on the density and kind of dopants used, and through proper design radiation tolerance can be extended considerably.

The radiation is concentrated in the Van Allen belts by the earth's magnetic field and includes the components and energy ranges shown in Table 7.V. The flux density ϕ and energy is a complex function of position as suggested by Fig. 7.20 and may not be isotropic. In addition, solar flares

Table 7.V

Solar Radiation Characteristics[a]

Type of radiation	Energy	Flux ($cm^{-2}\ sec^{-1}$)	K_L
Protons	1–5 MeV	10^8	10^{-6}
	20–200 MeV	10^4	$10^{-7}-10^{-5}$
Electrons	1–2 MeV	10^5	$10^{-11}-10^{-9}$
	1–100 keV	10^8	$10^{-12}-10^{-11}$

[a] α-particles stopped by cover glass.

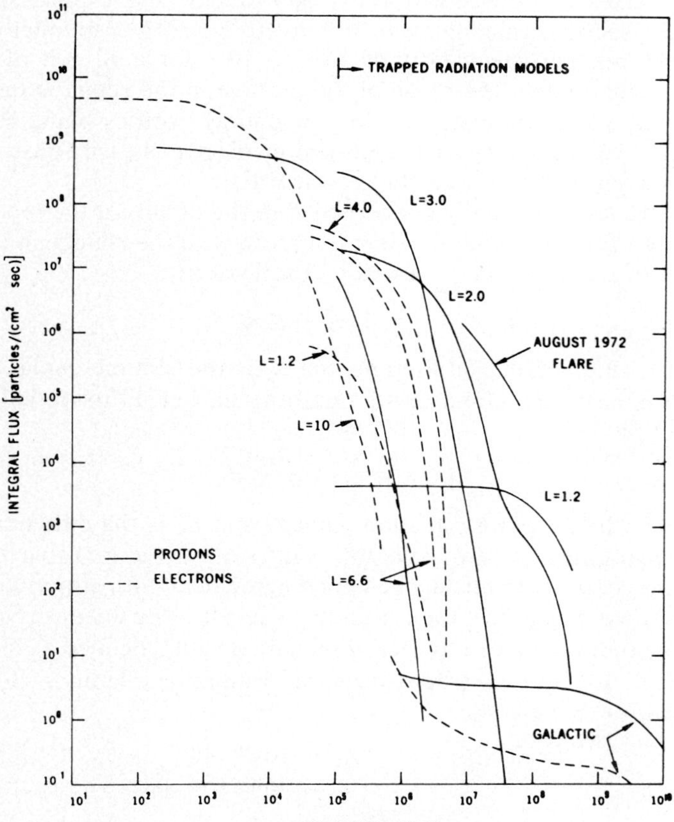

Fig. 7.20. Integral spectra of protons and electrons in the inner magnetosphere (out to about 10 earth radii). The *L* parameter is called McIlwain's shell and it is of the order of the earth radius. [From D. M. Sawyer, K. W. Chan, M. J. Teague, and J. J. Schofield, Jr., *Proc. 13th IEEE Photovoltaic Specialists Conf.* (1978), p. 545. © 1978 IEEE.]

can generate short, intense bursts of radiation. The solar flare event of August 1972 reduced efficiencies of typical solar panels by some 2–5% of their preflare efficiency.

The penetration of a semiconductor by a high-energy particle causes considerable lattice damage such as vacancies, interstitials, defect clusters, and various vacancy–impurity complexes. To quantify the effects of radiation damage the number density of radiation-induced defects acting as effective recombination centers is assumed to vary linearly with the dose: $N_r = K_0\Phi$, where K_0 is the number of centers per centimeter of path length produced by each particle and Φ is the integrated flux over time called the *fluence*. Depending on the character and the energy of the radiation, K_0 values are between 10^{-3} cm^{-1} and 10^2 cm^{-1} corresponding to collision cross sections ranging from 10^{-28} to 10^{-23} cm^2 (or a number of damage events per particle transit of 10^{-6} to 10^{-1} for a Si cell of typical thickness). Of course, the range of the particle in the semiconductor depends on its mass and energy, with low-energy protons being absorbed within a few micrometers while high-energy electrons easily pass through the cell leaving a trail of defects in their wake.

Since we are not usually concerned with the details of the recombination kinetics for each type of defect but more with the change in lifetime, the effect of radiation on τ is usually described by

$$1/\tau = 1/\tau_0 + K_\tau\Phi, \tag{7.6}$$

where τ_0 is the undamaged lifetime and K_τ is the lifetime damage coefficient.† Similarly, the effect on the minority carrier diffusion length L is described by

$$1/L^2 = 1/L_0^2 + K_L\Phi, \tag{7.7}$$

where L_0 is the L before radiation damage and K_L is the diffusion length damage coefficient. A few typical K_L values are listed in Table 7.V.

The preceding formalism gives good agreement with experimental results as shown in Fig. 7.21 for n/p and p/n Si cells. The fluence required to reduce the output of a cell to 75% of the initial value, defined as the "critical fluence" Φ_c, is a convenient measure of radiation tolerance. Some typical values for Si cells are

for electrons: $K_L \simeq 10^{-10}$–10^{-9},
$\Phi_c \simeq 10^{13}$–10^{15} cm^{-2},

for protons: $K_L \simeq 10^{-7}$–10^{-5},
$\Phi_c \simeq 10^{9}$–10^{12} cm^{-2}.

† For a single-level center with a recombination cross section σ and a probability of occupancy P_r, $K_\tau = \sigma v_{th} P_r K_0$.

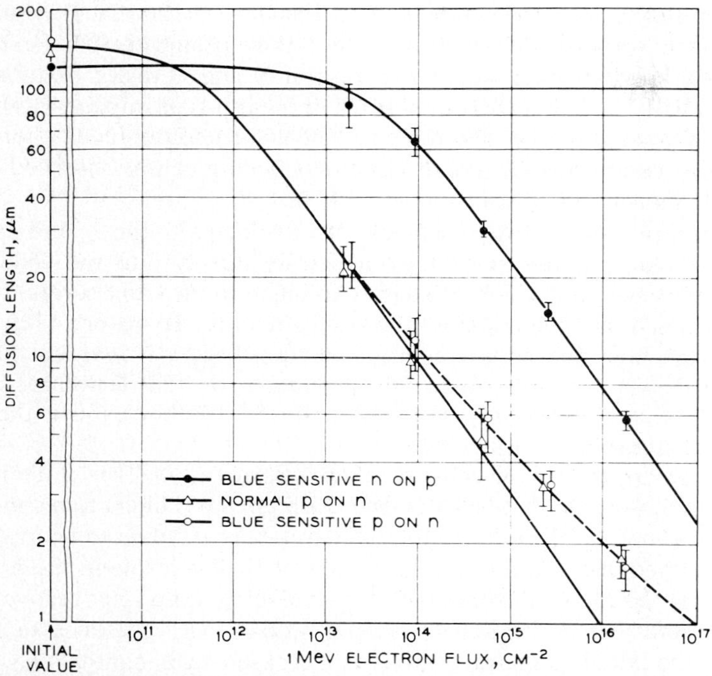

Fig. 7.21. Diffusion length of minority carriers versus radiation fluence of 1-MeV electrons. [After W. Rosenzweig, H. K. Gummel, and F. M. Smits, *Bell Syst. Tech. J.* **42,** 399 (1963). Copyright 1963, AT&T.]

In recent investigations the relation

$$L_0^2/L^2 - 1 = (K_L'\Phi)^n \tag{7.8}$$

with $n \simeq 0.7$ gives better fits to radiation damage data (Bernard and Mottet, 1976).

Since μ_n is greater than μ_p, giving a larger diffusion length for photogenerated electrons in p-type materials, the n/p cells tend to be more radiation tolerant than p/n cells. In addition, radiation tolerance is reduced as the base doping density is increased. It is thought that the dopant associates readily with the radiation-induced defects to form active recombination centers and increasing the doping density increases the probability for forming the centers. In this regard Al-doped p-layers appear to be somewhat more radiation tolerant than B-doped layers.

Differences in FZ- and CZ-grown material are also noticed and have been correlated with the relatively high O and C concentrations in the CZ material. Although the FZ material produces higher initial η_s, the Φ_c for

CZ material is larger (Pschunder and Fischer, 1976). The FZ material experiences "photon degradation" after a short period (24 hr) of exposure to light following irradiation.† Addition of O and C to FZ material (Pschunder and Fischer, 1976) or using Ga or Al as *p*-type dopants (Rahilly *et al.*, 1976) appear to alleviate the effect of these photoinduced changes in the defect complexes. A similar effect has been recently observed in terrestrial Si solar cells (Weizer *et al.*, 1978).

In general, high-energy particles affect principally the J_L and particularly the red response since they penetrate deeply into the cell. Low-energy particles, for which damage is confined to the front layers, tend to lower V_{oc} and the blue response more significantly. To absorb a large portion of the high flux of low-energy radiation before it reaches the cell, "cover glasses" of glass, fused SiO_2, sapphire, or FEP Teflon are almost universally used (typically $\simeq 0.15$ mm thick). Without these the cells would be unusable in a few weeks.

The Li-doped p/n Si cell (ca. 1966) (Berman, 1972) is a device designed to increase radiation tolerance. Lithium has a high atomic mobility and drifts (at $T > 20-50°C$) to the radiation-caused defect to neutralize its electrical activity. Larger concentrations of O play a role in the process, increasing the recovery time, but also producing more total recovery and more stability than for low O concentrations. The presence of Li decreases the initial η_s slightly, but Φ_c is increased significantly. The details of Li-induced recovery are still not well understood, the recovery parameters having been related to the gradient of the Li concentration (Faith, 1972) and to the difference between the bulk Li concentration and the concentration of defects. Li-doped cells with η_s up to 12.8% have been produced.

Drift fields (see Section 7.5.2) have yielded some improvement in radiation tolerance but at the expense of initial efficiency.

The radiation resistance of cells made from direct gap materials such as GaAs and Cu_xS/CdS is considerably better than that of Si because of the much shorter absorption length for photons for these materials. Damage occurring more than $\simeq 6$ μm from the surface for GaAs and $\simeq 2$ μm for Cu_xS/CdS has virtually no effect on efficiency. Of course, low-energy particle protection using cover glasses is even more crucial with these cells. Cu_xS/CdS cells have been shown to have a Φ_c for 100 keV electrons about two decades greater than that for Si cells (Van Aerschodt *et al.*,

† Irradiation in the dark appears to make the Si susceptible to a process that is completed by the action of light and causes the lifetime to be reduced. The effect appears to be similar to "photochemical reactions" occurring in CdS and other materials (Tscholl, 1968; Im *et al.*, 1970; Fahrenbruch and Bube, 1974).

1971). Unfortunately the Cu_xS/CdS cell has never reached high enough efficiencies to be practical in space. Despite the high efficiency of the GaAs cell, space use has been restricted to powering a few Russian satellites and some experimental testing at this writing.

7.5.2 Electric Fields to Increase Efficiency

In Section 4.5.1 we showed that an electric field could be used to effectively increase the minority carrier diffusion length L in a bulk semiconductor. The effective diffusion length was given by $L_{eff} \simeq L_0(\mathscr{E}/\mathscr{E}_c + 1)$, where $\mathscr{E}_c = kT/qL_0$ and L_0 is the zero field L. Electric fields have been used in three principal ways to improve the efficiency of Si cells:

(1) drift fields to increase L_{eff} in the bulk of the absorber (base) layer;

(2) back-surface fields (BSF) to drive carriers away from the back surface and to reduce the effective surface recombination velocity S there; and

(3) front-surface fields to accelerate carriers generated in the front layer toward the junction, decreasing the effective S there and also reducing the recombination in the bulk of the thin front-surface layer.

Drift fields are useful for material with short L such as low-cost Si or polycrystalline material and may have application to radiation-tolerant cells. The fields are induced by gradients in the doping density $N_A(x)$ for the example of a p-type absorber

$$q\mathscr{E} = (dE_c/dx) = (dE_v/dx) = (kT/N_A)(dN_A/dx), \qquad (7.9)$$

and the usual approximation is an exponential doping profile so that the field is constant. The selection of the best magnitude for the field depends on a compromise between reducing V_{oc} and increasing J_L due to the increased L. For example, if we start from the back of the cell and establish a doping gradient to produce the built-in field, then at the junction the doping density must be somewhat less than for an optimal zero-field cell. This increases J_0 and decreases V_{oc} relative to the zero-field cell. Using computer solutions, Van Overstraeten and Nuyts (1969) showed that the drift field in the 10–30 μm of the absorber layer (base) closest to the junction was the most important for increasing the efficiency. Computer modeling of such devices made from high-lifetime Si indicates only a slight improvement in initial efficiency but considerable improvement in radiation tolerance (Baraona and Brandhorst, 1976).

Not only are characteristics of such devices difficult to assess theoretically since μ, τ, and $\mathscr{E}$ are all usually changing with distance, but the devices are difficult to fabricate in some cases as well. A particularly successful method is by epitaxial growth of Si layers so that N_A can be varied during growth.

Back-surface fields (BSF) are of particular importance with regard to the trend toward thinner absorber layers. The BSF acts as a mirror for photogenerated carriers, effectively decreasing the back-surface recombination velocity S by the ratio $\simeq N_A/N_A^+$ (for $y^+ \ll L^+$), where N_A, N_A^+; y_p, y^+; and L_p, L^+ are the doping levels, thicknesses, and diffusion lengths in the bulk absorber and p^+-layers, respectively. In addition, the BSF confines the forward-bias injected carriers particularly for $y_p/L_p \lessapprox 1.5$, tending to decrease J_0 in the ratio N_A/N_A^+ and increasing V_{oc} substantially for thin cells. A more exact formulation is given by Godlewski *et al.* (1973) in which the surface recombination velocity S is replaced by S_{eff} given by (S^+ is the surface recombination velocity of the p^+-layer)

$$S_{eff} = \frac{N_A D^+}{N_{A^+} L^+}\left[\frac{(S^+L^+/D^+)\cosh(y^+/L^+) + \sinh(y^+/L^+)}{(S^+L^+/D^+)\sinh(y^+/L^+) + \cosh(y^+/L^+)}\right], \quad (7.10)$$

where D^+ is the diffusion coefficient in the p^+-layer [cf. Eq. (5.15)]. When the p^+-layer is somewhat thicker, the configuration is sometimes called a high–low junction. For small absorber thicknesses the cell behaves like a p^+/n junction with accordingly large V_{oc}. For cell thickness such that $y_p/L_p \gtrsim 2.0$, the results converge to the infinite thickness values and no effect of a BSF is seen even for $S \rightarrow \infty$. The BSF may be thought of as a way to obtain the high V_{oc} inherent in cells with *high* N_A adjacent to the junction without sacrificing the higher J_L generating capacity of *low* N_A material. The BSF field cell has been treated by computer modeling by von Roos (1978) and BSF theory is discussed by Fossum *et al.* (1980).

In practical cells utilizing BSF, increases in red response are seen as well as increased radiation tolerance due to thinner Si layers. Increases in V_{oc} of 20–80 meV have been reported (Mandelkorn and Lamneck, 1972). The fabrication of BSF cells allows simplified "alloy-through" techniques to be used in which the p-type wafer is n-diffused, both front and back; then an evaporated Al film is diffused through the back n-layer to form a p^+-layer about 1 μm thick. Epitaxial layers and B diffusions have also been used to obtain the p^+-layer (Mandelkorn *et al.*, 1973).

High V_{oc} values have also been seen in cases with a thick absorber layer where $y_p/L_p \gtrsim 3$, thus requiring a different explanation. In these cases the increase in V_{oc} has been attributed to the small V_{oc} component generated at the high–low junction between the p- and the p^+-layers (Mandelkorn and Lamneck, 1975). In general, the increases in V_{oc} are probably made up of contributions from all three mechanisms mentioned earlier.

In principal a front-surface field (FSF) should offer the same advantages as the BSF. The FSF probably already exists by virtue of the doping profile in modern high-efficiency cells with very thin front layers. However, the front-surface (diffused) layer is usually so thin and the doping levels so high with correspondingly short lifetimes that other considerations take precedence. Indeed, Godlewski *et al.* (1973, p. 32) have proposed that for very high doping levels a retrograde field may exist at the surface that lowers efficiency by a small fraction.

Lindholm *et al.* (1978) have demonstrated significantly increased V_{oc} (to 0.640 V) using the high–low junction configuration for the front layer (called HLE for high–low emitter). The reverse saturation current J_{0E} for the front layer of these cells has been reduced to 10^{-14} A cm^{-2} and V_{oc} = 0.70 V was predicted.

7.5.3 Optimization of "Conventional" Cells

There continues to be much interest in raising the solar efficiency of "conventional" Si cells toward the 22% practical maximum predicted by Wolf (1971). Back-surface fields are now quite standard for high-efficiency cells. Increases in base-layer lifetime have almost reached their practical limits in optimizing J_L,† reflectance loss has been minimized in the CNR cell, and further increases in J_L must come from an increase in blue response. Increases in the ff and V_{oc} have been pursued by optimization of base- and diffused-layer carrier concentrations and concentration profiles. Aside from the thickness, these are perhaps the easiest parameters to modify in the construction of a cell. In the usual high-efficiency cell the doping of the diffused emitter layer (usually *n*-type; N_D) is made as high as possible to reduce R_s and to increase V_{oc}. The base layer doping N_A can be varied. Most of the possibilities mentioned earlier (increasing blue response, increases in ff, and V_{oc}) lead to reconsideration of the diffused *n*-layer.

Wolf's (1971) early theory predicted that as the carrier concentration on both sides of the p/n junction was increased, the reverse saturation current J_0 would continue to decrease and η_s would reach a maximum of $\simeq$22% at AM0 for $N_A = 3.5 \times 10^{18}$ cm^{-3}. Of course, when the acceptor concentration N_A (for a *p*-type base) becomes too high, decreases in μ_n and τ_n will decrease J_L, beginning to have an appreciable effect on the efficiency for $N_A > 10^{18}$ cm^{-3} ($\rho \simeq 0.05\ \Omega$ cm). An extension of Wolf's theory by Brandhorst (1972) and co-workers predicts a practical maximum η_s of 18% for a base acceptor density N_A of $\simeq 4 \times 10^{18}$ cm^{-3}. How-

† Although the portion of the loss in J_L that occurs in the long wavelength region is still not negligible.

ever, later experimental work showed the maximum η_s to lie at a somewhat lower value, $N_A \simeq 4 \times 10^{17}$ cm^{-3} (Iles and Soclof, 1975; Godlewski *et al.*, 1975). In particular, V_{oc} is smaller, reaching a broad maximum at $\simeq 0.62$ V between $N_A = 10^{17}$ and 10^{18} cm^{-3} and slowly falling for larger values of N_A. By comparison simple theory predicts a continual increase of V_{oc} with N_A and a value of $\simeq 0.7$ V for $N_A = 10^{18}$ cm^{-3} (see Fig. 11.3). The J_L values usually lie within the predicted range or can be accounted for by low τ_n. The discrepancy between measured and predicted V_{oc} was pointed out by Brandhorst (1972) and others in 1972 and it became a subject of great interest by 1975–1977 when several theories were advanced.

The dark, reverse-bias saturation current J_0 consists of two components: the electron current injected into the p-type base J_{0n} (controlled by μ_n, τ_n, and S_n in the base layer for the example of a n^+/p cell) and the hole current injected into the n-type emitter layer J_{0p} (μ_p, τ_p, and S_p). The base component J_{0n} behaves essentially as expected and dominates the emitter current J_{0p} for thin n-layers (0.2–0.5 μm) with longer τ_p ($>10^{-8}$ sec) and lower S_p($<10^4$ cm sec^{-1}). However, a change in any or all of these emitter parameters may cause J_{0p} to contribute strongly to J_0 and thus decrease V_{oc}. If J_{0n} is reduced sufficiently, as is desirable to obtain higher V_{oc}, then J_{0p} controls J_0 as shown schematically by curves c and d in Fig. 7.22. Because of the thinness of the n-layer and its extremely high N_D (10^{19}–10^{20} cm^{-3}) as well as the almost certain presence of large gradients in N_D, τ_p, and μ_p, these parameters as well as S_p are difficult to measure precisely and are even more difficult to model satisfyingly. For example, Lindholm *et al.* (1975) suggest a linearly graded N_D with $dN_D/dx = 10^{23}$ cm^{-4}.

Most of the present theories for the saturaton of V_{oc} point to J_{0p} being much larger than predicted (curve b in Fig. 7.22). Reasons advanced for this revolve around various so-called heavy doping effects such as

(1) impurity clustering leading to a junction that is inhomogeneous in a three-dimensional sense and enhanced tunneling through depletion layer defects, pointed out by Lindholm *et al.* (1975),

(2) band-gap narrowing in the highly doped n-layer due to impurity banding, band tailing, and strain effects as suggested by Dunbar and Hauser (1975, 1977),

(3) a retrograde field established by a decreasing carrier density as a function of the increasing N_D, which increases the effective surface recombination as proposed by Godlewski *et al.* (1975).

(4) band-gap narrowing due to high electric fields, the thermal band-gap equivalent of the Franz–Keldysh effect, and the effect on the generation/recombination current through the junction as proposed by Rittner (1977), and

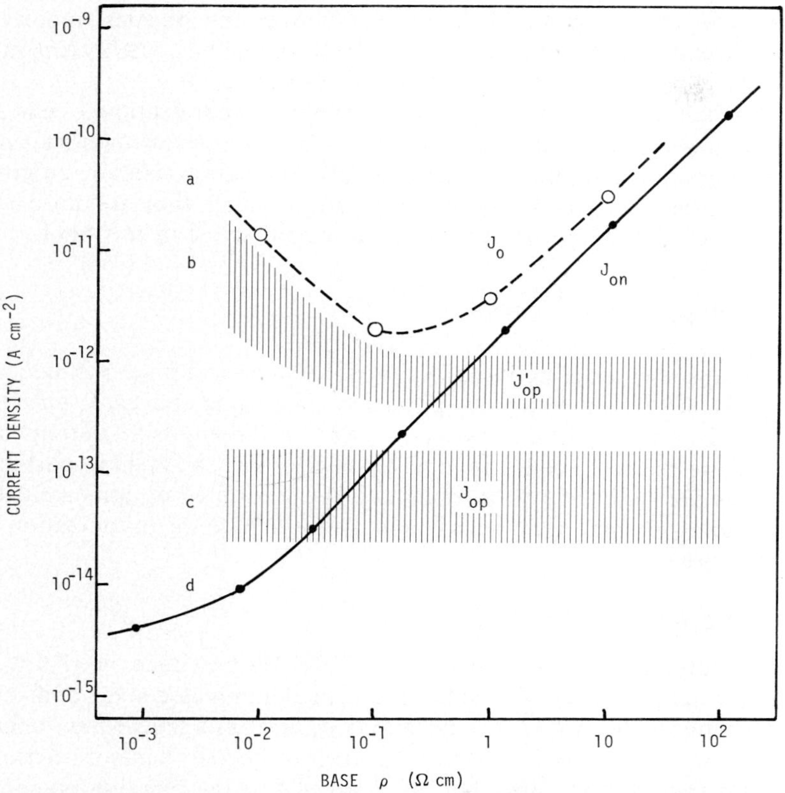

Fig. 7.22. Diode saturation current versus base resistivity for high-efficiency solar cells. (a) Total J_0 obtained from measured values of V_{oc} and J_{sc}. (b) Range of J_{0p} including heavy doping effects. (c) Range of calculated J_{0p} without heavy doping effects. (d) J_{0n} calculated from measured values of τ and L_n in good-quality Si.

(5) the effects of the emitter surface recombination combined with band-gap narrowing (Fossum *et al.*, 1979; Shibib and Fossum, 1981).

Most of these considerations have been incorporated into a computer modeling study by Hauser and Dunbar (1977), who predict that η_s as high as 17.5 to 18.5% could be obtained (AM0, total area basis, including R_s and grid coverage losses).

The importance of the heavy doping effects is emphasized by consideration of the inversion-layer cell of Blakers and Green (1981), where the high carrier density in the *n*-layer is induced by oxide charge rather than by doping. These cells produce V_{oc} values of 0.678 V (see Fig. 11.3).

There was still considerable controversy over the physics underlying heavy doping effects (Shibib *et al.*, 1979; Fossum *et al.*, 1979; Amantea, 1980; Lanyon, 1981; and Redfield, 1980, 1981).

Over the years considerable improvement in conventional cells has been obtained by various geometrical and optical improvements such as optimization of collecting grids, contacts, and antireflection coatings. Although these increases in η_s are by no means small, they are more technological in nature and lie outside the principal scope of this book.

7.6 NOVEL APPROACHES

Perhaps the most powerful methods of increasing η_s and decreasing costs are by innovative approaches—changing the basic structure of the cell. This is particularly true in view of the intensive optimization already applied to Si cells. Several novel approaches aimed at particular purposes are described to demonstrate the wealth of options available. These relatively new approaches should also yield to the optimization discussed earlier.

7.6.1 Vertical Multijunction Cells

The term *vertical multijunction cell* (VMJ) has been used for two somewhat different types of cells: (1) an edge-illuminated stack of discrete p/n junction cells whose V_{oc} are additive, and (2) a single p/n junction with a convoluted junction plane. In both of these cells a major fraction of the light traverses the absorber layer parallel to the junction plane. We refer to the former as a *stacked* VMJ and the latter as a *convoluted* VMJ.

The stacked VMJ is designed to increase device V_{oc} for concentrating systems. The configuration is shown in Fig. 7.23. The cell is fabricated by first making many conventional $p^+/n/n^+$ cells each with continuous contacts covering the entire n and p faces. These contacts are "tinned" and the cells are stacked 50–100 layers deep. The stack is soldered (or brazed) in a furnace and then sliced perpendicular to the junction planes to form each VMJ cell. Polishing the side where the light enters and applying an antireflection coating completes the cell. Since light enters parallel to the junction the η_Q is high and both red response and radiation tolerance are improved. There is essentially no grid loss and only the width of the solder joints is lost to incoming light, amounting to $<5-7\%$ of the area.

The surface recombination velocity on the light entrant side of the cell must, of course, be low but the n-layer can now be thicker and can be doped to optimize V_{oc} without regard for sheet resistance. Since V_{oc} for the stack is high (30–60 V), the current density is low and the contact resistances are minimized by the large area contacted, so these cells appear

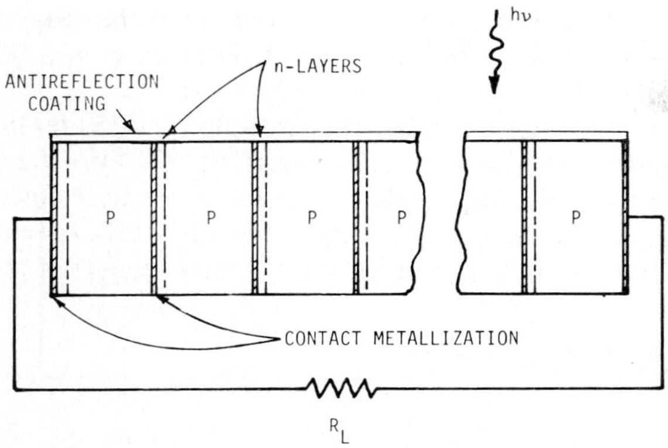

Fig. 7.23. Cross section of stacked vertical multijunction cell. Typical cells comprise 50–100 such individual cell elements.

to be good choices for concentrator systems. Despite the promise of this concept, η_s had reached only 8% (AM0) in 1976 (Goradia and Goradia, 1976), because of nonoptimal design parameters. With optimization its efficiency at high concentration ratios is expected to far surpass conventional cells. One difficulty is that the minority carrier diffusion length must be greater than the width of each of the *n*-layers to avoid excessive recombination losses. This dictates the use of somewhat higher resistivity Si than would be used if V_{oc} were to be maximized. Since the total current of the stack is limited by the lowest current generated by any of the elements, the illumination must be uniform; this is a potential difficulty in concentrator systems.

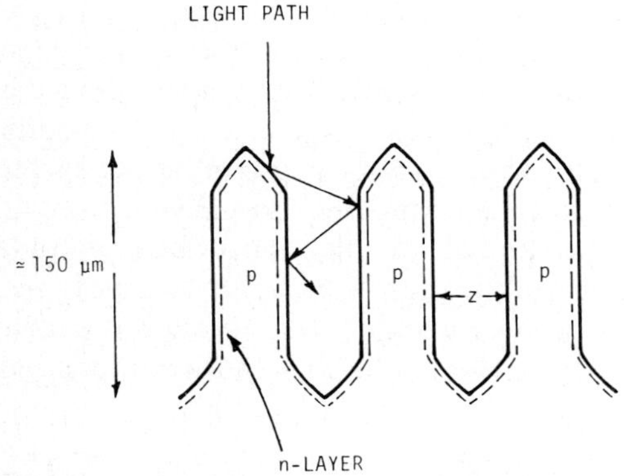

Fig. 7.24. Cross section of a small portion of a convoluted vertical multijunction cell. The widths of the grooves and lands z is between 7 and 50 μm.

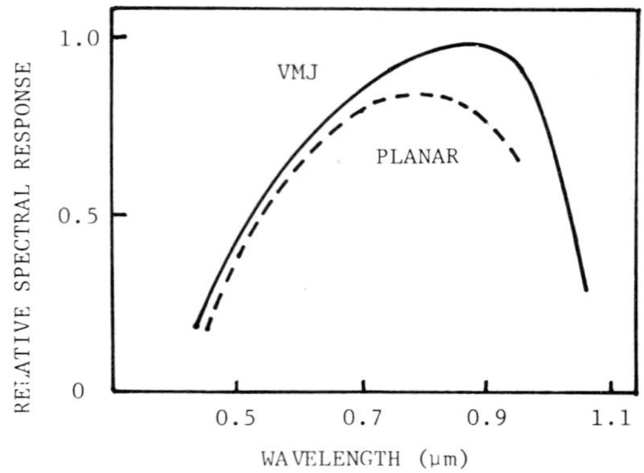

Fig. 7.25. Comparison of measured relative spectral response for planar and VMJ cells processed similarly. [Redrawn from W. W. Lloyd, *Proc. 11th IEEE Photovoltaic Specialists Conf.*, (1975), p. 349. © 1975 IEEE.]

The convoluted VMJ design [also called the etched-multiple-vertical-junction cell (EMVJ)] is specific to increased radiation tolerance and to use in concentrating systems (Frank and Goodrich, 1980). In the cross section shown in Fig. 7.24, the grooves (7–50 μm wide × 150 μm deep) are fabricated by etching and the angled bottoms of the troughs trap the light to automatically produce antireflection properties similar to the CNR cell. Solar efficiencies of $\simeq 13\%$ have been measured with J_{sc} = 40 mA cm^{-2} at AM0 (Lindmayer and Wrigley, 1976, p. 30). Figure 7.25 compares the increased red response of these cells with that of conventional Si cells. Of course, R_s and gridding remain problems.

7.6.2 Tandem Junction Cell†

An interesting design is the so-called thin tandem junction cell (TTJC) developed by Chiang *et al.* (1978a, 1978b). In this device both n and p contacts are alternated on the back of the cell with an additional n^+-layer on the front of the cell, which may or may not be connected (shown in Fig. 7.26). The front surface is texturized and a BSF is used at the rear surface contact. In addition, the optical reflection at the rear contact is maximized by adjusting contact fabrication parameters (see next section) so that the remaining light can make another pass through the absorber layer. The back collection largely eliminates back-surface recombination losses.

† The term *tandem junction* has been applied to both the cell configuration described here and to the multi-bandgap, stacked configuration described in Chapter 12.

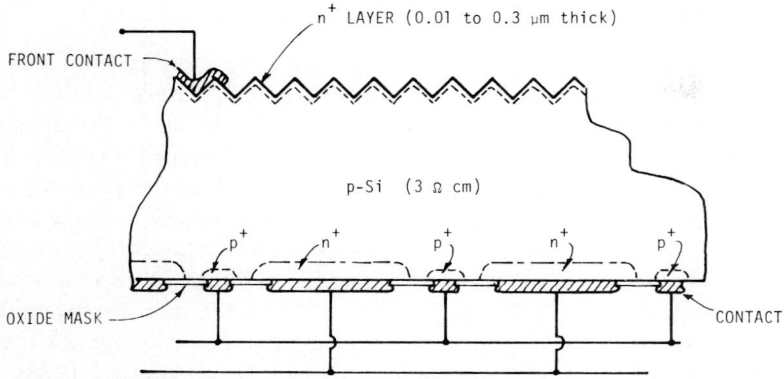

Fig. 7.26. Tandem junction cell. Back contact coverage: n^+, 80% and p^+, 15%. [Redrawn from S. Y. Chiang, B. G. Carbajal, and G. F. Wakefield, *Proc. 13th IEEE Photovoltaic Specialists Conf.* (1978), p. 1290. © 1978 IEEE.]

Although the greatest efficiencies have been realized with both front and back contacts connected, the front contact can be left floating at V_{oc}, which reduces the effective surface recombination there (Landis, 1981). The total recombination current in the absorber layer is given by $\int_x^{x'} U(x)\,dx$ and, partially because of the reflection of light at the back surface, the J_L of this cell *increases* with decreasing cell thickness W, in contrast to the asymptotic saturation at large W seen in conventional cells. The AM0 parameters for preliminary 2.7-Ω cm cells are given in Table 7.VI. The potential advantages of the TTJC are the absence of front grid loss making the cell ideal for concentration systems, a greater power-to-weight ratio, and simplification of assembly into panels, since the connections are on one side.

Table 7.VI

Tandem Junction Cell Characteristics

	J_{sc} (mA cm^{-2})	V_{oc} (V)	η_s (%)
Both front and back connected (measured) (W = 90 μm)	46	0.57	—
Back only connected (measured) (W = 110 μm)	43	0.572	13.5
Practical maximum (calculated) (W = 50 μm)	47	0.73	20.0

7.6.3 Thin Cells

In these designs the power-to-weight ratio is optimized. Cells are thinned from sawed wafers by etching in a hot NaOH solution to a thickness of 50–60 μm. The time and temperature of the back-surface Al alloy step can be manipulated to yield optimal reflectance for long-wavelength light passing through the cell, giving a second pass. Devices utilizing this mode are called back-surface reflection (BSR) cells (Rasch *et al.*, 1980; Chai, 1980). Reflectance values up to $\simeq 0.8$ can be obtained although lower values are chosen in practice because of other considerations (Wrigley, 1977). Efficiencies of 14.3% at AM0 have been obtained in pilot plant production and power-to-weight ratios are approximately 1250 W_p kg^{-1} (Lindmayer and Wrigley, 1976; Scott-Monck, 1978). Because the recombination loss region is thinner, those cells also have an increased resistance to high-energy radiation.

7.6.4 Other Approaches

There are of course other approaches that are more divergent from the conventional single-crystal Si cell than those discussed previously. To reduce fabrication costs polycrystalline structures can be used. They have lower efficiency in general but eliminate many fabrication steps. This technology is not limited to Si and is discussed in detail in Chapter 9. Configurations such as Schottky, MOS, and MIS junctions can have higher blue response and eliminate the diffusion step in fabrication (which is made more complex for the case of polycrystalline materials by the grain boundaries). These are discussed in more detail in Chapters 5 and 11. However, a few approaches closer to the single-crystal Si field are mentioned here.

To reduce the cost of the Si crystal, directional solidification casting has been used rather than the conventional crystal growth techniques. In this case polycrystalline material is produced but the grains are fibrous and when the material is sliced perpendicularly to the fiber axis, and when the grains are sufficiently large, the resulting cells are much like their single-crystal cousins. Fischer and Pschunder (1976; Fischer, 1978) reported such cells having $\eta_s \lesssim 12\%$ when the average grain size was above several millimeters (AM0). Fischer presents an interesting curve of the relation of solar efficiency to grain size shown in Fig. 7.27. Lindmayer (1978a) has reported polycrystalline cell efficiencies up to 16% (AM1).

Further in the direction of polycrystalline material are the efforts of Chu *et al.* to grow *p*-Si layers on large-grained metallurgical grade Si, ceramic, or metal substrates by CVD and then to complete the cells with an epitaxially grown *n*-layer. Efficiencies of 6–10.6% (AM1) were ob-

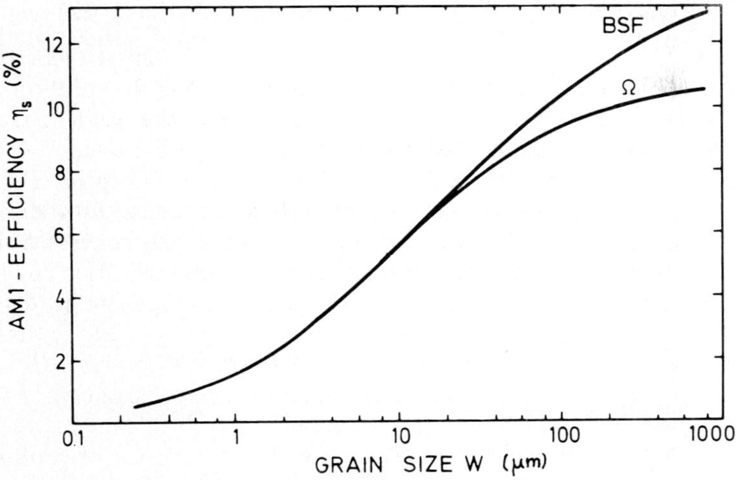

Fig. 7.27. Solar conversion efficiency as a function of grain size for a ''nonsingle-crystalline'' Si cell (calculated for an n^+/p cell with $\rho_p = 5\ \Omega$ cm). The curve marked BSF is for cells with back-surface fields whereas Ω donotes a back-surface recombination velocity approaching infinity. [From H. Fischer, *Proc. 1st EC Photovoltaic Solar Energy Conf.*, Luxembourg, 1977, p. 52. D. Reidel Publ., Dordrecht, Netherlands, 1978.]

tained for these devices (Chu *et al.*, 1978; Robinson *et al.*, 1978).

In polycrystalline material the fabrication of the thin, diffused layer is complicated by rapid diffusion of dopant along grain boundaries. If all grain boundaries are perpendicular to the plane of the cell, this situation may be used to advantage since a sort of vertical multijunction cell could be made (cf. Fig. 7.24). However, if the orientation of the grain boundaries is more random, other means of creating the junction may be more favorable. These include deposition of transparent metal (and insulator) layers to form Schottky barrier or MIS cells and the use of ion implantation doping so that little grain boundary diffusion occurs. MIS cells offer ease of fabrication and good blue response, and solar efficiencies up to 10.5% (AM1) were achieved in 1978 with polycrystalline Si (Green *et al.*, 1978).

Another approach is the design of single-crystal Si cells for systems using moderate concentration ratios ($\lesssim$300). Such designs incorporate modifications to decrease series resistance (more efficient grid designs, lower contact and bulk resistivities), improve heat conduction, and maximize efficiencies (even at the expense of higher cost, since system cost varies inversely with cell efficiency). Back-surface minority carrier confinement is particularly important in order to realize the potential of such cells (Hu and Drowley, 1978). Si concentrator cells have demonstrated $\eta_s = 17\%$ at 400 suns and O'Donnell *et al.* (1978) believe that 20% at 500 suns can be obtained.

7.7 ECONOMICS AND INNOVATION

Further reduction in the cost of photovoltaics is crucial for their widespread use in terrestrial solar conversion. There are several possible broad areas of cost reduction:

(1) raw materials,
(2) purification,
(3) fabrication, and
(4) innovation.

The first three items yield most readily to the experience of production and the accumulation of enough volume is a critical factor in reducing the cost. For this reason the United States Department of Energy and the European Common Market have artificially stimulated research and development as well as certain markets for solar cells, such as prototype solar generation plants and field demonstrations. Despite its unpredictability, innovation has the potential to make even more dramatic reductions in cost.

An interesting perspective was presented by Maycock and Wakefield (1975) with a "learning curve" graphing the cost of Si cells in revenue per peak watt versus the industry accumulated volume of cells (Fig. 7.28.) The electronic calculator market has enjoyed a 70% slope for several orders of magnitude but virtually no device has experienced such a decrease over the six to eight orders required for widespread use of solar photovoltaics. The solid curve shows the result of the experience of some 25 years on the price of Si cells and the dashed curve represents the impact that the DOE† programs hope to have on this cost. Figure 7.28 suggests that these programs, which began to have appreciable funding in 1972, are quite successful. The policy of DOE has relied heavily on the innovative aspect of cost reduction and the programs reflect this in funding both basic and applied research in many different areas of photovoltaics involving Si as well as many less-well-known materials. The goal is such that the cost will permit a large-scale intrusion into the total electrical power needs of the United States that will expand to about 3% by the year 2000—saving 10^6 barrels of oil per day!

Because of the increased awareness that energy and raw materials are not inexhaustible, some new considerations have appeared. How long

† Early programs were funded through NSF and ERDA.

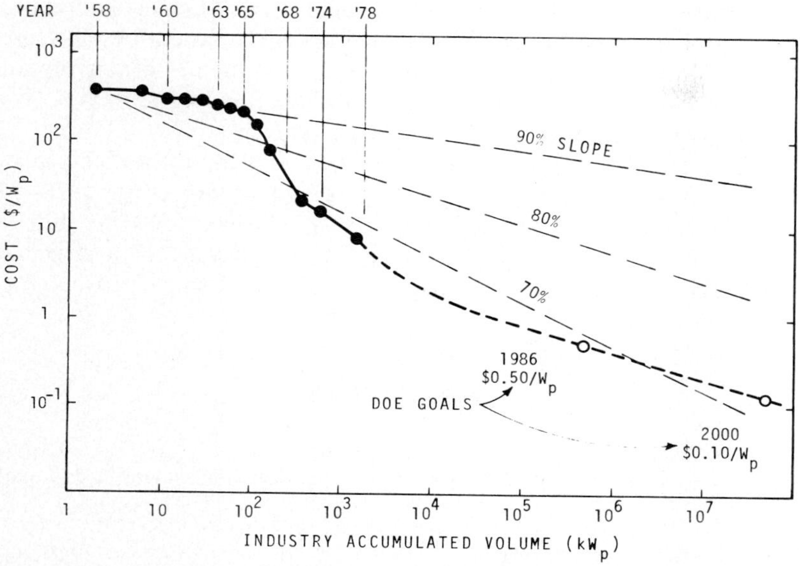

Fig. 7.28. Learning curve for Si solar cell production. A 70% slope means a 30% reduction in consumer cost for each doubling in volume. The ERDA-DOE goal for the year 2000, a production of 5×10^7 kW$_p$ per year, represents about 3% of the total expected electrical energy demand for the US in that year. [Redrawn from P. B. Maycock and G. F. Wakefield, *Proc. 11th IEEE Photovoltaic Specialists Conf.*, (1975), p. 252. © 1975 IEEE, and other sources.]

does it take for a solar cell system (or any other power-generating system for that matter) to pay back the energy required to produce the system itself (Lindmayer, 1978b)? Are there enough raw materials to produce an appreciable contribution to the total power-generating capacity? The answers to these questions rely heavily on experience gained in Si cell production and utilization.

Although Si is inherently abundant and cheap, purification makes it expensive in terms of both energy and other raw materials (as suggested by Fig. 7.2). A great deal of effort is going into finding cheaper ways of producing purified Si. Fundamental to this task is defining a solar grade of Si: What impurities can be tolerated and which are highly detrimental to solar efficiency (Wakefield *et al.*, 1975; Hill *et al.*, 1976; Davis *et al.*, 1978)? An exhaustive study by Hunt (1975) of Dow-Corning selects about 20 chemical "routes," out of some 150 possible reactions, to produce \$10 kg^{-1} solar grade Si.

The cost of Si wafers to the solar cell fabricator (about \$1.50–\$2.50 per peak watt) represented a rather minor part of the total array cost of \$8–\$15 per peak watt in 1978. Thus other studies have concentrated on low-cost fabrication techniques such as Ralph's (1975) investigation of the feasibility of automation, the fabrication of cell contacts by silk screen processes (Ralph, 1975; Frisson *et al.*, 1978), and the use of ion implantation with subsequent laser annealing to produce thin emitter layers (Kirkpatrick *et al.*, 1978; Muller *et al.*, 1978).

Several studies of the effect of production scale-up indicate a possible array cost of $2 per peak watt in large quantities, using 1978–1980 technology (Grenon and Coleman, 1978; Carbajal, 1978). For example, Bickler *et al.* (1978) give a good cost analysis for a large-scale (20 MW per year) fabrication process for terrestrial Si cell arrays.

REFERENCES

Amantea, R. (1980). *IEEE Trans. Electron Devices* **ED-27,** 1231.

Baraona, C. R. and Brandhorst, H. W., Jr. (1976). *Proc. 12th IEEE Photovoltaic Specialists Conf.,* p. 9.

Berman, P. A. (1972). *Proc. 9th IEEE Photovoltaic Specialists Conf.,* p. 281.

Bernard, J., and Mottet, S. (1976). *Proc. 12th IEEE Photovoltaic Specialists Conf.,* p. 262.

Bickler, D. G., Gallagher, B. D., and Sanchez, L. E. (1978). *Proc. 13th IEEE Photovoltaic Specialists Conf.,* p. 241.

Blakers, A. W., and Green, M. A. (1981). *Appl. Phys. Lett.* **39,** 483.

Brandhorst, H. W., Jr. (1972). *Proc. 9th IEEE Photovoltaic Specialists Conf.,* p. 37.

Brandhorst, H. W., Jr. (1977). *Jpn. J. Appl. Phys.* **16,** Suppl. 16-1, 399.

Cabajal, B. G. (1978). *Proc. 13th IEEE Photovoltaic Specialists Conf.,* p. 252.

Chai, A-T. (1980). *Proc. 14th IEEE Photovoltaic Specialists Conf.,* p. 156.

Chandler, T. C., Jr., Hillborn, R. B., Jr., and Faust, J. W., Jr. (1976). *Proc. 12th IEEE Photovoltaics Specialists Conf.,* p. 282.

Chapin, D. M., Fuller, C. S., and Pearson, G. L. (1954). *J. Appl. Phys.* **25,** 676.

J.-W. Chen and A. G. Milnes (1980). Energy levels in silicon. *In* "Annual Review of Materials Science" (R. A. Huggins *et al.,* eds.), Vol. 10, p. 157. Annual Reviews, Inc., Palo Alto, Calif.

Chiang, S. Y., Carbajal, B. G., and Wakefield, G. F. (1978a). *Proc. 13th Photovoltaic Specialists Conf.,* p. 1290.

Chiang, S. Y., Carbajal, B. G., and Wakefield, G. F. (1978b). *IEEE Trans Electron Devices* **ED-25,** 1405.

Chu, T. L., Chu, S. S., Duh, K. Y., and Yoo, H. I. (1976). *Proc. 12th IEEE Photovoltaic Specialists Conf.,* p. 74.

Chu, T. L., Chu, S. S., Stokes, E. D., Lin, C. L., and Abderrassoul, R. (1978). *Proc. 13th IEEE Photovoltaic Specialists Conf.,* p. 1106.

Ciszek, T. F. (1972). *Mater. Res. Bull.* **7,** 731.

Ciszek, T. (1982). *Proc. 16th IEEE Photovoltaic Specialists Conf.* (to be published).

Ciszek, T. F., and Schwuttke, G. H. (1975). *Phys. Status Solidi A* **27,** 231.

Dale, B., and Rudenberg, H. (1960). *Proc. 14th Annu. Power Sources Conf.*

Davis, J. R., Rai-Choudhury, P., Blais, P. D., Hopkins, R. H., and McCormick, J. R. (1976). *Proc. 12th IEEE Photovoltaic Specialists Conf.,* p. 106.

Davis, J. R., Rohatgi, A., Rai-Choudhury, P., Blais, P. D., and Hopkins, R. H. (1978). *Proc. 13th IEEE Photovoltaic Specialists Conf.,* p. 490.

Dermatis, S. N., Faust, J. W., Jr., and John, H. F. (1965). *J. Electrochem. Soc.* **112,** 792.
Dunbar, P. M., and Hauser, J. R. (1975). *Proc. 11th IEEE Photovoltaic Specialists Conf.,* p. 13.
Dunbar, P. M., and Hauser, J. R. (1977). *Solid-State Electron.* **20,** 697.
Duncan, C. S., Seidensticker, R. G., McHugh, J. P., Hopkins, R. H., Skutch, M. E., Driggers, J. M., and Hill, F. E. (1980). *Proc. 14th IEEE Photovoltaics Specialists Conf.,* p. 25.
Fahrenbruch, A., and Bube, R. H. ((1974). *J. Appl. Phys.* **45,** 1264.
Faith, T. J. (1972). *Proc. 9th IEEE Photovoltaics Specialists Conf.,* p. 292.
Feldman, C., Blum, N. A., and Satkiewicz, F. G. (1980). *Proc. 14th IEEE Photovoltaic Specialists Conf.,* p. 391.
Fischer, H. (1978). *Proc. EC Photovoltaic Sol. Energy Conf., 1st, Luxembourg, 1977* p. 52. Reidel Publ., Dordrecht, Netherlands.
Fischer, H., and Pschunder, W. (1975). *Proc. 11th IEEE Photovoltaic Specialists Conf.,* p. 25.
Fischer, H. and Pschunder, W. (1976). *Proc. 12th IEEE Photovoltaic Specialists Conf.,* p. 86.
Fossum, J. G., Lindholm, F. A., and Shibib, M. A. (1979). *IEEE Trans. Electron Devices* **ED-26,** 1294.
Fossum, J. G., Nasby, R. D., and Pao, S. C., (1980). *IEEE Trans. Electron Devices* **ED-27,** 785.
Frank, R. J., and Goodrich, J. L. (1980). *Proc. 14th IEEE Photovoltaic Specialists Conf.,* p. 423.
Frisson, L., Lauwers, P., Bulteel, P., De Smet, L., Mertens, R., Govaerts, R., and Van Overstraeten, R. (1978). *Proc. 13th IEEE Photovoltaics Specialists Conf.,* p. 590.
Godfrey, R. B., and Green, M. A. (1979). *Appl. Phys. Lett.* **34,** 790.
Godlewski, M., Baraona, C. R., and Brandhorst, H. W., Jr. (1973). *Proc. 10th IEEE Photovoltaic Specialists Conf.,* p. 40.
Godlewski, M. P., Brandhorst, H. W., Jr., and Baraona, C. R. (1975). *Proc. 11th IEEE Photovoltaic Specialists Conf.,* p. 32.
Goradia, C., and Goradia, M. G. (1976). *Proc. 12th IEEE Photovoltaic Specialists Conf.,* p. 789.
Graff, K., and Fischer, H. (1979). Carrier lifetime in silicon and its impact on solar cell characteristics. *In* "Solar Energy Conversion—Solid State Physics Aspects" (B. O. Seraphin, ed.), *Topics in Applied Physics,* vol. 31, p. 204. Springer-Verlag, Berlin and New York.
Green, M. A., Godfrey, R. B., and Davies, L. W. (1978). *Proc. 13th IEEE Photovoltaic Specialists Conf.,* p. 651.
Grenon, L. A., and Coleman, M. G. (1978). *Proc. 13th IEEE Photovoltaic Specialists Conf.,* p. 246.
Grove, A. S. (1967). "Physics and Technology of Semiconductor Devices," p. 25. Wiley, New York.
Hall, R. N. (1981). *Solid-State Electron.* **24,** 595.
Hauser, J. R., and Dunbar, P. M. (1977). *IEEE Trans. Electron Devices* **ED-24,** 305.
Heaps, J. D., Zook, J. D., Grung, B. L., and Butter, C. D. (1980). *Proc. 14th IEEE Photovoltaic Specialists Conf.,* p. 39.
Hill, D. E., Gutsche, H. W., Wang, M. S., Gupta, K. P., Tucker, W. F., Dowdy, J. D., and Crepin, R. J. (1976). *Proc. 12th IEEE Photovoltaic Specialists Conf.,* p. 112.
Hoffman, H. (1981). *J. Appl. Phys.* **52,** 4070.
Hu, C. and Drowley, C. (1978). *Proc. 13th IEEE Photovoltaic Specialists Conf.,* p. 786.
Hunt, L. P. (1975). *Proc. 11th IEEE Photovoltaic Specialists Conf.,* p. 259.

Hunt, L. P., Dosaj, W. D., McCormick, J. R., and Crossman, L. D. (1976). *Proc. 12th IEEE Photovoltaic Specialists Conf.*, p. 125.
Iles, P. A., and Soclof, S. I. (1975). *Proc. 11th IEEE Photovoltaic Specialists Conf.*, p. 19.
Im, H.-B., Matthews, H. E., and Bube, R. H. (1970). *J. Appl. Phys.* **41,** 2581.
Jewett, D. N. (1978). *Proc. 13th IEEE Photovoltaic Specialists Conf.*, p. 376.
Kalejs, J. P., Mackintosh, B. H., Sachs, E. M., and Wald, F. V. (1980). *Proc. 14th IEEE Photovoltaic Specialists Conf.*, p. 13.
Khattak, C. P., and Schmid, F. (1980). *Proc. 14th IEEE Photovoltaics Specialists Conf.*, p. 484.
Kirkpatrick, A. R., Minnucci, J. A., Shaughnessy, T. S., and Greenwald, A. C. (1976). *Proc. 12th IEEE Photovoltaics Specialists Conf.*, p. 299.
Kirkpatrick, A. R., Minnucci, J. A., Greenwald, A. C., and Josephs, R. H. (1978). *Proc. 13th IEEE Photovoltaic Specialists Conf.*, p. 706.
Kressel, H., D'Aiello, R. V., and Robinson, P. H. (1976). *Appl. Phys. Lett.* **28,** 157.
LaBelle, H. E., Jr., and Mlavsky, A. I. (1971). *Mater. Res. Bull.* **6,** 571.
Landis, G. A. (1981). *Solar Cells* **3,** 91.
Lanyon, H. P. D. (1981). *Solar Cells* **3,** 289.
Lesk, I. A., Baghdadi, A., Gurtler, R. W., Ellis, R. J., Wise, J. A., and Coleman, M. G. (1976). *Proc. 12th IEEE Photovoltaic Specialists Conf.*, p. 173.
Lindholm, F. A., Li, S. S., and Sah, C. T. (1975). *Proc. 11th IEEE Photovoltaic Specialists Conf.*, p. 3.
Lindholm, F. A., Neugroschel, A., Pao, S. D., Fossum, J. G., and Sah, C. T. (1978). *Proc. 13th IEEE Photovoltaics Specialist Conf.*, p. 1300.
Lindmayer, J. (1978a). *Proc. 13th IEEE Photovoltaic Specialists Conf.*, p. 1096.
Lindmayer, J. (1978b). *Proc. EC Photovoltaic Sol. Energy Conf. 1st, Luxembourg, 1977*, p. 825. Reidel Publ., Dordrecht, Netherlands.
Lindmayer, J., and Allison, J. F. (1973). *COMSAT Tech. Rev.* **3,** 1. [Reprinted *in* "Solar Cells" (C. E. Backus, ed.), p. 209. IEEE Press, New York, 1976.]
Lindmayer, J., and Wrigley, C. (1976). *Proc. 12th IEEE Photovoltaic Specialists Conf.*, p. 53.
Mandelkorn, J., and Lamneck, J. H. (1972). *Proc. 9th IEEE Photovoltaic Specialists Conf.*, p. 66.
Mandelkorn, J., and Lamneck, J. H. (1975). *Proc. 11th IEEE Photovoltaic Specialists Conf.*, p. 36.
Mandelkorn, J., Lamneck, J. H., and Scudder, L. R. (1973). *Proc. 10th IEEE Photovoltaics Specialists Conf.*, p. 207.
Matlock, J. H. (1979). *Semiconductor International* **2,** 33 (October).
Maycock, P. B., and Wakefield, G. F. (1975). *Proc. 11th IEEE Photovoltaic Specialists Conf.*, p. 252.
Mlavsky, A. I., Serreze, H. B., Stormont, R. W., and Taylor, A. S. (1976). *Proc. 12th IEEE Photovoltaics Specialists Conf.*, p. 160.
Muller, J. C., Grob, A., Grob, J. J., Stuck, R., and Seffert, P. (1978). *Proc. 13th IEEE Photovoltaic Specialists Conf.*, p. 711.
Noel, G. T., Kulkarni, S., Wolf, M., Pope, D. P., and Graham, C. D., Jr. (1976). *Proc. 12th IEEE Photovoltaic Specialists Conf.*, p. 168.
O'Donnell, D. T., Robb, S. P., Rule, T. T., Sanderson, R. W., and Backus, C. E. (1978). *Proc. 13th IEEE Photovoltaic Specialists Conf.*, p. 804.
Pschunder, W., and Fischer, H. (1976). *Proc. 12th IEEE Photovoltaic Specialists Conf.*, p. 270.
Rahilly, W. P., Scott-Monck, J., Anspaugh, B., and Locker, D. (1976). *Proc. 12th IEEE Photovoltaic Specialists Conf.*, p. 276.

Ralph, E. L. (1975). *Proc. 11th IEEE Photovoltaic Specialists Conf.*, p. 315.

Rasch, K. D., Roy, K., Schilling, R., and Fischer, H. (1980). *Proc. 14th IEEE Photovoltaic Specialists Conf.*, p. 141.

Ravi, K. V., Serreze, H. E., Bates, H. E., Morrison, A. D., Jewett, D. N., and Ho, J. C. T. (1975). *Proc. 11th IEEE Photovoltaic Specialists Conf.*, p. 280.

Redfield, D. (1980). *IEEE Trans. Electron Devices* **ED-27,** 766.

Redfield, D. (1981). *Solar Cells* **3,** 313.

Riel, R. K. (1973). Status of silicon web solar cells. *In* "Proc. Photovoltaic Conversion of Solar Energy for Terrestrial Applications, Vol. 2," NSF-RANN Workshop, Cherry Hill, New Jersey, NSF-RANN-74-013, p. 36.

Rittner, E. S. (1977). *J. Energy* **1,** 9.

Robinson, P. H., D'Aiello, R. V., Richman, D., and Faughnan, B. W. (1978). *Proc. 13th IEEE Photovoltaic Specialists Conf.*, p. 1111.

Runyan, W. R. (1965). "Silicon Semiconductor Technology." McGraw-Hill, New York.

Scott-Monck, J. (1978). *Proc. 13th IEEE Photovoltaic Specialists Conf.*, p. 421.

Schwuttke, G. H. (1979). *Proc. 2nd EC Photovoltaic Solar Energy Conf.*, Berlin, 1979 (R. Van Overstraeten and W. Palz, ed.), p. 130. D. Reidel Publ., Dordrecht, Netherlands.

Seidensticker, R. G., Scudder, L., and Brandhorst, H. W., Jr. (1975). *Proc. 11th IEEE Photovoltaic Specialists Conf.*, p. 299.

Seidensticker, R. G., Kothmann, R. E., McHugh, J. P., Duncan, C. S., Hopkins, R. H., Blais, P. D., Davis, J. R., and Rohatgi, A. (1978). *Proc. 13th IEEE Photovoltaics Specialists Conf.*, p. 358.

Shah, P., and Fuller, C. B. (1976). *Proc. 12th IEEE Photovoltaic Specialists Conf.*, p. 286.

Shewmon, P. G. (1963). "Diffusion in Solids." McGraw-Hill, New York.

Shibib, M. A., and Fossum, J. G. (1981). *J. Appl. Phys.* **52,** 1072.

Shibib, M. A., Lindholm, F. A., and Therez, F. (1979). *IEEE Trans. Electron Devices* **ED-26,** 959.

Siekhaus, W. (1976). Personal communication.

Smits, F. M. (1976). *IEEE Trans. Electron Devices* **ED-23,** 640.

Storti, G., Wrigley, C., Wohlgemuth, J., Whitehouse, D., and Scheinine, A. (1980). *Proc. 14th IEEE Photovoltaic Specialists Conf.*, p. 137.

Tscholl, E. (1968). *Philips Res. Rep. Suppl.* **6,** 1.

Tsuya, N., *et al.* (1980). *Jpn. J. Appl. Phys.* **19,** Suppl. 19-1, 655. [Also see K. I. Arai, N. Tsuya, and T. Takeuchi, *Proc. 14th IEEE Photovoltaics Specialists Conf.*, (1980) p. 31.]

Van Aerschodt, A. E., Capart, J. J., David, K. H., Fabbricotti, M., Heffels, K. H., Loferski, J. J., and Reinhartz, K. K. (1971). *IEEE Trans. Electron Devices* **ED-18,** 471.

Van Overstraeten, R., and Nuyts, W. (1969). *IEEE Trans. Electron Devices* **ED-16,** 632.

Varma, R., and Swerdling, S. (1976). *Proc. 12th IEEE Photovoltaic Specialists Conf.*, p. 293.

von Roos, O. (1978). *J. Appl. Phys.* **49,** 3510. [Also see *J. Appl. Phys.* **50,** 5371 (1979).]

Wakefield, G. F., Maycock, P. D., and Chu, T. L. (1975). *Proc. 11th IEEE Photovoltaic Specialists Conf.*, p. 49.

Weizer, V. G., Brandhorst, H. W., Broder, J. D., Hart, R. E., and Lamneck, J. H. (1978). *Proc. 13th IEEE Photovoltaic Specialists Conf.*, p. 1327.

Westbrook, R. D., and Polgreen, T. L. (1976). *Proc. 12th IEEE Photovoltaic Specialists Conf.*, p. 67.

Wolf, M. (1971). *Energy Convers.* **11,** 63. [Reprinted *in* "Solar Cells" (C. E. Backus, ed.), p. 191. IEEE Press, New York 1976.]

Wolf, M. (1972). *Proc. Annu. Power Sources Symp., 25th,* p. 120. [Reprinted *in* "Solar Cells" (C. E. Backus, ed.), p. 38. Wiley (Interscience), New York, 1976.]

Wolf, M. (1975). *Proc. 11th IEEE Photovoltaic Specialists Conf.*, p. 306.
Wolf, M. (1980). *IEEE Trans. Electron Devices* **ED-27,** 751.
Wolf, M., MacDiarmid, A. G., Noel, G. T., and Thallum, K. (1976). *Proc. 12th IEEE Photovoltaic Specialists Conf.*, p. 137.
Wrigley, C. (1977). *Proc. 12th IEEE Photovoltaic Specialists Conf.*, p. 343.

CHAPTER 8

HETEROJUNCTION AND HETEROFACE STRUCTURE CELLS

In this chapter we discuss the first-order criteria for the selection of heterojunction (HJ) solar cell materials. We develop a figure of merit and compare some of the many possible candidates for HJ couples. By using a low lattice-mismatch, isotype HJ to provide a window for a homojunction (i.e., a heteroface structure), a considerable reduction of surface recombination loss for direct band-gap materials can be obtained. The AlGaAs/GaAs cell is discussed as an example of such a structure and as an example of the design of solar cells for concentrator systems. Finally, the CdS/InP cell is considered as an example of an anisotype HJ in which the properties of low lattice mismatch appear to be directly advantageous to junction transport.

8.1 CHOICE OF HETEROJUNCTION SOLAR CELL COMPONENTS

For the selection of a HJ pair the following first-order criteria can be considered.

(1) The absence of conduction (or valence) band spikes that could impede photogenerated carrier transport. This condition is suggested if

$\Delta E_c = \chi_1 - \chi_2 < 0$ and $\Delta E_v = \chi_2 - \chi_1 + E_{g2} - E_{g1} > 0$ (see Sections 5.3.1 and 5.3.2).

(2) ΔE_c (or ΔE_v for n-type absorbers) as close to zero as possible to maximize the diffusion voltage V_d and V_{oc} for the cell.

(3) E_{g1} close to 1.4–1.6 eV to take advantage of the maximum in solar efficiency of the absorber.

(4) E_{g2} as large as possible to pass as much of the solar spectrum as possible, while still allowing a low-resistance material.

(5) Small lattice mismatch.

(6) Small thermal expansion mismatch. Since most junctions are prepared at elevated temperatures, the thermal expansion coefficients must be matched to some degree.

The subject of lattice mismatch is complex. Although small lattice mismatch does not assure large η_s and large mismatch does not prevent large η_s, the evidence does point to low mismatch as being beneficial. An example of the former situation is the ZnSe/GaAs junction, which has a low lattice mismatch and, according to the preceding criteria, should have an extremely high efficiency. Thus far the ZnSe/GaAs cell has not realized this promise because of insufficient photogenerated current collection.† The effects of lattice mismatch and interfacial defects are discussed in Chapter 5 and by Fahrenbruch and Aranovich (1979). A mismatch strain $\epsilon_{mf} < 1\%$ appears to be generally beneficial, but for $\epsilon_{mf} > 1\%$ there appears to be no strong dependence of the junction properties on the magnitude of ϵ_{mf}. In many cases large lattice mismatch does not decrease the quantum efficiency appreciably, but leads to current paths that decrease V_{oc} and ff.

Ternary compounds such as $Al_xGa_{1-x}As$ and $Zn_xCd_{1-x}S$ allow an additional degree of freedom in the selection of either the band-gap or the lattice constant. Quaternary compounds such as $Al_xGa_{1-x}As_yP_{1-y}$ and $Cu_xIn_{1-x}Se_yTe_{1-y}$ yield two degrees of freedom, and ideally both the band gap *and* lattice constant can be adjusted to give optimal junction properties. Of course the price of this freedom is the additional complexity of the system. Several such systems are described in Chapter 11.

Since the fill factor is relatively constant for efficient cells (ff $\simeq$ 0.75–0.90), a simple figure of merit for HJ couples is the product of J_{sc} and V_{oc}. By assuming $\eta_Q(\lambda) = 1$, J_{sc} can be determined approximately from $(E_{g2} - E_{g1})$ using the integrated photon flux shown in Fig. 6.3. The open-circuit voltage is assumed to be proportional to the diffusion voltage $V_d = E_{g1} - \delta_n - \delta_p + \Delta E_c$, where $\Delta E_c \leq 0$, $\delta_n = E_c - E_F$,

† Gaugash *et al.* (1976) report 8 to 9% efficiency. See also Balch and Anderson (1972).

$\delta_p = E_F - E_v$, and junctions with appreciable $\Delta E_c > 0$ are not included. A brief list of HJ candidates with parameters is shown in Table 8.I with the figure of merit expressed as a "merit efficiency" η_{merit} by assuming the common values ff = 0.75 and $V_{oc}/V_d = 0.65$.†

Other such lists have been compiled by Milnes and Feucht (1972), and Fahrenbruch (1977). Bucher (1978) gives a comprehensive listing of both homojunctions and HJs that have been fabricated and their photovoltaic parameters.

There are, of course, other considerations for the formation of HJs such as interdiffusion, compound formation at the interface, resistivity, interfacial band pinning, and oxide layers formed prior to or during growth.

At the temperature of fabrication, interdiffusion of the component species can occur at either a small or large scale. In the small-scale case we are considering transport of only sufficient material to cause changes in doping. The diffusion of Cu into CdS during fabrication of the Cu_xS/CdS HJ is a prime example here and the extent of Cu doping dominates the properties of this junction.

Interdiffusion on a larger scale can lead to the formation of intervening, third-compound layers at the metallurgical junction. An example of this is provided by the spectral response curves shown in Fig. 8.1 for two *p*-CdTe/*n*-ZnSe cells produced at rather high temperatures (~600°C) by CSVT using two slightly different methods (Buch *et al.*, 1977). Cell M-1 shows an interlayer of *n*-CdSe ($E_g \simeq 1.74$ eV) that absorbs photons above its band gap, but not usefully. The spectral response for cell M-21 indicates an interlayer of *p*-ZnTe ($E_g \simeq 2.3$ eV) that has a relatively good η_Q itself but prohibits all but hot photogenerated carriers from the CdTe from being collected.

An additional consideration is optical mismatch between the components of a HJ that can add to the reflection loss. The effect is small for most compound semiconductors ($\lesssim 1\%$) but can approach 5% for HJs that include metal oxide films (ZnO, SnO_x). Of course this reflection loss might be eliminated by making the window layer part of an antireflection sandwich or by making textured interfaces.

In large-band-gap materials, particularly the II–VI compounds, the tendency toward self-compensation is large, and production of sufficiently low-resistivity material is difficult. This is particularly the case for polycrystalline thin films since the self-compensation is most severe in regions of disorder such as grain boundaries. For example, *n*-$Zn_xCd_{1-x}S$

† Measured V_{oc}/V_d values at AM1 for common cells are Si, 0.59; AlGaAs/GaAs, 0.71; Cu_xS/CdS, 0.58; CdS/InP, 0.71; and CdS/CdTe, 0.50.

Table 8.1

Selected Heterojunction Solar Cells—Figure of Merit

Heterojunction	E_{g2} (eV)	E_{g1} (eV)	χ_2 (eV)	χ_1 (eV)	$\Delta a_0/\bar{a}_0$ (%)[a]	ΔE_c (eV)	δ_{n2} (eV)[d]	V_d (V)	J_L (mA cm^{-2})	η_{merit}[b] (%)
Deg. n-ITO/p-InP	3.35[c]	1.35	4.5	4.38	≈15%	−0.12	−0.3	1.43	29	24.0
n-ITO/p-InP	3.05[c]	1.35	4.5	4.38	≈15	−0.12	0	1.13	28	18.7
Deg. n-ITO/p-CdTe	3.35	1.50	4.5	4.28	≈25	−0.22	−0.3	1.48	24	21.3
n-ITO/p-CdTe	3.05	1.50	4.5	4.28	≈25	−0.22	0	1.18	24	16.4
Deg. n-ITO/p-Si	3.35	1.12	4.5	4.01	≈7	−0.49	−0.3	0.83	36	17.6
n-ZnSe/p-Ge	2.67	0.66	4.09	4.13	0.15	+0.04	0.04	0.51	49	14.8
n-ZnSe/p-Si	2.67	1.12	4.09	4.01	4.3	−0.08	0.05	0.89	33	17.5
n-CdS/p-Zn_3P_2	2.42	1.4	4.5	3.6	2.26[e]	−0.70	0.05	0.65	22	8.5
n-CdS/p-InP	2.42	1.35	4.5	4.38	0.31	−0.12	0.05	1.08	24	15.0
n-ZnO/p-Zn_3P_2	3.3	1.4	4.35	3.6	22[e]	−0.75	0	0.55	27	8.8
n-CdS/p-CdTe	2.42	1.50	4.5	4.28	10.2	−0.22	0.05	1.13	19	12.8
n-ZnO/p-CdTe	3.3	1.5	4.35	4.28	34.0	−0.07	0	1.33	24	18.9
n-ZnSe/p-GaAs	2.67	1.43	4.09	4.07	0.23	−0.02	0.05	1.26	24	17.5
n-ZnSe/p-Zn_3P_2	2.67	1.4	4.09	3.6	0.93	−0.49	0.05	0.76	25	11.0

[a] a_0 for hexagonal materials is given as $2^{1/2}a_0$.

[b] $$\eta_{merit} = \left[q \int_{E_g 1}^{E_g 2} (d\Gamma/dE)\, dE\right] (0.65V_d)\, ff/P_s,$$ with ff = 0.75 and P_s = 82.3 mW cm^{-1}.

[c] ITO band gap changes for degenerate material because of Burstein shift.

[d] $\delta_{n2} = E_c - E_F$.

[e] Zn_3P_2 is pseudocubic with a = 2.86 Å; $2a$ is used in these lattice match calculations.

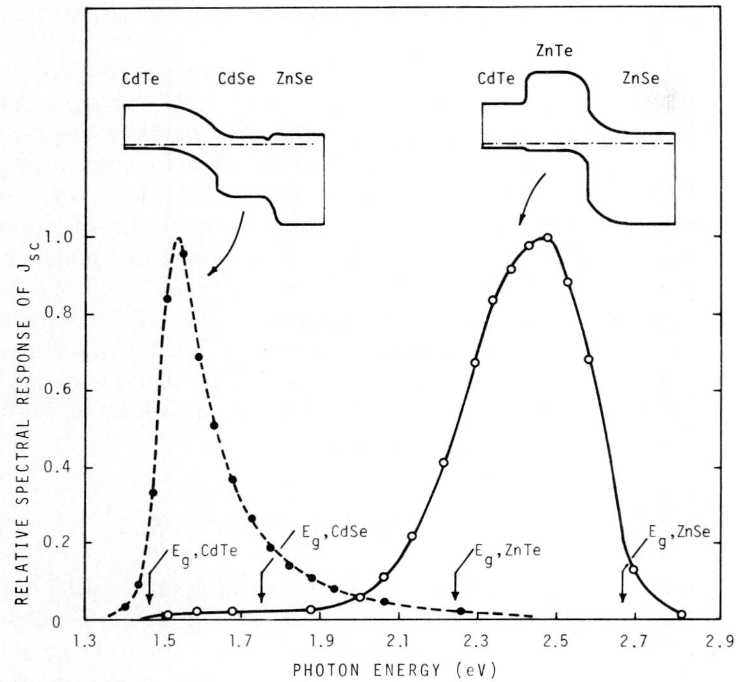

Fig. 8.1. Spectral response of two *p*-CdTe/*n*-ZnSe heterojunctions prepared by CSVT. Proposed band diagrams of each cell are also shown. [Redrawn from F. Buch, Ph.D. Thesis, Dep. Mat. Sci. and Eng., Stanford Univ., Stanford, California, 1976.]

films with compositions of $x > 20\%$ ($E_g > 2.65$ eV) cannot easily be made with $\rho < 1\ \Omega$ cm (Feigelsen *et al.*, 1977; Chynoweth and Bube, 1980).

As noted in Chapter 5, the surfaces of most semiconductors take on a different potential than the bulk. Surface states dominate the electronic properties of the surfaces of covalent semiconductors and pin E_F at the surface (i.e., the "$\frac{2}{3}$ rule"). For some materials the surface state effects appear to be strong enough to invert the surface. For example, an atomically clean surface of *p*-InP readily inverts to *n*-type on exposure to O_2 at even moderate temperatures (Weider, 1979) or with sputter cleaning in Ar (Tsai *et al.*, 1980). Exposure of the "pure" surface (e.g., cleaved) of a semiconductor to air very rapidly produces a thin oxide film (~15 Å for GaAs or InP within seconds at room temperature) that also affects junction transport, sometimes beneficially. These considerations are of critical importance in the fabrication of HJs and prompt such techniques as the *in situ* etching required for stable results with the CdS/InP HJ cell, and the melt-back techniques sometimes used for preparation of AlGaAs/GaAs heteroface cells.

As was pointed out in Chapter 5 and by Fahrenbruch and Aranovich (1979), the band configuration at the interface of a HJ and the electrical transport there is considerably more complex than we have suggested by considering just a simple difference in electron affinities across the junction. Despite the basic nature of this difference, it may play a minor role in junction transport. Indeed, other effects may dominate the situation: (1) microscopic pinning of the Fermi level at the interface (as influenced by chemical reactivity of the HJ components and the resulting charged interface states and dipoles), (2) the defect structure in the vicinity of the interface and the preferential segregation of dopants and impurities there, and (3) the presence of intervening insulating layers, formed during junction fabrication, which may contain trapped charge and can act as buffering layers to accommodate lattice mismatch.

8.2 THE AlGaAs/GaAs HETEROFACE SOLAR CELL

The band gap of GaAs (1.43 eV) is near the optimum for solar conversion, indicating a theoretical efficiency of 26–29% at AM1. Because the high mobility of GaAs allows the fabrication of very-high-frequency devices, because its band gap usefully complements that of Si, and because of the ease of forming lattice-matched ternary compounds, GaAs rivals Si as the most investigated semiconductor material. GaAs has a direct band gap and a large optical absorption coefficient, absorbing 97% of the AM1 photons within about 2 μm. The near perfect lattice match and absence of interface state recombination of the AlGaAs/GaAs isotype junction has been used to advantage to remove the front surface recombination loss from the heteroface structure cell and to yield the highest efficiencies of any solar cell type in 1983. Because of the high cost of the material and fabrication, the present goals of the GaAs-based solar cell technology have been strongly directed toward concentrator systems and space use.

8.2.1 The Materials GaAs and AlGaAs

Although relatively abundant in the earth's crust (18 ppm as compared to 64 ppm for Cu), the supply of Ga is not large in terms of present-day refining practice. In 1980 the cost for highly purified Ga was some \$3000 kg^{-1} in small lots. However, the presence of Ga in fly ash from the burning of coal promises a large future source should the need warrant setting up recovery methods. Although arsenic is also quite abundant and easily

Table 8.II

Properties of GaAs and AlAs (all at 300°K)

	GaAs	AlAs
Physical properties		
Structure	Zincblende cubic	Zincblende cubic
Melting point	1238°	1740°C
Density	5.316	3.73 g cm^{-3}
Thermal expansion	5.9×10^{-6} °C^{-1}	5.2×10^{-6} °C^{-1}
Lattice constant	5.654 Å	5.661 Å
Electronic properties		
n_i (297°K)	1.8×10^{6} cm^{-3}	
N_c	4.27×10^{17} cm^{-3} [a]	
N_v	8.19×10^{18} cm^{-3} [a]	
χ	4.07 eV	≈3.5 eV
E_g, direct	1.424 eV [b]	3.018 eV
E_g, indirect	$E_g^L = 1.708$ eV	$E_g^L = 2.3$ eV
	$E_g^x = 1.900$ eV	$E_g^x = 2.168$ eV
μ_n ($N_D = 10^{17}$ cm^{-3})	4000 cm^2 V^{-1} sec^{-1}	280 cm^2 V^{-1} sec^{-1}
μ_p ($N_A = 10^{17}$ cm^{-3})	250 cm^2 V^{-1} sec^{-1}	
τ_n, typical	10^{-9} sec	
maximum	6×10^{-8} sec	
τ_p, typical	3×10^{-9} sec	
maximum	8×10^{-9} sec	
L_n, typical	4–8 μm	
maximum	23 μm	
L_p, typical	4 μm	
maximum	5–6 μm	

[a] From H. C. Casey, Jr., and M. B. Panish, "Heterostructure Lasers, Part A," p. 207. Academic Press, New York, 1978.

[b] $E_g(T) = 1.519 - 5.405 \times 10^{-4}T^2/(204 + T)$. [From C. D. Thurmond, *J. Electrochem. Soc.* **122,** 1133 (1975).]

won, its chemistry and the inefficiency of refining the small lots used at present makes semiconductor grade As expensive (\$500 kg^{-1}) as well.

GaAs absorbs strongly in the visible and has a zincblende cubic structure. A brief list of its physical and electronic properties is given in Table 8.II.† Although minority carrier lifetimes in GaAs are quite short ($\sim 10^{-9}$ sec) because of the direct band-gap nature of the material, the large mobility gives diffusion lengths quite sufficient for high quantum efficiencies (6–8 μm in LPE GaAs:Ge). Optical absorption data are shown

† An excellent source for the fundamental electronic properties of GaAs and AlGaAs alloys and junctions is Casey and Panish (1978a, Chapter 4).

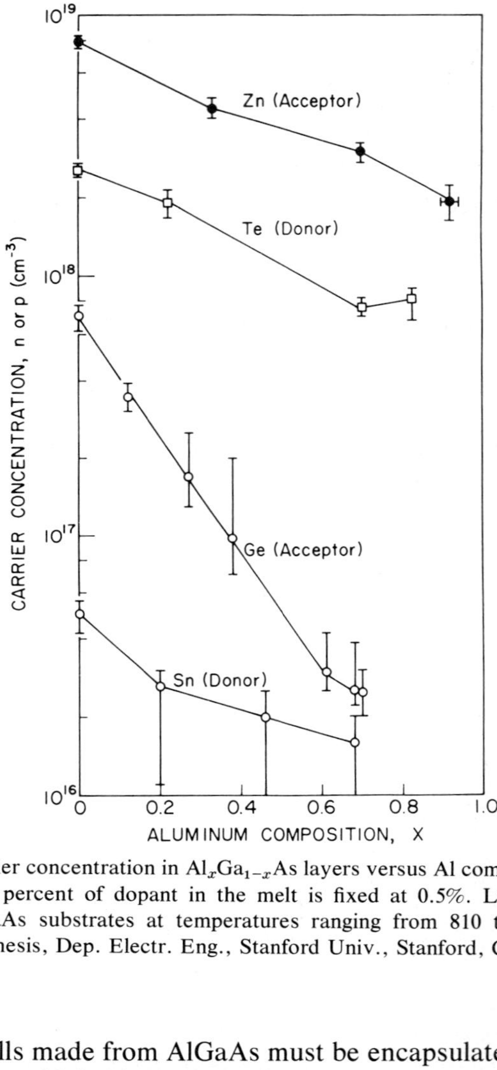

Fig. 8.2. Carrier concentration in $Al_xGa_{1-x}As$ layers versus Al composition for four dopants. The mole percent of dopant in the melt is fixed at 0.5%. Layers are grown on (100)-oriented GaAs substrates at temperatures ranging from 810 to 840°C. [From D. Cheung, Ph.D. Thesis, Dep. Electr. Eng., Stanford Univ., Stanford, California, 1975.]

in Fig. 3.8. Cells made from AlGaAs must be encapsulated since AlGaAs is hygroscopic at high Al concentrations.

Crystal growth of the material is usually by the Bridgman technique, although the Czochralski method can also be used. Synthesis from the elements must be done carefully to avoid explosion. Because of the large difference in the vapor pressure ($P_{Ga} = 3 \times 10^{-4}$ atm at 1240°C and $P_{As} = 1$ atm at 615°C over the elements), an arsenic overpressure must be used during growth to preserve stoichiometry.

Crystal growth can also be accomplished by chemical vapor deposition and by growth from solution; these are the most common means of fabrication of solar cells from these materials.

Dopants include S, Se, Te, Sn, Si, C, and Ge as shallow donors and Zn, Be, Mg, Cd, Si, Ge, and C as shallow acceptors. The column IV elements—C, Si, Ge, and Sn—are amphoteric in GaAs, and their electrical activity in the crystal depends on growth conditions. For example, Ge is substitutional on an As site, acting as an acceptor for growth near 900°C, whereas for growth near the melting point Ge substitutes on a Ga site as a donor. Zn is volatile and fast diffusing and thus easy to introduce in diffused junctions, but Ge and Be appear to give the highest lifetimes among the common acceptor dopants. Te appears to be the dopant of choice as a donor impurity. The low atomic diffusivity of both Ge and Te make it necessary to introduce them during the growth of layers rather than by postgrowth diffusion. Dopants generally become less electrically active with increasing Al content in AlGaAs, as shown in Fig. 8.2.

Recombination centers active in GaAs are similar to those in Si (Cr, Fe, Ni, Cu, and Ag) but also include oxygen.

$Al_xGa_{1-x}As$ fulfills the need for an adjustable band-gap counterpart for GaAs with an exceptionally good lattice match (only 0.16% mismatch between AlAs and GaAs), and growth methods involving AlGaAs have the ability to form clean interfaces. Figure 8.3 shows the energy band gap

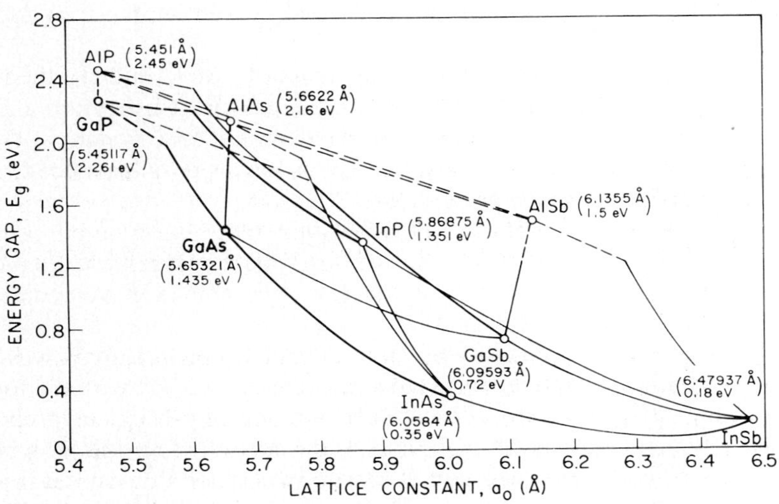

Fig. 8.3. Energy gap versus lattice constant of various III–V binary and ternary compound systems. Direct band gaps are shown by the solid line and indirect band gaps by the dashed line. [From D. Cheung, Ph.D. Thesis, Dep. Electr. Eng., Stanford Univ., Stanford, California, 1975.]

versus lattice constant for Al, Ga, In, P, As, Sb compounds. At a composition of $x \simeq 0.44$, the indirect and direct band gaps of $Al_xGa_{1-x}As$ coincide; for higher x, the indirect band gap of the ternary is lower, as shown in Fig. 3.9.

8.2.2 Fabrication by LPE and CVD

The fabrication of layered structures by liquid phase epitaxy LPE depends on the solubility of As in liquid Ga and AlGa alloys and the subsequent precipitation of $Al_xGa_{1-x}As$ layers onto crystalline substrates, which determine the crystallographic orientation of the layers.† The growth apparatus shown in Fig. 8.4 is fabricated of high-purity carbon and quartz, and growth occurs in an ambient of high-purity H_2. The top portion, containing the Ga melt, slides over the bottom, which contains both the substrate and a GaAs source. After the melt has been saturated at $T \simeq 900°C$ over the GaAs source, the slider carries the Ga melt over the substrate and the temperature is lowered at a rate of 0.1–0.5°C min^{-1}. The solubility-versus-temperature data are shown in Fig. 8.5. Growth rates of 0.1–0.5 $\mu m\ min^{-1}$ are obtained. During deposition the relatively large volume of Ga acts as a sink for segregated impurities. Since the growth proceeds very near thermal equilibrium, the resulting epitaxial layers are of very high quality. To stabilize the growth, temperature gradients are sometimes established across the freezing front by cooling devices, producing layers of very uniform thickness and preventing the formation of hillocks.

Since the growth takes place in a system of C, SiO_2, and H_2, the resistivity of the resulting layers depends on the growth temperature and the resulting activity of the C, which acts as an amphoteric dopant in GaAs. The oxygen partial pressure must be kept as low as possible since O is an active recombination center in GaAs.

During the initial moments of deposition there is, in some cases, a small dissolution of the GaAs substrate, which ensures an atomically clean interface and accounts for the low recombination velocities observed at the metallurgical interface.

By adding a small quantity of Zn to the Al, Ga melt, a shallow p-type layer can be formed in the GaAs substrate simultaneously with the formation of the AlGaAs layer. Alternatively, a separate p-GaAs layer can be grown from a Ge-doped Ga melt prior to the growth of the AlGaAs layer. The diffused layer and the grown layer are two major alternatives for this sort of device. Naively, one would expect greater crystalline perfection and longer lifetime in the grown layers.

† A review of LPE is given by Dawson (1972).

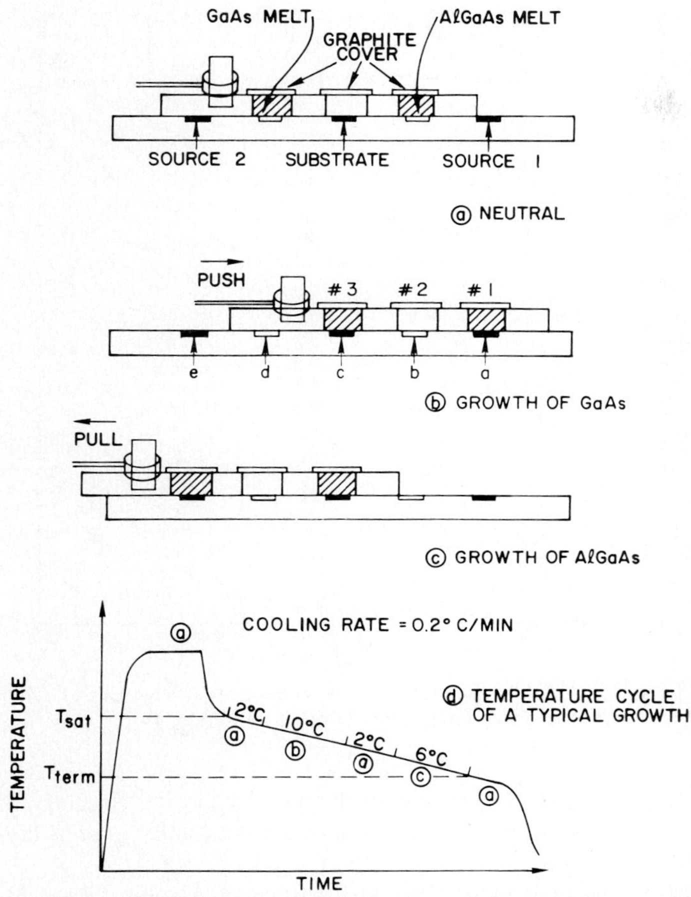

Fig. 8.4. Cross section of LPE growth system and temperature–time profile during growth of GaAs and AlGaAs layers. The growth system is situated within a quartz tube carrying a flow of high purity H_2 at atmospheric pressure. [From D. Cheung, Ph.D. Thesis, Dep. Electr. Eng., Stanford Univ. Stanford, California, 1975.]

Several variations on LPE growth are used to enhance the properties of the devices. Deposition can take place using undersaturated, saturated, or supersaturated melts, the latter typically giving smaller minority carrier diffusion length L in the grown layer (Woodall and Hovel, 1977). "Leaching" or gettering of impurities from the underlying p or n substrate layers to enhance L can be accomplished by annealing the substrate in the Ga, Al melt before growth (Woodall and Hovel, 1975). An "etchback-regrowth" method, also used by Woodall and Hovel (1977), accomplishes

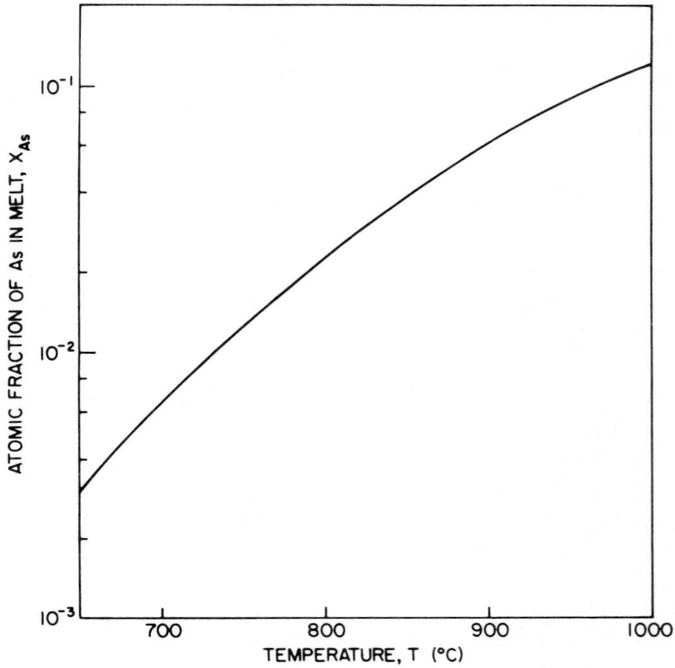

Fig. 8.5. Solubility of As in liquid Ga versus temperature. [From L. C.-C. Shen, Ph.D. Thesis, Dep. Electr. Eng., Stanford Univ., Stanford, California, 1975.]

both gettering and growth of a compositionally graded Al_xGa_{1-x} layer in a single step. The advantages of such compositionally graded layers are discussed in Section 8.2.6.

The other major growth method for films of $Al_xGa_{1-x}As$ is chemical vapor deposition CVD [or, more restrictively, vapor phase epitaxy (VPE)] by synthesis from gaseous Ga and As compounds. By reaction of $GaCl_3$ and arsine (AsH_3) on a substrate at 600–800°C, growth rates for GaAs of >10 μm hr^{-1} can be achieved (Johnston and Callahan, 1976). Deposition can also be accomplished by pyrolysis reaction of organometallic Ga compounds (e.g., trimethylgallium) and AsH_3 at an rf-heated substrate (MO-CVD) (Dapkus *et al.*, 1978; Dupuis *et al.*, 1977). A particular advantage of the latter technique with regard to future scale-up is that it is done in a cold wall reactor and material is deposited only on heated areas. Both processes utilize H_2 as a carrier gas at atmospheric pressure.

Lifetimes are comparable in layers grown by both LPE and CVD. Both methods can be also used to deposit polycrystalline GaAs films on amorphous substrates.

Because of the extreme difference in vapor pressures of the elements (a ratio of 10^{11} at 400°C) stoichiometry is impossible to maintain in standard vacuum evaporation of GaAs from the compound. High quality layers can be grown by molecular beam epitaxy (MBE), however. A review is given by Casey and Panish (1978b, p. 132).

8.2.3 The GaAs-Based Solar Cell—Basic Structure and Preliminary Optimization

Except for an isolated report of $\eta_s = 4\%$ for a p/n homojunction by Jenny *et al.* (1956), the history of the GaAs cell is much more recent than that of the Si cell; one of the early reports of a GaAs p/n junction cell was by Gobat *et al.* (1962), who quoted $\eta_s = 11\%$ and $J_L = 17$ mA cm^{-2} at AM1, with a large V_{oc}. In 1970 Alferov *et al.* (1971) reported the first p-AlGaAs/n-GaAs heterojunction structure with an AM0 efficiency of 10–11% and a strongly increased short-wavelength response with respect to the p/n homojunction. A large jump in efficiency occurred in 1972 with the heteroface structure, apparently first reported by Woodall and Hovel (1972), who obtained $\eta_s = 15.3\%$ at AM1 and 19.1% at AM2 with a p-$Al_{0.7}Ga_{0.3}As$/p-GaAs/n-GaAs device fabricated by LPE. Diffusion of Zn during the AlGaAs deposition formed the intermediate p-type layer. Because of its high efficiency, the GaAs-based cell was soon associated with solar concentration where the high cell cost could be absorbed in the cost of the concentrator system.

In 1978 Fan and Bozler (1978) reported a $n^+/p/p^+$ homojunction GaAs cell with a structure much like that of the standard BSF Si cell. The p and n^+ layers of the cell were fabricated by CVD and the thin n^+ layer was reduced to its final thickness (0.045 μm) by anodization. The anodic oxide layer was left on to function as an antireflection coating for the cell. This cell had an AM1 efficiency of 20% with no concentration. Heteroface structures still have a somewhat higher η_s with reports of 24.7% at AM1 ($C = 180$) by Sahai *et al.* (1978) for a heteroface structure with a graded AlGaAs layer, $\eta_s = 21\%$ at AM1.4 (no concentration) by James and Moon (1975), and of 21.9% at AM1 (no concentration) by Woodall and Hovel (1977).

Woodall and Hovel's cell configuration, similar to that shown in Fig. 8.6, has been changed little for most of the present designs except for variations in thickness and composition of the layers. Typical values of the parameters of an experimental cell (without concentration) are listed in Table 8.III.

At the light-incident surface x'_{p^+} a considerable portion of the photogenerated carriers is lost to surface recombination ($S = 10^6$ cm sec^{-1}). However, because of the large indirect band gap of AlGaAs, only a small

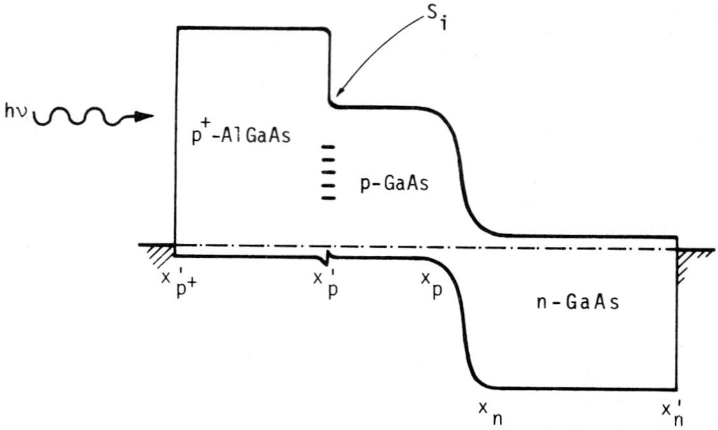

Fig. 8.6. Typical heteroface AlGaAs/GaAs solar cell structure.

Table 8.III

AlGaAs/GaAs Heteroface Cell Parameters[a]

p^+-$Al_xGa_{1-x}As$ window layer	
Dopant	Ge
$\lvert x'_p - x'_{p^+}\rvert$	7 μm
Composition, x	0.7
N_A	10^{18} cm^{-3}
E_g, indirect	2.08 eV
E_g, direct	2.41 eV
p-GaAs layer	
Dopant	Ge
$\lvert x_p - x'_p\rvert$	1.2 μm
L_n	5–8 μm
N_A	9×10^{17} cm^{-3}
n-GaAs substrate	
Dopant	Te
$\lvert x'_n - x_n\rvert$	≃200 μm
N_D	1.6×10^{18} cm^{-3}
Cell	
$W_d = \lvert x_n - x_p\rvert$	0.04 μm
J_0	10^{-10} A cm^{-2}
A factor	≃2
V_{oc} (AM1)	0.88 V
J_{sc} (AM1)	0.02 A cm^{-2}
ff	0.75
η_s	14.8%

[a] L. C.-C. Shen, Ph.D. Thesis, Dep. Electr. Eng., Stanford Univ. Stanford, California, 1976.

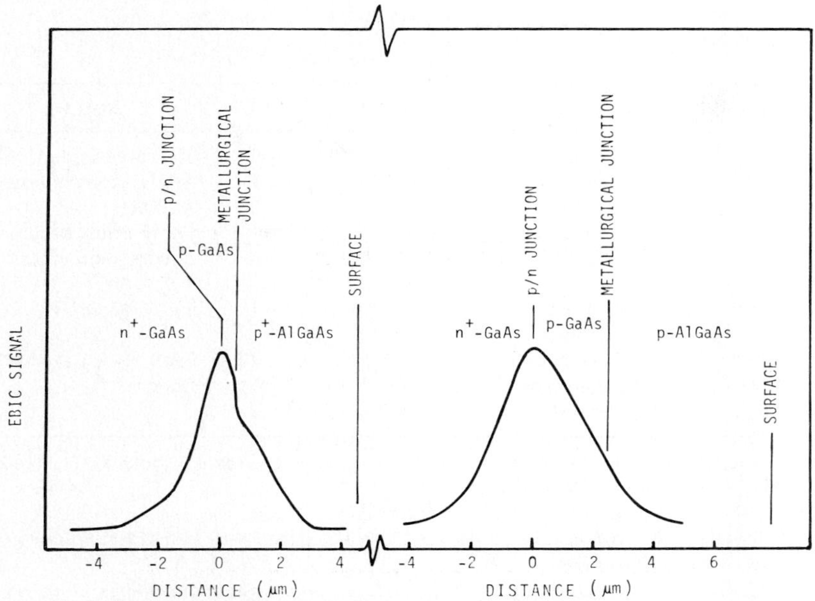

Fig. 8.7. Electron-beam-induced current versus distance for (left) a structure with a *p*-GaAs/Zn layer formed by diffusion during AlGaAs layer growth and (right) a structure with a *p*-GaAs/Ge layer grown by LPE with subsequent growth of the AlGaAs layer. [Redrawn from L. C.-C. Shen, Ph.D. Thesis, Dep. Electr. Eng., Stanford Univ., Stanford, California, 1976.]

fraction of the total J_L is generated in this window layer, and so the relative loss is small. Because of good lattice matching and because the fabrication processes can produce clean interfaces, the recombination loss at the AlGaAs/GaAs interface is small. Measurements by Ettenberg and Kressel (1976) indicate values of $S_i < 10^4$ cm sec.$^{-1}$.

Because the p^+-layer also serves for lateral current collection in most cases (see Section 8.2.4 for an exception), the resistivity of the thin p-layer, where most of the useful absorption occurs, can be optimized with regard for L_n rather than for current collection considerations. Shen (1976) compared Zn-diffused junctions with Ge-grown junctions, finding greater L_n and effective collection for the latter case, as illustrated by the EBIC plots of Fig. 8.7. Other EBIC measurements indicated $L_n = 6.7\ \mu$m for Ge doping at a doping level of 2×10^{17} cm^{-3}.

The diode current–voltage characteristics of GaAs-based cells are represented remarkably well by the Shockley injection and recombination/generation models. For current levels corresponding to unconcentrated AM1 illumination either injection or recombination/generation

Table 8.IV

Diode Parameters of Selected GaAs-Based Solar Cells

Reference	Type	J_0	A	Details
Alferov *et al.*[a]	Homojunction	$0.25–1.7 \times 10^{-12}$	$\simeq 2$	Measured in dark
Van der Plas *et al.*[b]	Heteroface		1	At high concentration
			2	At AM1
Ewan *et al.*[c]	Heteroface	2.08×10^{-19}	$\simeq 1$	Curve fitting of illuminated data for $200 \lesssim C \lesssim 10^3$
Fan *et al.*[d]	n^+/p Homojunction	1.4×10^{-17}	1.1	$\log J_{sc}$ versus V_{0c}
Loo *et al.*[e]	Heteroface	1.6×10^{-9}	2.28	Dark; $C \simeq 1$ type cell
Shen[f]	Heteroface	1.1×10^{-10}	$\simeq 2$	Dark
Calculated[g]	Heteroface	2×10^{-18}	1	

[a] Zh. K. Alferov, U. M. Andreev, M. B. Kagan, I. I. Protasov, and V. G. Trofim, *Sov. Phys.—Semicond.* **4,** 2047 (1971).

[b] H. Van der Plas, personal communication, 1979.

[c] J. Ewan, R. C. Knechtli, R. Loo, and G. S. Kamath, *Proc. 13th IEEE Photovoltaic Specialists Conf.* (1978) p. 941.

[d] J. C.-C. Fan and C. O. Bozler, *Proc. 13th IEEE Photovoltaic Specialists Conf.* (1978) p. 953.

[e] R. Loo, L. Goldhammer, B. Anspaugh, R. C. Knechtli, and G. S. Kamath, *Proc. 13th IEEE Photovoltaic Specialists Conf.* (1978) p. 562.

[f] L. C.-C. Shen, Ph.D. Thesis, Dep. Electr. Eng., Stanford Univ., Stanford, California, 1976.

[g] Ideal, $A = 1$ homojunction current calculated using the material properties and dimensions of Table 8.III.

modes may dominate, but under high concentration most cells show the $A = 1$ injection behavior with corresponding low J_0. Table 8.IV lists the diode A factors and J_0 values of several cells, along with a J_0 calculated from the materials parameters of Shen's cell for comparison. Hovel (1973) comments on the junction transport mechanisms, using the Sah–Noyce–Shockley and Choo models to calculate ultimate efficiencies for GaAs-based cells.

The heteroface cell may be optimized in several ways:

(1) *Decreasing the window* (p^+)*-layer thickness.* Figure 8.8 shows the calculated effect of decreasing the window thickness $|x'_{p+} - x'_p|$ for AM0 conditions. The results suggest making the window as thin as possible consistent with its spreading resistance. These results are, of course, strongly dependent on the Al/Ga composition chosen and the air mass number. The effect on η_s is shown in Fig. 8.9.

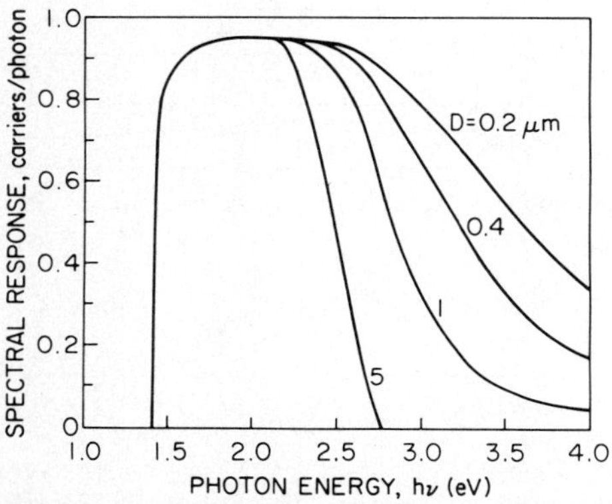

Fig. 8.8. Calculated spectral response of an $Al_xGa_{1-x}As$/GaAs cell with the $Al_xGa_{1-x}As$ thickness D as the independent variable. The parameters used here are

$$d = |x_p - x_p'| = 1\ \mu\text{m}, \quad L_n = 0.5\ \mu\text{m for AlGaAs},$$
$$W_d = 0.01\ \mu\text{m}, \quad L_n = 5.0\ \mu\text{m for } p\text{-GaAs},$$
$$S(x_{p+}') = 10^6\ \text{cm sec}^{-1}, \quad L_p = 1.0\ \mu\text{m for } n\text{-GaAs},$$
$$S(x_p') = 10^4\ \text{cm sec}^{-1}, \quad x = 0.86.$$

[From L. C.-C. Shen, Ph.D. Thesis, Dep. Electr. Eng., Stanford Univ., Stanford, California, 1976.]

(2) *Optimization of the p-layer thickness.* The effect of changing the p-layer thickness $d = |x_{p+}' - x_p|$ to maximize carrier generation and collection is shown in Fig. 8.10, which indicates a broad maximum for $d \simeq 1\ \mu$m for the larger L_n values using the same conditions as the pre-

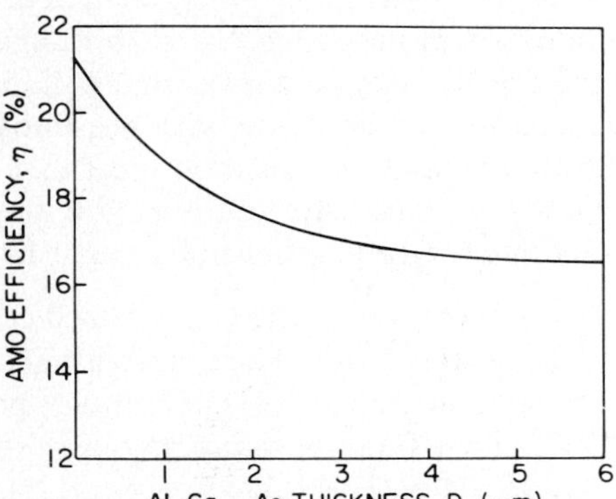

Fig. 8.9. Calculated AM0 efficiency as a function of the AlGaAs thickness D. Parameters used are the same as for Fig. 8.8. $V_{oc} = 1.0$ V and ff = 0.8 are assumed. [From L. C.-C. Shen, Ph.D. Thesis, Dep. Electr. Eng., Stanford Univ., Stanford, California, 1976.]

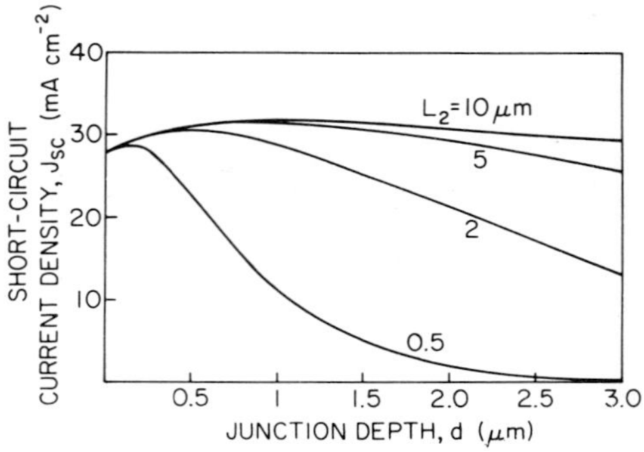

Fig. 8.10. Calculated J_{sc} at AM0 as a function of the *p*-GaAs layer thickness for several values of L_n. Parameters used are the same as those of Fig. 8.8. [L. C.-C. Shen, Ph.D. Thesis, Dep. Electr. Eng., Stanford Univ., Stanford, California, 1976.]

ceding.† Van der Plas *et al.* (1978), optimized the total efficiency versus *p*-layer thickness in concentrator cells and found similar broad maxima for $d \simeq L_n/4$.

(3) *Optimization of doping level.* The doping of the *n* and *p* regions can be varied to maximize V_{oc} while retaining a long L_n in the *p*-layer. In junctions where transport is controlled by recombination/generation this optimization involves the adjustment of W_d with consideration for the effect of doping level in the quasi-neutral regions on the lifetime in the depletion layer. The dependence of η_s on the *n*-type quasi-neutral region doping level for cells controlled by recombination/generation is discussed by Sekela *et al.* (1977), who assumed that the lifetime in the depletion layer is controlled by τ_p in the *n*-type quasi-neutral region. Their calculations indicate that an optimum exists in the range $N_D = 10^{16}$ to 10^{17} cm^{-3} but with no strong dependence on N_D. They estimate that cells with $\eta_s \lesssim 22\%$ at AM0 with no concentration can be made with presently available materials (assuming no grid coverage or reflection losses). For cells in which injection and diffusion dominate, the optimization variables are more directly accessible and higher doping densities might be favored.

One of the principal advantages of the GaAs-based cell is the small decrease of efficiency with temperature. Most measurements show a monotonic decrease of percent efficiency of 0.033 percentage points per °C (e.g., from 20.000% to 19.967% for 1°C temperature increase) such as that shown in Fig. 6.23.

† For these conditions the J_0 increase is approximately linear with *p*-layer thickness (if transport is dominated by J_{0n}) as indicated by Fig. 5.6 and so the change in V_{oc} with thickness is rather small.

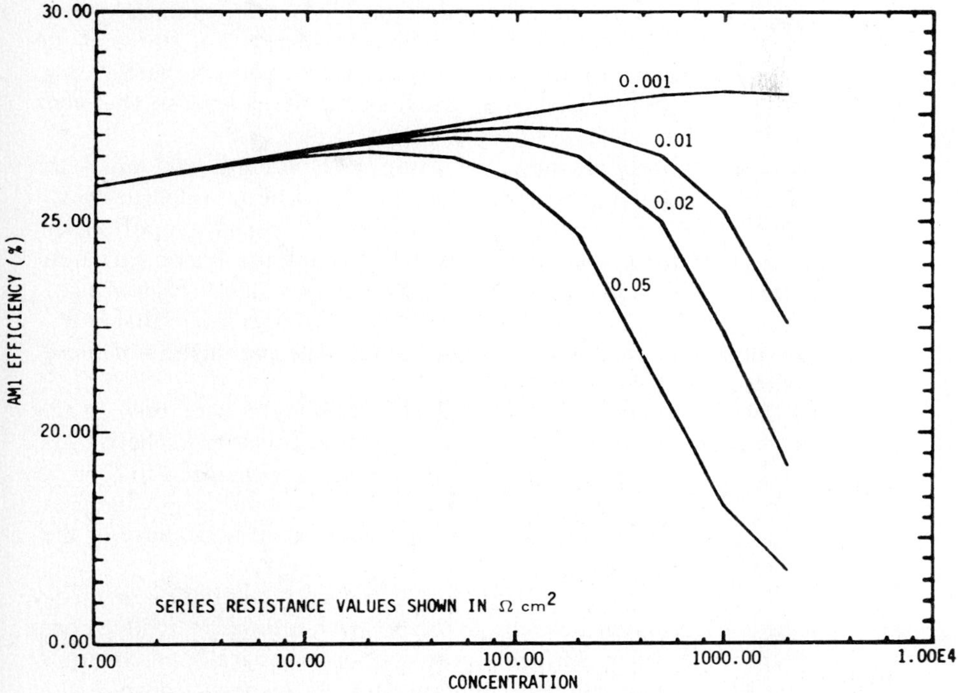

Fig. 8.11. Calculated solar efficiency as a function of concentration ratio for an AlGaAs/GaAs LPE cell for various values of lumped series resistance. [Courtesy of R. Sahai, D. D. Edwall, and J. S. Harris, Jr., *Proc. 13th IEEE Photovoltaic Specialists Conf.* (1978), p. 946. © 1978 IEEE.]

8.2.4 AlGaAs/GaAs Concentrator Cells

At the high concentration ratios favored for terrestrial concentration systems using GaAs-based cells, other optimization procedures are required. Since efficiency generally increases with intensity as shown in Fig. 8.11, concentration ratios of 10^3 and above are favored. The major limitation at these high ratios arises from current collection, requiring total R_s values of less than 10^{-3} Ω cm.† The window layer plays a major role in current collection, so some sort of compromise between AlGaAs sheet resistance (thickness and resistivity), grid spacing, and the amount of light transmitted by the AlGaAs layer must be made. The design of the grid itself is critical, and every aspect—contact resistivity; finger width, thickness, and spacing; and bus resistance—must be considered in proper optimization.

† Thermal considerations at high concentration ratios are discussed in Section 12.2.2.

For GaAs-based cells the Dember voltage has only a small effect ($\lesssim$1 mV), since the photogenerated carrier density ($\approx 2 \times 10^{14}$ cm^3 at $C = 10^3$) is small compared to the majority carrier density. Nevertheless, the high-intensity designs have light incident on the p-side so that the Dember voltage adds to V_{oc}.

Three contemporary designs, all using LPE growth and all with $\eta_s > 20\%$ under high concentration, are considered here: a design by L. James reported by Van der Plas *et al.* (1978) and designs reported by Ewan *et al.* (1978) and Sahai *et al.* (1978). Perhaps the major variation among them is in the thickness of the AlGaAs window: 1.2 μm (Sahai *et al.*, der Plas *et al.*, 1978), 10 μm (Ewan *et al.*, 1978), and 0.05 μm (Sahai *et al.*, 1978). Compositions of $x \lesssim 0.9$ are used in all. The parameters of these cells are summarized in Table 8.V.

In the design by Sahai the 0.05 μm AlGaAs layer is used only to reduce surface recombination rather than for current collection. These cells show a considerably larger J_{sc} ($\approx$31 mA cm^{-2} when normalized to $C = 1$, 100 mW cm^{-2}) than do the cells designed by Ewan ($\approx$23.4 mA cm^{-2} at $C = 1$) or James ($\simeq$25.7 mA cm^{-2} at $C = 1$), presumably because of the difference in AlGaAs thickness.

In the design reported by Sahai *et al.* the grid contracts are made *through* the AlGaAs layer, directly to the p-GaAs in order to reduce contact resistance. However, surface recombination is generally much larger at metal/p-GaAs interfaces, possibly resulting in higher J_0 despite the small contact area. This may be responsible for the somewhat lower V_{oc} seen in these cells as compared with those in which contact is made to the AlGaAs layer.

The p-layers of these cells are more uniform in thickness. Doping is by Be, Mg, or Zn and $L_n = 4$–5 μm.

The lowest effective series resistance is for the cell designed by James, which utilizes a p^+-GaAs layer between the contact metallization and the AlGaAs window to obtain contact resistivities in the 10^{-5}–10^{-4} Ω cm^2 range and an overall R_s of 4×10^{-5} to 4×10^{-4} Ω cm^2.

All of these cells, particularly the one by James, showed almost exact agreement with the elementary theory utilized in the computer simulation used for their design.

8.2.5 GaAs-Based Cells for Space Power Systems

A promising use for GaAs solar cells is in space, where the power-to-weight ratio and radiation resistance are of primary importance. By 1980 several experimental cell test groups had undergone testing in simulated space radiation environments and other groups have been tested in space. GaAs cells appear to have significantly better radiation resistance

Table 8.V

Parameters for Selected AlGaAs/GaAs Concentrator Cells

	Van der Plas[a]	Ewan[b]	Sahai[c]
$Al_xGa_{1-x}As$ Layer			
x	0.93	>0.90	—
D (μm)	1.2	10	0.05
N_A (cm^{-3})	5×10^{17}	1.5×10^{18}	3×10^{18}
Dopant	Mg	Be	Be
p-GaAs Layer	Diffused	Diffused	LPE grown
d (μm)	1.2	—	2
N_A (cm^{-3})	7×10^{17}	—	2×10^{18}
Dopant	Mg	Be	Zn
L_n (μm)	4	—	5
n-GaAs Base			
N_D (cm^{-3})	7×10^{17}	2×10^{17}	—
Dopant	Sn	Sn	—
L_p (μm)	4	2 (calculated)	—
Device			
R_s (Ω)	—	0.01	0.02
Grid spacing (μm)	62	250	90–100
Photovoltaic parameters			
Concentration ratio	945 (AM1.5)	1006 (AM1)	178 (AM2)
J_{sc} (A cm^{-2})	20.7	23.7	4.35
V_{oc} (V)	1.14	1.19	1.07
η_s (%)	23	20	24.7
Temperature (°C)	50	—	50
J_{sc} (mA cm^{-2}), normalized to $C = 1$, 100 mW cm^{-2}	25.6	23.6	≈31
η_s (%), at other concentration ratios	—	22.3 (at $C = 197$)	20.3 (at $C = 440$)

[a] H. A. Van der Plas, L. W. James, R. L. Moon, and N. J. Nelson, *Proc. 13th IEEE Photovoltaic Specialists Conf.* (1978), p. 934.

[b] J. Ewan, R. C. Knechtli, R. Loo, and G. S. Kamath, *Proc. 13th IEEE Photovoltaic Specialists Conf.* (1978), p. 941.

[c] R. Sahai, D. D. Edwall, and J. S. Harris, Jr., *Proc. 13th IEEE Photovoltaic Specialists Conf.* (1978), p. 946.

than Si cells, although this resistance appears to be more strongly dependent on fabrication variables than that for Si (Loo *et al.*, 1978). The decreases in efficiency on exposure to radiation appear to fit the models set up to quantify radiation effects in Si cells [e.g., Eq. (7.7)].

Several methods exist for minimizing or alleviating radiation damage. Postexposure annealing (e.g., 30 h at 200°C) has been used to restore the major part of lost efficiency (Walker and Conway, 1978). Graded junction structures (discussed in the next section) utilizing built-in electric fields can reduce the effects of the radiation-lowered L_n. Hutchby's (1978) detailed computer analysis for a doubly graded structure indicates substantially increased radiation resistance. This is particularly appealing for the AlGaAs system since grading is rather easy to accomplish during fabrication by CVD or LPE.

8.2.6 GaAs-Based Graded Band Gap and Schottky Barrier Cells

Two alternative GaAs-based structures are discussed briefly here: compositionally graded structures and Schottky barrier cells. Polycrystalline, thin-film GaAs cells are discussed in Chapters 9 and 11 and by Hovel (1975).

Introduction of a band-gap gradient by compositional grading has sprung from three somewhat different but interrelated goals:

1. Grading is a natural extension of the window concept for optimizing both the admission of light and the absorption of light in a layer backed by a p/n junction.
2. Grading and consequent high electric fields near the incident surface act to reflect carriers away from the surface and reduce the surface recombination loss.
3. High electric fields throughout the generation volume can substantially increase the effective L_n there [$L_{\text{eff}} \simeq L_{n0}(\mathscr{E}qL_{n0}/kT)$ as in Eq. (4.28)].

These ideas have been considered for a long time, and the AlGaAs–GaAs system appears to be ideal for exploring their effects. For example, Tauc (1957) apparently first proposed the reduction of the surface recombination velocity by a gradient-induced electric field in 1957. Tauc also showed that a potential is generated across an illuminated graded band-gap semiconductor because of the gradients there.

Marfaing and Chevallier (1971) extended the theory of Tauc to graded structures in which band gap, mobilities, lifetimes, carrier densities, and effective masses varied with position and compared their calculations with experimental $Hg_xCd_{1-x}Te$ structures, finding good agreement. They found that a small photovoltage of $\lesssim$1.5 mV could be generated by the gradients at very high illumination levels. However, for all practical solar cell purposes the voltage generated by the gradients is completely negligible and the primary effects of junction grading are those listed at the beginning of this section.

Hutchby and Fudurich (1976) have made a very detailed (and quite conservative) computer simulation of n-AlGaAs/p-GaAs graded struc-

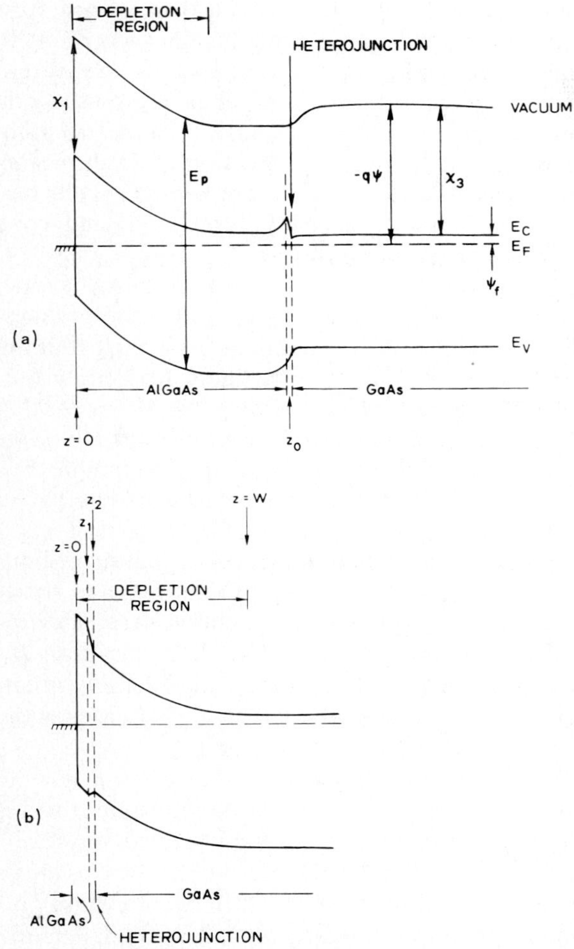

Fig. 8.12. Energy band diagram of Au/n-AlGaAs/n-GaAs Schottky barrier solar cells. The thickness of the AlGaAs layer is greater than the depletion layer width in part (a) and smaller than the depletion layer width in part (b). [From Y.-D. Shen, Ph.D. Thesis, Stanford Univ., Stanford, California, 1978.]

tures, using parameters for available materials as far as known and considering all known loss mechanisms, even to the antireflection coating. Their optimal cell has an $Al_xGa_{1-x}As$ layer graded over 1.0 μm from $x = 0.35$ to $x = 0$, yielding a solar efficiency of 17.7% at AM0. The grading provides increased resistance to degradation of L_n and S in radiation environments.

Woodall and Hovel (1977) used an "isothermal etchback-regrowth" technique for growth of p-$Al_xGa_{1-x}As$ layers graded from $x \approx 0.9$ to $x = 0$ over a distance of 0.2–0.4 μm. The Ga–Al–Zn melt was slightly undersaturated when positioned over the n-GaAs substrate. This first dissolves a portion of the substrate and then regrows a graded layer of AlGaAs as a result of the concentration gradients set up in the melt during the dissolution part of the process. Diffusion of Zn during the process results in a 0.8–2 μm p-GaAs:Zn layer. The resulting cells have 21.8% efficiency at AM1 (η_s = 18.4% at AM0, simulated) and considerably improved short-wavelength response.

Kordos *et al.* (1979) prepared 0.2 μm AlGaAs layers by a similar method. Using Auger depth profiling to verify the grading profile, they found good agreement with calculations assuming aluminum diffusion through a concentration gradient in the liquid adjacent to the growth interface. They also found that the graded layer depth (or etchback thickness) could be controlled by the degree of undersaturation within a range of 0.15–0.25 μm. Solar cells made using this technique by Kordos and Pearson (1981) suggested strongly improved high-energy response could be obtained using these methods.

Schottky and MIS structures on GaAs are discussed in detail in Section 11.1.2, but an interesting Au/n-AlGaAs/n-GaAs structure reported by Shen and Pearson (1979) deserves mention here. Theoretical analyses predict that this structure, shown in Fig. 8.12, can have J_{sc} as large as a conventional Au/n-GaAs cell and that a considerably greater V_{oc} can be obtained. Experimental results show that V_{oc} increases with Al mole fraction x from 0.53 V ($x = 0$) to 0.70 V ($x = 0.5$).

Since a band-gap difference ΔE_g ($\approx$0.4 eV for $x = 0.3$) exists that could impede hole transport to the Au (as indicated by Fig. 8.12a), the AlGaAs layer must be thin enough so that both AlGaAs and GaAs are depleted at the HJ and so that the HJ interface occurs in a high electric field region. The result of variation of the AlGaAs thickness on the measured spectral response is shown dramatically in Fig. 8.13. A family of curves of similar appearance is generated by variation of reverse bias V_R in a cell with a larger thickness (e.g., d = 2000 Å) as increasing V_R reduces the effect of ΔE_g.

Some junction grading at the HJ interface may also reduce the effective ΔE_g, as discussed in Section 5.3.3. Such measurements do much to restore one's faith in the concept of HJ band discontinuities.

An unoptimized Au/n-AlGaAs/n-GaAs cell recorded V_{oc} = 0.88 V and η_s = 10.5% (AM1 simulation, with no antireflection coating, and a AlGaAs layer thickness of 300 Å) (Yang *et al.*, 1980). A more general treatment of these double layer Schottky barrier devices is given by Lee and Pearson (1980).

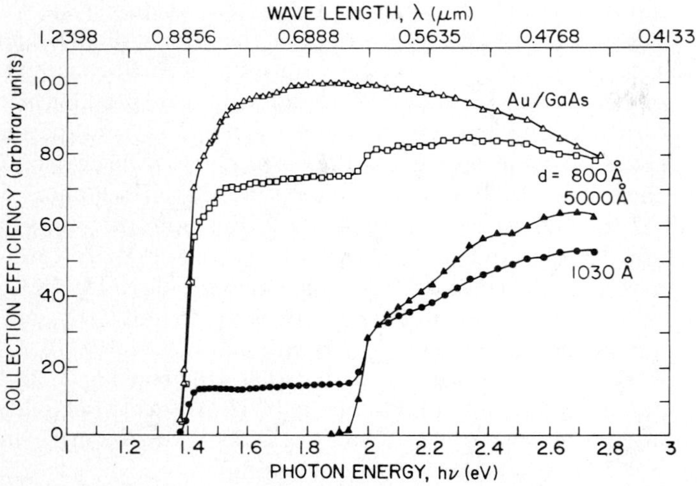

Fig. 8.13. Normalized collection efficiency versus photon energy of Au/n–$Al_{0.43}Ga_{0.57}As$/n–GaAs Schottky barrier solar cell. The independent variable is the AlGaAs layer thickness d. The characteristic of a Au/GaAs cell is shown for comparison purposes. [From Y.-D. Shen, Ph.D. Thesis, Stanford Univ. Stanford, California, 1978.]

8.3 InP-BASED CELLS

The CdS/InP system is discussed as an example of a true HJ in which excellent lattice matching and interfacial ''cleanliness'' appear to play an important role in the junction current transport. Highly efficient cells have been fabricated by vacuum evaporation, chemical vapor deposition, and close-spaced vapor transport (CSVT) methods.

InP is a direct-band-gap material with a band gap of 1.34 eV, close to the optimum for solar energy conversion. Like other homojunctions of direct-band-gap materials, the efficiency InP homojunction appears to be limited by surface recombination at the illuminated surface (Galavnov *et al.*, 1967). However, in a heterojunction configuration the potential advantages of InP can be realized.

Indium is a rather expensive metal (\$2000 kg^{-1} in small lots in 1980) with a natural abundance of 0.14 ppm (as compared to 64 ppm for Cu) in the earth's crust. Phosphorus, although abundant and inexpensive in its impure forms, is expensive to purify (\$4000 kg^{-1} in small lots of 6n purity). InP crystals are grown by the Czochralski method at high pressures or by using a liquid encapsulation technique to preserve stoichiometry.

Research on InP heterojunctions did not start until about 1974, probably because of unavailability of high-quality single crystals of p-type InP.

The first reported results on n-CdS/p-InP heterojunctions, by Wagner *et al.* (1975), described a 12.5% efficient cell. The choice of CdS appears to be particularly well suited for such a heterojunction. Because the lattice constant of zincblende InP is 5.869 Å and the corresponding parameter ($2^{1/2}a$) of wurtzite CdS is 5.850 Å, there is only 0.32% lattice mismatch between the (111) plane of InP and the basal plane of hexagonal CdS. In addition, the tetrahedral atomic distance is 2.533 Å in InP and 2.532 Å in CdS, indicating that the orientation of the InP is not critical, as has been experimentaly borne out. Wagner's first cells were prepared by vacuum deposition of CdS on 0.4 Ω cm InP single crystals doped with Cd. They showed a quantum efficiency of 0.70 between 0.55 and 0.91 μm, and recombination/generation type transport with a low $J_0 = 5 \times 10^{-9}$ A cm^{-2}, suggestive of a good lattice match. The solar efficiency $\eta_s = 12.5\%$ was obtained at a relatively low solar intensity of 53 mW cm^{-2} (a cloudy day in New Jersey), which gave $V_{oc} = 0.63$ V, $J_{sc} = 15$ mA cm^{-2}, and a fill factor ff = 0.71 (with an antireflection coating).

It was discovered that a heat treatment at $T \simeq 600°$C in a nonoxidizing atmosphere improves this type of CdS/InP cell, yielding $V_{oc} = 0.72$ V and a solar efficiency of 14% (Shay *et al.*, 1976). No change in junction properties was observed for a subsequent 15-min anneal at temperatures less than 550°C, indicating good operating stability of such a cell. By assuming that optimum junction quality corresponds to that of an InP homojunction, extrapolated maximum theoretical values for the CdS/InP junction of $V_{oc} = 0.85$ V, $J_{sc} = 20$ mA cm^{-2}, ff = 0.75, and $\eta_s = 17.2\%$ were estimated.

Further improvement of actual cell performance was achieved by going to a chemical vapor deposition of CdS on InP using an open-tube H_2S/H_2 flow system (Shay *et al.*, 1977; Bettini *et al.*, 1977, 1978). At an appropriate concentration of H_2S (about 2 mol%) in the gas flow, the surface of the InP is continuously etched by the formation and sublimation of indium sulfides; CdS nucleates on this clean surface and prevents further attack of the InP by the H_2S. With single-crystal InP, values were obtained as follows: $V_{oc} = 0.79$ V, $J_{sc} = 18.7$ mA cm^{-2}, ff = 0.735, and $\eta_s = 15.0\%$ at AM2. Capacitance data on these cells again indicated an abrupt junction but with a diffusion voltage of 1.00 ± 0.15 V, close to the value of 1.25 V for the diffusion voltage of an InP homojunction.

CdS/InP junction cells have also been prepared by the vacuum evaporation of CdS onto single-crystal p-type InP homoepitaxial layers, which were, in turn, grown by a chemical vapor deposition method on p^+ single-crystal InP substrates.† Cells showing between 8 and 12% effi-

† In these cells, the surface of the CdS layer is slightly textured due to faceting during growth, resulting in some antireflective properties (Manasevit *et al.*, 1978).

ciency were prepared, with the best cell having $V_{oc} = 0.72$ V, $J_{sc} =$ 23.2 mA cm^{-2}, ff = 0.63, and $\eta_s = 11.9\%$ (without an antireflection coating, AM $\approx$ 1 illumination of 94 mW cm^{-2}, active area basis). Extrapolation of the open-circuit voltage to $T = 0°K$ yielded a diffusion voltage of 1.06 eV in this case.

The highest V_{oc} observed to date was reported in an excellent work by Yoshikawa and Sakai (1977) for a cell prepared by close-spaced-vapor-transport. The photovoltaic parameters were $V_{oc} = 0.807$ V, $J_{sc} =$ 18.6 mA cm^{-2}, ff = 0.74, and $\eta_s = 14.4\%$ (AM2, 77 mW cm^{-2}). In contrast with the cells previously mentioned, these were grown on the (110) face rather than the (111) face. The growth temperature was $\approx$710°C. These researchers found that $V_D = 1.14$ eV, a value in agreement with the V_{oc} values seen by them and other workers as well as being in almost exact agreement with published electron affinity values for CdS and InP. Capacitance data on these cells indicated an abrupt junction corresponding to the band diagram shown in Fig. 8.14.

In view of the preceding comments on the degree of lattice matching between CdS and InP and the implied relevance of this match for the good cell properties observed, it is something of a surprise to find that essentially equally good cells can be made with ITO/InP junctions for which the lattice mismatch is large. Ion-beam-deposited ITO layers with resistivity of 10^{-3} Ω cm on Zn-doped p-type single crystals of InP yield solar cells with $V_{oc} = 0.76$ V, $J_{sc} = 21.5$ mA cm^{-2}, ff = 0.65, and $\eta_s = 14.4\%$

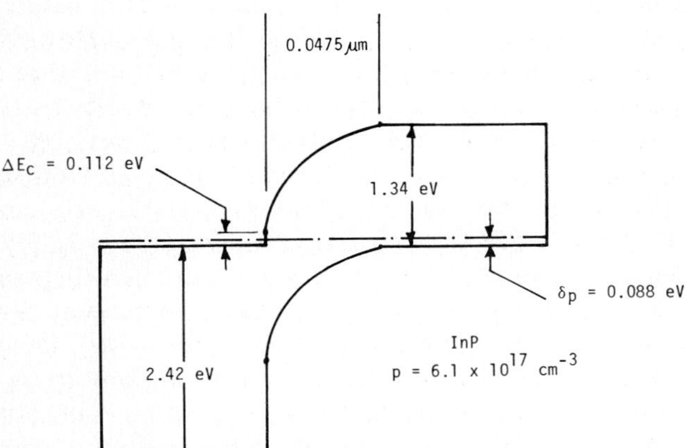

Fig. 8.14. Energy band diagram for CdS/InP heterojunction prepared by close-spaced vapor transport of CdS film onto single crystal InP. [Redrawn from A. Yoshikawa and Y. S. Sakai, *Solid State Electron.* **20,** 133 (1977). Copyright (1977), Pergamon Press, Ltd.]

(using a MgF_2 antireflection coating) (Sree Harsha *et al.*, 1977). Similar results have been obtained for sputtered ITO layers on homoepitaxial *p*-InP layers grown by chemical vapor deposition on p^+-InP substrates (Manasevit *et al.*, 1978). The best cell reported for this process had $V_{oc} = 0.69$ V, $J_{sc} = 23.4$ mA cm^{-2}, ff = 0.65, and $\eta_s = 12.4\%$ (~AM 1.5 illumination at 85 mW cm^{-2} without an antireflection coating). Extrapolation of the open-circuit voltage to $T = 0°$K gives a diffusion voltage of 1.09 V, in good agreement with the values cited previously. In the latter case some heat treatment is required to produce the maximum efficiency. The success of the ITO/InP junctions in spite of lattice mismatch can probably be attributable to the existence of a buried homojunction in the InP.

Evidence for a buried homojunction has been obtained by measurements of rf sputtering deposited ITO/InP cells by Bachmann *et al.* (1979). Ion probe analyses showed the presence of sufficient Sn penetration into the *p*-InP substrate to change the conductivity type for nominal substrate temperatures of 250°C. When the ITO was deposited at a nominal 27°C substrate temperature, no evidence of Sn penetration was found, although this does not preclude the existence of a very thin homojunction.

Tsai *et al.* (1980) found that buried homojunctions in the ITO/*p*-InP system were produced by the action of sputter deposition. Schottky diodes were formed by vacuum evaporation of thin films of Au on *p*-InP substrates. When these substrates had been previously sputter cleaned in Ar under the same conditions as those for the ITO deposition, spectral response curves with sharply reduced high-energy response were obtained, typical of homojunction behavior. These spectral responses were nearly the same as those for the ITO/*p*-InP cells. However, when Au/*p*-InP Schottky barriers were formed on substrates that were not sputter cleaned after etching, the high-energy spectral response was much larger, typical of that seen for good Schottky barrier devices.

These last conclusions may be further compared with those for CdS/GaAs heterojunctions fabricated by Bettini *et al.* (1978) and by Yoshikawa and Sakai (1975) by CVD (lattice mismatch $\simeq 3.5\%$). Surface etching and heat treatment are again required to obtain the best diodes. Although diodes with moderately good diode characteristics were obtained (e.g., $J_0 = 10^{-8}$ A cm^{-2} with $A = 1.80$; Bettini *et al.*, 1978), both η_Q and V_{oc} for the CdS/GaAs heterojunctions were too low to be competitive. Indium-tin-oxide/GaAs cells made by Bachmann *et al.* (1979) using rf and magnetron sputtering and ion beam deposition show considerably lower J_{sc}, V_{oc}, and η_s ($\lesssim 5\%$) than do similar ITO/InP cells ($\eta_s \gtrsim 14\%$). Apparently the details of the junction formation permit good η_Q and V_{oc} in the presence of large mismatch in the case of the ITO/InP heterojunction but not in the case of CdS/GaAs or ITO/GaAs.

8.4 SUMMARY

The GaAs heteroface structure, the GaAs $n^+/p/p^+$ homojunction, and the Au/AlGaAs/GaAs Schottky diode cells are all in remarkable agreement with solar cell theory. In particular, J_0 can be calculated with reasonable accuracy from measured materials properties.

The CdS/InP cell also gives a close to theoretical efficiency, but these cells frequently show tunneling transport rather than recombination/generation transport, for reasons that are not completely understood.

In addition to the fortuitous lattice matching of the CdS/InP junction, the experiments reported here indicate that either heat treatment (at $T > 500°C$) or *in situ* etching is required to realize the highest and most uniform efficiencies. This suggests one or more of the following possibilities.

(1) Interfacial debris is removed or inactivated by these treatments.

(2) Some sort of compound formation occurs at the interface that neutralizes the interfacial recombination.

(3) A heteroface structure is produced that effectively moves the p/n junction away from the metallurgical interface.

Two of the strongest general guiding principles in the historical development of heterojunction cells have been the use of a heteroface structure to avoid surface recombination loss, and the demand for excellent lattice match. Thus it is rather ironic to consider the existence of important counter examples for both these criteria: the p^+/n GaAs homojunction with 20% efficiency reported by Fan and Bozler (1978) (Section 8.2.3) and the ITO/InP heterojunction with 14.4% solar efficiency reported by Sree Harsha *et al.* (1977). The first example shows that surface recombination loss *can* be controlled by suitable fabrication techniques, in this case by making the n^+-layer extremely thin (and by the *possible* advantageous effect of passivation of the n^+-GaAs surface recombination by anodization).† Thus other direct band-gap materials may prove to be highly efficient in the homojunction configuration, provided suitable fabrication methods can be found.

Although the second counterexample, ITO/p-InP, is explained by the formation of a buried homojunction, our understanding of the details of the heterojunction interface still do not permit us to predict the formation of the buried homojunction or its effect on carrier transport. For example, why does the ITO/p-GaAs analog apparently not produce good cells?

† On the other hand, these thin n^+-layer cells do appear to be much more difficult to fabricate than the heteroface AlGaAs/GaAs cells.

Thus still other guiding principles, in particular those involving the microscopic chemistry of the interface region, must be added to the list beginning this chapter.

A number of other heterojunctions and heteroface cells, including both single crystal and polycrystalline materials are considered in Chapter 11.

REFERENCES

Alferov, Z. I., Andreev, U. M., Kagan, M. B.,Protasov, I. I., and Trofim, V. G. (1971). *Sov. Phys.—Semicond.* **4,** 2047. [*Fiz. Tekh. Poluprovodn.* **4,** 2378 (1970).]

Bachmann, K. J., Schreiber, H., Jr., Sinclair, W. R., Schmidt, P. H., Thiel, F. A., Spencer, E. G., Pasteur, G. Feldmann, W. L., and Sree Harsha, K. S. (1979). *J. Appl. Phys.* **50,** 3441.

Balch, J. W., and Anderson, W. W. (1972). *Phys. Status Solidi A* **9,** 567.

Bettini, M., Bachmann, K. J., Buehler, E., Shay, J. L., and Wagner, S. (1977). *J. Appl. Phys.* **48,** 1603.

Bettini, M., Bachmann, K. J., and Shay, J. L. (1978). *J. Appl. Phys.* **49,** 865.

Buch, F., Fahrenbruch, A. L., and Bube, R. H. (1977). *J. Appl. Phys.* **48,** 1596.

Bucher, E. (1978). *Appl. Phys.* **17,** 1.

Casey, H. C., Jr., and Panish, M. B. (1978a). "Heterostructure Lasers, Part A." Academic Press, New York.

Casey, H. C., Jr., and Panish, M. B. (1978b). "Heterostructure Lasers, Part B." Academic Press, New York.

Chynoweth, T. A., and Bube, R. H. (1980). *J. Appl. Phys.* **51,** 1844.

Dapkus, P. D., Dupuis, R. D., Yingling, R. D., Yang, J. J., Simpson, W. I., Moudy, L. A., Johnson, R. E., Campbell, A. G., Manasevit, H. M., and Ruth, R. P. (1978). *Proc. 13th IEEE Photovoltaic Specialists Conf.*, p. 960.

Dawson, L. R. (1972). *In* "Progress in Solid-State Chemistry" (H. Reiss and J. O. McCaldin, eds.), p. 117. Pergamon, Oxford.

Dupuis, R. D., Dapkus, P. D., Yingling, R. D., and Moudy, L. A. (1977). *Appl. Phys. Lett.* **31,** 201.

Ettenberg, M., and Kressel, H. (1976). *J. Appl. Phys.* **47,** 1538.

Ewan, J., Knechtli, R. C., Loo, R., and Kamath, G. S. (1978). *Proc. 13th IEEE Photovoltaic Specialists Conf.*, p. 941.

Fahrenbruch, A. L. (1977). *J. Cryst. Growth* **39,** 73.

Fahrenbruch, A. L., and Aranovich, J. (1979). Heterojunction phenomena and interfacial defects in photovoltaic converters. *In* "Solar Energy Conversion—Solid State Physics Aspects" (B. O. Seraphin, ed.), Topics in Applied Physics, Vol. 31, p. 257. Springer-Verlag, Berlin and New York.

Fan, J. C.-C., and Bozler, C. O. (1978). *Proc. 13th IEEE Photovoltaic Specialists Conf.*, p. 953. [Also see J. C.-C. Fan, C. O. Bozler, and R. L. Chapman, *Appl. Phys. Lett.* **32,** 390 (1978).]

Feigelsen, R. S., N'Diaye, A., Yin, S.-Y., and Bube, R. H. (1977). *J. Appl. Phys.* **48,** 3162.

Galavanov, V. V., Kundukhov, R. M., and Nasledov, D. N. (1967). *Sov. Phys.—Solid State* **8,** 2723.

Gaugash, P. V., Kas'yan, V. A., Kovol'kov, V. I., and Rakhimov, N. R. (1976). *Sov. Phys.—Semicond.* **9,** 1239.

Gobat, A. R., Lamorte, M. F., and McIver, G. W., (1962). *IRE Trans. Mil. Electron.* **6,** 20.

Hovel, H. J. (1973). *Proc. 10th IEEE Photovoltaic Specialists Conf.*, p. 34.
Hovel, H. J. (1975). "Solar Cells," Semiconductors and Semimetals, Vol. 11. Academic Press, New York.
Hutchby, J. A. (1978). *Proc. 13th IEEE Photovoltaic Specialists Conf.*, p. 571.
Hutchby, J. A., and Fudurich, R. L. (1976). *J. Appl. Phys.* **47,** 3140, 3152.
James, L. W., and Moon, R. L. (1975). *Appl. Phys. Lett.* **26,** 476.
Jenny, D. A., Loferski, J. J., and Rappaport, P. (1956). *Phys. Rev.* **101,** 1208.
Johnston, W. D., and Callahan, W. M. (1976). *Appl. Phys. Lett.* **28,** 150.
Kordos, P., and Pearson, G. L. (1980). *Solid-State Electron* **23,** 399.
Kordos, P., Powell, R. A., Spicer, W. E., Pearson, G. L., and Panish, M. B. (1979). *Appl. Phys. Lett.* **34,** 366.
Lee, S. C., and Pearson, G. L. (1980). *IEEE Trans. Electron Devices* **ED-27,** 844.
Loo, R., Goldhammer, L., Anspaugh, B., Knechtli, R. C., and Kamuth, G. S. (1978). *Proc. 13th IEEE Photovoltaic Specialists Conf.*, p. 562.
Manasevit, H. M., Hess, K. L., Dapkus, P. D., Ruth, R. P., Yang, J. J., Campbell, A. G., Johnson, R. E., Moudy, L. A., Bube, R. H., Fabick, L. B., Fahrenbruch, A. L., and Tsai, M.-J. (1978). *Proc. 13th IEEE Photovoltaic Specialists Conf.*, p. 165.
Marfaing, Y., and Chevallier, J. (1971). *IEEE Trans. Electron Devices* **ED-18,** 465.
Milnes, A. G., and Feucht, D. L. (1972). "Heterojunctions and Metal-Semiconductor Junctions," p. 8. Academic Press, New York.
Sahai, R., Edwall, D. D., and Harris, J. S., Jr. (1978). *Proc. 13th IEEE Photovoltaic Specialists Conf.*, p. 946.
Sekela, A. M., Feucht, D. L., and Milnes, A. G. (1977). *IEEE Trans. Electron Devices* **ED-24,** 373.
Shay, J. L., Wagner, S., Bachmann, K. J., and Buehler, E. (1976). *J. Appl. Phys.* **47,** 614.
Shay, J. L., Wagner, S., Bettini, M., Bachmann, K. J., and Buehler, E. (1977). *IEEE Trans. Electron Devices* **ED-24,** 483.
Shen, L. C.-C. (1976). Ph.D. Thesis, Dept. Electr. Eng., Stanford Univ., Stanford, California.
Shen, Y. D., and Pearson, G. L. (1979). *Solar Energy Materials* **2,** 31.
Sree Harsha, K. S., Bachmann, K. J., Schmidt, P. H., Spencer, E. G., and Thiel. F. A. (1977). *Appl. Phys. Lett.* **30,** 645.
Tsai, M.-J., Fahrenbruch, A. L., and Bube, R. H. (1980). *J. Appl. Phys.* **51,** 2696.
Tauc, J. (1957). *Rev. Mod. Phys.* **29,** 308.
Van der Plas, H., James, L. W., Moon, R. L., and Nelson, N. J. (1978). *Proc. 13th IEEE Photovoltaic Specialists Conf.*, p. 934.
Wagner, S., Shay, J. L., Bachmann, K. J., and Buehler, E. (1975). *Appl. Phys. Lett.* **26,** 229.
Walker, G. H., and Conway, E. J. (1978). *Proc. 13th IEEE Photovoltaic Specialists Conf.*, p. 575.
Weider, H. H. (1979). *In* "Insulating Films on Semiconductors, 1979" (G. G. Roberts and M. J. Morant, eds.), p. 234. Inst. Phys. Conf. Series No. 50, Bristol, U.K.
Woodall, J. M., and Hovel, H. J. (1972). *Appl. Phys. Lett.* **21,** 379.
Woodall, J. M., and Hovel, H. J. (1975). *Appl. Phys. Lett.* **27,** 447.
Woodall, J. M., and Hovel, H. J. (1977). *Appl. Phys. Lett.* **30,** 492.
Yang, H. T., Shen, Y.-D., Edwall, D., Miller, D. L., and Harris, J. S. (1980). *IEEE Trans. Electron Devices* **ED-27,** 851.
Yoshikawa, A., and Sakai, Y. (1975). *Jpn. J. Appl. Phys.* **14,** 1547.
Yoshikawa, A., and Sakai, Y. (1977). *Solid-State Electron.* **20,** 133.

CHAPTER 9

POLYCRYSTALLINE THIN FILMS FOR SOLAR CELLS

9.1 INTRODUCTION

There are three types of photovoltaic converters for large-scale terrestrial use—single-crystal cells, concentrating systems, and thin-film devices. Although the basic physics of thin-film devices is the most complex, they offer two important advantages: (1) because the required thickness of the active layers is on the order of two or three times the optical absorption length, the material costs remain a small part of the total cell cost,† and (2) the thin-film configuration lends itself to large area, continuous flow processing. The tasks of developing large-scale procedures and building the machines to produce thin-film solar arrays in a continuous process are formidable, but once in action these machines should be able to provide enormous output and consequently low unit cost. To illustrate the potential for scale-up, Fig. 9.1 shows a large vacuum evaporation facility capable of coating 10 × 12 ft^2 sheets of architectural float glass at the rate of one sheet per minute. The machine has been operating since

† The approximate materials costs for the minimum required thicknesses of the absorber and window layers for various cells are illustrated by the following comparison: Si *p*/*n*, \$0.09 W_p^{-1}; CdS/CdTe, \$0.08 W_p^{-1}; Cu_xS/CdS, \$0.08 W_p^{-1}; $CuInSe_2$/CdS, \$0.21 W_p^{-1}; CdS/InP, \$0.32 W_p^{-1}, and Au/GaAs, \$0.25 W_p^{-1}. The assumptions are η_s = 10%, AM1.5, and 1981 *small-lot* (1 kg) prices for 5N elements. No fabrication or substrate costs are included. These costs can probably be divided by a factor of 3 to 4 for large-lot pricing.

Fig. 9.1. Semicontinuous vacuum evaporation coating facility designed by Airco Temescal for a Libbey-Owens-Ford architectural glass making plant. The system routinely coats 10 × 12 ft^2 substrates with single- or multiple-layer films on a one-minute cycle. The machine utilizes 24 electron beam evaporation sources, each operating at 150 kW. Vacuum is 10^{-6} Torr, established by 19 48-inch diffusion pumps. [Photo courtesy of LOF Glass, a division of Libbey-Owens-Ford, Toledo, Ohio.]

1966. If such a machine were producing 8% efficient solar cells continuously, its output would translate to over 1 MW_p per day.

Although much of the past research on thin-film cells has been done with epitaxial films deposited on single-crystal or large-grain polycrystalline substrates, the real advantages of the thin-film approach come from devices fabricated on low-cost amorphous or polycrystalline (PX) substrates. In almost all cases, such thin films are polycrystalline. The individual crystallites are generally of good quality, but these are interrupted by numerous grain boundaries and perhaps occasional voids. The grain boundaries are regions of increased disorder with both structural defects and segregated impurities in large densities. Thus they are generally regions of increased recombination. In addition, charge trapping at grain boundary energy levels can form potential barriers that impede carrier transport.

What then are the baseline optical, electronic, and structural constraints for thin-film photovoltaics? Roughly speaking, we must have the following:

(1) The film thickness of the absorber layer t_a must be sufficient to absorb most of the solar illumination, i.e., about two absorption lengths, or $t_a \gtrsim 2/\alpha$ [e.g., for ≈90% absorption of the available $(h\nu > E_g)$AM1.5 solar spectrum in GaAs $t_a \simeq 2$ μm is required and in Si, $t_a \simeq 100$ μm is required].

(2) Considerable loss of photogenerated carriers can occur at grain boundaries. The situation within the grains can be viewed simply by using a treatment due to Rothwarf (1976) if we assume for the moment that the grain boundary recombination velocity S_{gh} is infinite. We further assume that, at the surface of the "recombination cone" of Fig. 9.2, carriers are equally likely to diffuse to the grain boundaries and be lost or to the junc-

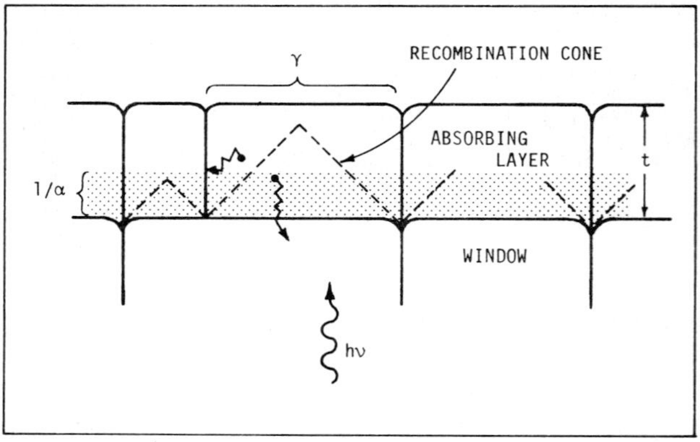

Fig. 9.2. Simple model of the effects of grain boundary recombination in a PX absorber layer. It is assumed that $S_{gb} \rightarrow \infty$ and that most of the carriers photogenerated outside the recombination cone are lost to the grain boundary whereas those inside it diffuse to the junction and are collected.

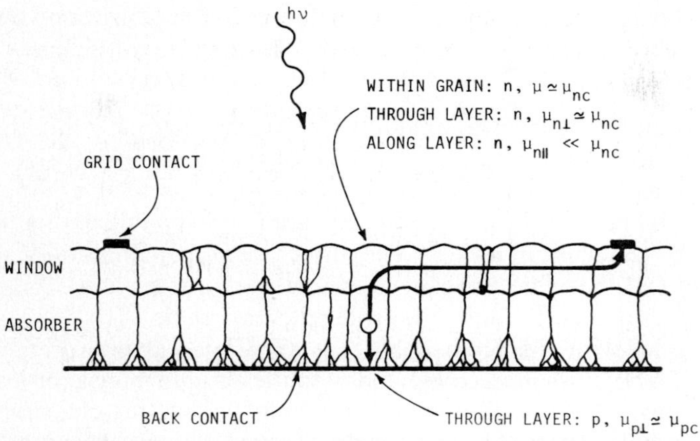

Fig. 9.3. Schematic diagram of carrier flow in a typical PX heterojunction solar cell with different parallel and transverse resistivities and mobilities. The carrier densities are n and p and the mobility values for bulk, single-crystal material are μ_{nc} and μ_{pc}.

tion to be usefully collected. Then in grains that are small with respect to the absorption length $1/\alpha$, a large proportion of the photogeneration volume lies outside the cone where carriers are likely to be lost. Consideration of Fig. 9.2 suggests that the grain size γ should be several times the mean optical absorption length $\overline{1/\alpha}$ for solar photons to avoid significant loss.†

(3) The electronic conduction properties of thin PX films are in general anisotropic, following the anisotropy of the grain structure. Thus we consider mobilities along the film (parallel to the film plane) and across the film plane (transverse), the former usually being considerably smaller. In the conventional absorber/window/grid configuration shown in Fig. 9.3, the parallel resistivity $\rho_{\parallel}$ of the window layer contributes the largest part of the total series resistance, and, if we assume the distributed window resistance is $R_{\rm sw} \lesssim 0.5\ \Omega\ {\rm cm}^2$, then by Eq. (6.18), the value of $\rho_{\parallel}$ must be roughly such that $\rho_{\parallel} \lesssim 600 t_{\rm w}$, Ω cm, where the window thickness $t_{\rm w}$ is expressed in centimeters.

(4) Generally (but certainly not always), growth in thin films produces a fibrous structure with the fiber axis perpendicular to the film plane and with the grain diameter smaller than its dimension along the fiber axis t (or $\gamma/t \approx 0.2$–0.5). Thus the absorber thickness must be great enough to provide the requisite grain size. Several options, such as increased sub-

† Although these results are in general agreement with more careful analyses (Section 9.4.4), the assumptions are not strictly correct. They are used here for their intuitive value.

strate temperature during film growth or recrystallization after film growth are available to increase the ratio of grain size to thickness, as discussed in Sections 9.2.3 and 9.2.4.

The grain boundaries dominate the performance of PX thin-film solar cells in most cases, and the importance of the role of grain size is illustrated dramatically in Fig. 9.4, which shows η_s as a function of γ for experimental cells. The data for Si cells are in reasonable agreement with theoretical calculations. For some direct band-gap materials, notably Cu_xS and $CuInSe_2$, the grain sizes can be over two orders of magnitude smaller than that of Si, and still produce comparable η_s.

In keeping with the philosophy that experimental interpretation and theory must be intimately connected with the actual material preparation, an overview of PX thin-film growth methods begins this chapter. Rather than describe each of the many methods in detail, we assume that the reader is familiar with most of them and we focus on their common aspects. Next we discuss the optical and electronic properties of PX films in

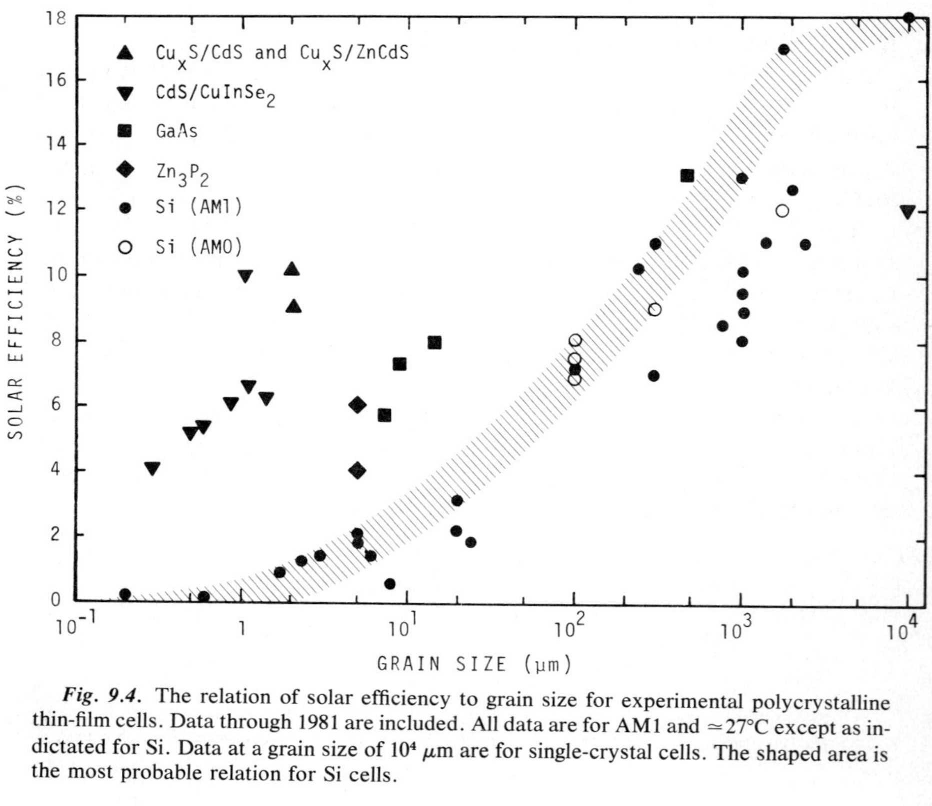

Fig. 9.4. The relation of solar efficiency to grain size for experimental polycrystalline thin-film cells. Data through 1981 are included. All data are for AM1 and ≈27°C except as indictated for Si. Data at a grain size of 10^4 μm are for single-crystal cells. The shaped area is the most probable relation for Si cells.

Sections 9.3 and 9.4. The concepts described in the previous sections are integrated in a discussion of the principles of PX solar cells in Section 9.5.

The deposition and properties of thin PX films have been the topics of tens of thousands of articles and hundreds of books. In this chapter we restrict our discussion to those properties that are directly relevant to photovoltaic cells, and even then we must be quite brief. We recommend an excellent resource book by Maissel and Glang (1970) as well as the list of general references at the end of the chapter. Grain boundaries are discussed by Leamy *et al.* (1982).

9.2 GROWTH OF THIN FILMS

The diversity of growth methods available represents a real flowering of the ingenuity of scientists and engineers. A representative listing is given in Table 9.I. Of these we have chosen vacuum evaporation (VE) as a representative or "baseline" method upon which to expand in order to obtain an intuitive perspective on the common aspects of film growth.

9.2.1 Common Aspects of Film Growth

Film growth can be thought of in terms of three basic themes:

(1) *the flux of atoms or molecules* to the surface of the substrate, which is determined, for example, by the source temperature for vacuum evaporation or by diffusion in the carrier gas for chemical vapor deposition (CVD),

(2) *the energetics of the incoming species* on the substrate, which are determined, for example, by the thermal energy of the atoms as they arrive, by the substrate temperature T_s, and by the sort of chemical bonding that takes place there, i.e., by the heats of condensation and/or formation, and

(3) *the organizational influences* that are present at the substrate. These arise from the periodicity of the substrate and film lattices, the topography of the growing surface, the crystal growth habit of the film, and the chemical affinities that may exist.

The ability of the incoming molecules to accommodate these organizational forces depends on the atomic mobility at the growing surface, which is a strong function of the energy of the atoms or molecules, and the impurities and the structural defects present on the surface. The better the accommodation on the substrate the closer the layer is to being a single crystal.

The growth of thin films is similar in many ways to the settling of immigrants in a new land. If the incoming numbers are large or the

Table 9.1

Common Means of Thin-Film Deposition

A. Atomic and Molecular Deposition in Vacuum

Name	Acronym	Typical values: Rate†	Typical values: Ambient pressure (Torr)	Details and references to deposition method
Vacuum evaporation	VE	Medium	10^{-8}–10^{-5}	Expensive equipment; poor material utilization[a]
Molecular beam epitaxy	MBE	Low	10^{-11}–10^{-8}	Very expensive equipment; excellent impurity and stoichiometry control[b]
Flash VE	FVE	High	10^{-8}–10^{-5}	Retains stoichiometry for incongruently evaporating compounds[a]
Reactive VE	RVE	Medium	10^{-6}–10^{-2}	Stoichiometry control is possible[a]
Activated RVE	ARVE	High	10^{-6}–10^{-2}	Enhances difficult reactions;[c,d] similar to ion beam plating[e]
Ion beam deposition	IB	Medium	$\lesssim 10^{-6}$	Ion energy is variable (also called ion condensation)[d,f]
Hot wall VE	HWVE	Low to medium	10^{-8}–10^{-5}	Closer to thermodynamic equilibrium; good material utilization (pressure in quasi-closed hot wall chamber, $\sim 10^{-3}$ Torr)[g]
Sputtering	dcSp rfSp	Low	10^{-3}–10^{-1}	Depositing atoms have high energy (4–40 eV)[a]
Magnetron sputtering	MSp	Medium	10^{-3}–10^{-1}	Plasma confined away from substrate[h]
Triode sputtering	TSp	Medium	10^{-5}–10^{-3}	Lower background pressure at substrate[i]
Reactive sputtering	RSp	Low	10^{-3}–10^{-1}	Stoichiometry control possible[a,j]
Ion beam sputtering	IBSp	Medium	$\sim 10^{-5}$	Additional control; lower pressure[d,k]
Planar reactive deposition	PRD	Low	$\sim 10^{-4}$	Good material utilization; stoichiometry control possible[l]
Plasma CVD	PCVD	Medium	$\sim 10^{-3}$	Enhances reactions[m]

B. Atomic and Molecular Deposition from a Carrier Gas or Liquid Near Atmospheric Pressure‡

Name	Acronym	Typical values: Source temp. (°C)	Typical values: Substrate temp. (°C)	Rate	Details and references to deposition method
Chemical vapor deposition	CVD	~850	~700	Medium to high	Good stoichiometry and doping control[n,o]
Metal organic CVD	MO-CVD	~25	550–750	Medium	Good materials utilization with rf heated substrate[p,q]

† Rates $\lesssim 0.1\ \mu m\ min^{-1}$, low; rates $\gtrsim 1\ \mu m\ min^{-1}$, high.

‡ CVD and CSVT are done at reduced pressure in some cases.

[a] L. I. Maissel and R. Glang, "Handbook of Thin Film Technology." McGraw-Hill, New York, 1970.

[b] A. Y. Cho and J. R. Arthur, *Prog. Solid State Chem.* **10,** 157 (1975).

[c] A. C. Raghuram and R. F. Bunshaw, *J. Vac. Sci. Technol.* **9,** 1389 (1972); R. F. Bunshaw and A. C. Raghuram, *J. Vac. Sci. Technol.* **9,** 1385 (1972).

[d] C. Weissmantel, H. J. Erler, and G. Reisse, *Surf. Sci.* **86,** 207 (1979).

[e] D. Mattox, *J. App. Phys.* **34,** 2493 (1963).

[f] T. Takagi, I. Yamada, and A. Sasaki, *Proc. 12th IEEE Photovoltaic Specialists Conf.* (1976), p. 55.

[g] A. Lopez-Otero, *Thin Solid Films* **49,** 3 (1978).

[h] J. A. Thornton, *J. Vac. Sci. Technol.* **11,** 666 (1974).

[i] T. Van Vouros, *Optical Spectra* **11,**(11), 30 (1977).

[j] A. D. Jonath, W. W. Anderson, J. A. Thornton, and D. G. Cornog, *J. Vac. Sci. Technol.* **16,** 200 (1979).

[k] D. E. Burk, J. B. DuBow, and J. R. Sites, *Proc. 12th IEEE Photovoltaic Specialists Conf.* (1976), p. 971.

[l] L. M. Fraas and K. Zanio, *J. Electron. Mater.* **7,** 211 (1978).

[m] R. P. H. Chang, A. J. Polak, D. L. Allara, and C. C. Chang, *J. Vac. Sci. Technol.* **16,** 888 (1979).

[n] J. J. Tietjen, Chemical vapor deposition of electronic materials. *Ann. Rev. Mater. Sci.* **3,** 317 (1973).

[o] K. J. Bachmann, E. Buehler, J. L. Shay, S. Wagner, and M. Bettini, *J. Electrochem. Soc.* **123,** 1509 (1976).

[p] H. M. Manasevit and W. I. Simpson, *J. Electrochem. Soc.* **111,** 180 (1964).

[q] P. D. Dapkus, R. D. Dupuis, R. D. Yingling, J. J. Yang, W. I. Simpson, L. A. Moudy, R. E. Johnson, A. G. Campbell, H. M. Manasevit, and R. P. Ruth, *Proc. 13th IEEE Photovoltaic Specialists Conf.* (1978), p. 960.

(*continued*)

B. Atomic and Molecular Deposition from a Carrier Gas or Liquid Near Atmospheric Pressure‡ (*continued*)

Name	Acronym	Typical values: Source temp. (°C)	Typical values: Substrate temp. (°C)	Rate	Details and references to deposition method
Close-spaced vapor transport	CSVT	~750	~400	High	Planar geometry; good materials utilization[r,s]
Liquid phase epitaxy	LPE	~800	~800	Medium	Expensive set up; excellent quality layers[t]
Spray pyrolysis	SP	~25	~450	Low	Inexpensive, simple equipment; small grain size[u,v]
Chemplating (chemical reduction plating)	CP	25–90	25–90	Low	Inexpensive, simple equipment; small grain size[w]
Electroplating	EP	25–90	25–90	Low	Inexpensive, simple equipment; small grain size; stoichiometry control possible with two anodes[w,x]
Ion exchange	IEX	45–100	45–100	Medium	Good materials utilization; good interface cleanliness; simple equipment[y,z]
Anodization	AN	~25	~25	Low	Controlled growth of thin (<1 μm) oxide films[a]

C. Particulate Deposition§

Name	Acronym	Typical values: Application temp. (°C)	Typical values: Minimum thickness (μm)	Details and references to deposition method
Spray (airbrush)	SPR	25–400	~5	Very inexpensive equipment; easy scale up.
Brush painting	BP	~25	20–40	Inexpensive but scale up difficult
Screen print (seriography)	SER	~25	~20	Inexpensive equipment; easy scale up and automation; good definition of layer area[aa]
Doctor blade	DB	~25	~10	Inexpensive equipment; easy scale up
Sedimentation	SED	~25	~5	Inexpensive process; used for CRT phosphors[bb]

Plasma spray	PS	25–800	~5	Used for refractory coatings (flame temperature >2000°C)[cc]
Electrophoresis	EL	25	5	Used for phosphor application[dd,ee]

§ All processes at atmospheric pressure. Generally, layers produced by these processes (except PS) must be sintered at 500–900°C in inert atmospheres. Rates are generally high except for SED.

[r] J. Saraie, M. Akiyama, and T. Tanaka, *Jpn. J. Appl. Phys.* **11,** 1758 (1972).

[s] K. W. Mitchell, A. L. Fahrenbruch, and R. H. Bube, *J. Vac. Sci. Technol.* **12,** 909 (1975).

[t] H. C. Casey, Jr. and M. B. Panish, "Heterostructure Lasers, Part B," p. 109. Academic Press, New York (1978).

[u] R. S. Berg, R. D. Nasby, and C. Lampkin, *J. Vac. Sci. Technol.* **15,** 359 (1978).

[v] J. Aranovich, A. Ortiz, and R. H. Bube, *J. Vac. Sci. Technol.* **16,** 994 (1979).

[w] F. A. Lowenheim, "Modern Electroplating." Wiley, New York, 1963.

[x] F. A. Kröger, *J. Electrochem. Soc.* **125,** 2028 (1978); M. P. R. Panicker, M. Knaster, and F. A. Kröger, *J. Electrochem. Soc.* **125,** 566 (1978).

[y] T. S. TeVelde, *Energy Conversion* **14,** 111 (1975).

[z] F. A. Shirland, *Adv. Energy Conversion* **6,** 201 (1966).

[aa] N. Nakayama, H. Matsumoto, K. Yamaguchi, S. Ikegami, and Y. Hioki, *Jpn. J. Appl. Phys.* **15,** 2281 (1976).

[bb] F. de Boer and H. Emmens, *Philips Tech. Rev.* **16,** 232 (1955).

[cc] J. B. Huffadine and A. G. Thomas, *Powder Met.* **7,** 290 (1964).

[dd] R. Audubert and S. de Mende, "The Principles of Electrophoresis" (A. J. Pomerans, trans.). Hutchinson and Co., Ltd., London, 1959.

[ee] E. W. Williams, K. Jones, A. J. Griffiths, D. J. Roughley, J. M. Bell, J. H. Steven, M. J. Huson, M. Rhodes, and T. Costich, *Solar Cells* **1,** 357 (1980).

resources (energy) are limited, the resulting settlement will likely be chaotic. On the other hand, if the influx is reasonably slow and the resources ample, an orderly (epitaxial) settlement can result after some rearrangement.

Vacuum evaporation (VE) is shown schematically in Fig. 9.5, with a Knudsen cell source. To make the example more concrete, we consider CdS, which dissociates into Cd and S_2 upon evaporation. The flux of Cd atoms and S_2 dimers from the source is determined by the pressure inside the cell $P(T)$, which is in turn a strong function of the cell temperature. The flux at a distance r from the source is given approximately by

$$\Gamma = A_k P(T)/2\pi r^2 (2\pi MkT)^{1/2} \quad \text{atoms cm}^{-2}\ \text{sec}^{-1}, \tag{9.1}$$

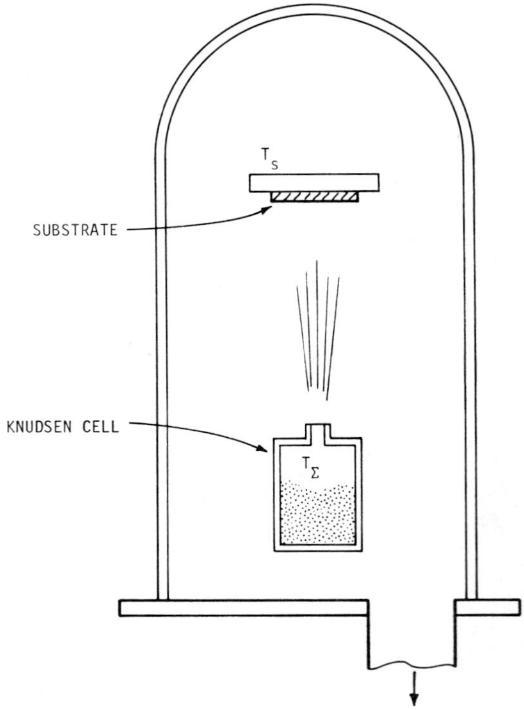

Fig. 9.5. Schematic of VE system using a Knudsen cell source. Typical parameters for evaporation of CdS at about 0.2 μm min^{-1} are source temperature $T_\Sigma \simeq 820°C$ (giving an internal cell pressure of $\simeq 10^{-1}$ Torr) and substrate temperature $T_s = 200-400°C$. Cd is present in the vapor flux primarily as a monomer whereas S is present principally as a dimer, S_2. The Cd flux at the substrate is $\Gamma \simeq 3.8 \times 10^{15}$ atoms cm^{-2} sec^{-1}; an equivalent pressure of Cd of 2.1×10^{-5} Torr near the substrate would give about the same flux. The flux of molecules from the background pressure, assumed to be 5×10^{-7} Torr, is ~2×10^{14} molecules cm^{-2} sec^{-1}.

where A_K is the source aperture area, P is the vapor pressure, T is the absolute temperature, and M is the mass of each atom or molecule of the evaporant.†

Typically, a pressure of 10^{-1} Torr exists within the Knudsen cell during evaporation. In the vapor stream Cd atoms and S_2 dimers have 0.1–0.2 eV of thermal energy in addition to their potential energy of formation at the substrate. Near the substrate, situated about 15 cm away in this example, the equivalent partial pressure of the Cd and S_2 is about 1.5×10^{-5} Torr for a film growth rate of 0.2 μm min^{-1}. In addition to the Cd and S_2 impinging on the substrate, the background pressure in the vacuum system provides a constant flux of impurity atoms. For example, a background partial pressure of 10^{-7} Torr of O_2 would be sufficient to produce a monolayer of O_2 in 30 sec if the sticking coefficient at the growth surface were unity.‡

In this deposition of CdS, the substrate temperature T_s is typically 200–400°C. The vapor pressure of Cd and S_2 over CdS is many orders of magnitude smaller than that of Cd or S_2 over the respective elements so any excess Cd or S will immediately be reevaporated and stoichiometry is preserved automatically (as in the three-temperature method of Guenther, 1966). The elevated temperature is also sufficient to give enough atomic mobility on the growing surface to form polycrystalline material with fairly large grains (0.5–3 μm in diameter) and, at the higher temperatures, to permit epitaxy on suitable single-crystal substrates. At T_s above about 400°C, reevaporation of CdS from the substrate competes with deposition and so the growth rate is reduced.

Some materials evaporate without dissociation or as dimers. These include the elements and some compounds:

$$\text{Au} \rightarrow \text{Au} \qquad \text{Sb} \rightarrow \text{Sb}_2 \qquad \text{SiO} \rightarrow \text{SiO} \qquad \text{Se} \rightarrow \text{Se}_2$$

Other compounds MX that dissociate on evaporation do so congruently, i.e., for every M atom leaving the evaporant, an X atom also leaves so that the source stoichiometry is preserved. For this to happen, the temperature of thermal dissociation of MX must be greater than the temperature required for appreciable evaporation of either of the elements over the compound.§

$$\text{CdSe} \rightarrow 2\text{Cd} + \text{Se}_2 \qquad \text{MgO} \rightarrow \text{Mg} + \text{O} \qquad \text{Zn}_3\text{P}_2 \rightarrow 3\text{Zn} + 2\text{P}$$

† Providing that $A_K^{1/2}$ is less than the mean free path of the atoms in the cell at P and that the total area of the evaporant within the cell is $\gg A_K$.

‡ The sticking coefficient is considerably less than one unless the material being deposited has a great affinity for O_2. For example, Si, Mg, and Ti have such a large affinity for O_2, whereas for Ag, Au, or In the sticking coefficients are much smaller than one.

§ For CdSe at 600°C, $P_{Cd} = 10^2$ Torr and $P_{Se_2} = 2 \times 10^2$ Torr over the elements and $P_{Cd} = 2 \times 10^{-4}$ Torr and $P_{Se_2} = 10^{-4}$ Torr over the compound.

These materials are easily evaporated from the compounds and stoichiometry of the deposited material can be maintained on a heated substrate.

In the third mode of evaporation, incongruent dissociation, the material tends to decompose with one of the elements evaporating faster than the other. Incongruent evaporation is favored when the vapor pressures of the elements are widely disparate;† then stoichiometry is difficult to maintain for VE.

$$\mathrm{InP} \rightarrow \mathrm{P\ (vapor)} + \mathrm{In\ (liquid)}$$
$$\mathrm{GaAs} \rightarrow \mathrm{As\ (vapor)} + \mathrm{Ga\ (liquid)}$$

To obtain stoichiometric layers, flash evaporation has been used with some success. Evaporation from separate M and X sources has been used to produce high-quality layers, but the method requires extremely fine flux control. CVD has been one of the favored methods of deposition for these compounds.

Here we discuss several modifications to simple VE, which illustrate the commonality of many of the growth methods. These modifications are shown schematically in Fig. 9.6.

1. *Hot-Wall Evaporation* (HWVE). To increase the flux of atoms to the substrate and to bring the deposition closer to a thermal equilibrium situation, hot walls (450–600°C) may be added that surround the vapor flux region within the vacuum chamber. Atoms hitting the walls are "reflected" with a temperature near the source temperature. The substrate temperature can be increased substantially since reevaporation loss is compensated by the large flux. The system approaches the situation for growth in a closed tube with a low ambient pressure. Significant advantages are conservation of material‡ and increased grain size. An excellent review of HWVE is given by Lopez-Otero (1978).

2. *Reactive Evaporation* (RVE). In this case one of the elements of the compound to be deposited is part of the gas ambient. For example, Si can be evaporated in a small partial pressure of O_2 to produce SiO_x or Zn in O_2 to produce ZnO (Brodie *et al.*, 1980). This situation is similar to that for CVD except that pressures in this case are much lower and the flow is molecular rather than laminar with diffusion control.

3. *Activated Reactive Evaporation* (ARVE). To add more energy to the impinging atoms and to increase chemical reactivity, an rf plasma or electron flood beam can be established within the chamber to ionize or activate the gaseous species. Such systems have been used for the deposi-

† For InAs at 800°C, we have $P_{\mathrm{In}} \simeq 4 \times 10^{-4}$ Torr and $P_{\mathrm{As}} \simeq 10^{+4}$ Torr over the elements and $P_{\mathrm{In}} \simeq 4 \times 10^{-4}$ Torr and $P_{\mathrm{As}} \simeq 2 \times 10^{-3}$ Torr over the compound.

‡ In a conventional laboratory-scale VE system, typically less than 5% of the evaporant is utilized on the substrate. The rest is lost to the walls or shields.

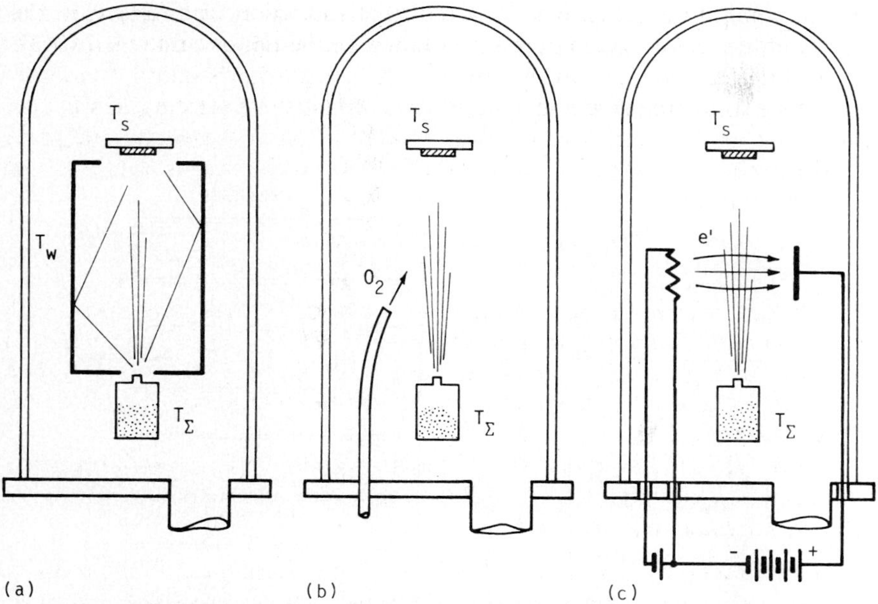

Fig. 9.6. Hot-wall, reactive, and activated evaporation modifications. (a) Hot wall at temperature T_w surrounds evaporation region. (b) Oxygen partial pressure is maintained to form metal oxide by reaction at substrate. (c) Electron beam crosses vapor stream to activate molecular species. Accelerating and/or retarding grids are usually added to control ion energy.

tion of hard refractory coatings (HfC, TaC, etc.) (Raghuran and Bunshah, 1972).

Molecular beam epitaxy (MBE) is an extension of multiple source VE in which the fluxes of the constituents can be separately controlled (Casey and Panish, 1978, p. 132). The process is carried out under ultrahigh vacuum conditions ($10^{-11}-10^{-9}$ Torr) and the sources, usually of the Knudsen cell type, are surrounded by shields at liquid N_2 temperature that serve to trap impurities and also to increase the temperature stability of the sources. The technique is capable of producing films of single-crystal quality (e.g., GaAs, InP), coevaporation doping, and deposition of subsequent layers 10 to 20 Å or less thick. Although MBE is an important research tool for photovoltaics, it may be too costly and too slow for high volume production of solar cells.

If we choose VE as a base point, many of the other deposition methods can be thought of as variations on three basic themes:

(a) adding energy to the depositing species (e.g., sputtering, reactive evaporation),

(b) increasing the pressure with the introduction of diffusion in the gas or liquid as an important consideration in the deposition (e.g., CVD, CSVT, LPE), and

(c) introducing reacting species (e.g., reactive sputtering, CVD).

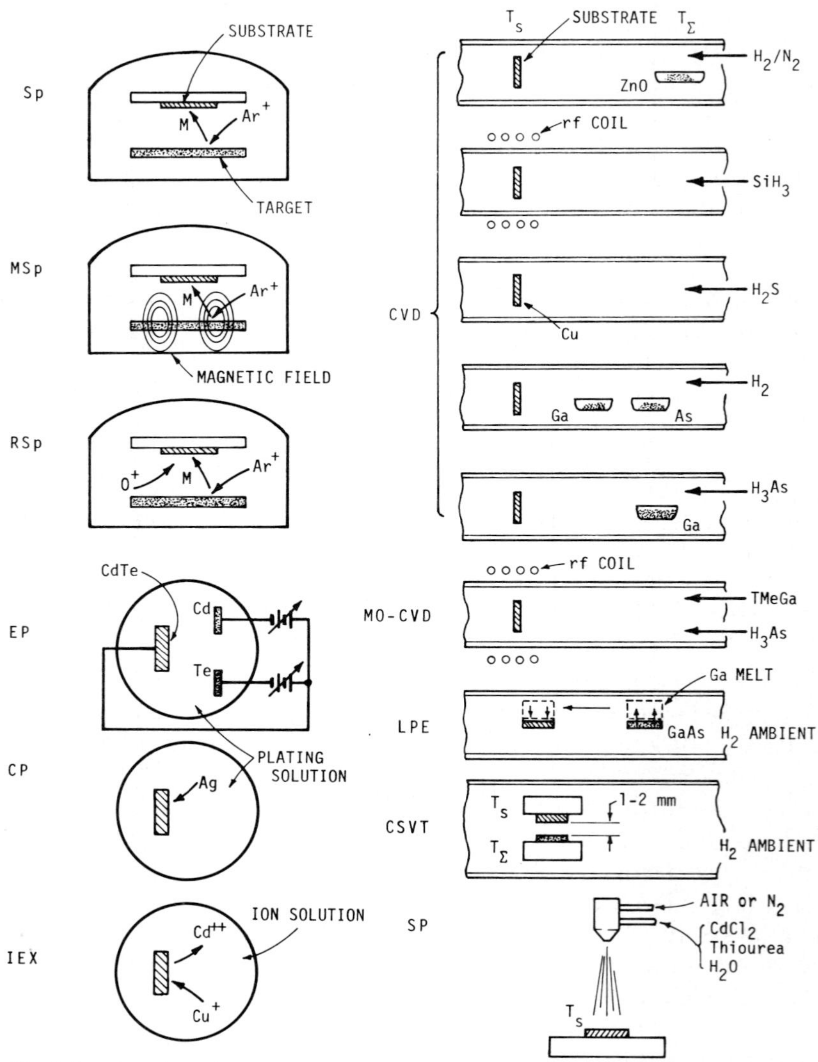

Fig. 9.7. Atomic and molecular deposition processes. Key to acronyms is found in Table 9.I.

Several other of the atomic or molecular deposition methods are summarized in Fig. 9.7. In CVD, LPE, and chemplating, the reacting species are carried to the substrate by a gas or liquid, usually at atmospheric pressure. The growth rate is controlled by concentration, carrier gas flow rate, substrate temperature, reaction rate at the substrate, and diffusion within the carrier gas or liquid. Electroplating might be considered in a broad sense to be the "activated" equivalent of chemplating.

In ion exchange growth, ions from a liquid solution *replace* atoms of the substrate to form the desired compound layer. An example is the topotaxial growth of Cu_xS from CdS in a hot solution of Cu^+ ions (see Chapter 10).

In CSVT the distance between the source and the substrate is made very small (1–3 mm) and since the carrier gas is stagnant and at atmospheric pressure, diffusion is the predominant means of species transport.

In sputtering (Sp) and reactive sputtering (RSp) atoms or molecules are removed from the source (target) by impinging high-energy noble gas atoms provided by a high voltage dc or rf glow discharge. The sputtered atoms have a very high energy (5–10 eV) compared with thermally evaporated atoms. Magnetic fields are sometimes used to increase deposition rate, permit lower pressures, and confine the plasma away from the substrate region.

In spray pyrolysis, the reacting species are carried to the substrate in droplets of liquid sprayed on a hot substrate. The liquid is, for example, an organic complex containing the material to be deposited, diluted in H_2O and/or alcohol. Substrate temperatures are quite high (350–500°C) and the resulting films are complex in structure. The relative simplicity of the apparatus makes the procedure potentially very inexpensive and capable of high throughput.

In addition, there is a class of deposition methods, shown in Fig. 9.8, in which the compound to be deposited is applied to the substrate in the form of small particles, suspended in a suitable vehicle that is later removed by heating. The particles are then fused together by high-temperature sintering in an inert or reducing atmosphere, sometimes with the aid of a fluxing agent. These methods include painting (with a brush or by spray), screen printing, doctor blade application,† and sedimentation from a slurry. In plasma and flame spray, the material is carried to the substrate as fine droplets in a very-high-temperature (2000–5000°C) flame or plasma jet, usually close to atmospheric pressure.

† In doctor blade application the material is spread onto the substrate using a carefully controlled blade much like a knife blade.

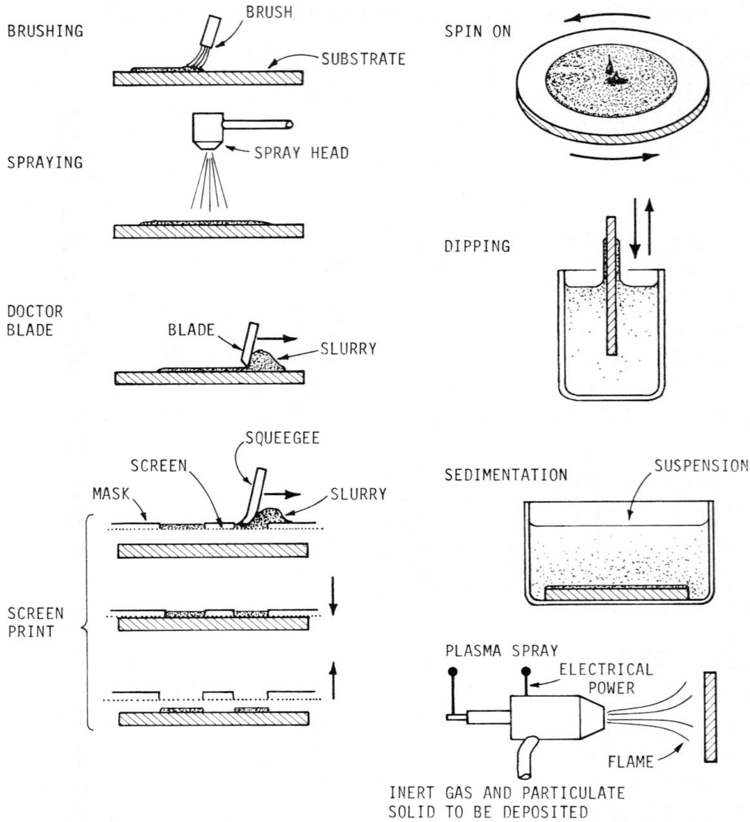

Fig. 9.8. Particulate deposition methods.

Finally, thin PX layers can be formed directly from the liquid phase as illustrated by the following examples:

(1) dipping ceramic substrates into molten Si (Zook *et al.*, 1977, 1978),

(2) edge-fed-growth (EFG) of Si ribbon from the melt (see Chapter 7),

(3) dendritic web growth of Si (Chapter 7),

(4) directional solidification of ribbon Si using laser or electron beam heating (RTR) (Lesk *et al.*), and

(5) growth of Si ribbon from a slot in the bottom of a crucible containing molten Si (the inverted Stepanov technique) (Kim *et al.*, 1977).

A good overview and comparison of the techniques for growth of thin Si sheet is given by Schwuttke (1979).

9.2.2 Atomic Surface Mobility, Nucleation, and Growth

In this section we follow the stages of atomic or molecular deposition of a thin film from *nucleation*, in which isolated nuclei are formed on the substrate; through *agglomeration*, in which these nuclei grow in area until they touch; and finally to the coalescence stages when islands develop

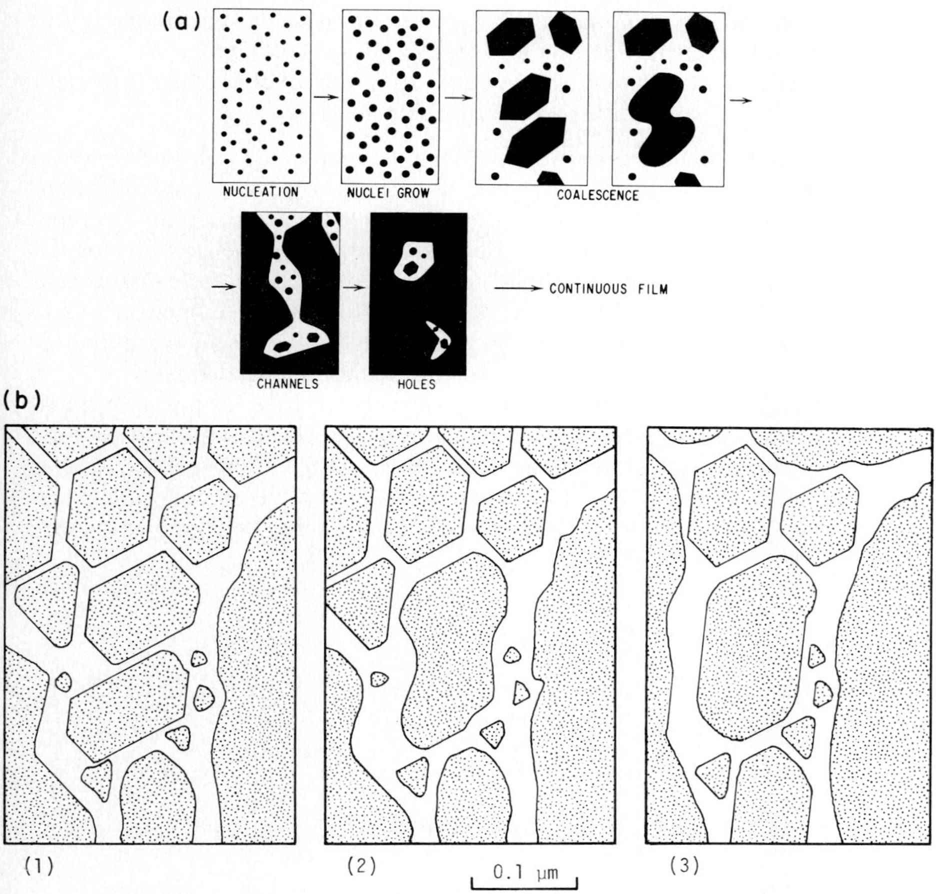

Fig. 9.9. (a) Schematic of the stages of film growth. [From C. I. Neugebauer, Condensation, nucleation, and growth of thin films. *In* "Handbook of Thin Film Technology" (L. I. Maissel and R. Glang, eds.), p. 8/33. McGraw-Hill, New York, 1970. Copyright © 1970 McGraw-Hill. Used with the permission of McGraw-Hill Book Company.] (b) Tracings of successive electron micrograph images of Au grown on MoS_2 at 400°C, showing the change in shape of islands during and after coalescence. (1) Arbitrary time zero, (2) 1–2 sec, and (3) 60 sec. [Drawn from microphotographs by D. W. Pashley, M. J. Stowell, M. H. Jacobs, and T. J. Law, *Phil. Mag.*, **10**, 127 (1964).]

and coalesce and the voids are filled. These processes, shown in Fig. 9.9, are substantially complete before appreciable growth in thickness occurs.

An idealized atomic picture of a substrate surface during deposition is shown in Fig. 9.10. The attractive forces that hold the impinging atoms onto the substrate are of three varieties:

(1) *Van der Waals forces* due to polarization of atoms (without electrostatic interaction) with typical binding energies of ≈ 0.1 eV;

(2) *metallic binding forces* due to reduction of the kinetic energy of the electron sea, with binding energies of about 2 eV; and

(3) *chemical binding forces* due to ionic exchange or covalent sharing of electrons, with 1–3 eV of binding energy.

Atoms on the "flat" substrate surface have two degrees of freedom without overcoming the attractive forces that prevent reevaporation. Atoms at steps (one degree of freedom) or at kinks have, on the average, less energy than those on the "flat" surface. Thus the mobility of the atoms or molecules on the surface decreases as the atoms give up their energy to the substrate. The effect of elevated substrate temperature is to permit the atoms to retain sufficient energy to make the movements necessary for accommodation to the substrate and among themselves.

Nucleation involves a balance between the surface (or perimeter) and volume energy of the developing nuclei. Beyond a critical nucleus size (on the order of 10–100 Å), the large nuclei grow at the expense of the smaller ones, and so the number of nuclei on the substrate is continually decreasing during growth. The initial nucleation is enhanced by the presence

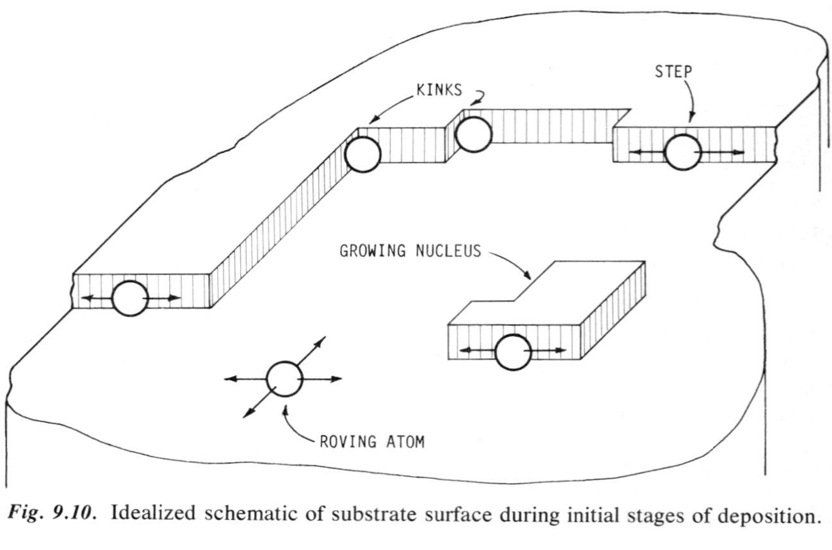

Fig. 9.10. Idealized schematic of substrate surface during initial stages of deposition.

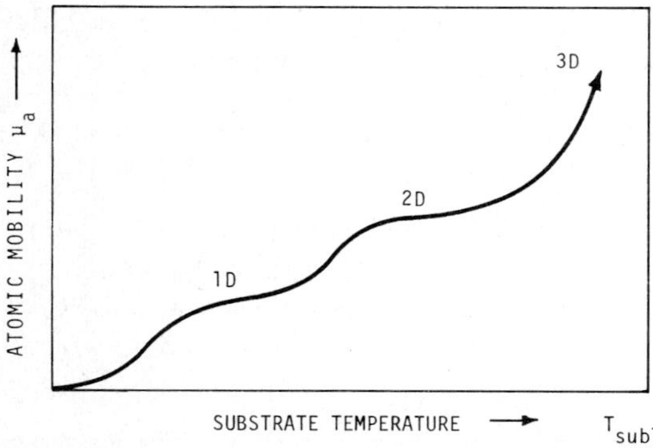

Fig. 9.11. A simplistic view of the average atomic mobility of atoms being deposited on a crystalline substrate. One degree of freedom corresponds to movement along a step (1D) while two degrees of freedom are present in free movement on the surface (2D). Increasing reevaporation is represented by (3D), and T_{subl} is the temperature at which the evaporation rate from the substrate exceeds the incident flux.

of defects on the substrate. At this stage the individual crystallites (about 100 Å in diameter) are quite perfect structurally. The average mobility of the atoms on the substrate is shown schematically in Fig. 9.11 as a function of the substrate temperature. At higher temperatures the atoms have enough energy for reevaporation from the substrate.

The coalescence phase is critical for the formation of grain boundaries and dislocations. As the larger nuclei (perhaps several thousand atoms in size) coalesce, the amount of disorder at the merging boundary depends on the orientation of each nucleus before contact. If the nuclei are aligned (as in epitaxial growth on a well-matched substrate lattice), either a twin boundary or none at all may be formed. If the islands are not aligned before contact (as in growth on an amorphous substrate or growth at low temperature), a grain boundary is formed. The misorientation at the merging boundary may be rotational, around a normal to the growth plane, and/or translational, because of lattice mismatch (Seto, 1972; Holt, 1974). As the nuclei become large, the energy of the aggregate becomes smaller so the larger nuclei have less ability to accommodate to each other as they coalesce.

As a result of the balance between the surface and volume energies and the attractive forces, both within the emerging films and from the substrate, the total energy in the film decreases as the degree of ordering increases, as shown in Fig. 9.12. The potential energy barriers between the regions of ordering represent the energy required to reorient the coalescing nuclei.

As the process of coalescence continues, the holes remaining in the film become reservoirs of virtual dislocations arising both from the lattice

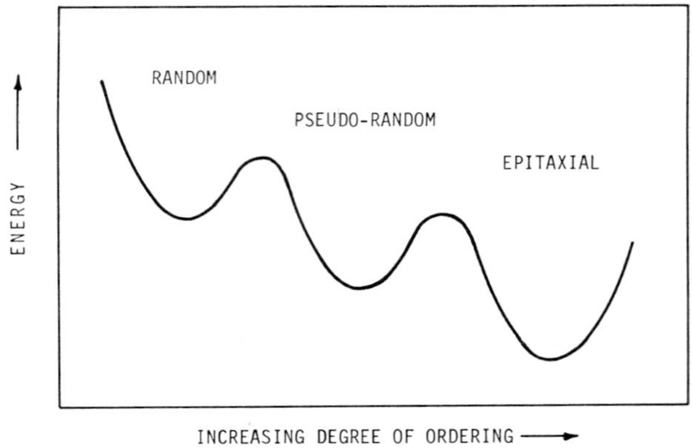

Fig. 9.12. A qualitative diagram of the energy of the aggregate of atoms in a film versus the degree of ordering in the film.

mismatch and from the effects of stresses developed within the islands during their growth. If the flux is large or the substrate temperature is low, voids may be left during the coalescence phases.

As the film continues to grow, now in thickness, the orientational forces inherent in the deposited material itself make themselves felt. For example, many semiconductors have a preferred growth orientation, even on completely amorphous substrates, and frequently cubic materials prefer to grow with the [111] axis normal to the substrate and hexagonal materials with the [002] axis normal to the substrate. The differences in growth rate of the different crystal faces produce faceting on each of the grains as the film becomes thicker. Now the large crystallites grow at the expense of the small ones by crowding them out as columnar grains develop as shown in Fig. 9.13a, producing a fibrous texture in the film.

Finally, the surface of the film has a texture of dimensions on the order of the grain size due to faceting during growth. Such a surface texture may act to reduce the reflectivity of the film with respect to an optically flat surface, and thus be beneficial.

On completion of growth the cross section of the film may show the characteristic structure illustrated in Fig. 9.13a. On amorphous substrates the grains are very often oriented with the [111] or hexagonal axis perpendicular to the film plane with a random or pseudo-random orientation around the normal to the surface (Hasson *et al.*, 1972), as for example in CdS, shown in Fig. 9.13b. Individual grains may also contain subgrain stacking faults and twins. Amith (1978) has studied the structure of VE CdS films such as those used for Cu_xS/CdS cells in great detail. The complex structure of films deposited by sputtering is shown in Fig. 9.14.

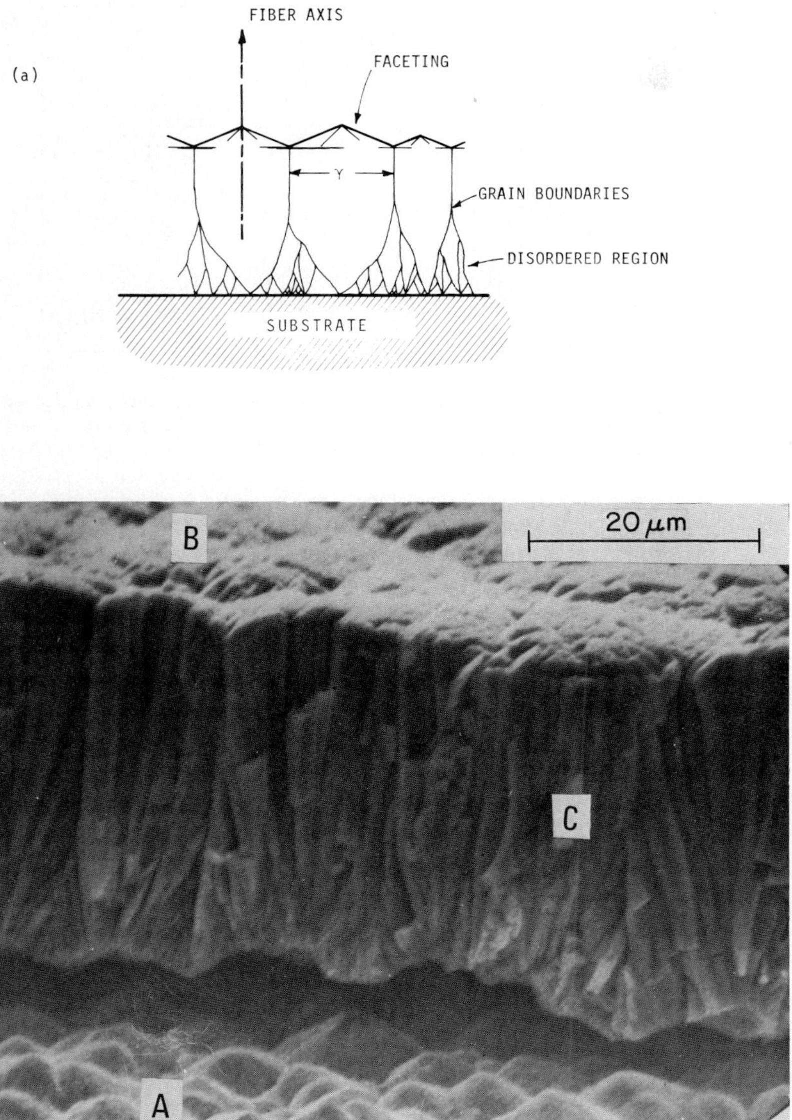

Fig. 9.13. (a) Schematic cross section of PX thin film grown on an amorphous substrate. Grains may also contain subgrain stacking faults and/or antiphase domain boundaries. (b) SEM micrograph of a vacuum-evaporated CdS film approximately 30 μm in thickness. Region A is the Zn-coated Cu substrate, B is the surface of the CdS, and C is the fracture surface of the CdS. [Micrograph courtesy of John Meakin, Institute of Energy Conversion, Univ. of Delaware, Newark, Delaware.]

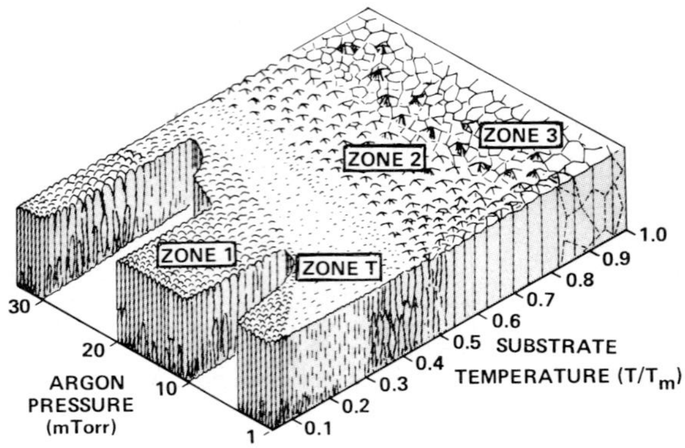

Fig. 9.14. Schematic representation of the dependence of the structure of sputtered coatings on substrate temperature (expressed as a ratio with the melting temperature of the layer). The energy of the sputtered atoms generally increases with decreased Ar pressure. Several zones can be identified: Zone 1, with densely packed fibrous grains; Zone 2, somewhat larger, faceted, columnar grains; Zone 3, larger, equiaxed grains; and Zone T, a transition between Zones 1 and 2. [After J. A. Thornton, *J. Vac. Sci. Technol.* **11,** 666 (1974). Figuse courtesy of the author.]

Normally the section of the film closest to the substrate is the most disordered, since this is where lattice mismatch and impurity defects on the substrate are accommodated (Aleksandrov, 1972, 1979). The grain diameter is the largest in the sections farthest from the substrate. Thus it may be advantageous to have the region of photogeneration near the growth surface of the film rather than near the substrate.

The grains are separated by a variety of different types of grain boundaries including: (1) coherent twin boundaries (containing no defect states), (2) low-angle grain boundaries (with misfit dislocations), and (3) incoherent twin planes and high-angle grain boundaries (with high densities of dislocations and segregated impurities) in order of increasing electrical activity. During the growth of each single grain, impurities and defects are segregated out and left at the boundary. The situation is much like the zone refining that takes place in the growth of single crystals, except that the imperfections remain trapped in the film rather than being cut off the end of the ingot.

9.2.3 Enhancement of Grain Size during Growth

Given the discussion at the beginning of the chapter, we wish the grain size to be as large as possible to obtain the greatest solar efficiency.†

† Whereas large grain size is desirable, *absolute* grain size may be less important for direct band-gap materials (see Fig. 9.4).

Methods to increase grain size, both during and after growth, are described in this and the following section. The essence of these methods is to increase the atomic mobility enough so that the rather weak inherent organizational forces within the film (and perhaps in a crystalline substrate) can act.

Grain size is a strong function of the growth parameters. Given sufficient surface mobility, the grain size tends to increase as growth proceeds, in order to lower the total energy of the film by decreasing the grain boundary area. This leads to the qualitative relationship between grain size γ and substrate temperature T_s shown in Fig. 9.15c. As the film grows in thickness and grains become larger, the driving force toward larger grains becomes less, however, and the grain size tends to saturate, as in Fig. 9.15a. These considerations are true for both crystalline and amorphous substrates. There is a useful relationship between the microstructure of such films and the ratio of T_s to the melting point T_m of the film. For example, in the epitaxial growth of Cd_3As_2 by VE, the films are amorphous or small grained for $T_s/T_m < 0.4$; for $0.4 < T_s/T_m < 0.5$, columnar growth is observed; and for $T_s/T_m > 0.5$, large equiaxed crystallites are obtained (Juruzik and Zdanowicz, 1980). The observation of three growth morphology zones with respect to T_s/T_m appears to be quite general (but not universal), having been seen in sputtered films (Thornton, 1974), VE

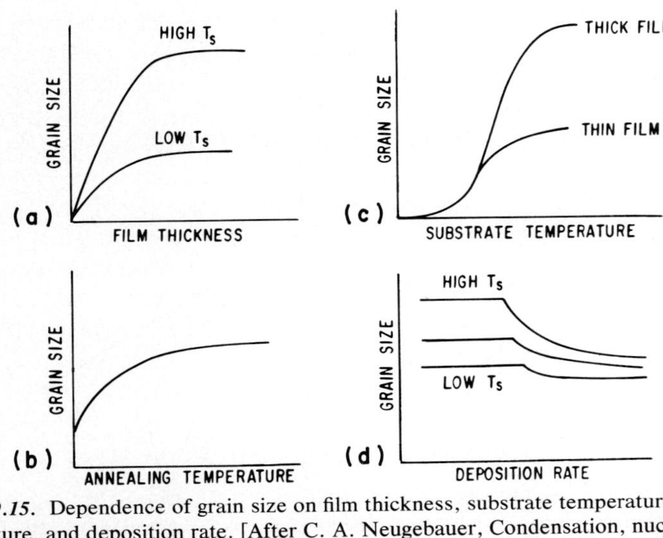

Fig. 9.15. Dependence of grain size on film thickness, substrate temperature, annealing temperature, and deposition rate. [After C. A. Neugebauer, Condensation, nucleation, and growth of thin films. *In* "Handbook of Thin Film Technology" (L. I. Maissel and R. Glang, eds.), p. 8-41. McGraw-Hill, New York 1970. Copyright © 1970 McGraw-Hill. Used with permission of McGraw-Hill Book Company.]

metal and metal oxide films (Movchan and Demchishin, 1969), and in the semiconductor Cd_3As_2 films.

With larger fluxes, the accommodation mechanisms can become saturated, leading to a decrease in grain size for increasing Γ, as in Fig. 9.15d. This effect can be viewed in terms of a supersaturation ratio, Γ/Γ_0, i.e., the ratio between the flux of atoms arriving at the substrate Γ and the flux leaving the substrate by evaporation at the substrate temperature Γ_0 (T_s), which is given by

$$\Gamma_0(T_s) = P(T_s)/[2\pi MkT_s]^{1/2}. \qquad (9.2)$$

The relation between grain size and supersaturation ratio for Si is illustrated by the data of Fig. 9.16.

An empirical relationship between grain size, substrate temperature, growth rate dy/dt, and film thickness y has been obtained by Lopez-Otero (1977) for PbTe films grown epitaxially on BaF_2 by HWVE:

$$\gamma = C_1 y(dy/dt)^{-C_2} \exp(-\Delta E/kT_s),$$

where C_1 and C_2 are constants and ΔE is the activation energy for the process. For PbTe there are two well-defined branches of the relationship: for $T_s < 450°C$, $C_2 = 0.5$ and $\Delta E = 0.64$ eV; for $T_s > 450°C$, $C_2 = 0.67$ and $\Delta E = 1.80$ eV.

For many semiconductor material depositions at reasonable growth

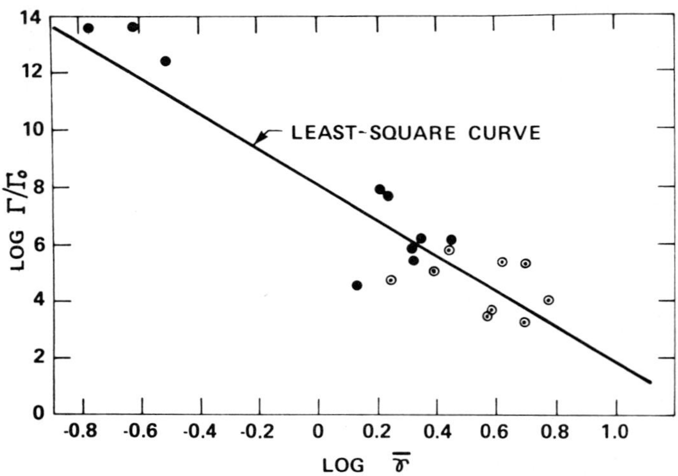

Fig. 9.16. Supersaturation ratio versus grain size for Si deposited by electron beam vacuum evaporation onto sapphire substrates. The incident flux is determined directly from film thickness rate, assuming all Si atoms stick. [After C. Feldman, F. G. Satkiewicz, and N. A. Blum, *Thin Solid Films* **90**, 461 (1982).]

rates, there is a moderately well-defined temperature range for which epitaxy is possible on a particular crystalline substrate. The lower bound is called the epitaxial temperature T_e. Of course T_e depends strongly on the substrate and the deposition conditions (i.e., rate, ambient pressure, residual gases present, etc.), but the situation is shown qualitatively in Fig. 9.17. The T_e ranges reported for several materials are: CdTe, 200–600°C; CdS, 100–600°C; and GaAs, 520–825°C (Manasevit, 1974; Holt, 1974).

A method for increasing grain size for epitaxial growth, called "temperature profiling," utilizes variation of both flux and T_s during the initial phases of growth. The method, as applied to CSVT, is shown in Fig. 9.18. During the initial phase of growth T_s is high and Γ is low, so that few nuclei form, and atomic mobility is high so that the nuclei can rearrange easily. As growth proceeds, T_s is lowered so that reevaporation is decreased and the flux is increased to permit faster growth. There are some similarities of this crossover profile method to the vapor growth of single crystals, where the seed end of the crystal is pulled through a temperature gradient (Scholz and Kluckhow, 1967).

Increased grain sizes and growth rates can also be obtained by rheotaxy (Milnes and Feucht, 1976; Rasmanis, 1963, 1964). In this method, a thin film of material in which the compound to be deposited is almost insoluble (usually a metal) is applied to the substrate before growth. When the substrate is held near the melting point of the metal, the atomic mobility of the deposited atoms is increased strongly and large grains result. Romeo (1981) developed the technique of "quasi-rheotaxy," in which

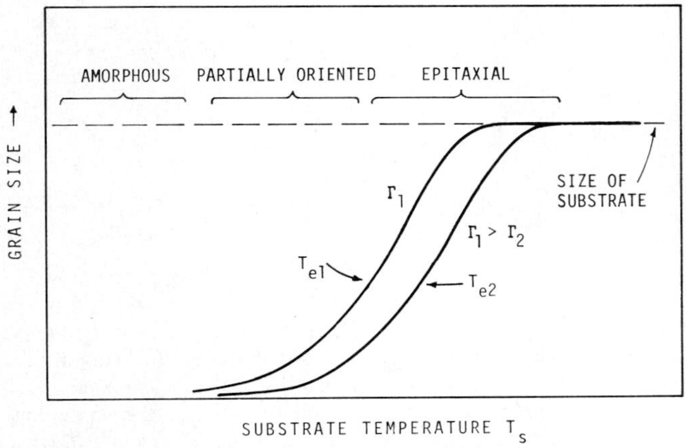

Fig. 9.17. Qualitative plot of grain size versus substrate temperature for two values of incident flux Γ showing epitaxial temperature T_e.

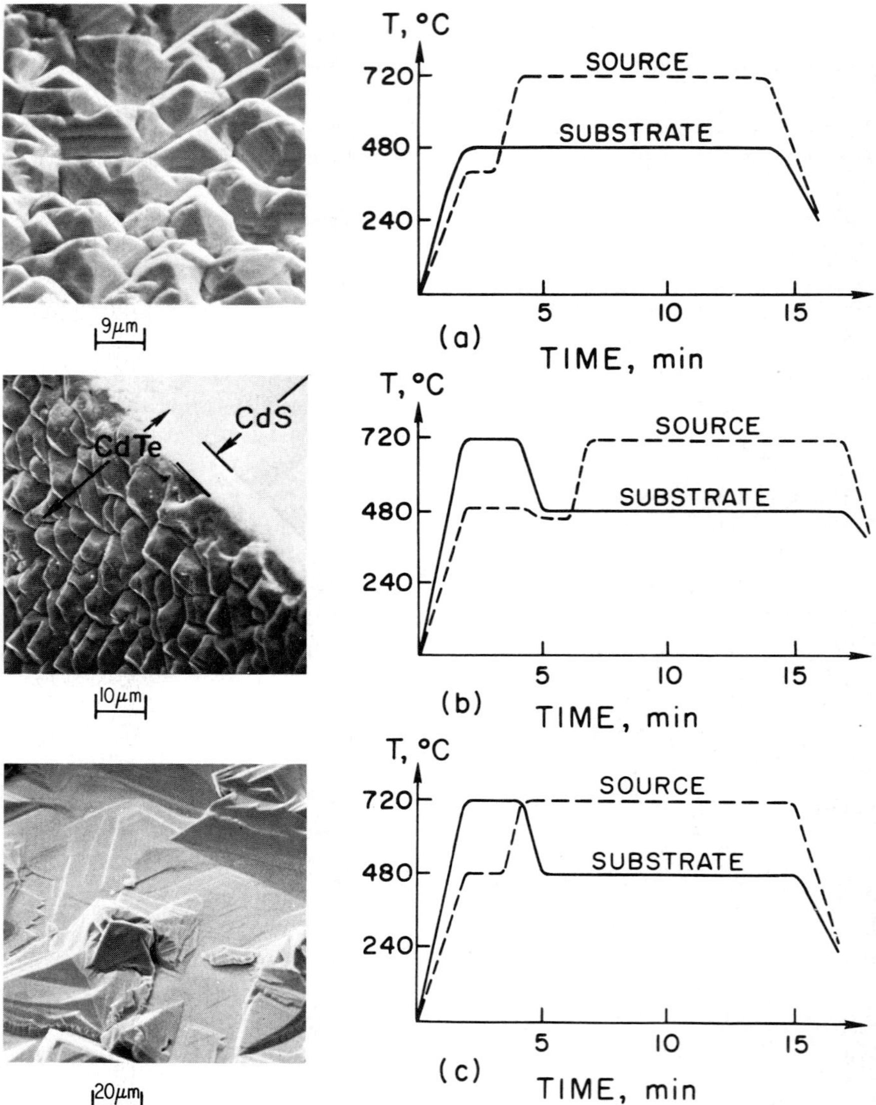

Fig. 9.18. Temperature–time profiles for source and substrate in CSVT deposition of CdTe on CdS single-crystal substrates and the corresponding film structures as shown by SEM photographs for three deposition schemes: (a) "normal" temperature profile, (b) thermal etch followed by normal deposition, and (c) "crossover" temperature profile. [After K. Mitchell, A. L. Fahrenbruch, and R. H. Bube, *J. Vac. Sci. Technol.* **12,** 909 (1975).]

semiconductor layers are grown on thin metal films held slightly *below* their melting point (i.e., quasi-liquid metal films). Romeo demonstrated growth of CdS, CdTe, Zn_3P_2, $CuIn_xGa_{1-x}Se_2$, and $CuInSe_yTe_{1-y}$ on films of Sn, Pb, Bi, or Tl deposited on amorphous sapphire or stainless steel substrates, obtaining grain sizes larger than 10 μm. In another variation of rheotaxy, Si layers can be grown on substrates that have been coated with several monolayers of Pt. The Pt film apparently "rides" on top of the Si layer as it grows and can be removed after growth is completed. Grain size is enhanced and defect density is reduced (Siekhaus, 1976).

Large-grain polycrystalline Si has been grown by Bloem *et al.* (1979; Graef *et al.*, 1977) on graphite substrates using a thin layer of Sn. In this case the Sn is in the molten state and the deposition proceeds by a "vapor–liquid–solid" mechanism; Si is dissolved in the Sn and then precipitates because of supersaturation. The Sn evaporates during the initial part of the deposition and when it is gone, the grain boundaries propagate in a direction normal to the substrate. The presence of intense illumination (Kumagawa *et al.*, 1968; see also Dresner and Stringfellow, 1968) or electric fields (Chopra and Kahn, 1967) at the substrate during deposition can also be used to enhance grain size, but the effects appear to be weaker than those obtained by rheotaxy or "vapor–liquid–solid" growth.

9.2.4 Postgrowth Film Modification

In addition to using high substrate temperatures and low rates to obtain large grains during growth, there are a number of methods for increasing grain size after growth is completed.

Annealing or sintering the film permits an increased mobility condition to be maintained for much longer times than would be possible during growth (Fig. 9.15b). The sintering ambient is, of course, critical, to avoid changing the stoichiometry or vacancy concentration in the film.

The presence of certain fluxing agents strongly increases the atomic mobility during annealing, decreasing the time and temperature required. This technique is usually called *recrystallization* and it has been most often applied to CdS. Gilles and Van Cakenberghe (1958); using Ag and Cu as the fluxing agents, sintered CdS films on amorphous quartz substrates at 500–600°C to obtain grain sizes up to 12 mm. In, Pb, Al, Zn, and Bi have a similar effect, but with considerably smaller magnitude. Sintering of screen printed CdS + $CdCl_2$ layers apparently uses the same principles. Recrystallization without fluxing has been observed but at temperatures above 700°C (Kahle and Berger, 1970).

By annealing CdS films on Cu doped glass substrates in a H_2S/H_2 atmosphere, Fraas *et al.* (1975) have obtained grain sizes up to 800 μm (using 490°C for 40 min). For 12 μm thick CdS films on ITO coated glass, they obtained crystallite sizes up to 30 μm by the same process. Initial crystallite sizes in both cases were <1 μm. The films made by this process show an almost single-crystal optical transmission as shown in Fig. 9.19. The latter technique would be quite amenable to solar cell fabrication since no undesirable impurities are left in the film by the fluxing operation. The CdS resistivity is decreased to 10–20 Ω cm by subsequent anneals in Cd overpressure. Krajenbrink *et al.* (1981) have used recrystallized CdS thin films made by this technique on ITO-coated glass as substrates for the epitaxial growth of CdTe and InP films. The CdTe (~30 μm thick) was deposited by physical vapor deposition and the InP (~1 μm thick) was formed by planar reactive deposition. Both CdTe and InP films had grain sizes up to ~40 μm.

Recrystallization has been accomplished in Si, increasing grain sizes from 0.1 to 10 μm up to ≃50 μm, and, with the addition of small amounts

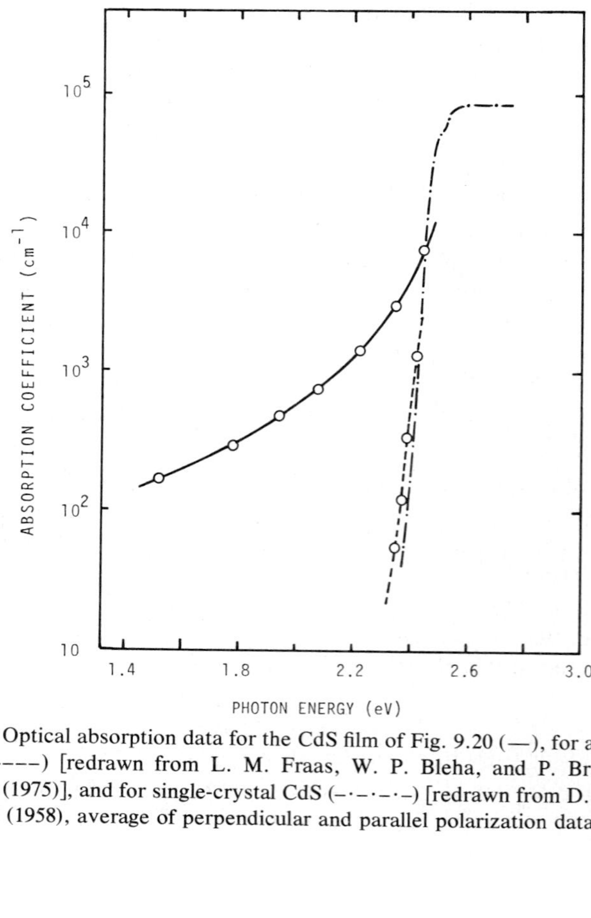

Fig. 9.19. Optical absorption data for the CdS film of Fig. 9.20 (—), for a recrystallized CdS film (-----) [redrawn from L. M. Fraas, W. P. Bleha, and P. Braatz, *J. Appl. Phys.* **46,** 491 (1975)], and for single-crystal CdS (–·–·–·–) [redrawn from D. Dutton, *Phys. Rev.* **112,** 785 (1958), average of perpendicular and parallel polarization data].

of B or Al during recrystallization, to >100 μm (Schins *et al.*, 1979). However, the temperatures required are quite high, 1200–1420°C. Favorable orientation of the grain boundaries (normal to the substrate) is also aided by the presence of B or Al.

The methods just described call for general heating of the film. Localized heating, by scanning laser or electron beams over the layers, has also been used successfully to increase grain size and enhance crystallinity. Local temperatures can be considerably higher than in the methods described and may exceed the melting point of the material. The localized heating method has some similarity to zone refining of single crystals. Lest *et al.* (1976) used an intense scanned laser beam to recrystallize EFG Si ribbon. In this work a molten zone was produced and the ribbon was transported with respect to the zone. Another approach is the use of a recrystallized layer on which to grow the active layers of the solar cell. For example, Ge films on metal or graphite substrates are recrystallized by a scanned laser beam. Then a GaAs layer is grown epitaxially on the Ge yielding large-grain material for Schottky barrier cells (Yeh *et al.*, 1978).

Gold *et al.* (1979) have used laser-induced radial recrystallization to produce large-grain (~ 100 μm) PX films of Si and Ge. In this case the laser beam initiates the crystallization in amorphous layers of Si or Ge that are held at an elevated temperature (but below the temperature for the spontaneous amorphous-to-crystalline transition). The latent heat of crystallization provides the driving force to continue the growth at extremely rapid rates (>100 cm sec^{-1}). Grain sizes produced by slow heating (homogeneous nucleation) in similar films are only on the order of 0.1 μm.

9.3 OPTICAL EFFECTS IN THIN FILMS

Optical effects in thin films are outlined here very briefly in order to point out a few items of importance to photovoltaics. A text by Heavens (1965) and a review by Clark (1980) are recommended.

A typical specular optical transmission versus wavelength curve for a thin CdS film on glass illustrates several of the effects of interest (Fig. 9.20). In this case the reflection coefficient for $h\nu > E_g$, about 19%, is close to the value predicted by the index of refraction:

$$R = (n_r - 1)^2/(n_r + 1)^2. \tag{9.3}$$

A film thickness of $t = 3.7$ μm is deduced from the wavelength difference between the interference maxima (or minima).† The absorption coeffi-

† The interference fringes at $\lambda = 2n_r t$ are also typically seen in J_{sc} versus wavelength data when such films are used as the window layers in heterojunction solar cells.

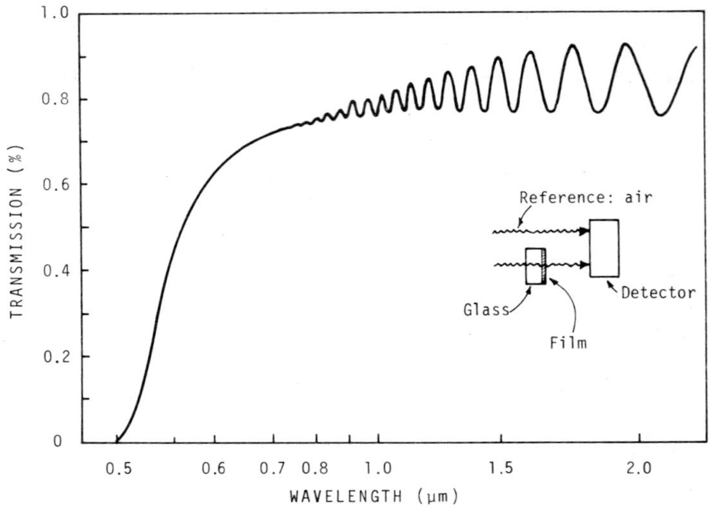

Fig. 9.20. Optical transmission of an *n*-CdS:In film deposited by VE on soda-lime glass. Data were taken with a standard Beckman DK-2 spectrophotometer (specular). The film appears clear and transparent to the naked eye.

cient data for the same film (Fig. 9.19), show a considerably "softer" edge than that for a single crystal. This is due to a combination of effects:

(1) scattering, with altered reflection and transmission, at the textured (faceted) front surface of the film, especially when the grain size is on the order of the wavelength of the incident light (Szczyrbowski and Czapla, 1977);

(2) optical scattering at voids and grain boundaries within the film (Filinsky, 1972);

(3) extrinsic absorption of light at the high density of imperfection states present at the grain boundaries (Bagley *et al.*, 1979); and

(4) possible changes in the optical absorption coefficient arising from the high electric fields at grain boundaries (the Franz–Keldysh effect) (Bujatti and Marcelja, 1972; Dow and Redfield, 1972).

Thus the softening of the absorption curve is extremely dependent on film deposition parameters.

Because of the inherent scattering and nonspecular reflection and transmission character of PX thin films, optical measurements on these materials are difficult. Standard spectrophotometers, which utilize small-aperture detector geometries ($\approx f/10$) yield data on only the specular component of transmission, whereas a solar cell utilizing such a film would see the scattered component as well. Use of a large-aperture detector geometry or an integrating sphere can overcome these problems to obtain a true photon flux assessment (Fagen, 1979).

Spectroscopic ellipsometry is becoming an important tool for the precise determination of optical constants and optical film thickness for PX materials. In particular, the high accuracy obtainable is valuable for studying the sometimes small perturbations on the optical constants, due to grain boundary effects (Aspnes, 1976; Bashara and Azzam, 1976).

9.4 ELECTRICAL TRANSPORT PROPERTIES

The electrical transport properties of polycrystalline PX materials play a significant role in determining the efficiencies of solar cells involving these materials. The basic theory of electrical transport in homogeneous single-crystal material has been discussed in Chapter 3. However, in most instances the behavior of PX thin films cannot be predicted directly from the known and understood properties of the corresponding bulk materials. The electrical characteristics of these PX layers are strongly influenced by the bounding surfaces of the individual crystallites, especially those at the internal surfaces or grain boundaries (e.g., see Leamy *et al.* 1982). Three aspects of grain boundary phenomena in PX semiconductor materials can be identified as being important for photovoltaics:

Resistivity In PX layers a distinction can be made between the microscopic mobility and carrier density within each grain, versus the macroscopic effective mobility and average carrier density for the layer as a whole. The potential energy barriers associated with the grain boundaries dominate the carrier mobility in the layer in many cases and thus control the resistivity. In these cases the conductivity does not depend linearly on the carrier density; the influence of carrier density is manifested most strongly through its effect on barrier height. For the case of fibrous grains oriented normal to the layer, the carrier transport in PX layers can be separated into components along the PX layer (parallel) and through the layer (transverse), as suggested by Fig. 9.3. The parallel resistivity is thus strongly affected by the grain boundaries whereas the transverse resistivity is affected very little. In a solar cell, increased resistivity primarily results in a loss in ff with little effect on J_{sc} or V_{oc}.

Recombination in the quasi-neutral bulk absorber layer The defect and impurity states present at the external and internal surfaces of the crystallites cause recombination of excess carriers, ultimately reducing the effective minority carrier lifetime in an inhomogeneous manner. In most films suitable for use as absorber layers in solar cells, this results in a moderate loss in J_{sc} and V_{oc}. This is treated in Section 9.5.

Grain boundary penetration of the junction The interiors of the grain boundaries are disordered regions that may be associated with reduced band gap and enhanced two-dimensional conductivity parallel to the boundary surface. Thus the penetration of the depletion layers of the junction by such grain boundaries can lead to shunting conductance and to increased diode current, mainly reducing V_{oc} and ff. Values of these parameters in most PX-based cells (particularly those involving direct band-gap materials) are considerably lower than those observed in the corresponding single or pseudo-single-crystal cells. This is considered in Section 9.5.3.

Reduction of the series resistance loss due to polycrystallinity should not usually be a problem, given conditions such that the other losses are minimized. For example, the effects of high series resistance parallel to PX layers can be almost entirely overcome by using a transparent conducting coating such as ITO or SnO_x. Percentage losses in J_{sc} and V_{oc} due to increased recombination appear to be about equal in magnitude for most Si cells, while the loss in ff is somewhat greater. For direct band-gap materials however, the losses in V_{oc} and ff predominate with η_Q remaining quite high. The effect of grain boundary penetration of the junction is perhaps the least understood of the loss mechanisms (Fossum *et al.*, 1980).

After a discussion of film thickness effects, the remainder of this section focuses on the barrier properties of grain boundaries and their effect on mobility, mobile carrier density, and resistivity.

9.4.1 Effects of Film Thickness

When the thickness of a semiconductor layer (or the crystallite size) becomes comparable to the mean free path of the carriers in the film, then scattering of the carriers at the film surfaces begins to decrease the mobility. For a bulk mobility of 10^3 cm^2 V sec^{-1}, for example, and a carrier effective mass of 0.1 m_0, Eq. (3.20) shows that the scattering relaxation time is about 5×10^{-14} sec, giving a mean free path of about 150 Å. This indicates the order of magnitude of film thickness that is important for thickness scattering effects. For films more than 1000 Å thick such effects are usually negligible.

A very simple treatment of surface scattering is often adequate. If τ_b is the bulk scattering relaxation time and τ_s is the relaxation time for surface scattering, then the total relaxation time τ is given approximately by

$$\tau^{-1} = \tau_b^{-1} + \tau_s^{-1}. \tag{9.4}$$

If we take $\tau_s \simeq (t/\lambda)\tau_b$, where t is the film thickness and λ is the mean free path, then the mobility in the film μ_f is given in terms of the mobility in the bulk μ_b by

$$\mu_f/\mu_b = (1 + \lambda/t)^{-1} \simeq 1 - \lambda/t. \tag{9.5}$$

This result is explicitly for the case of total diffuse scattering from the surfaces, and for the absence of band bending at the surface. The effect of surface scattering is reduced if some fraction of the surface scattering is specular rather than diffuse. For majority carrier mobility, scattering is decreased by a large factor if there is a depletion layer at the surface of the film (e.g., by surface state pinning) that excludes carriers from the surface. For minority carrier mobility such a depletion layer attracts carriers to the surface and increases the effect of scattering.

More detailed calculations (Anderson, 1970) show that surface scattering is larger in nondegenerate semiconductors than it is in degenerate semiconductors or metals. For a nondegenerate semiconductor the mobility in the film is reduced to about one-half that in the bulk when the film thickness is less than about twice the mean free path, in good agreement with the simple result of Eq. (9.5), for totally diffuse scattering.

Kazmerski *et al.* have made a thorough study of the combined effects of scattering and the grain boundary potentials on the mobility and conductivity of CdS (Kazmerski *et al.*, 1972), and $CuInTe_2$ (Kazmerski and Juang, 1977) films.

9.4.2 Grain Boundaries in PX Layers

The grain boundaries that separate the crystal grains in a PX layer arise from misorientation of neighboring crystallites during the coalescence stages of growth. These boundaries are generally regions with a high density of defects and impurities that have been segregated from the grains during growth. For example, direct evidence of a high density of segregated impurities has been obtained by Kazmerski *et al.* (1980) using scanning Auger and SIMS analysis on cast and directionally solidified PX Si samples. Helmreich and Seiter (1979) have made a comparison of the morphology and defect segregation in Czochralski-grown cast and CVD-grown layers of PX Si, finding appreciable segregation of slowly diffusing elements at the grain boundaries. Redfield (1981) suggests that the charged grain boundary states in several varieties of large-grain PX Si arise mainly from oxygen segregation during post-growth anneals.

Grain boundaries can be separated into three classes:

(1) *Coherent twins.* These have negligible electrical activity.
(2) *Low-angle grain boundaries.* These include a net of edge dis-

locations accompanied by lattice strain fields. Electrical activity is moderate.

(3) *High-angle grain boundaries and incoherent twinning.* This class includes high densities of dislocations (some of which may extend hundreds of angstroms outward from the plane of the grain boundary into the bulk of the grains e.g., Fig. 5.20), much lattice strain, and considerable impurity segregation. They show strong electrical activity. The dislocations present at such boundaries produce new energy states both as a result of the dilatation or strain field (by splitting of existing energy levels) and at dangling bonds.

Electrical charge present at these defect and impurity energy levels produces several interrelated, but distinct, effects at the grain boundary interface:

(a) almost "metallic" conduction *along* the grain boundary interface in some materials;
(b) charged interface states that produce potential barriers, much like isotype homojunctions, within the layer that may dominate electrical conduction parallel to the layer; and
(c) energy levels that act as recombination centers for minority carriers.

The one-dimensional periodicity along the dislocation cores causes banding of the dislocation energy levels. Thus they can behave as one-dimensional, almost metallic channels for charge carriers. This situation is described by Elbaum (1974) and Kamieniecki (1976) and has been evidenced by strongly anisotropic conduction in plastically deformed CdS (Elbaum, 1974) and anomalous optical absorption bands associated with screw dislocations at 0.04 and 0.27 eV above the CdS valence band (Merchant and Elbaum, 1976, 1979). In a two-dimensional dislocation net at a grain boundary, such banding can lead to a high two-dimensional conductivity along the interface. For example, in n-Ge, Mataré (1971, p. 272) and others observed appreciable conduction along grain boundaries that appear to be degenerate p-type. However, it has apparently not been observed for other materials.

If the plane of interface states in an n-type layer has an acceptorlike nature, for example,† compensating space charge regions extend into the adjacent grains and produce a potential barrier of height ϕ_{gb} as shown in Fig. 9.21a. This situation is very similar to the pinning of the Fermi level at semiconductor surfaces. In principle, the semiconductor adjacent to

† Oxygen adsorbed at grain boundaries in CdS gives such electron traps (Wu and Bube, 1974).

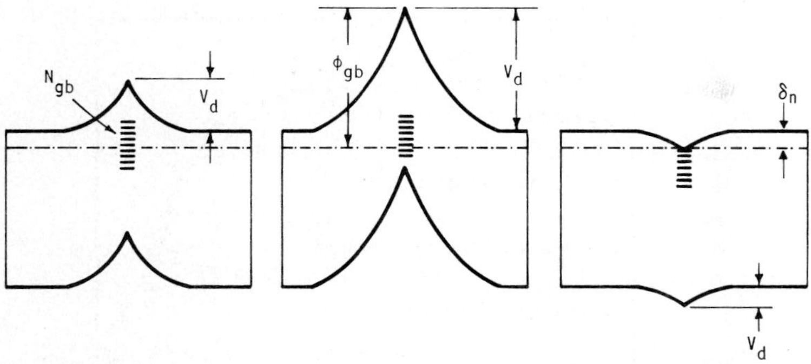

Fig. 9.21. Three different types of grain boundary potential barriers. Note that we use V_d here rather than ϕ_b, which is commonly used in grain boundary potential literature. We do this to differentiate it from the ϕ_b used in Schottky barrier literature. The barrier height referred to the equilibrium Fermi level is denoted by ϕ_{gb}.

the grain boundary can be depleted, inverted, or accumulated, depending on the nature and energy distribution dN_{gb}/dE of the grain boundary states. The depletion situation appears to be the most common by far for intrinsic, undoped grain boundaries. Doped, extrinsic grain boundaries, can be accumulated.

The formation of potential barriers at grain boundaries is well substantiated and Mataré (1971) has reported extensive studies on their behavior in bicrystals. The electrical properties of grain boundaries in Si were also studied by Sosnowski (1959), who used light spot scanning and voltage probing to show conclusive evidence of isotype homojunction behavior and of increased recombination at such boundaries. Also see Leamy *et al.* (1982).

The diffusion potential V_d can be related to the grain boundary charge and the bulk majority carrier density by the charge conservation relation

$$N_{gb}^{*} = 2N_D W_d, \tag{9.6}$$

where N_{gb}^{*} is the area density of *charged* interface states, N_D is the effective donor density in the bulk of the grain, and W_d is the depletion layer width on either side of the grain boundary. Integration of Poisson's equation then yields

$$V_d = qN_D W_d^2/2\epsilon_s = qN_{gb}^{*2}/8\epsilon_s N_D = \phi_{gb} - \delta_n/q, \tag{9.7}$$

where ϵ_s is the permittivity of the semiconductor and $\delta_n = E_c - E_F$. The grain boundary depletion layer widths in bicrystals can be determined by capacitance measurements and this provides a convenient way of measuring N_{gb}^{*} and ϕ_{gb}. If the grains are fully depleted, then W_d must be re-

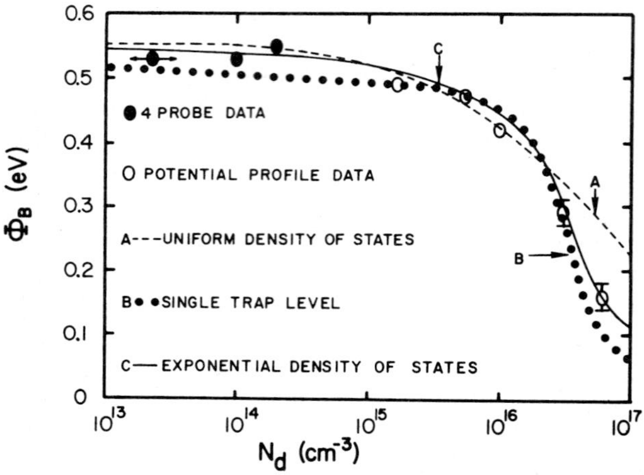

Fig. 9.22. The largest experimental values of grain-boundary barrier height V_d (extrapolated to $T = 0$°K) as a function of dopant density. The *large-grain n*-Si ($\gamma \simeq 25\ \mu$m) had been neutron-transmutation doped with P. Unfilled circles are data from small-area potential profile experiments, and filled circles are from four-point probe bulk resistivity data at $\simeq$270°K. At a doping level of 1.67×10^{15} cm^{-3} these two methods give identical results; below this dopant density, the potential probe data were difficult obtain because of the sample resistance, and hence only the four-point probe data are shown. The three curves shown are the results of model calculations. Note that Φ_B in the figure is called V_d in this book. [After C. H. Seager and T. G. Castner, *J. Appl. Phys.* **49,** 3879 (1978).]

placed by approximately $\gamma/2$,† and the diffusion voltage is then given by

$$V_d = q\gamma^2 N_D / 8\epsilon_s . \tag{9.7a}$$

In this case the QNRs no longer exist to control the position of the Fermi level and it moves farther from the conduction band (for the example of an *n*-type material) to a position set by the occupation of N_{gb} (i.e., so that ϕ_{gb} is such that $N_{gb}^* = N_D\gamma$).‡

The value of N_{gb}^* depends on ϕ_{gb} and the energy distribution of grain boundary states dN_{gb}/dE. A number of different forms for this distribution have been modeled, including linear, exponentially increasing or decreasing, and discrete distributions. Seager and Castner (1978) predict the relation between N_D and V_d for Si shown in Fig. 9.22 for *large-grain* material (i.e., such that $W_d \ll \gamma$), with experimental points for the dark case. The results are not strongly dependent on the form of dN_{gb}/dE. In sub-

† This depends on a geometric factor close to unity depending on the shape of the grains.

‡ Note that it is common to express the barrier height in terms of ϕ_{gb} *and/or* V_d. Whereas V_d can depend directly on bulk carrier density $n \simeq N_D$, the relation of ϕ_{gb} to N_D is more subtle and, in some cases, ϕ_{gb} remains almost constant as N_D is varied.

sequent experiments Seager and Pike (1979) found the distribution for Si shown in Fig. 9.23 with experimental data.

Baccarani *et al.* (1978) studied the more general case that includes the possibility of the grains being completely depleted. This yields the relation for V_d and N_D shown in Fig. 9.24a, for the dark case. Note that a maximum in V_d occurs when the grains (assumed to be of uniform size) are just fully depleted. Baccarani *et al.* found the best fit for their experimental data using an assumed delta function distribution of 4×10^{12} cm^{-2} traps located at midgap. The point at which the grains are just fully depleted also corresponds to an almost discontinuous rise in the ratio of free carrier to introduced-dopant density. For N_D smaller than this value the ratio may be several orders of magnitude less than one; for greater N_D values, the ratio goes to unity.

This discrepancy between the measured carrier density and the introduced dopant density at low densities was originally interpreted by Cowher and Sedgwick (1972) and by Fripp (1975a) as segregation of dopant to the grain boundaries where it passivated dangling bonds. The dopant segregation model was later questioned by analysis of conductivity activation energy measurements (e.g., Rai-Choudhury and Hower, 1973).

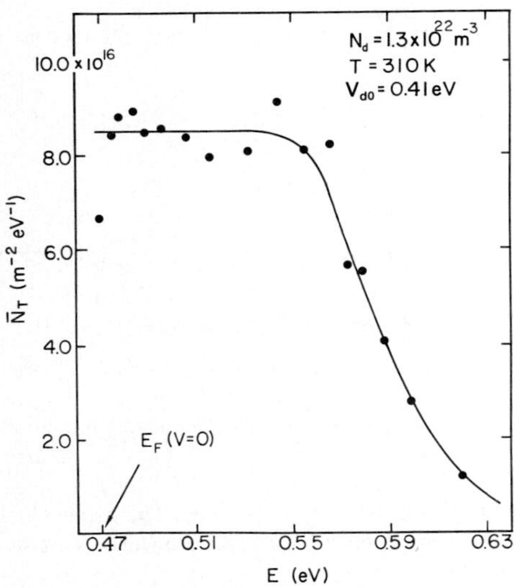

Fig. 9.23. Energy distribution of grain boundary states in a *n*-Si:P (transmutation-doped) bicrystal. The density of states in the grain boundary deduced from dc $I-V$ data is plotted as a function of energy measured from the valence band. Information about dN_{gb}/dE is available only at energies above the equilibrium Fermi level (at 0.47 eV). [After C. H. Seager and G. E. Pike, *Appl. Phys. Lett.* **35,** 709 (1979); errata, *Appl. Phys. Lett.* **37,** 251 (1980).]

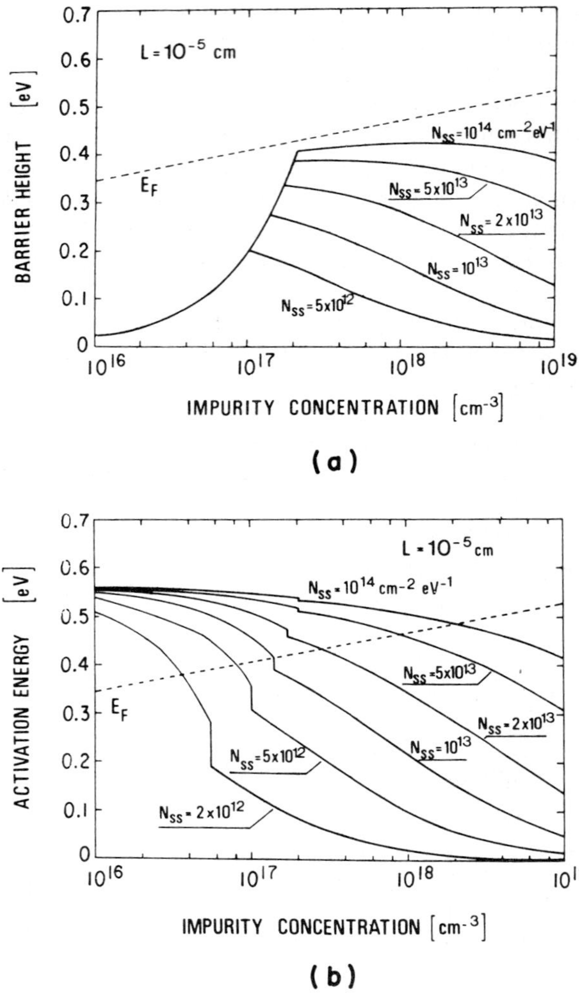

Fig. 9.24. Parameters for a one-dimensional PX Si layer model with a *small grain size* of 10^{-5} cm. Calculated values of (a) barrier height V_d and (b) conductivity activation energy ΔE_σ versus donor density. The running variable is the interface state density, assumed to be a linear distribution with respect to energy above E_i. The dashed line shows the position of E_F with respect to the intrinsic level E_i. [After G. Baccarani, B. Ricco, and G. Spadini, *J. Appl. Phys.* **49,** 5565 (1978).]

In general, V_d is a function of dN_{gb}/dE, the temperature, the donor or acceptor density within the grains, and the illumination intensity. Thus recombination kinetics must be accounted for in the determinaton of V_d. Theoretical results have been obtained for the illuminated case by Card and Yang (1977), who developed a model that predicted that in general the barriers are substantially lowered by carrier trapping under illumination to maximize the recombination rate. For example, in an *n*-type material, the grain boundary traps are negative when filled. Under illumination, minority carriers (holes) are attracted to the grain boundaries by the depletion layer fields, neutralizing the grain boundary traps and strongly reducing the barrier height so that more electrons are available. In the case modeled by Card and Yang (1977), V_d was reduced from ~0.5 eV in the dark to ~0.07 eV under 1-sun illumination for $N_{gb}^* = 10^{13}$ cm^2 and $N_D = 10^{16}$ cm^{-3}. Maruska *et al.* (1980) found that the effect of the grain boundary potential barriers on the conductivity was completely eliminated by 1-sun illumination in large grain PX *n*-Si.

Representative values of barrier height are given in Table 9.II for some semiconductor materials of interest. The effect of the potential barriers on majority carrier mobility is discussed in the next section, and the recombination of minority carriers at grain boundaries is discussed in Section 9.5.1.

The grain boundary potential profiles can be measured using an ultrafine contact probe or they can be inferred from capacitance or beam-induced current measurements. An interesting variation is the use of a contactless voltage probe by Cohen *et al.* (1979). They measured the shift in absolute energy of an Auger signal arising from a surface impurity (C), to obtain the potential profile for grain boundaries in GaAs bicrystals.

9.4.3 Electrical Conduction in PX Material

The intergrain potential barriers present obstacles to carrier transport through PX layers and may reduce the carrier mobility by many orders of magnitude from that found in single-crystal material. For thin films with a thickness comparable to the grain size, such intergrain barriers are clearly much more important for parallel conductivity than for transverse conductivity. If a thin-film component of a heterojunction solar cell is covered by a conducting coating so that current flow can proceed through the thickness of a fibrously oriented film, the effects of polycrystallinity need not play a large role; if, however, parallel flow along the film is needed for current collection, an appreciable contribution to the series resistance can arise (Fig. 9.3).

Even when the potential barrier height is effectively zero (e.g., at coherent twin boundaries), such intergrain boundaries may still act as diffuse internal scattering surfaces. A simplified treatment for a nondegenerate semiconductor follows. If the mean free path of carriers between collisions at intergrain boundaries is equal to the grain size γ, then the

Table 9.II

Selected Grain Boundary Potential Barrier Parameters

Material	Grain boundary type	N_{gb} (cm^{-2})	n_n or p_p (cm^{-3})	ΔE_σ (eV)†	V_d (V)	Remarks‡ and references
Any	Low angle	$<10^{11}$ eV^{-1}	—	—	—	—
	Medium angle	$10^{11}-10^{13}$ eV^{-1}	—	—	—	—
	High angle	$>10^{13}$ eV^{-1}	—	—	—	—
n-Ge	High angle	$\simeq 10^{12}$	—	0.52	0.29	273°K, *a*
n-Ge	—	—	—	—	0.31	*b*
n-Ge	—	10^{13} eV^{-1}		0.71	—	*c*
p-Ge	—	—		~0	~0	Apparently grain boundaries are p^+, *c*
n-Si	High angle	7×10^{12} eV^{-1}	10^{16}	0.62	0.41	310°K, *d*
n-Si	PX	—	2.3×10^{19}	0.56	0.44	Grains completely depleted, *e*
n-Si	PX	—	—	—	0.22	Large grain, *f*
p-Si	PX	—	10^{16}	0.52	—	*g*
		—	5×10^{17}	0.42	—	*g*, *h*
		3.4×10^{12}	10^{19}	0.025	—	*g*
n-CdS	PX	—	—	~0.3	0.11–0.22	*i*
n-CdS	PX	—	—	—	0.07–0.13	*j*
n-$CuInTe_2$	PX	—	4×10^{17}	0.16	0.055	*k*
p-$CuInTe_2$	PX	—	5×10^{16}	0.12	0.031	*k*
p-CdTe	PX	—	5×10^{15}	—	0.6	5 μm grains, *l*
n-InP	PX	—	4×10^{15}	—	0.08	epitaxial film, *m*
n-InSb	PX	—	10^{17}	0.11	0.04	*n*
n-GaAs	PX	—	10^{17}	0.4	0.4	*o*
p-GaAs	PX	—	10^{17}	0.4	0.4	*o*

† $\Delta E_\sigma = qV_d + \delta_n + kT - qT(d\phi_{gb}/dT)$, as in Eq. (9.17).

‡ All measurements in dark and at ≃300°K except as noted.

[a] W. E. Taylor, N. H. Odell, and H. Y. Fan, *Phys. Rev.* **88,** 867 (1952).

[b] Y. Hamakawa and J. Yamaguchi, *Jpn. J. Appl. Phys.* **1,** 334 (1962).

[c] R. K. Mueller, *J. Appl. Phys.* **32,** 640 (1962).

[d] C. H. Seager and G. E. Pike, *Appl. Phys. Lett.* **35,** 709 (1979); errata, *Appl. Phys. Lett.* **37,** 251 (1980).

[e] M. L. Tarng, *J. Appl. Phys.* **49,** 4069 (1978).

[f] H. P. Maruska, A. K. Ghosh, A. Rose, and T. Feng, *Appl. Phys. Lett.* **36,** 381 (1980).

[g] J. Y. W. Seto, *J. Appl. Phys.* **46,** 5247 (1975).

[h] Both Seto (footnote *g*) and Baccarani *et al.* [G. Baccarani, B. Ricco, and G. Spadini, *J. Appl. Phys.* **49,** 5565 (1978)] find that E_σ collapses at about $p_p \simeq 10^{18}$ cm^{-3} in PX Si.

[i] C. Wu and R. H. Bube, *J. Appl. Phys.* **45,** 648 (1974).

[j] L. L. Kazmerski, W. B. Berry, and C. W. Allen, *J. Appl. Phys.* **43,** 3515 (1972).

time between scattering events is approximately $\tau = \gamma/v$ with $v = (2kT/m^*)^{1/2}$ and from $\mu = q\tau/m^*$ we get

$$\mu = q\gamma/(2kTm^*)^{1/2}. \tag{9.8}$$

Thus the mobility should vary as $T^{-1/2}$. For $\gamma = 0.1\ \mu\text{m}$ and $m^* = 0.1$ m, Eq. (9.8) gives $\mu = 5800\ \text{cm}^2\ \text{V}^{-1}\ \text{sec}^{-1}$; thus for reasonable grain sizes, the zero-barrier-height grain boundaries should have negligible effect.

In most PX materials, however, the intergranular boundary potentials are large enough to affect transport significantly and in many cases to dominate it. A simple treatment of carrier transport parallel to a layer with transverse grain boundaries is given here as an illustration. It assumes that (1) the grains are only partially depleted; (2) scattering within the depletion layers can be neglected (as in the Bethe approach to Schottky diodes); (3) transport within grain boundaries that are parallel to the applied electric field is neglected, reducing the problem to one dimension;† and (4) V_d does not change with bias voltage. The voltage drop across each grain $\Delta V'$ is given by

$$\Delta V' = \Delta V + J\rho_g(\gamma - 2W_d),$$

where the internal resistivity of each grain is $\rho_g = 1/qn\mu_g$, as shown in Fig. 9.25. For the whole film (of length y) the voltage drop is

$$V = \sum^{m} \Delta V' = \sum^{m} [\Delta V + J\rho_g(\gamma - 2W_d)] = m\,\Delta V + J\rho_g y[1 - 2W_d/\gamma],$$

where there are $m = y/\gamma$ grains. If we wish to define an effective conductivity mobility μ_{eff} for the whole film, then the voltage across the entire film is

$$V = Jy/qn\mu_{\text{eff}} = y\,\Delta V/\gamma + Jy[1 - 2W_d/\gamma]/qn\mu_g \tag{9.9}$$

† In any real PX material, conduction along grain boundaries may be a consideration, giving a two- or three-dimensional grid problem. These cases are treated by Blount *et al.* (1970) and by Ghosh *et al.* (1980a).

---

[k] L. L. Kazmerski and Y. J. Juang, *J. Vac. Sci. Technol.* **14,** 769 (1977).

[l] K. W. Mitchell, A. L. Fahrenbruch, and R. H. Bube, *J. Vac. Sci. Technol.* **12,** 909 (1975).

[m] See analysis of Fig. 9.36.

[n] J. E. Greene and C. E. Wickersham, *J. Appl. Phys.* **47,** 3630 (1976).

[o] J. J. J. Yang, P. D. Dapkus, R. D. Dupuis, and R. D. Yingling, *J. Appl. Phys.* **51,** 3794 (1980).

[l] G. Baccarani, B. Ricco, and G. Spadini, *J. Appl. Phys.* **49,** 5565 (1978).

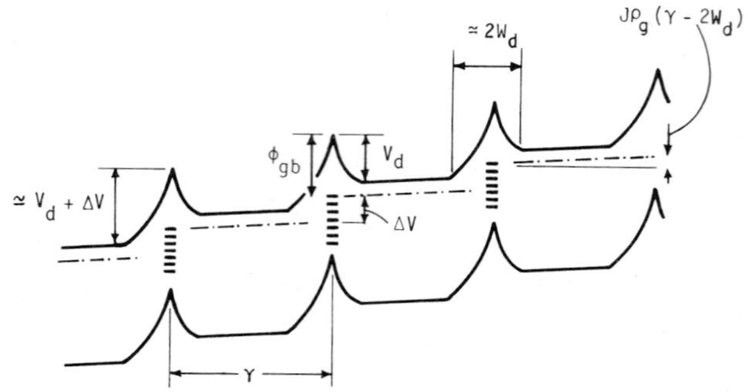

Fig. 9.25. One-dimensional model of a PX material under bias. The situation shown is when most of the bias ΔV on each grain boundary is sustained across the reverse-biased (left) side of each barrier.

so that

$$\mu_{\text{eff}} = \frac{1}{(qn\ \Delta V/J\gamma) + [1 - (2W_{\text{d}}/\gamma)]/\mu_{\text{g}}}. \tag{9.10}$$

In consideration of the voltage drop ΔV across each grain boundary, there are two broadly opposite models: (1) a model in which the charged interface state density N^*_{gb} is unchanged by moderately small applied bias and (2) a model in which N^*_{gb} increases with applied bias, pinning the

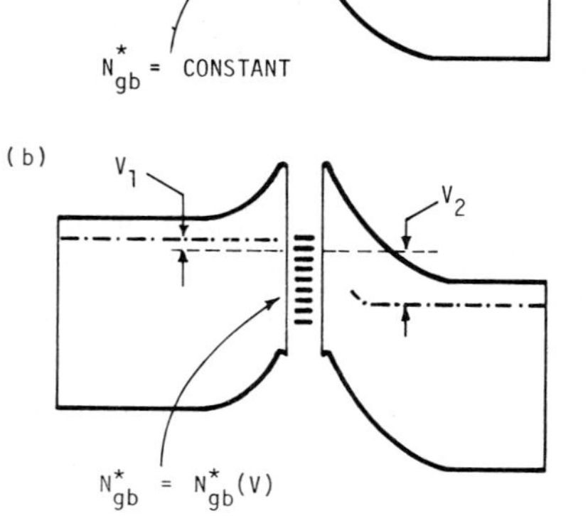

Fig. 9.26. Single grain boundary potential barriers at applied bias $V = V_1 + V_2$ for (a) the case where N^*_{gb} is constant and carriers are not scattered at the boundary and (b) the case where $N^*_{\text{gb}} = N^*_{\text{gb}}(V)$ and carriers come to equilibrium with grain boundary states.

barrier height near its zero-applied-bias value until a large bias is applied. For small applied bias ($\Delta V \lesssim kT/q$) across a single grain boundary, both of these models lead to the same result. This would be the case for an efficient solar cell with negligible series resistance. Because of the value of diagnostic study of grain boundary conduction at higher bias voltages, both cases are described here briefly.

Constant N_{gb}^* Assuming the charged grain boundary state density N_{gb}^* is unchanged by applied bias, charge conservation requires that the bias voltage V be split between the two sides of the grain boundary barrier (Fig. 9.26a) according to

$$V_2 = 4V_d\{1 - [1 - (V_1/V_d)]^{1/2}\} - V_1. \tag{9.11}$$

Note that the application of $V \geq 4V_d$ causes complete collapse ($V_1 = V_d$, flat band) of the barrier; μ_{eff} would then be dominated by the bulk grain mobility. For applied bias such that $V_1/V_d \leq 0.4$, we have $V_1 \simeq V_2 \simeq V/2$, and, assuming current transport by thermionic emission (Bethe type), the $J-V$ relation for each boundary is given by (Taylor *et al.*, 1952†; Tarng, 1978; Korsh and Muller, 1978)

$$J = A^*T^2 \exp(-q\phi_{gb}/kT)[2 \sinh(qV/2kT)]. \tag{9.12}$$

This assumes that carriers are emitted across the boundary without scattering and, in particular, without coming to equilibrium with the grain boundary states.‡ In a PX material with many grain boundaries the bias across each boundary ΔV is small, $q\,\Delta V/kT \lesssim 1$, and sinh is approximately equal to its argument so that

$$J \simeq A^*T^2 \exp(-q\phi_{gb}/kT)[q\Delta V/kT]. \tag{9.13}$$

N_{gb}^* *Increasing with Bias* The opposite situation is described by Mueller (1961), Pike and Seager (1979), Seager and Castner (1978), Baccarani *et al.* (1978), and others in which carriers proceed over the first part of the barrier, come to equilibrium with the grain boundary states,§ and then are thermionically emitted over the second, reverse-biased barrier (Fig. 9.26b). In this situation N_{gb}^* increases with bias to maintain the barrier height and most of the applied bias voltage is dropped across the reverse-biased barrier. Using the relative voltages across two back-to-

† A very good article!

‡ The result of Eq. (9.12) is for a bicrystal. For a series of m potential barriers it would read $J \simeq A^*T^2 \exp(-\phi_{gb}/kT)[2 \sinh(qV/2mkT)]$.

§ The thickness of a large-angle grain boundary has been taken as up to 50 or 100 Å by various workers. Therefore carrier transport over a grain boundary is likely to be accompanied by more or less complete equilibration of the carriers with the grain boundary states.

back Schottky diodes passing the same current, it is easy to show that $V_1 \lesssim (kT/q) \ln 2$ and $V_2 \simeq V$ for moderate biases. The exact relationship of V, V_1, and V_2 is obtained using the solution of Poisson's equation and the particular charged interface state distribution dN_{gb}^*/dE that is a function of V_1. At small applied bias $\Delta V \lesssim kT/q$, the current is again given by Eq. (9.13). At moderate bias, the increasing N_{gb}^* maintains the barrier height and the current saturates near the reverse-bias saturation current of diode 2 or varies more slowly than linearly with voltage. In some cases single barriers can withstand bias voltages up to 100 V. At large enough bias, depending on the values of N_{gb}^* and N_D, the barrier begins to collapse and the current increases exponentially. Using this model, Seager and Pike (1979) have been able to deconvolute the $J(V,T)$ behavior to yield the distribution of grain boundary states shown in Fig. 9.23, for an n-Si bicrystal.

Returning now to the calculation of small-bias mobility of PX layers with partially depleted grains, we use Eq. (9.13) in Eq. (9.10) and assume $W_d \ll \gamma$ giving

$$\mu_{eff} = \{1/\mu_g + [nk/\gamma A^* T] \exp(q\phi_{gb}/kT)\}^{-1}. \tag{9.14}$$

If the grain boundary barriers dominate the mobility, then

$$\mu_{eff} \simeq [\gamma A^* T/nk] \exp(-q\phi_{gb}/kT). \tag{9.15}$$

Finally, if we substitute for A^* and $N_c/n = \exp(\delta_n/kT)$ with $\phi_{gb} = V_d + \delta_n/q$, we obtain

$$\mu_{eff} = [(q\gamma/kT)(kT/2\pi m^*)^{1/2}] \exp(-qV_d/kT) = \mu_0 \exp(-qV_d/kT), \tag{9.16}$$

in agreement with the Petritz (1956, 1958) theory for PX material mobility. The effects of image forces and tunneling are likely to be important in some high carrier density solar cell materials, but have been neglected here. For high carrier densities, the barriers decrease in height and width and tunneling becomes important, particularly at low temperatures. The relative contributions of thermionic emission and tunneling to the mobility in PX layers are considered by Martinez and Piqueras (1980) and Wu and Bube (1974).

Measurements of the temperature dependence of the conductivity of a PX layer give an activation energy ΔE_σ from the slope of the ln σ versus $1/T$ plot which can be related to the barrier height. Using $\sigma = qn\mu_{eff}$ and assuming ϕ_{gb} is pinned by grain boundary states, Eq. (9.15) gives

$$-d(\ln \sigma)/d(1/T) = \Delta E_\sigma/k \simeq q\phi_{gb}/k + T - qT(d\phi_{gb}/dT)$$

so that

$$\Delta E_\sigma \simeq q\phi_{\text{gb}} + kT - qT(d\phi_{\text{gb}}/dT) = qV_{\text{d}} + \delta_n + kT + qT(d\phi_{\text{gb}}/dT), \tag{9.17}$$

for partially depleted grains.† For lack of detailed knowledge, in some cases $d\phi_{\text{gb}}/dT$ is either taken to be zero or to have the same temperature dependence as the band-gap energy (Mueller, 1961, p. 635). The differences between the activation energy and the barrier height V_{d} are illustrated by the calculations of Fig. 9.24.

Experimentally the measured effective mobilities of PX samples follow the exponential relationships of Eqs. (9.15) or (9.16), but usually indicate a value of A^*, which is 10^2-10^3 times smaller than that for a single-crystal/metal junction.‡

The meaning of Hall effect measurements on such PX films is most easily seen by going back to the basic equations. If $W_{\text{d}} \ll \gamma$ and we neglect scattering terms, the induced Hall field $\mathscr{E}_y$ counterbalances the force exerted on the moving majority carriers by the magnetic induction B_z:

$$q\mathscr{E}_y \simeq 10^{-8}\, qB_z v_x,$$

where $\mathscr{E}_y$ is in volts per centimeter, B_z in gauss, and v_x in centimeters per second. The drift velocity of the carriers is given either by $v_x = \mu\mathscr{E}_x$ or by $v_x = J_x/nq$. Since $\mathscr{E}_x$ is supported mainly by grain boundary potential barriers having no relation to the mobility inside the grains μ_{g}, then $v_x = \mu_{\text{g}}\mathscr{E}_x$ is not a good description. Since J is continuous throughout the layer (neglecting conduction along grain boundaries), the second description is appropriate giving

$$R_{\text{h}} = 10^{-8}\, \mathscr{E}_y/B_z J_x = 1/qn, \tag{9.18}$$

where n is the carrier density inside the grains. Then the mobility as calculated from $\rho = J_x\mathscr{E}_x = 1/qn\mu$ is the effective mobility μ_{eff} given by Eq. (9.16). Maruska *et al.* (1980)§ made Hall measurements on PX n-Si for

† For the particular case in which all the charged grain boundary states N_{gb}^* are considerably below E_{F} (n-type), then N_{gb}^* is constant with variation of bias and temperature. Therefore V_{d} = constant, and ϕ_{gb} varies as E_{F} moves with a change in temperature so that $d\phi_{\text{gb}}/dT = (1/q)d\delta_n/dT$. In this case Eq. (9.17) gives $\Delta E_\sigma = qV_{\text{d}} + kT$.

‡ In cases of substantial depletion it is probably more meaningful to use the Schottky theory of junction transport (Section 5.7.1), which introduces the drift velocity $\mu_{\text{g}}\mathscr{E}_{\text{max}}$ into the preexponential factor in Eq. (9.16) (i.e., $\mu_{\text{g}}\mathscr{E}_{\text{max}}$, rather than $(kT/2\pi m^*)^{1/2}$, where $\mathscr{E}_{\text{max}}$ is the maximum electric field at the barrier and μ_{g} is the grain-interior mobility). This was done, for example by Taylor *et al.* (1952) and by Stratton (1956).

§ These authors found a conductivity activation energy of 0.216 eV that reduced the conductivity by several orders of magnitude with respect to the illuminated case at temperatures below 200°K. See also Ghosh *et al.* (1980a).

which solar intensity illumination effectively removed the effect of grain boundary barriers on the conductivity. They found that the carrier density n, as measured by Eq. (9.18), was essentially the same in either light or dark, showing that the preceding identification is valid, at least in their regime. When the depletion layers are large with respect to the grain size, the meaning of Hall effect measurements is more elusive. The situation is discussed by Volger (1950), Petritz (1958), Blount *et al.* (1970), and Kamins (1971).

Other, more nearly complete models of carrier transport in thin PX layers have been offered by a number of workers. Up until a few years ago most of the interest centered around varistors, electroluminescent and photoconductive devices.

Seager and Castner (1978) and Pike and Seager (1979) developed a conductivity model for the dark case that includes thermal carrier traffic and a number of different trap distributions (Fig. 9.22) and found excellent agreement with experimental measurements on Si bicrystals and large-grain PX slices. Baccarani *et al.* (1978) and Seto (1975) included the situation in which the depletion layers fill the interior of the grains. Their analysis predicts a rather discontinuous change in the conductivity activation energy when conditions go from partially depleted to fully depleted grains as shown in Fig. 9.24b. Seto (1975) and Cowher and Sedgwick (1972) both observed strong minima in the effective mobility at the doping density at which the grains go from fully depleted to partially depleted (see Fig. 9.33a). This corresponds directly to a maximum in the barrier height V_d (such as that shown in Fig. 9.24a). Graef *et al.* (1979) have applied the theories of Baccarani and Seto to experimental measurements on CVD-grown PX layers of p-Si for a variety of acceptor densities and grain sizes with good agreement, indicating a continuous distribution of traps over the band gap. All the measurements mentioned here were done on unilluminated samples.

Card and Yang (1977) modeled the illuminated case using the Shockley–Read–Hall recombination model, finding that under solar illumination (100 mW cm^{-2}) the diffusion potential V_d was reduced to values (Fig. 9.27) that would affect the conductivity to a much smaller extent. This and other research dealing with the illuminated case are discussed in the next sections since recombination is a determining factor for the barrier height.

Other causes of scattering may be important in polycrystalline thin films under a variety of specific circumstances. These include deformation (Dexter and Seitz, 1952) and/or charged-imperfection-like scattering from dislocations, effects of mechanical stress because of different coeffi-

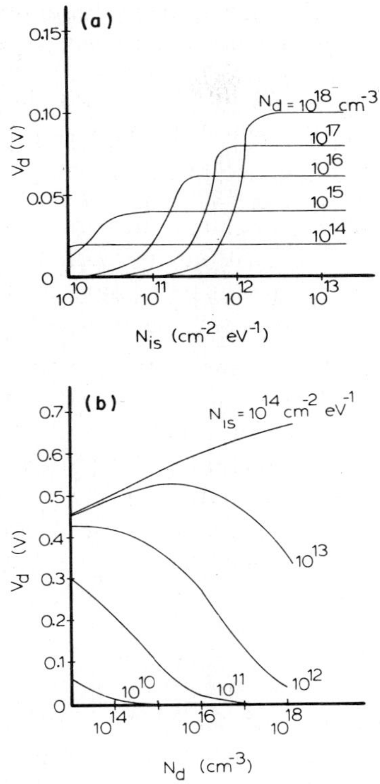

Fig. 9.27. Calculated diffusion potential at a grain boundary in n-Si and its dependence on interface state density N_{is} and doping concentration N_D (300°K). (a) Under solar illumination (100 mW cm^{-2}). Saturation is due to establishment of condition $p(0) \simeq n(0)$, at the grain boundary ($x = 0$). (b) Under dark conditions. [Both figures after H. C. Card and E. S. Yang, *IEEE Trans. Electron Devices* **ED-24,** 397 (1977). © 1977 IEEE.]

cients of thermal expansion between the film and substrate, and scattering from stacking faults (Kazmerski *et al.*, 1972), which can be treated as potential barriers like intergrain barriers.

In this section, theories and experimental work on the effect of grain boundaries on the conductivity and mobility of PX films have been discussed. Under dark conditions the grain boundary barriers most often dominate the conductivity of PX material. Illumination reduces the grain boundary potentials producing a strong increase in conductivity and, at solar illumination intensities, the effect of the barriers on the conductivity of PX Si samples, for example, appears to be almost negligible in some cases. There has been little research on the magnitudes of these effects as a function of illumination wavelength and intensity, however. Recombination is a determining factor in this relation; in general, a three-dimensional transport problem must be solved to obtain a theoretical result. Recombination at grain boundaries is discussed in Section 9.5.1.

9.4.4 Examples of Polycrystalline Films

To illustrate these phenomena more specifically, we describe some observations for PX films of different types.

PX films of CdS are *n*-type, whether deposited by VE, SP, or other methods. Figure 9.28 shows the temperature dependence of the electron density and the electron mobility for VE films of CdS, as determined by measurements of thermoelectric power and conductivity. The variation of mobility with temperature can be described by Eq. (9.16), the three films shown having mobility activation energies of 0.11, 0.15, and 0.19 eV. The electron density in these films is about 10^{16} cm^{-3} and is associated with excess Cd. The increase of V_d with increasing electron density suggests

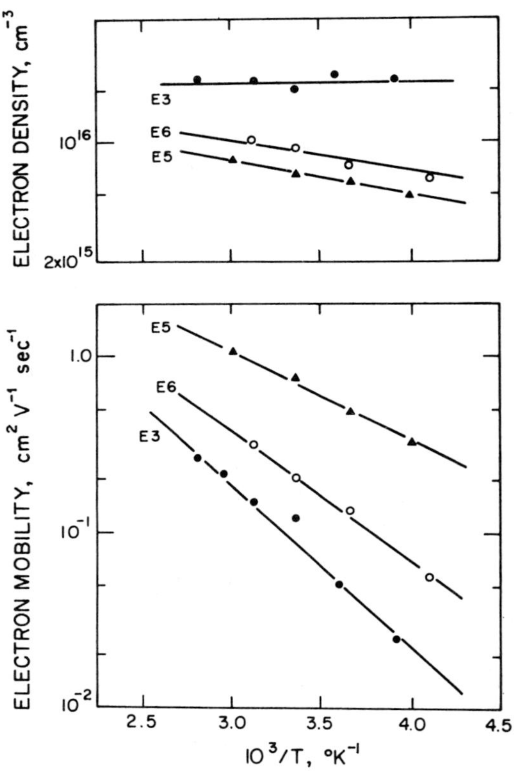

Fig. 9.28. Dark electron density and mobility as a function of temperature for three different CdS films vacuum-evaporated on rutile substrates at different substrate temperatures: E_3, T_s = 200°C; E_6, 220°C; and E_5, 260°C. Grain size is ≥0.1 μm. [After C. Wu and R. H. Bube, *J. Appl. Phys.* **45,** 648 (1974).]

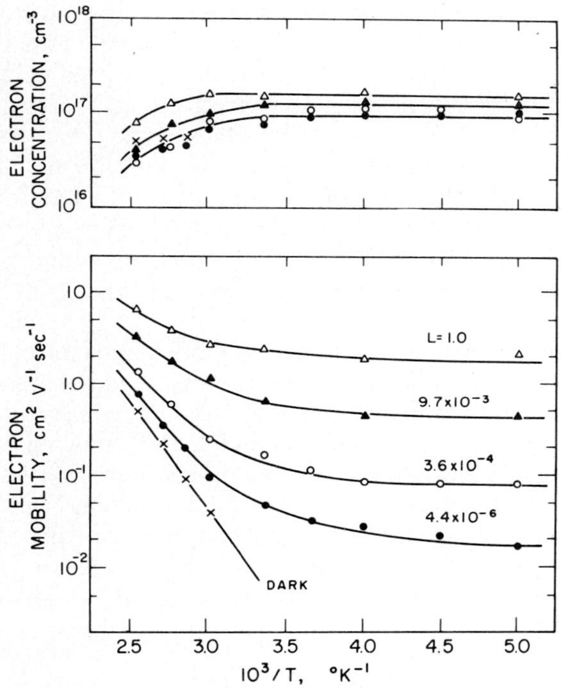

Fig. 9.29. Electron density and mobility as a function of temperature for a CdS film deposited by spray pyrolysis, for different values of illumination intensity. $L = 1$ corresponds to tungsten illumination of 50 mW cm^{-2}. These authors were able to fit the low-temperature mobility values with a tunneling model. [After C. Wu and R. H. Bube, *J. Appl. Phys.* **45,** 648 (1974).]

that many of the grains are close to being fully depleted, in agreement with the estimated values of grain size and depletion layer width. A survey of measurements on CdS films reported in the literature between 1964 and 1973 indicates that CdS films made in different ways have a range of barrier heights from 0.07 to 0.50 eV, with most values lying between 0.10 and 0.20 eV, and a range of μ_0 values from 1 to 10^5 cm^2 V^{-1} sec^{-1}, with most values on the order of a few hundred cm^2 V^{-1} sec^{-1}.

Barrier heights of as-deposited CdS films can be modified by a variety of means, such as by illumination or the field effect, which induce temporary changes, or by heat treatment in various ambients (e.g., O_2, H_2), which induces quasi-stable changes in V_d.

Figure 9.29 shows the effect of photoexcitation on the electron density and mobility in a film of CdS deposited by spray pyrolysis. As with most CdS films of this type having high electron densities in the dark, the primary effect of photoexcitation is to increase the electron mobility, particularly in the low-temperature range where tunneling dominates.

The mobility in PX films of CdS can also be varied using the field effect: the application of a field voltage effectively decreases the value of V_d for the intergrain barriers by inducing excess charge that counteracts the charge depletion at the barriers. Figure 9.30 shows the mobility as a function of temperature for a VE CdS film, for different values of applied field voltage. The value of V_d is decreased from 0.095 eV to 0.069 eV by application of a 2-V field voltage. This effect is potentially important for the situation in which grain boundaries pass through the depletion layers of solar cell junctions.

Heat treatment can be used to induce more long-lasting changes in film properties. If oxygen is adsorbed on the surface and in the grain boundaries of a CdS film, it removes electrons from the conduction band of the CdS in order to become chemically bound, thus building up a depletion layer at the surface and at the grain boundaries. This adsorbed oxygen can be removed by low-temperature annealing in vacuum. Typical results are shown in Fig. 9.31, where removal of adsorbed oxygen decreases the

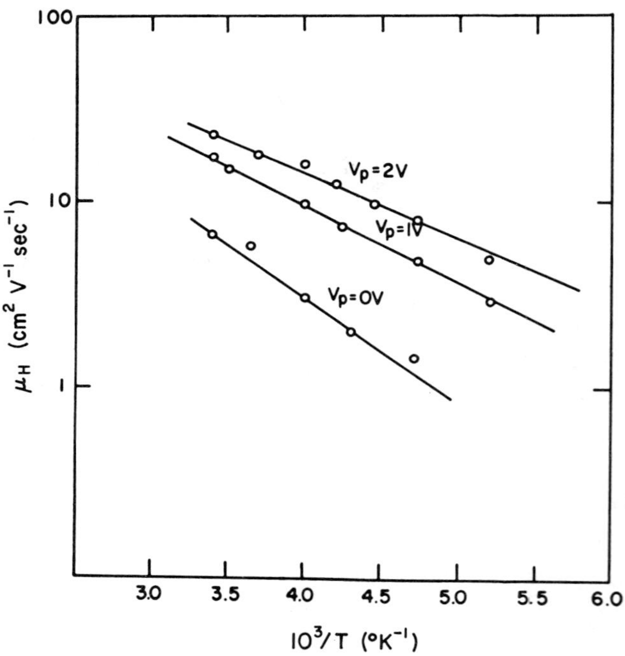

Fig. 30. Variation of Hall mobility with temperature for a vacuum-evaporated polycrystalline CdS film for three values of applied field voltage V_p. [After A. Waxman, V. E. Henrich, F. V. Shallcross, H. Borkan, and P. K. Weimer, *J. Appl. Phys.* **36**, 168 (1965).]

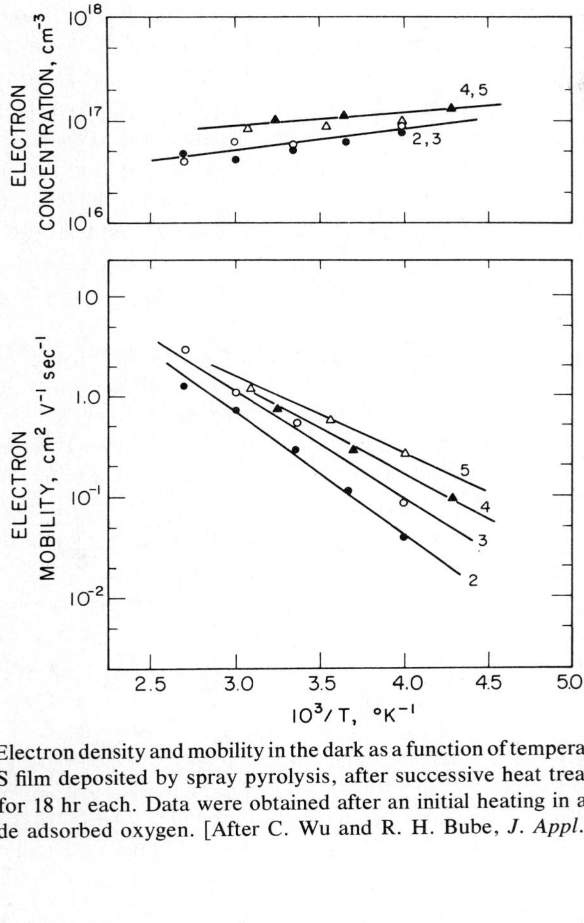

Fig. 9.31. Electron density and mobility in the dark as a function of temperature for a polycrystalline CdS film deposited by spray pyrolysis, after successive heat treatments in vacuum at 120°C for 18 hr each. Data were obtained after an initial heating in air at 100°C for 18 hr to provide adsorbed oxygen. [After C. Wu and R. H. Bube, *J. Appl. Phys.* **45**, 648 (1974).]

value of V_d from 0.22 to 0.15 eV and increases the average free electron density slightly.

The resistivity of PX CdS films can also be changed dramatically by heat treatment at higher temperatures, as shown in Fig. 9.32 for various ambients. Whereas the higher ranges of CdS film resistivities produced by heat treatment are somewhat unstable with respect to oxygen adsorption, the low-resistivity CdS films ($\leq 1\ \Omega$ cm) produced by heat treatment appear to be quite stable. The following mechanisms have been suggested for the heat treatment effects observed for higher temperatures:

(1) A decrease in the amount of oxygen adsorbed at grain boundaries;

(2) Adsorption of hydrogen within the grain boundary, acting to

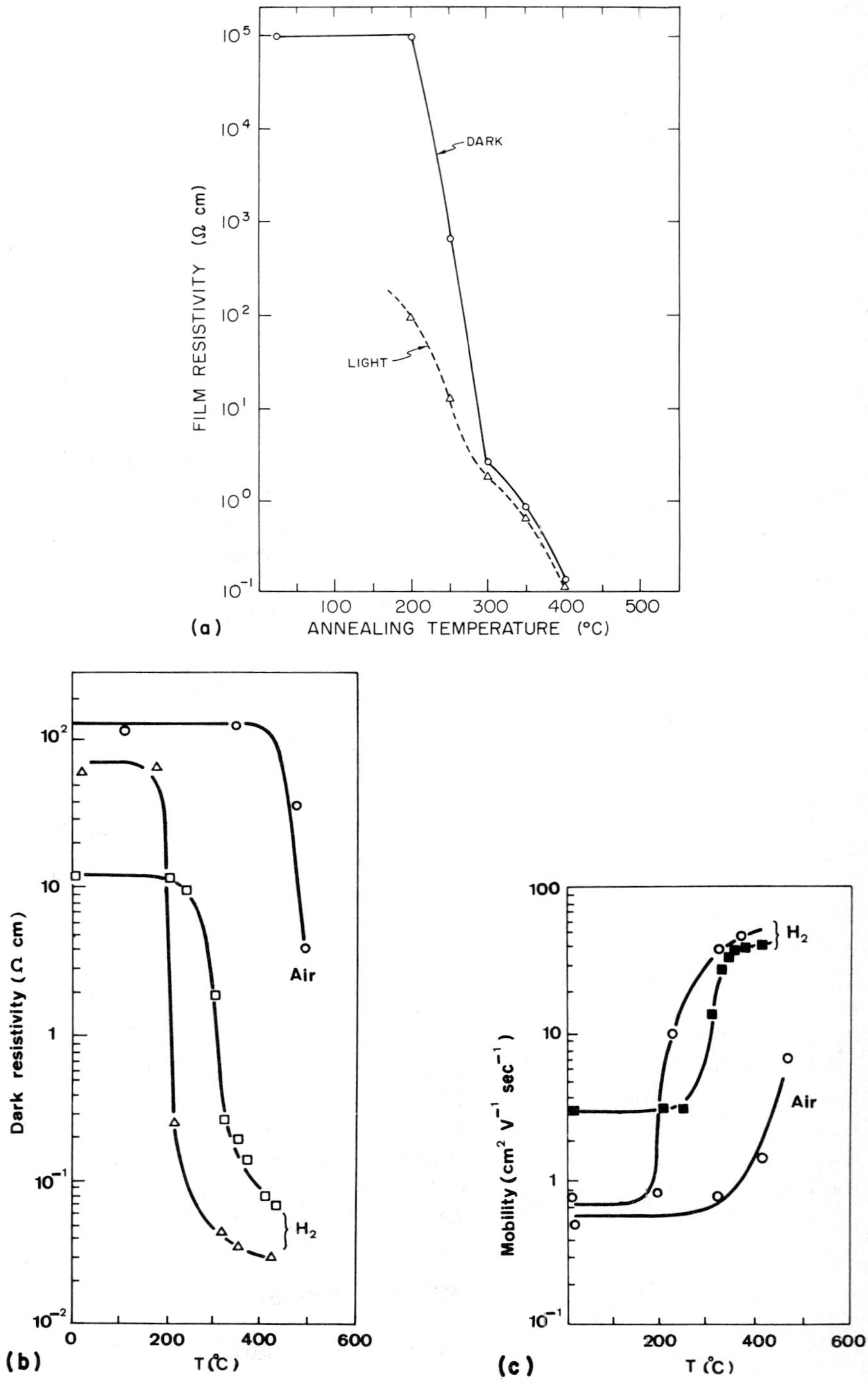

FILM RESISTIVITY (Ω cm)
10^5
10^4
10^3
10^2
10^1
10^0
10^-1
DARK
LIGHT
100
200
300
400
500
(a)
ANNEALING TEMPERATURE (°C)
Dark resistivity (Ω cm)
10^2
10
1
10^-1
10^-2
Air
H2
0
200
400
600
(b)
T(°C)
Mobility (cm^2 V^-1 sec^-1)
100
10
1
10^-1
H2
Air
200
400
600
(c)
T (°C)

annihilate acceptor grain boundary states (i.e., passivation of broken bonds).

(3) Partial recrystallization to increase grain size or anneal out dislocations; and

(4) Loss of sulfur from the interiors of the crystallites, causing increased N_D (interstitial Cd or S vacancies are donors).

The effect of grain boundary limited mobility in Si is illustrated by Fig. 9.33 where the resistivity parallel to the plane of VE Si films is plotted against measured carrier density. Only at high carrier density is the bulk mobility value approached for these films. The Hall mobility data for these films show a minimum slightly above a doping density of 10^{18} cm^{-3} (about 1 cm^2 V^{-1} sec^{-1}), where the barrier height V_d reaches its maximum. The variation of resistivity with free carrier density for MO-CVD deposited PX GaAs films (both *n*- and *p*-type) is given by the valuable article of J. J. J. Yang *et al.* (1980).

In most of the preceding examples the grain size of the films was small and the influence of the grain boundaries on transport dominated the bulk properties. Similar results can be seen on a much reduced scale even in films that are essentially epitaxial. An example of this effect is observed in epitaxial films of n-InP deposited by metallo-organic CVD on GaAs or InP substrates (Tsai and Bube, 1978). Figure 9.34 shows the variation of mobility in these films as a function of electron density, at room temperature. Below an electron density of about 3×10^{16} cm^{-3}, the mobility increases with increasing electron density, whereas for higher electron densities, the mobility decreases with increasing density. In the higher-density range, scattering by charged impurities dominates and the temperature dependence of the mobility for these samples is actually identical to that reported in the literature for single-crystal samples with the same electron density. In the lower-density range, however, the behavior is suggestive of a barrier-limited mobility, the height and width of the depletion layers around the barrier being inversely related to the electron density.

The detailed temperature dependence of the mobility for three of the

---

Fig. 9.32. Effect of heat treatments in various ambients on the electrical properties of CdS films. (a) Resistivity of a CdS film deposited by VE at a substrate temperature of 300°C as a function of annealing temperature in H_2. [R. H. Bube, "Applied Research on II–VI Compound Materials for Heterojunction Solar Cells." Progress Report NSF/RANN/SE/GI-38445X/PR/74/1 (1974).] (b) Resistivity and (c) mobility (at room temperature) of CdS films deposited by spray pyrolysis on glass at a substrate temperature of 340°C as a function of annealing temperature in H_2 and air. [M. Savelli and J. Bougnot, Problems of the Cu_2S/CdS cell. *In* "Solar Energy Conversion" (B. O. Seraphin, ed.), p. 213, Topics in Applied Physics, Vol. 31. Springer-Verlag, Berlin and New York, 1979.]

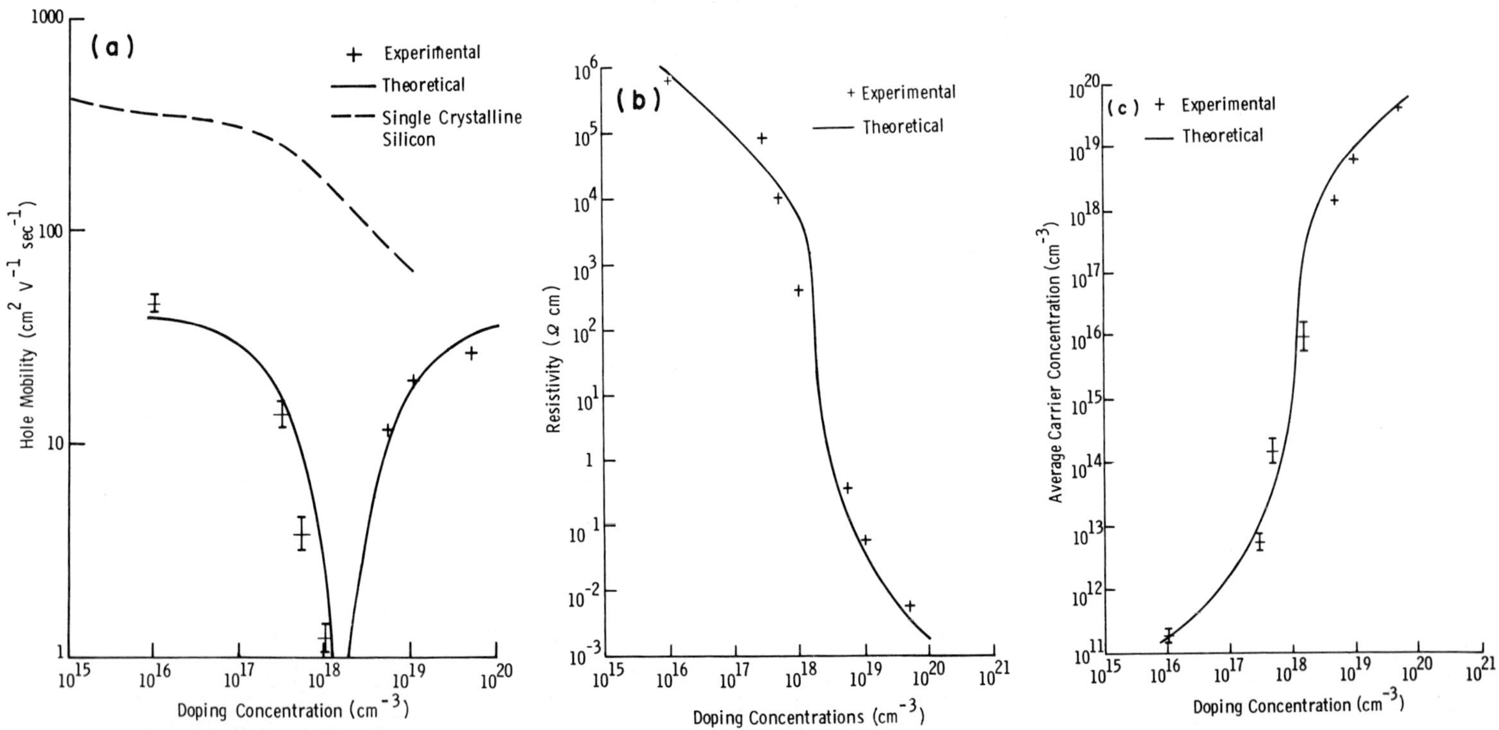

Fig. 9.33. (a) Hall mobility, (b) resistivity, and (c) carrier density versus doping density for Si films grown by CVD on SiO_2 and doped *n*-type by ion implantation with a subsequent anneal at 1100°C. Film thickness was 1 μm and average grain size was estimated at 200 Å by TEM. All measurements were made at room temperature. [After J. Y. W. Seto, *J. Appl. Phys.* **46,** 5247 (1975).]

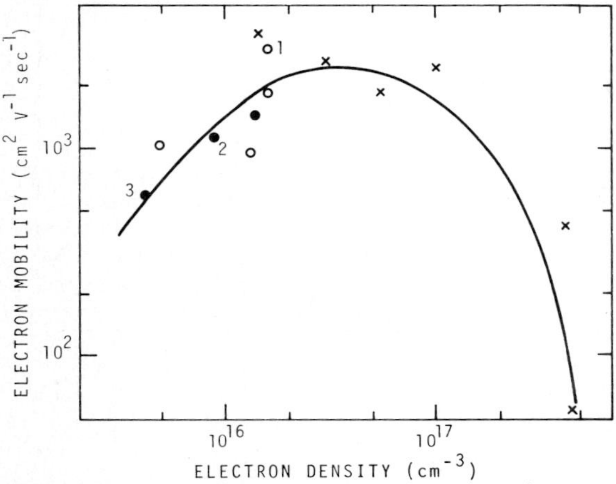

Fig. 9.34. Electron mobility as a function of electron density in undoped *n*-InP films deposited by metallo-organic CVD on insulating substrates. Substrates: (○) (100) GaAs:Cr; (●) (111a) GaAs:Cr; and (×) (100) InP:Fe. Numbered points refer to Fig. 9.35.

samples indicated in Fig. 9.34 is plotted in Fig. 9.35. For two of the samples, variations in the light are also included. The electron mobility in single-crystal InP at and below room temperature is dominated by optical-mode scattering so that the mobility increases rapidly with decreasing temperature. The mobility data for thin-film sample 1 are quite similar to those for single-crystal InP, whereas those for films 2 and particularly 3 (with the lowest carrier density) suggest that transport is modified by tunneling and thermal activation over grain boundary potential barriers. These data can be analyzed by rewriting Eq. (9.14), using the Schottky formulation for barrier transport in parallel with tunneling through the barrier, the latter dominating at low temperatures. The grains are assumed to be mostly undepleted.

$$\mu_{\text{eff}} = \frac{1}{\mu_{\text{g}}^{-1} + [(\mu_{\text{g}}\mathscr{E}_{\text{max}}\gamma q/kT)\exp(-qV_{\text{d}}/kT) + \mu_{\text{tun}}]^{-1}}. \tag{9.19}$$

The mobility of the grain interiors is μ_{g} and the grain boundary tunneling limited mobility is μ_{tun}. Equation (9.19) was used to extract the thermal excitation term for sample 3 by assuming that $\mu_{\text{tun}}/\mu_{\text{g}}$ does not depend on temperature and that the grain-interior mobility of film 3 is equal to the bulk mobility of film 1 (i.e., $\mu_{\text{g3}} \simeq \mu_1$).

$$\mu_{\text{eff,3}} = \frac{\mu_1}{1 + \{(\gamma q\mathscr{E}_{\text{max}}/kT)\exp(-qV_{\text{d}}/kT) + [\mu_{\text{tun}}/\mu_1]_{150^\circ\text{K}}\}^{-1}}. \tag{9.20}$$

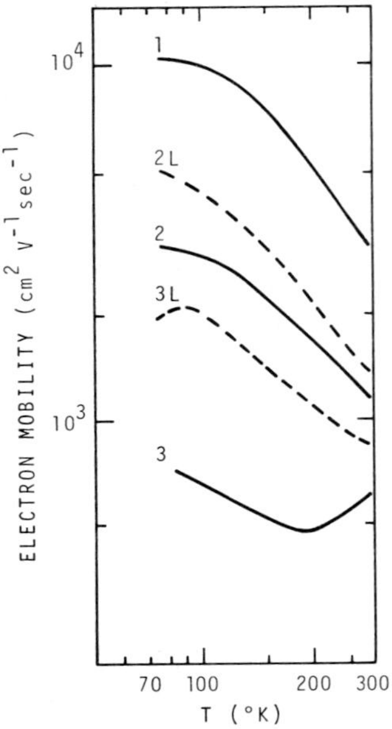

Fig. 9.35. Temperature dependence of the electron mobility of three undoped n-InP films. Numbers on curves correspond to the data points given in Fig. 9.34. Dashed curves 2L and 3L represent the mobility under photoexcitation.

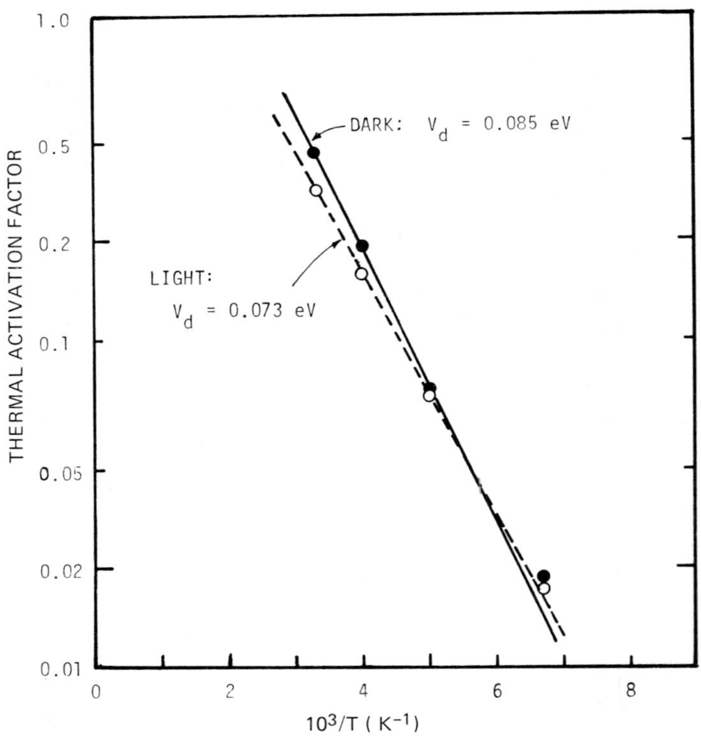

Fig. 9.36. Thermal activation factor of grain boundary dominated mobility for n-InP samples of Fig. 9.35, analyzed according to Eq. (9.20) for sample 3. The value of $[\mu_{tun}/\mu_1]$ at 150°K was used to fit these data, i.e., 0.075 in the dark and 0.223 in the illuminated case. The thermal activation factor is $(\gamma q \mathscr{E}_{max}/kT) \exp(-qV_d/kT)$.

The results, shown in Fig. 9.36, show a barrier height in the dark of 0.085 eV, which decreases slightly under illumination. The fit is excellent and the principal effect of illumination appears to be to increase the tunneling component. Overall, the grain boundary effects in these films are much smaller than those commonly found in the more grossly polycrystalline CdS and other films and indicate the relatively minor role played by such potential barriers in epitaxial films.

9.5 THE EFFECT OF GRAIN BOUNDARIES IN POLYCRYSTALLINE SOLAR CELLS

Because of the large densities of defects and segregated impurities at grain boundaries, and because of the local minority carrier collecting fields, the boundaries are regions of strong recombination. The reduction of the photovoltaic parameters, J_{sc}, V_{oc}, and ff by this recombination is strongly dependent on the grain size in the PX layers. The importance of these effects is highlighted by the beam-induced current image of Fig. 9.37, which show grain boundaries that are electrically active as dark contrast.

To gain an overall perspective, we consider a simple p/n junction in a PX material with columnar grains oriented normal to the junction plane as shown in Figs. 9.38a and 9.39. For the moment we assume zero bias voltage. Under illumination the photogenerated carrier density n_p is highest along the central axis of each columnar grain and decreases toward the grain boundry where the carriers recombine more rapidly. In the direction perpendicular to the junction plane there are the usual carrier density variations produced by the boundary conditions and the spatial variation of photogeneration (Fig. 9.38b). Thus the calculation of minority carrier density within the grains requires the solution of a three-dimensional transport equation.

What is the boundary condition at the grain boundary? Since the potential barrier height ϕ_{gb}, the effective recombination velocity S_{gb}, and the excess carrier density $n_p\ (x, y)$ are interdependent, the loss there cannot be expressed by a simple constant lifetime formalism.† Instead we have

$$S_{gb} = S_{gb}[n_p(x, y), p_p(x, y), dN_{gb}/dE, \tau_g, x], \qquad (9.21)$$

where the symbols have their usual meanings and τ_g is the lifetime of the grain's interior material. The region of recombinative influence of the grain boundary is roughly equal to the larger of either the grain interior

† The spatial coordinates used here are x = (distance away from the junction) and y = (distance away from the grain boundary).

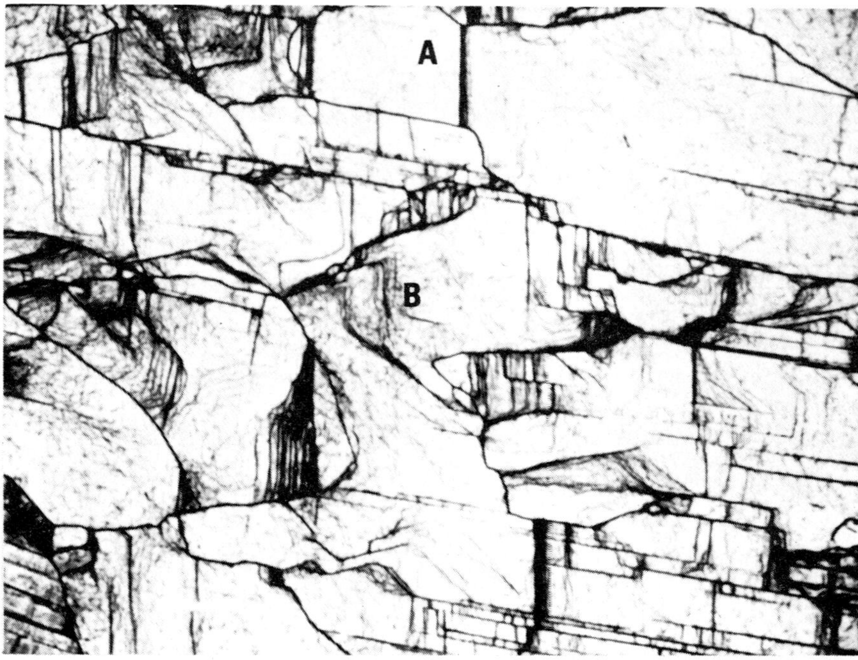

Fig. 9.37. Scanning EBIC response of an area of an ITO/Si solar cell fabricated with Monsanto zone-refined PX Si. The surface shown is almost featureless under an optical microscope. Subsequent etching revealed etch pits corresponding to the small dark spots in area B. The dark spots are caused by active dislocations in the intergrain regions. The portion of the sample imaged is about 1.1 mm in its longest dimension. [Photo courtesy of K. Jones. From N. Inoue, C. W. Wilmsen, and K. A. Jones, *Solar Cells* **3,** 35 (1981).]

diffusion length L_g or the grain boundary depletion layer width W_{dgb}. Thus the problem has not only three spatial dimensions but a position-dependent, nonlinear boundary condition (Lindholm and Fossum, 1981). Given the recombination parameters, the solution could presumably be obtained by a machine calculation. Because of the three-dimensional nature of the problem, it is, strictly speaking, not possible to assign a unique theoretical average lifetime or diffusion length to the PX aggregate, since it is crucial how and where that average is taken. We examine that consideration further in the next section.

The situation is even more complex at forward bias. For the sake of argument, assume that S_{gb} is very large and consider the value of $n_p(x, 0)$ along the grain boundary, represented by line $y = y_0$ in Fig. 9.38c. The large value of S_{gb} implies that $n_p(x, 0)$ approaches n_{p0} along the grain boundary, but this conflicts with the commonly assumed boundary condi-

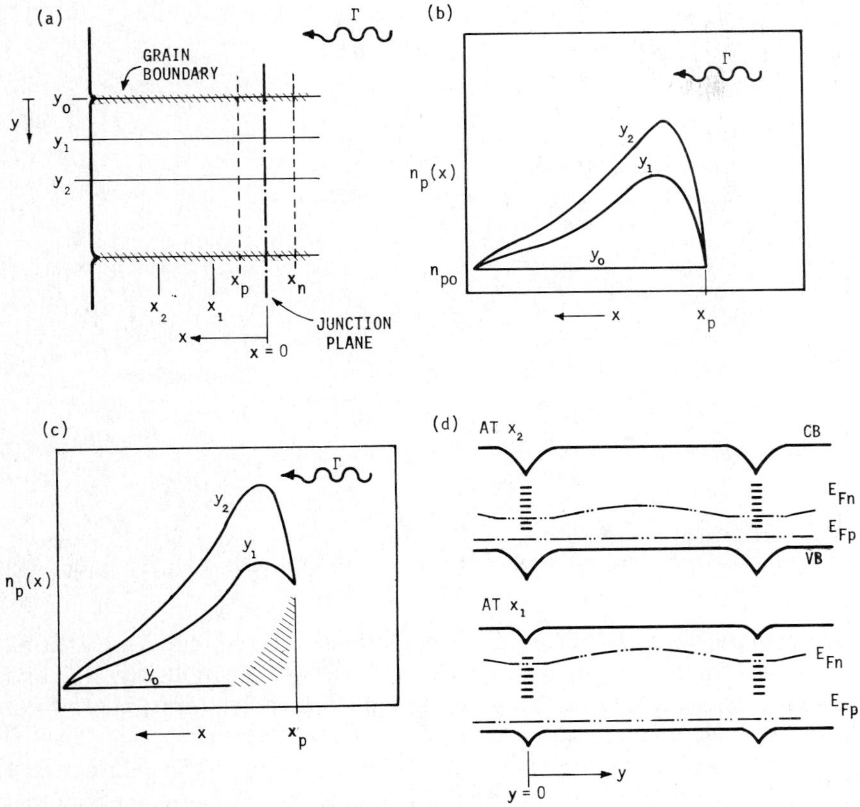

Fig. 9.38. Intersection of a grain boundary with a p/n homojunction.

(a) Plan view: junction is at $x = 0$, with junction depletion layer edges at x_p and x_n.

(b) Carrier density $n_p(x, y)$ under illumination for zero applied bias. Values of $n_p(x, y)$ are shown on two lines within the grain, y_1 and y_2, and on the grain boundary, y_0, assuming $S_{gb} \to \infty$.

(c) Carrier density $n_p(x, y)$ under illumination with forward bias injection. Shaded portion represents uncertainly in the description of $n_p(x, y)$ near x_p.

(d) Sections of the band diagram parallel to the junction plane at x_1 and x_2 under illumination, now assuming that $S_{gb} < V_{th}/4$.

tion at the junction depletion layer edge x_p, namely, that $n_p(x_p, y) = n_{po} \exp(qV/kT)$. Forward-bias-injected carriers are funneled into the grain boundary regions by the depletion layers intersecting at the grain boundary and at the junction. The carrier density in these regions, denoted by the shaded area in Fig. 9.38c, is difficult to model satisfactorily. Thus both lateral carrier transport *within* the grain and transport *along* the grain

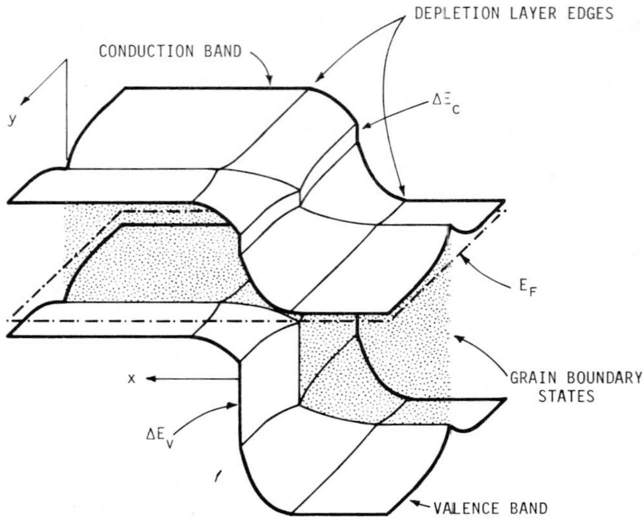

Fig. 9.39. Band diagram with two spatial dimensions for a heterojunction solar cell with a grain boundary traversing the junction depletion layers. The grain boundary region is depleted in both *n*- and *p*-type components.

boundary depletion layer by diffusion (and possibly conduction) must be considered for a complete solution. Shockley (1949, Sections 4.6, 4.7, and Appendix 5 therein), apparently discussed this problem first, but did not offer a complete solution.

Several workers have simplified the problem by finding an average lifetime $\bar{\tau}$ or diffusion length $\bar{L}$ characteristic of the PX aggregate, but this concept must be used with caution. Considering Fig. 9.2 for a moment, it can be seen, at least qualitatively, that the probability of recombination is larger for photogeneration that occurs deep in the absorber and/or close to the grain boundaries. Also because the grain boundary recombination rate depends on n_p in a nonlinear way, the value of $\bar{L}$ depends on the injection level as determined by forward bias and the intensity of the illumination at each point in the absorber. Thus $\bar{L} = \bar{L}[\gamma, \alpha(\lambda), d\Gamma/d\lambda, V, x]$, and the carrier loss from an elemental volume $dx\, dy\, dz$ depends on the excitation level as well as its position within the grain.

We begin the first part of this section by focusing on studies of recombination at boundaries in single crystals and bicrystals. In the next part theories relating the photovoltaic parameters to grain size in PX materials are explored. The following part examines some of the more esoteric considerations such as the possible effect of reduction of the band gap at the disordered grain boundary. Finally we cite some attempts at passivation of grain boundary recombination.

9.5.1 Recombination at Grain Boundaries—Bicrystal and Large-Grain Polycrystal Investigations

Under illumination photogenerated excess minority carriers are drawn to the grain boundaries by the depletion layer fields surrounding them. Since the supply of majority carriers at the maxima of the grain boundary potential barriers is initially small, minority carriers are trapped at the grain boundary states, reducing the charge there and lowering the potential barriers. Steady state is reached when the majority carrier density at the grain boundary barrier maximum is high enough so that $n_p S_{gbn} \simeq p_p S_{gbp}$ there (for a p-type material, where S_{gbn} and S_{gbp} are the grain boundary recombination velocities for electrons and holes, respectively). The number of charged grain boundary interface states N_{gb}^* and hence ϕ_{gb} is changed radically by illumination of solar intensity.

This section begins with a treatment of recombination at the boundaries of filamentary grains in which recombination is quantified as a grain boundary recombination velocity S_{gb} which does not vary with carrier density. This is followed by a description of experimental measurements of S_{gb} in bicrystals and large-grain polycrystals. The section ends with a description of several theories which relate grain boundary barrier height, recombination rate, and illumination intensity.

Recombination at grain boundaries has been under fundamental investigation for many years, without regard to solar cell applications. Shockley (1950) developed a theory for the time decay of excess carrier density by recombination in isolated semiconductor filaments with recombination velocity S_{gb}, for the time intervals following *uniform* excitation. He assumed a rectangular filament with dimensions of $y = \pm A$ and $z = +B$ and a length x that was effectively infinite. The diffusion constant is D and τ_g is the grain-interior lifetime. Solution of the transport equation in two dimensions yields an eigenfunction series with each succeeding term having a smaller characteristic lifetime. After excitation ceases, the higher-order terms decay rapidly and only the term with the longest lifetime remains, giving an approximate "filament lifetime" corresponding to the lowest-order eigenvalues η_0 and ζ_0†:

$$1/\tau_{fil} \simeq D[(\eta_0^2/A^2) + (\zeta_0^2/B^2)] + 1/\tau_g, \tag{9.22}$$

where $\eta_0 \tan \eta_0 = S_{gb}A/D$ and $\zeta_0 \tan \zeta_0 = S_{gb}B/D$. If S_{gb} is effectively infinite, then the filament lifetime reduces to

$$1/\tau_{fil} \simeq (\pi^2 D/4)[A^{-2} + B^{-2}] + 1/\tau_g. \tag{9.23}$$

† Note that for steady state but uniform excitation it may not be reasonable to throw away the higher-order terms since they represent loss of carriers generated close to the grain boundary.

For small S_{gb} the tangent can be approximated by its argument and

$$1/\tau_{fil} \simeq S_{gb}[A^{-1} + B^{-1}] + 1/\tau_g . \quad (9.24)$$

From these equations it can be seen that the smallest dimension of the filament dominates the lifetime. Since the analysis pertains essentially to the later part of a transient response after the cessation of uniform excitation, it may not always be appropriate to deduce a value of $\bar{\tau}$ or $\bar{L}$ for steady state, nonuniform excitation from it.

Shockley (1949, Section 4.6 and Appendix 5) also applied these ideas to flow across a biased p/n junction into such a filament (injection in the x direction into one end of the filament) and found that the use of τ_{fil} from Eq. (9.24) in the standard diode equation [i.e., Eq. (5.13)] is a reasonable approximation, *providing* that $S_{gb}A/D$ is small enough so that tan $\eta_0 \simeq \eta_0$ and $\eta_0^2 \sim S_{gb}A/D$. For PX solar cell systems this condition is generally not satisfied. For larger $S_{gb}A/D$ values a complex series solution is required since the terms do not converge rapidly. The depletion layer surrounding the grain boundary was not considered in either of these situations.

Other investigators concentrated on finding the experimental electronic properties of grain boundaries in bicrystals (Vogel *et al.*, 1954; Pearson, 1949; Morrison, 1956; Werthiem and Pearson, 1957). Vogel *et al.* (1954) set up a grain boundary model that regarded the closely spaced dislocation cores at low-angle grain boundaries as a sheet of recombination centers characterized by a recombination velocity S_{gb}. In their experiments with lineage boundaries (i.e., boundaries with such low angle that the dislocations may be on the order of micrometers apart) in n-Ge they found S_{gb} = 1200–2600 cm sec^{-1} by using an optical probe (essentially dark conditions).

Sosnowski (1959) reviewed early work on Si and Ge bicrystals, finding that large-angle grain boundries had considerably larger values of ϕ_{gb} and much greater photoelectronic activity, but, curiously, smaller recombination velocities. He also reported considerably greater conductivity along the grain boundaries in Ge and concluded that they were so highly charged as to be degenerate p-type. McKelvey (1957, 1961) developed a model for recombination at lineage and twin boundaries in Ge to obtain experimental values of S_{gb}, but neglected any effects of the potential barrier fields. Mataré (1971) has made a comprehensive theoretical study of all types of grain boundaries, from both a structural and electronic point of view. His experimental evidence, particularly with regard to electronic properties, is largely confined to Ge.

Daud *et al.* (1978) studied grain boundary recombination in p/n junctions made on p-Si bicrystals, using EBIC techniques with beveled geometry. They measured an effective diffusion length $L_{eff} \simeq 2\ \mu m$ at the grain boundary that increased almost linearly with distance away from the grain boundary until it saturated at the single-crystal value $L_g \simeq 12\ \mu m$. The region of degraded L_{eff} extended approximately L_g on either side of the grain boundary, a distance much greater than the grain boundary depletion layer with ($W_{dgb} \leq 1\ \mu m$ in this case).

Several workers have studied the recombination at grain boundaries in large-grain polycrystalline material by scanning light spot or electron beam methods (Inoue *et al.*, 1981; Hanoka, 1980; Itoh *et al.*, 1978). For example, Armstrong *et al.* (1980) related the crystallographic orientation of the grain boundaries with their recombination behavior for Si.

The work discussed thus far in this section is confined essentially to the dark case (sometimes with small-area probe illumination) and the recombination properties are usually quantified by a constant value of S_{gb} which does not depend on carrier density. The illuminated case has been studied by Card and Yang (1977), who considered the effect of illumination on the grain boundary potential barrier and the recombination there using the Shockley–Read–Hall model. For an n-type material under illumination, photogenerated holes neutralize the negatively charged grain boundary states to reduce the barrier height until the recombination currents for electrons and holes are equal. Card and Yang assumed that E_{Fn} and E_{Fp} were constant across the grain boundary depletion layer. Then, assuming equal recombination cross sections for electrons and holes, $\sigma_n = \sigma_p = \sigma$, a stable condition is reached when the nonequilibrium carrier densities $n_p(x, 0)$ and $p_p(x, 0)$ at the grain boundary are equal. This maximizes the recombination rate. Only the states between E_{Fn} and E_{Fp} contribute significantly to the recombination current, which is given by

$$U_{gb} \simeq q(dN_{gb}/dE)(E_{Fn} - E_{Fp})\sigma v_{th} p_p(x, 0)/2, \tag{9.25}$$

where v_{th} is the thermal velocity and dN_{gb}/dE ($cm^{-2}\ eV^{-1}$) is the interface state density distribution. The resulting relation between the illuminated grain boundary barrier diffusion voltage V_{dgb}, N_D, and dN_{gb}/dE under AM1 illumination is shown in Fig. 9.27† (also see Hwang *et al.*, 1981). The density of holes at the grain boundary is enhanced by the presence of the remaining grain boundary barrier so that the recombination velocity at the grain boundary depletion layer edge S_{gb} (W_{dgb}) is given by

$$S_{gb}(W_{dgb}) \simeq (\sigma v_{th}/4)(dN_{gb}/dE)(E_{Fn} - E_{Fp}) \exp(qV_{dgb}/kT). \tag{9.26}$$

† Our dN_{gb}/dE is equivalent to Card and Yang's N_{is}. Card and Yang assumed dN_{gb}/dE = constant for this analysis.

The resulting grain boundary recombination rate

$$U_{\mathrm{gb}} \simeq 2q[p_n(x, W_{\mathrm{dgb}}) - p_{n0}]S_{\mathrm{gb}}(W_{\mathrm{dgb}}) \tag{9.27}$$

is not linear in $p_n(x, W_{\mathrm{dgb}})$. The $(dN_{\mathrm{gb}}/dE)(E_{\mathrm{F}n} - E_{\mathrm{F}p})$ term is a minor source of nonlinearity in the range of interest. However, the exponential dependence on illuminated barrier height can result in a much greater nonlinearity since V_{dgb} also depends on $E_{\mathrm{F}n}$ and $E_{\mathrm{F}p}$. Thus the regions of the cell that are highly excited [large $p_n(x, W_{\mathrm{dgb}})$, small V_{dgb}] have proportionately weaker recombination than do portions where low excitation [small $p_n(x, W_{\mathrm{dgb}})$, large V_{dgb}] prevails.† Because of the relation of V_{dgb} to doping density N_{D} (Figs. 9.22 and 9.27a) there is also a fairly strong dependence of S_{gb} on N_{D} for larger values of dN_{gb}/dE.

Card and Yang calculated an average lifetime $\bar{\tau}$ for square columnar grains (γ by γ), assuming uniform photoexcitation throughout the area *and* length of the grain. To do this they treated the grain boundary recombination centers as if they were spread uniformly throughout the volume of PX material, with their effectiveness enhanced proportionately by the grain boundary depletion layer field. This gives

$$\bar{\tau} = 1/N_{\mathrm{gb}}\sigma v_{\mathrm{th}} = [\gamma \exp(-qV_{\mathrm{dgb}}/kT)]/[\sigma v_{\mathrm{th}}(dN_{\mathrm{gb}}/dE)(E_{\mathrm{F}n} - E_{\mathrm{F}p})], \tag{9.28}$$

and the relation of $\bar{\tau}$ to γ, dN_{gb}/dE, and N_{D} is shown in Fig. 9.40 for an n-type material.

The analysis by Fossum and Lindholm (1980a) adds considerable depth to the theoretical treatment of PX solar cells. Several important observations were made concerning the dependence of S_{gb} on illumination. For the case of lower values of S_{gb}, when $E_{\mathrm{F}n}$ and $E_{\mathrm{F}p}$ are constant across the grain boundary depletion layer, they found that the recombination rate there could be written

$$U_{\mathrm{gb}} \simeq (S_p S_n)^{1/2}(n_{\mathrm{i}}/2) \exp[(E_{\mathrm{F}n} - E_{\mathrm{F}p})/2kT] \quad (\mathrm{cm}^{-2}\ \mathrm{sec}^{-1}), \tag{9.29}$$

where $S_p = \sigma_p v_{\mathrm{th}} N_{\mathrm{gb}}$, $S_n = \sigma_n v_{\mathrm{th}} N_{\mathrm{gb}}$, and σ_n, σ_p are the recombination cross sections for electrons and holes at empty recombination sites. This result is true only when sufficient excitation exists such that $(E_{\mathrm{F}n} - E_{\mathrm{F}p}) > 4\,kT$ so that certain approximations in the Shockley–Read–Hall model can be made. The result is similar to that of Card and Yang [Eq. (9.25)] except that it is written in terms of $E_{\mathrm{F}n}$ and $E_{\mathrm{F}p}$ rather than V_{dgb}.‡

† For example, using the parameters assumed by Card and Yang shows that for $p_n(x, W_{\mathrm{dgb}}) \simeq 1.6 \times 10^{10}\ \mathrm{cm}^{-3}$, $S_{\mathrm{gb}}(W_{\mathrm{dgb}}) \simeq 5 \times 10^6$ cm sec^{-1}, whereas for $p_n(x, W_{\mathrm{dgb}}) \simeq 1.6 \times 10^{12}\ \mathrm{cm}^{-3}$, $S_{\mathrm{gb}}(W_{\mathrm{dgb}}) \simeq 0.7 \times 10^6$ cm sec^{-1}.

‡ For sufficient illumination the results are the same since $n_i \exp[(E_{\mathrm{F}n} - E_{\mathrm{F}p})/2kT] \simeq p_p(x, 0) \simeq n_p(x, 0)$.

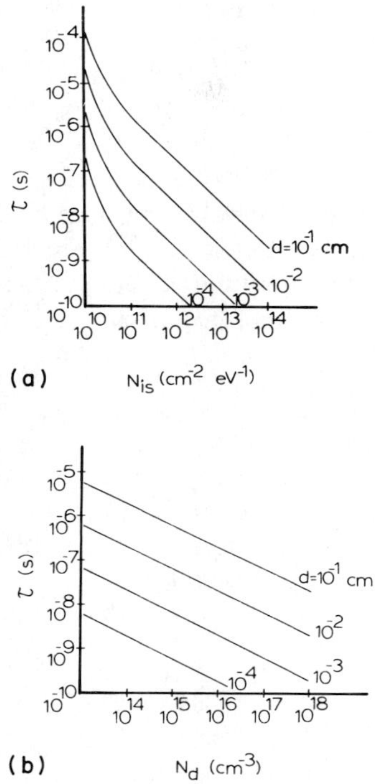

Fig. 9.40. (a) Calculated dependence of average minority carrier lifetime $\bar{\tau}$ in PX Si on grain size (called d here) and grain-boundary state density N_{is} (equals dN_{gb}/dE) for $N_D = 10^{16}\ \mathrm{cm}^{-3}$. Rapid initial decrease of $\bar{\tau}$ is due to increase in V_d (illuminated) with N_{is}. (b) Dependence of average minority carrier lifetime on N_D and grain size, for $N_{is} = 10^{12}\ \mathrm{cm}^{-2}\ \mathrm{eV}^{-1}$. [Both figures from H. C. Card and E. S. Yang, *IEEE Trans. Electron Devices* **ED-24,** 397 (1977). © 1977 IEEE.]

Equation (9.29) makes the striking assertion that the value of the potential barrier height ϕ_{gb} at the grain boundary has little or no effect on the recombination there, *provided* sufficient illumination exists.

For low-level injection in a p-type absorber, $p_p(x, y) \simeq N_A$ in the grain's interior so that $E_{Fp} \simeq E_F$ there and an effective S_{gb} can be written

$$S_{gb}^{eff} = (\tfrac{1}{4})(S_n S_p)^{1/2}[N_A/n_p(x, W_{dgb})]^{1/2}, \tag{9.30}$$

where $n_p(x, W_{dgb})$ is the carrier density at the grain boundary depletion layer edge under illumination. Note that this treatment, like that of Card and Yang, predicts that S_{gb}^{eff} decreases with increasing excitation level and increases with increasing doping level.

With regard to solution of the transport equation, Eq. (9.30) is a nonlinear boundary condition, dependent on injection level. However, for high injection levels $S_{gb}^{eff} \to (S_n S_p)^{1/2}/4$. Also when $(S_n S_p)^{1/2}$ is so high that $E_{Fn} - E_{Fp} \simeq 0$ in the grain boundary depletion layer, the barrier height becomes fixed at its dark value and $S_{gb}^{eff} \to v_{th}/4$. For these two cases the boundary condition is linear. The expressions of Eqs. (9.26) and (9.30) do much to lessen the uncertainty of description transport at the intersection of the grain boundary and the junction depletion layer edge (Fig. 9.38c), provided that conduction along the grain boundary core can be neglected.

Seager (1981) proposed a theory for recombination at single grain boundaries in n-Si (e.g., bicrystals) for a wide range of illumination intensities. Using detailed balance arguments similar to those of Hall (1952) and of Shockley and Read (1952) in their treatment of surface recombination, Seager coupled these results with his earlier model for the conductivity and capacitance of grain boundaries in Si (Pike and Seager, 1979). To formulate his model Seager (1981) assumed that:

(1) The carrier densities and currents are controlled by thermal emission over the grain boundary barriers (Sections 9.4.2 and 9.4.3);

(2) The same states that are involved in recombination are also involved in the trapping of charge that controls V_{dgb};

(3) All the grain-boundary states are at a single energy level;

(4) Uniform excitation and low injection conditions prevail ($\Delta n \ll n$); and

(5) The capture and emission processes are unaffected by illumination.

One of the main differences between Seager's theory and those mentioned previously (Card and Yang, 1977; Fossum and Lindholm, 1980a,b) is that the quasi-Fermi levels are permitted to vary with distance across the grain-boundary depletion layer. With some only moderately restrictive assumptions, Seager was able to formulate a relation between the illuminated barrier height V_{dgb}, the illumination intensity Γ, the donor density N_{D}, the recombination cross section for holes σ_p, and the dark barrier height V_{dgbd}. Although the resulting transcendental equations had to be solved numerically, several conclusions were drawn:

(1) Whereas large grain-boundary barriers are sensitive to even very low illumination levels, barriers with small V_{dgbd} remain essentially unchanged.

(2) For large V_{dgbd} (e.g., $\sim$0.4 eV), illumination produces a substantial readjustment of barrier height that is, to first order, independent of the recombination cross section for holes σ_p, for $\sigma_p < 10^{-18}$ cm^2.

(3) For smaller values of V_{dgbd} ($\lesssim$0.25 eV), the effective grain-boundary recombination velocity is given by

$$S_{\mathrm{gb}}^{\mathrm{eff}} \simeq q^{-1} v_{\mathrm{th}} \sigma_p (2V_{\mathrm{dgb}} \epsilon_{\mathrm{s}} N_{\mathrm{D}})^{1/2} \exp(qV_{\mathrm{dgb}}/kT), \tag{9.30a}$$

thus implying that $S_{\mathrm{gb}}^{\mathrm{eff}}$ decreases rapidly as V_{dgb} becomes smaller. For example, Seager assumed the values $N_{\mathrm{D}} = 10^{16}$ cm^{-3}, $\sigma_p = 2.5 \times 10^{-16}$ cm^2, and $\Gamma = 0.3$ suns and his calculations showed that $S_{gb}^{eff} = 10^4$ cm sec^{-1} for $V_{dgb} = 0.1$ eV, whereas $S_{gb}^{eff} = 1.5 \times 10^6$ cm sec^{-1} for $V_{dgb} = 0.25$ eV.

(4) The value of S_{gb}^{eff} is insensitive to illumination level for small V_{dgb} ($\lesssim$0.25 eV). However, for larger V_{dgb} ($\gtrsim$0.25 eV) the value of S_{gb}^{eff} begins to decrease when the illumination level increases above ~0.3 suns.

(5) For small V_{dgb} and /or high illumination level ($\gtrsim$3 suns) the hole quasi-Fermi level becomes nearly constant within the depletion layer, as was assumed by Card and Yang and by Fossum and Lindholm.

Seager's calculations agree well with his measurements on neutron-transmutation doped n-Si over the range of intensities studied (dark to 0.3 suns).

9.5.2 Polycrystalline Solar Cell Theories

Because of the complexity of the three-dimensional transport problem in PX material, various simplifying assumptions have been made to obtain estimates for the relation of solar efficiency to PX grain size and other materials parameters. Several of these estimates are compared with experimental values in Fig. 9.41. All the theories discussed here assume that most of the grain-boundary volume is undepleted ($2W_{dgb} \ll \gamma$).

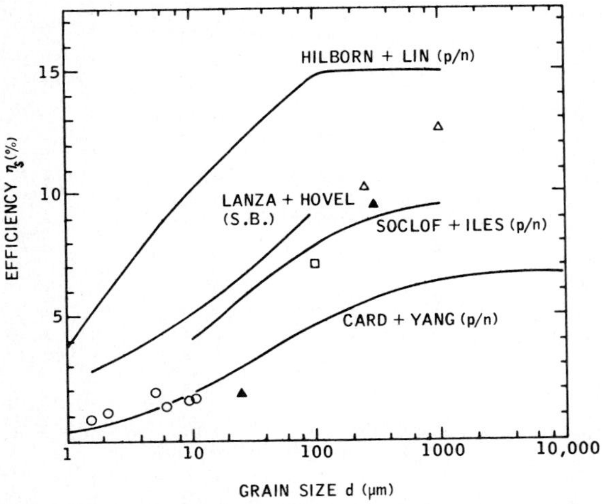

Fig. 9.41. Solar efficiency versus grain size for Si solar cells. The curves are theoretical results (Hilborn and Lin, 1976; Lanza and Hovel, 1977; Soclof and Isles, 1975; Card and Yang, 1977). The data points are experimental values obtaind by (○) Feldman *et al.* (1978), (▲) Chu *et al.* (1975) and Chu (1978), (△) Fischer and Pschunder (1976), and (□) Zook *et al.* (1977). The theoretical data of Card and Yang were estimated assuming ff = 0.70. [Figure from A. K. Ghosh, C. Fishman, and T. Feng, *J. Appl. Phys.* **51**, 446 (1980).]

Soclof and Iles (1975) assumed that $S_{gb} \to \infty$ and that any photogenerated carriers entering the grain-boundary depletion layer (of width W_{dgb}) were lost. Thus they subtracted the depletion layer widths from the grain size to obtain an effective grain size $\gamma_{eff} = \gamma - 2W_{dgb}$. An average diffusion length $\bar{L} = (D\tau_{fil})^{1/2}$ was then calculated from an equation such as Eq. (9.23) and used in the conventional expressions [such as Eq. (4.19)] to obtain J_{sc}. Their values of V_{oc} and ff were estimated from single-crystal cell data and the resulting relation of η_s to γ is shown in Fig. 9.41.

Lanza and Hovel (1977) used numerical methods to solve the three-dimensional transport equation within cylindrical grains. The value of J_L for the cell was taken as the sum of the currents generated in the quasi-neutral portion of the grains, where $S_{gb} \to \infty$, and in the *junction* depletion layer, where $S_{gb} = 0$ was assumed. Both Si and GaAs Schottky barrier solar cells were modeled with an AM1 illumination spectrum. An example of the resulting carrier density distribution is shown in Fig. 9.42. Lanza and Hovel assumed that forward biasing the junction did not affect the

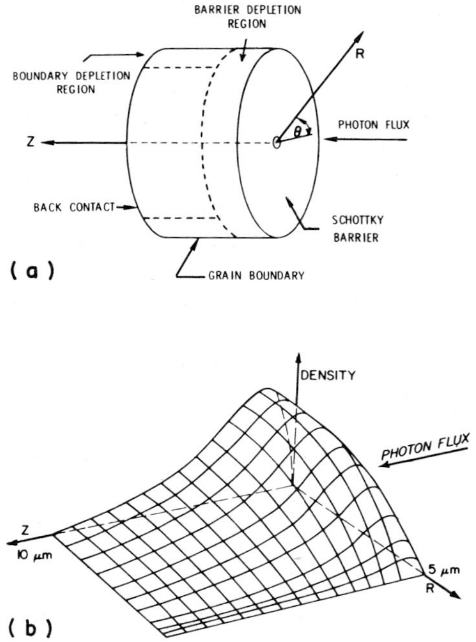

Fig. 9.42. (a) Grain model used for the calculations of Lanza and Hovel. (b) Excess carrier density as a function of position in a 10 μm thick grain of GaAs. The density goes to zero at the junction depletion layer edge. [After C. Lanza and H. J. Hovel, *IEEE Trans. Electron Devices* **ED-24**, 392 (1977). © 1977 IEEE.]

value of S_{gb}, although J_L was altered somewhat by the change in junction depletion layer width with bias voltage. The values of V_{oc} and ff for these Schottky barrier cells were calculated assuming that the junction properties were not affected by the presence of the grain boundaries. One of the Si results is shown in Fig. 9.41.†

Koliwad and Daud (1980; Koliwad *et al.*, 1979) used the empirical spatial variation of L_{eff} described in their earlier paper (Daud *et al.*, 1978) to integrate the contributions to the photogenerated current over each rectangular grain. The integration was performed over the solar spectrum as well, using equations such as Eqs. (4.19) and (4.26). This conceptual model enabled them to estimate the variation of J_{sc} with grain size. Subsequent measurements of J_{sc} for PX solar cells fabricated from Si with $\gamma >$ 100 μm gave good agreement with the theory. No attempt was made to model the dependence of V_{oc} and ff on L_{eff}, other than that introduced by changes in J_{sc}. From their experimental results Koliwad and Daud found the following percentage increases in the photovoltaic parameters in going from 100 to 1000 μm average grain size:

$$J_{sc}: \sim 8\%, \qquad V_{oc}: \sim 6\text{–}7\%, \qquad \text{ff}: \sim 12\text{–}13\%, \qquad \eta_s: \sim 30\text{–}32\%.$$

These relative changes among the photovoltaic parameters are fairly typical for PX Si cells. In particular the relative losses in J_{sc} and V_{oc} are similar in magnitude. These authors noted that the $A \simeq 2$ portion of the diode current increased strongly as grain size was decreased, but that this had a negligible effect on V_{oc} under AM1 conditions.

Card and Yang (1977) added several important refinements to the theory of PX solar cells. They allowed S_{gb} to be finite and considered the effect of illumination on the grain boundary potential barriers and the recombination there to calculate an average lifetime $\bar{\tau}$ for the PX Si absorber (as described in the previous section). Card and Yang's estimates for J_{sc} and V_{oc} are made by straightforward insertion of this $\bar{\tau}$ from Eq. (9.28) into the standard expressions: $\bar{L} = (D\bar{\tau})^{1/2}$, $J_{sc} \simeq qG\bar{L}$ with G, the generation rate, independent of position, and Eqs. (5.12) and (6.2a). Perhaps the most serious shortcoming of the Card and Yang theory is the assumption of uniform injection throughout the length of the grain. For this reason the theory gives good results for small grains (where activity is confined to thin layers, i.e., small x), but considerably underestimates η_s for larger grains (Fig. 9.41).

Ghosh *et al.*, (1980b) advanced a model that follows Card and Yang in averaging the grain boundary recombination centers as though they were spread throughout the volume of the semiconductor. However, these

† A similar analysis for p/n homojunction cells is given by Lanza and Hovel (1980).

authors do not consider the depletion layers at the grain boundaries or the effect of illumination on recombination there. For cubic grains of dimension γ on a side, they obtain

$$\bar{\tau} = \gamma/6\sigma v_{\text{th}} N_{\text{gb}}. \tag{9.31}$$

The corresponding empirical relationship $\bar{\tau} \simeq 5 \times 10^{-6}\ \gamma$ (sec), where γ is in cm, appears to fit a variety of Si experimental data for $0.2 < \gamma < 10^3\ \mu$m. Ghosh *et al.* (1980b) use this value of $\bar{\tau}$ with a semiempirical carrier mobility dependence on γ to obtain the J_L for the quasi-neutral absorber. The standard transport equation solutions [Eq. (4.22) with infinite back-surface recombination] are used. The contribution to J_L from the junction depletion layer ($\sim$8 mA cm^{-2}) is given by

$$J_{\text{L,dl}} \simeq q\Gamma_0[1 - \exp(-\alpha W_{\text{d}})]M, \tag{9.32}$$

where $M = 1$ for $\bar{\tau} \geqslant \tau_{\text{t}}$ and $M = \bar{\tau}/\tau_{\text{t}}$ for $\bar{\tau} < \tau_{\text{t}}$, and where τ_{t} is the transit time for carries crossing the depletion layer ($\tau_{\text{t}} \simeq W_{\text{d}}^2/V_{\text{d}}\mu$). The total J_{sc} is shown in Fig. 9.43a. To account for differences in J_{sc} versus γ from the preceding theory and experimental data at small γ, these authors assume a substantial shrinkage of the *junction* depletion layer under illumination [$W_{\text{d}} \simeq W_{\text{d}}(\Gamma)$] because of trapping in defect states there (Fig. 9.43b). The values of V_{oc} are calculated using the standard diode equations [e.g., Eqs. (5.12) and (5.27)] with the derived values of $\bar{\tau}$ from Eq. (9.31). The value of ff is obtained using the calculated values of J_{sc} and J_0. For small grain sizes recombination/generation dominates the J–V characteristics, whereas for larger γ, injection/diffusion controls and the appropriate J_0 values are calculated from $\bar{\tau}$. The results are shown in Fig. 9.44.

In their treatments of the PX solar cell Fossum and Lindholm (1980a,b) included the effects of grain boundaries within the *junction* depletion region. If $E_{\text{F}n}$ and $E_{\text{F}p}$ are nearly flat across the junction depletion layer, including the part occupied by the grain boundary, then Eq. (9.30) can be used directly to calculate the recombination current there. The situation involves a three-dimensional flow of carriers within W_{d}, but because of the high fields and diffusion currents present, the quasi-Fermi levels can be considered flat and then $(E_{\text{F}n} - E_{\text{F}p}) = qV$. The recombination rate [Eq. (9.29)]† must be multiplied by the "exposed" grain boundary area A_{gb} present in the depletion layer, i.e., for square columnar γ by γ grains $A_{\text{gb}} \simeq 4\gamma W_{\text{d}}$. The current density for each grain traversing the depletion layer is given by

$$J = I/\text{area} = (U_{\text{gb}}/2)(4\gamma W_{\text{d}})/\gamma^2 = q(S_n S_p)^{1/2}(W_{\text{d}} n_{\text{i}}/\gamma)\exp(qV/2kT). \tag{9.33}$$

† Since we are interested in recombination from only one side of the grain boundary here, U_{gb} is divided by 2.

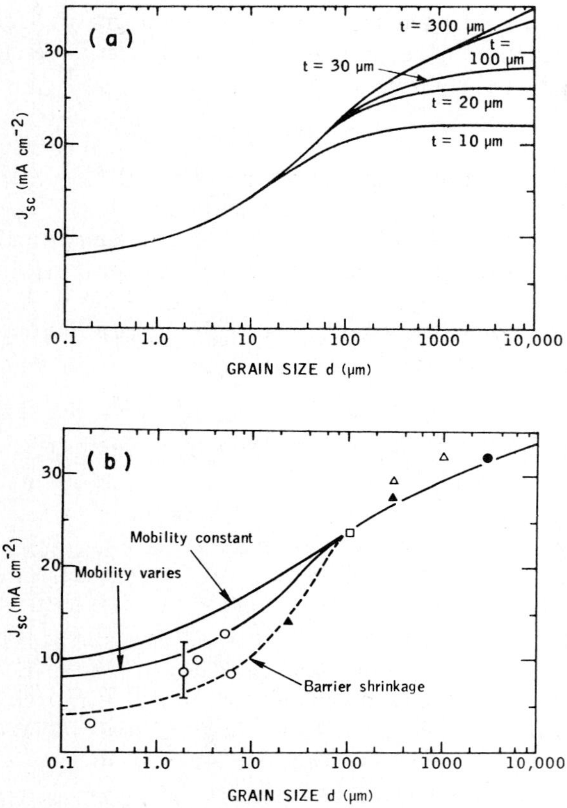

Fig. 9.43. (a) J_{sc} versus grain size for various PX Si film thicknesses. (b) Effect of barrier shrinkage and mobility assumptions on J_{sc}. Experimental point symbols are the same as for Fig. 9.41, except for (●) from the work cited here. All data are normalized to AM1, with some corrections for reflection and grid coverage losses where necessary. A decrease in W_d from 0.75 to 0.05 μm under illumination was assumed for $\gamma < 10$ μm. [After A. K. Ghosh, C. Fishman, and T. Feng, *J. Appl. Phys.* **51,** 446 (1980).]

Bulk recombination is neglected here. Fossum and Lindholm emphasized that the A factor of exactly 2 results from the assumptions of quasi-equilibrium and the steady-state conditions for maximum U_{gb} at the grain boundary (see also Lindholm and Fossum, 1981; Fossum *et al.*, 1980). Henry *et al.* (1977, 1978),† using different but equivalent reasoning, obtained the $\exp(qV/2kT)$ dependence for surface recombination at the periphery of $Al_xGa_{1-x}As$ heterojunctions (single crystal). Henry *et al.* used an "interaction length" rather than W_d in evaluating the exposed grain boundary area.

† The 1978 paper is highly recommended.

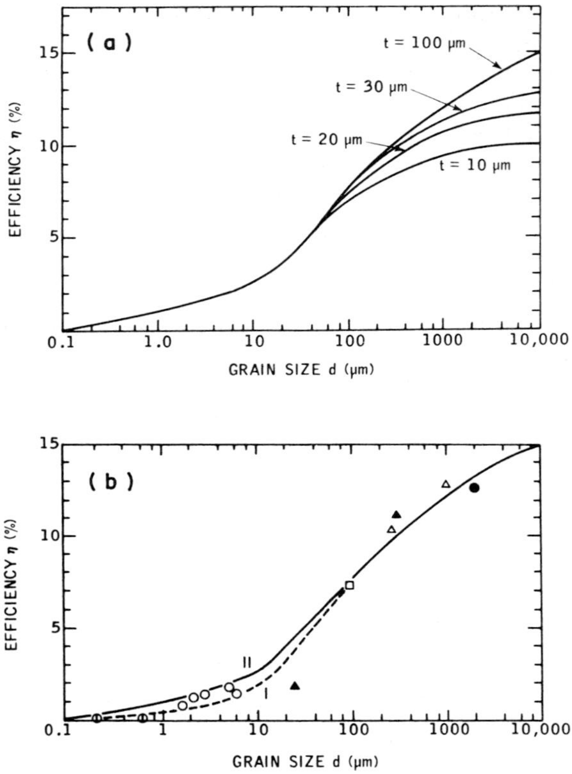

Fig. 9.44 Effect of grain size on Si solar efficiency. (a) For different PX Si layer thicknesses. (b) With (I) and without (II) junction barrier shrinkage effect. Point symbols are the same as in Fig. 9.43. [After A. K. Ghosh, C. Fishman, and T. Feng, *J. Appl. Phys.* **51,** 446 (1980).]

Comparison of the experimental data of Feldman *et al.* (1978) for PX p/n Si solar cells with Eq. (9.33) [using a value of $(S_nS_p)^{1/2} \simeq 10^6$ cm sec^{-1} and $W_d \simeq 0.2\ \mu$m] gives excellent agreement with the J_0 values of the $A = 2$ branches for the dark J–V curves of cells having $0.2 \leq \gamma \leq 6\ \mu$m.

Although these results are presented in terms of average grain size, it is evident that the important parameter is grain-boundary area. Thus in materials with a distribution of grain sizes, the smaller grains, with their greater surface-to-volume ratio, can considerably increase the value indicated by ''average'' grain size.

Makram-Ebeid (1979; Belovet *et al.*, 1979) did a computer calculation for the two-dimensional transport problem in a PX Si solar cell composed of planes of material separated by grain boundaries. They used a simplified form of the Shockley–Read–Hall expression to model the grain-boundary recombination velocity. Both *n*- and *p*-type layers were included with recombination at the front and back surfaces of the cell, in addition to the grain-boundary recombination. Temperature variation was accounted for and the calculation was further embellished by allowing the angle between the grain-boundary planes and the junction plane to vary. The effect of a Poisson distribution of distances between the grain boundary planes was also tested. The predicted spectral response $\eta_Q(\lambda)$ was in good agreement with data for PX Si cells.† In addition, the computer simulation predicted a variation of $\eta_Q(\lambda)$ with illumination intensity, in excellent qualitative agreement with experiment results that showed an 18% increase in η_Q at 0.875 μm when the cell was flooded with approximately 1 sun of AM1 illumination.

Other theoretical treatments of interest follow:

(1) Card and Hwang (1980) and E. S. Yang *et al.* (1981) focused on the recombinative effects of grain boundaries that are parallel to the junction plane, under dark and illuminated conditions. Such barriers also impede majority carrier flow.

(2) Kazmerski (1978, 1980a) expanded on the Card and Yang (1977) theory for direct band-gap materials, in particular $CuInSe_2$. Kazmerski (1980b) also presented a comprehensive overview of the status of PX solar cell research and development.

(3) Green (1978) used a novel three-dimensional equivalent circuit approach and focused on the role of grain boundaries in the junction depletion layer.

(4) Rothwarf (1976) used the assumption that all carriers generated closer to the cylindrical grain boundaries than to the junction are lost (i.e., the recombination cone concept of Fig. 9.2) to estimate the J_{sc} for the Cu_xS/CdS cell. Although admittedly an approximation, this theory does take account of the nonuniform generation rate [$d\Gamma/d\lambda$ and $\alpha(\lambda)$] in a real cell, which many of the other theories do not.

The three-dimensional transport problem in PX material is so complex as to practically defy complete solution.‡ Thus the transport models presented here are necessarily approximations and must be accompanied by some important qualifications:

(1) By their nature, averaging methods for $\bar{\tau}$ and $\bar{L}$ omit important aspects of the transport problem.

† Made by the Ribbon-Against-Drop process, in which molten Si is coated onto a pyrocarbon coating on a moving carbon ribbon (Belouet, 1980).

‡ A good overview of the problem is given by Lindholm and Fossum (1981).

(2) For most of the conditions encountered in solar cells the use of the filament lifetimes of Eq. (9.23) is questionable.

(3) The treatment of forward-bias injection into the QNRs and, in particular, the consideration of S_{gb} at the junction depletion layer edge under forward bias have been ignored in many of the models.

(4) The treatment of the parameters V_{oc} and ff (including the possible bias dependence of photogenerated current collection) has been generally superficial.

The effective grain-boundary recombination velocity $S_{gb}^{eff}(W_{dgb})$ has been considered for extreme cases: for the dark case and for the illuminated case where grain-boundary barriers are considerably reduced. For the illuminated case, S_{gb}^{eff} is either taken as infinite or calculated for the situation in which there is sufficient excitation so that the charging effects of the grain boundary states are effectively overwhelmed. A more rigorous treatment of the intermediate cases would be helpful, particularly for diagnostics of PX material.

There is reasonably good agreement between theory and experiment for PX silicon based cells, particularly with larger grain sizes. However, for cells based on direct band-gap materials, the V_{oc} and ff values are generally considerably smaller than predicted and more precise modeling appears to be necessary.

In spite of such criticisms, these theories have substantially increased our understanding of PX solar cells. For example, the dependence $S_{gb}^{eff} \sim [n_p(W_{dgb})]^{-1/2}$ from Eq. (9.30), suggests that considerable gains in solar efficiency might be obtained with solar concentration (using ratios of, say, 3 to 20). The combination of PX solar cells and low ratio concentrator systems has been largely ignored up to this time but might prove cost effective. Similarly the dependence $S_{gb}^{eff} \sim N_A^{1/2}$ suggests the use of lower doping densities in the absorber layer to improve J_{sc}, although a compromise with the variation of V_{oc} with N_A is required.

9.5.3 Other Considerations for PX Solar Cells

In this section we consider grain boundary core conduction and the effects of different doping level configurations and deposition sequences in heterojunctions.

Grain Boundary Sheet Conduction and Dislocation Banding Fraas (1978, 1980) and Fraas and Zonio (1980) have pointed out that, at least in some materials, the banding of the dense arrays of dislocation and dilatation states at the core of a grain boundary can lead to a significant decrease in the band gap there as well as parallel sheet conduction along the grain boundary. The effects of decreased band gap have been ob-

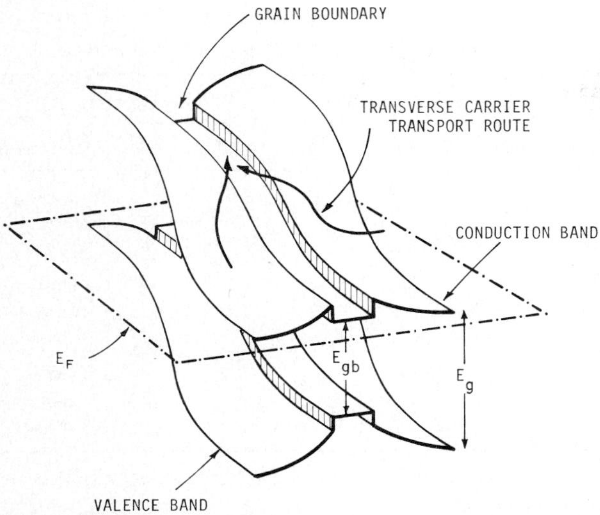

Fig. 9.45. Carrier transport in homojunction showing the effect of reduced band gap at grain boundary core.

served in Ge and CdS by optical absorption and electroluminescence. If such decreased core band gaps are present in solar cell, they should lead to the transport pathways shown in Fig. 9.45, and the activation energies in the diode J_0 term arising from grain boundary recombination should be significantly lowered.

Sheet conduction along grain boundaries has been observed in Ge (Mataré, 1971, p. 272) with values of 2000–10000 $\Omega\ \square^{-1}$ and in InSb (Shigeta *et al.*, 1976). Such a conductive grain boundary can be thought of qualitatively as another junction device in parallel with the junction comprising the bulk (interiors) of the grains, that shares the same (ohmic) contacts. Consider, for example, the case for which the Fermi level within the boundary material is pinned near the conduction band in the *n*-type side of the junction and near the valence band in the *p*-type side (Fig. 9.45). In this case a potential barrier of nearly E_{gb} still exists along the grain boundary to inhibit conduction. Grain-boundary conduction might be negligible in such a device. On the other hand, if the Fermi level were pinned close to the valence band on both sides of the junction (as could be the case for Ge, for which p^+ grain-boundary pinning is observed in both *n*- and *p*-type matrices), then no such barrier exists across the junction plane. In this case the potential barrier is at one of the cell contacts, which is more likely to be leaky than the bulk cell junction.† Thus conduction might

† The ohmic contact to the *n*-type side of the bulk junction acts as a tunneling contact to the grain boundary material in this case.

Table 9.III

Possible Grain Boundary Doping Configurations

	Absorber[a]		Window[a]		
Configuration	QNR	GB	QNR	GB	Comment
1	p	p^+	n	n^+	Low V_{oc}
2				n^-	Best
3				p	Low V_{oc}
4	p	p^-	n	n^+	Good
5				n^-	Good
6				p	Low V_{oc}
7	p	n	n	n^+	Low J_{sc}
8				n^-	Low J_{sc}
9				p	Low J_{sc}

[a] QNR is the grain interior quasi-neutral region; GB is the grain boundary.

occur along the grain boundary material between ohmic contacts on either side of the device giving a prohibitively low shunting resistance. Unfortunately, it is difficult to separate shunt resistance R_p from diode factor $A \simeq 2$ effects for forward $J-V$ characteristics, even in single-crystal cells, and grain boundary sheet resistance data for Si and other solar cell materials are difficult to obtain from existing data.† These effects have also been considered in a more general way by Shockley (1949, Sections 4.6 and 4.7).

Doping Configurations for Heterojunction Cells. Fraas (1978) also considered the effects of intentionally controlled grain boundary doping (with respect to grain interior values) as a means of optimizing photovoltaic parameters. Assuming that the alternatives of Fig. 9.21 are available, there are nine possibilities as shown in Table 9.III. A depleted or inverted grain boundary region (p^- or n) in a p-type absorber attracts minority carriers and increases S_{gb}^{eff}. In contrast, the presence of a p^+ grain boundary barrier could act much like a back-surface field in a conventional Si solar cell to decrease S_{gb}^{eff} and increase J_{sc}. Configurations 7, 8, and 9 are thus eliminated because of potentially high recombinaton losses. Configurations 1, 3, and 6 are eliminated because of potentially high J_0 (and thus low

† To our knowledge such grain boundary sheet conductance has not been observed unambiguously in Si or GaAs. However the work of Ghandi *et al.* (1979) suggests that grain boundaries in n-GaAs can be diffusion-doped to detrimental levels during CVD deposition on n^+-GaAs substrates.

V_{oc}) because of tunneling or low barrier height at the junction. This leaves configurations 2, 4, and 5 as promising candidates. Configuration 2, the closest to the optimum, is shown in Fig. 9.46. Since grain boundary diffusion of dopants is generally many orders of magnitude faster (Kamins, 1972) (and with lower activation energies) than bulk diffusion, the possibility of preferential doping of grain boundaries is certainly feasible. There are still little experimental data on the effect of preferential doping on S_{gb}, however.

Deposition Configurations At the onset of deposition of a layer, especially on an amorphous substrate, the material tends to be more disordered, becoming increasingly like a single crystal as the film thickness increases, as shown in Fig. 9.13 and discussed in Section 9.2.2. Since we wish most of the photogeneration to take place in the most well-ordered region, we are left with two possibilities shown in Fig. 9.47. The relative independence of the photovoltaic parameters on the properties of the window layer shown in some heterojunctions (e.g., large lattice mismatch in the ZnO/CdTe and ITO/Si cells) also reinforces this conclusion. Other factors aside, ease of cleaning and hermetic sealing argue for configuration b, whereas problems with current collection at the glass/window layer interface in b might favor configuration a.

Measurements in PX Material Layers Measurement of the electronic properties of PX semiconductor layers is complex because of the pres-

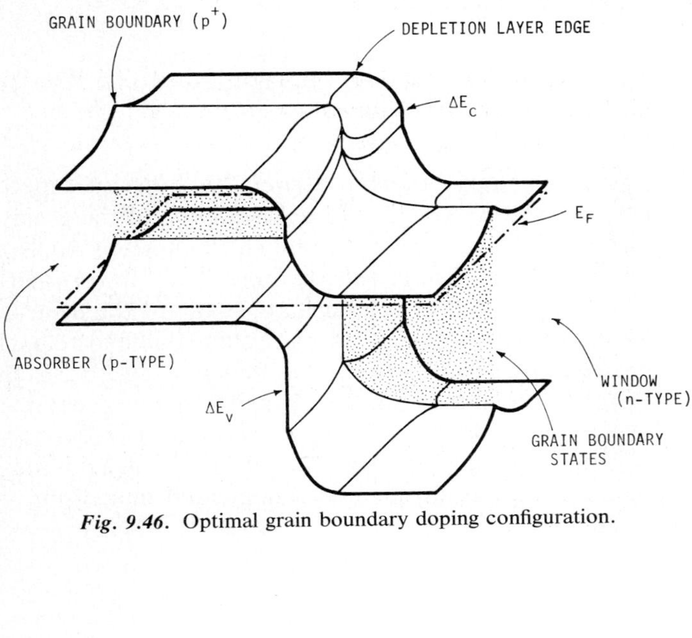

Fig. 9.46. Optimal grain boundary doping configuration.

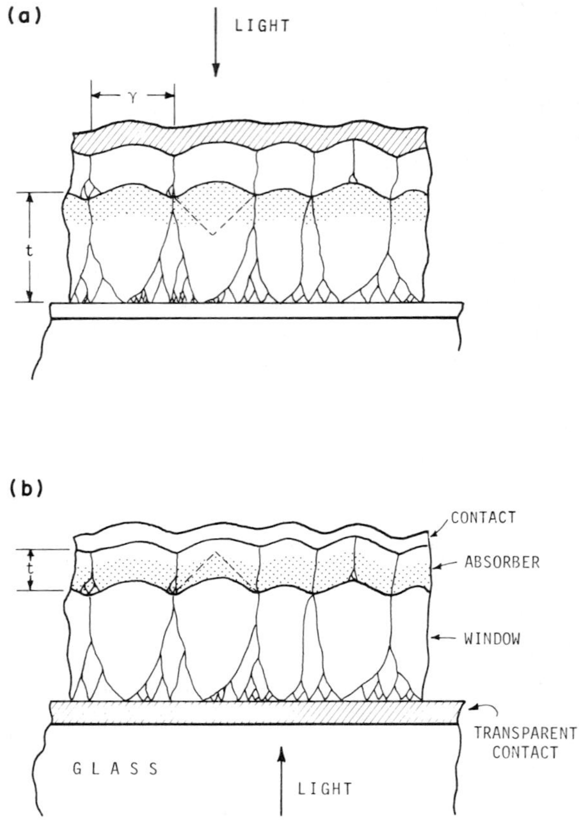

Fig. 9.47. Two possible configurations for heterojunction PX solar cells. Dot-shaded area is the principal photogeneration region of thickness $\sim\alpha^{-1}$.

ence of the grain boundary potential barriers, with height depending on doping density and the possibility of total depletion of the grains. Some measurements can be made by conventional techniques such as four-point resistance probing, Hall effect, thermoelectric power, and capacitance, but to be meaningful for solar cell applications, the measurements must be made under illumination of intensity comparable to actual use in a solar cell. For example, grain-boundary-dominated series resistance values that are prohibitive in the dark can change to quite acceptable values under AM1 conditions. Scanning EBIC is a powerful tool for assessing the effects of grain boundary structure (Fig. 9.37), but again it must be interpreted with care unless done under illuminated conditions. A large number of measurement techniques for PX materials are discussed in a 1979 SERI workshop.†

9.5.4 Passivation of Grain Boundary Recombination

There appear to be three main ways in which grain boundaries may be modified to reduce the losses there: (1) neutralization of the electrical activity of the grain boundary states by the addition of selected impurities, (2) preferentially "capping" the intersection of the grain boundary and the junction with an insulator, and (3) making the grain boundary into part of the junction by preferential diffusion doping of the grain boundary to the opposite type (e.g., an *n*-type boundary in a *p*-type absorber), thus producing an analog to the vertical multijunction cell (Fig. 7.24).

Neutralization of grain boundary states has been accomplished by a number of workers. Seager and Ginley (1979; Seager *et al.*, 1980) used anneals in atomic H to increase the conductivity across grain boundaries in Si bicrystals (Seager and Ginley, 1979), effectively removing all measurable boundary potential barriers (measured at 300°K), to a depth of 0.2–0.5 mm. Modest improvements in J_{sc}, V_{oc}, and ff in PX Si cells were indicated by preliminary experiments ($\gamma = 0.1–1$ mm). These p/n junction cells were fabricated by conventional diffusion processing and annealed in atomic H after contacting. Other passivation experiments have been done using atomic H (Robinson and D'Aiello, 1981; Makino *et al.*, 1980) and using Li (Miller and Orr, 1980; Young *et al.*, 1981).

Daud and Koliwad (1978) found that the addition of Cu apparently partially reduced grain boundary losses to increase η_s in PX Si solar cells. The Cu was introduced by diffusion from the surface in Wacker-cast *p*-Si and during crystal growth of CZ material, before junction formation by diffusion of P. Although the Cu substantially increased the η_s of small-grain material ($\bar{\gamma} \simeq 0.1$ mm, $\Delta P_m \simeq +15\%$), the opposite effect was noted for large-grain material ($\bar{\gamma} \simeq 1.0$ mm, $\Delta P_m \simeq -18\%$). The greatest effect of the addition of Cu was on the values of ff and V_{oc}. In particular, the low-voltage, $A \simeq 2$ branch of the dark ln J versus V curve was most affected, as would be expected from the theory of Fossum and Lindholm.

Less than a monolayer of Ru on *n*-GaAs has been shown by Nelson *et al.* (1980) to decrease surface recombination velocity by at least an order of magnitude. This might be expected to have similar effects at grain boundaries in this material.

† Photovoltaic Material and Device Measurements Workshop, Focus on Polycrystalline Thin Film Cells, SERI/TP-49-185, Arlington, Virginia (1979). Most of the articles have been reprinted in *Solar Cells*, **1,** Nos. 2 and 3 (1980).

The approach of electrically isolating the grain boundaries at the junction with an insulator was followed by Ghandi *et al.* (1979) for the Au/n-GaAs/n^+-GaAs/Mo Schottky cell. A thick oxide was formed on the surface of the n-GaAs by selective anodization and then judiciously etched with HCl to provide a thin oxide layer over the centers of the grains (0–20 Å), but to leave a thick layer ($\approx$ 100 Å) over the grain boundaries ("Ghandi caps"). The Au Schottky barrier was then deposited over all. Without the oxide caps the grain boundaries behave as n^+ material (possibly by diffusion from n^+-GaAs layer during CVD), effectively shorting the cell. Leakage currents are reduced by factors of 10^5–10^6 by the oxide passivation treatment.

The third approach to overcoming the effects of polycrystallinity in a solar cell is to utilize the boundaries of columnar grains as part of the junction, in effect producing a vertical multijunction cell (Fig. 7.24). In the approach by DiStefano and Cuomo (1977), P dopant is preferentially introduced into the grain boundaries of p-type silicon by a low-temperature diffusion process. A subsequent high-temperature diffusion forms a highly n-doped, convoluted skin covering a portion of each grain. The resulting junction around each grain collects carriers that might otherwise recombine at the grain boundaries. The grain boundary still intercepts the junction, although the intersection is no longer in a strongly illuminated region. The process must be used judiciously, of course, to avoid shorting to the back contact. The PX thin film Cu_xS/CdS cell (Fig. 10.1a) has this geometry.

9.6 EPILOGUE

Polycrystalline thin-film solar cells have great potential for large-scale use provided that adequate solar efficiency and stability can be achieved. Cells based on large-grain PX Si† have demonstrated that high η_s and the presence of grain boundaries are not mutually exclusive, as was commonly believed earlier. However, since this material must be sliced, it does not offer significant cost advantages. The real advantage of PX material lies in utilization of high throughput, thin-film deposition technology and the reduction in thickness of the active volumes of the cell by transfering the role of mechanical support to a cheap substrate material.

For large-grained PX Si cells, there is a small loss in J_{sc}, ff, and V_{oc}

† For example, Wacker SILSO™, HEM-cast silicon.

with respect to single-crystal cells. Cells based on direct band-gap PX materials have been shown to maintain their J_{sc} despite small grain size. However, for all small-grained materials there is a considerable loss in V_{oc} and ff, and therein lies the major challenge in increasing η_s for these devices.

The success of PX thin-film solar cells depends largely on reducing losses at the grain boundaries. This can be done by (1) reducing their density by modificaton during and/or after deposition to increase grain size; (2) passivation of their effect by removing or deactivating the recombination pathways there; and (3) use of materials in which the grain boundaries appear to be inherently inactive.

The high quantum efficiencies observed in films of Cu_xS and $CuInSe_2$ with small grain sizes and/or thicknesses suggests that grain boundaries in these materials are inherently inactive, at least with respect to generation of J_{sc}. However, their interfaces with heterojunction partners are still lossy, reducing the obtainable V_{oc} and ff. These materials may offer important clues as to the mechanisms involved in grain-boundary passivation in other semiconductors and at heterojunction interfaces.

GENERAL REFERENCES

Anderson, J. C. (1970). Conduction in thin semiconductor films. *Adv. Phys.* **19,** 311.

Bube, R. H. (1975). Electronic transport in polycrystalline films. *Annu. Rev. Mater. Sci.* **5,** 201.

Chopra, K. L. (1969). "Thin Film Phenomena." McGraw-Hill, New York.

Coutts, T. J., ed. (1978). "Active and Passive Thin Film Devices." Academic Press, New York.

Cullen, G. W., and Wang, C. C., eds. (1978). "Heteroepitaxial Semiconductors for Electronic Devices." Springer-Verlag, Berlin and New York.

Greene, R. F. (1964). Surface transport. *Surf Sci* **2,** 101.

Kazmerski, L. L., ed. (1980). "Polycrystalline and Amorphous Thin Film Devices." Academic Press, New York.

Leamy, H. J., Pike, G. E., and Seager, C. H., eds. (1982). "Grain Boundaries in Semiconductors" (*Proc. Mater. Res. Soc. Meet., Boston, 1981*). North-Holland Publ., Amsterdam.

Maissel, L. I., and Glang, R., eds. (1970). "Handbook of Thin Film Technology." McGraw-Hill, New York. An excellent resource book.

Milstein, J. B., Tsuo, Y. S., Hardy, R. W., Surek, T. (1981). The influence of grain boundaries on solar cell performance. *Proc. 15th Photovoltaic Specialists Conf.*, p. 1399. A good overview for Si devices.

Poate, J. M., Tu, K. N., and Mayer, J. W., eds. (1978). "Thin Films—Interdiffusion and Reactions." Wiley (Interscience), New York.

Smith, R. A. (1953). Infrared photoconductors. *Adv. Phys.* **2,** 321.

Surek, T., Ariotedjo, A. P., Cheek, G. C., Hardy, R. W., Milstein, J. B., and Tsuo, Y. S.

(1981). Thin-film polycrystalline silicon solar cells: Progress and problems. *Proc. 15th IEEE Photovoltaic Specialists Conf.*, p. 1251. An overview of growth methods and results.

REFERENCES

Aleksandrov, L. N. (1972). Structure and properties of transition layers formed in the epitaxy process. *Phys. Status Solidi A* **11,** 9.

Aleksandrov, L. N. (1979). Epitaxial interfaces between semiconductors. *Surf. Sci.* **86,** 144.

Amith, A. (1978). *J. Vac. Sci. Technol.* **15,** 353.

Anderson, J. C. (1970). *Adv. Phys.* **19,** 311.

Armstrong, R. W., Taylor, M. E., Storti, G. M., and Johnson, S. M. (1980). *Proc. 14th IEEE Photovoltaic Specialists Conf.*, p. 196.

Aspnes, D. E. (1976), "Optical Properties of Solids—New Developments" (B. O. Seriphin, ed.), p. 799. North-Holland Publ., Amsterdam.

Baccarani, G., Ricco, B., and Spadini, G. (1978). *J. Appl. Phys.* **49,** 5565.

Bagley, B. G., Aspnes, D. E., and Mogab, C. J. (1979). *Bull. Am. Phys. Soc.* **24,** 363.

Bashara, N. M., and Assam, S. C., eds. (1976). "Ellipsometry," p. 37. North-Holland Publ., Amsterdam.

Belouet, C. (1980). *J. Cryst. Growth* **50,** 279.

Belouet, C., Fabre, E., Makram-Edeid, S. M., Phuoc, N.-T., and Texier, C. (1979). *Proc. EC Photovoltaic Sol. Energy Conf., 2nd, Berlin* p. 114. D. Reidel Publ., Dordrecht, Netherlands.

Bloem, J., Giling, L. J., Graef, M. W. M., and de Moor, H. H. C. (1979). *Proc. EC Photovoltaic Sol. Energy Conf., 2nd, Berlin,* p. 759. D. Reidel Publ., Dordrecht, Netherlands.

Blount, G. H., Bube, R. H., and Robinson, A. L. (1970). *J. Appl. Phys.* **41,** 2190.

Brodie, D. E., Singh, R., Morgan, J. H., Leslie, J. D., Moore, C. J., and Dixon, A. E. (1980). *Proc. 14th IEEE Photovoltaic Specialists Conf.*, p. 468.

Bube, R. H. (1975). Electronic Transport in polycrystalline films. *Annu. Rev. Mater. Sci.* **5,** 201.

Bujatti, M., and Marcelja, F. (1972). *Thin Solid Films* **11,** 249.

Card, H. C., and Hwang, W. (1980). *IEEE Trans. Electron Devices* **ED-27,** 700.

Card, H. C., and Yang, E. S. (1977). *IEEE Trans. Electron Devices* **ED-24,** 397.

Casey, H. C., Jr., and Panish, M. B. (1978). "Heterostructure Lasers, Part B." Academic Press, New York.

Chopra, K. L., and Kahn, I. H. (1967). *Surf. Sci.* **6,** 33.

Chu, T. L. (1978). "SERI Semi-Annual Silicon Project Review Meeting." Unpublished.

Chu, T. L. Mollenkopf, H. C., Singh, K. N., Chu, S. S., and Wu, I. C. (1975). *Proc. 11th IEEE Photovoltaic Specialists Conf.*, p. 303.

Clark, A. H. (1980). Optical properties of polycrystalline semiconductor films. *In* "Polycrystalline and Amorphous Thin Films and Devices" (L. L. Kazmerski, ed.), p. 135. Academic Press, New York.

Cohen, M. J., Paul, M. D., Miller, D. L., Waldrop, J. R., and Harris, J. S., Jr. (1980). *J. Vac. Sci. Technol.* **17,** 899.

Cowher, M. E., and Sedgwick, T. O. (1972). *J. Electrochem. Soc.* **119,** 1565.

Daud, T., and Koliwad, K. M. (1978). *Proc. 13th IEEE Photovoltaic Specialists Conf.*, p. 503.

Daud, T., Koliwad, K. M., and Allen, F. G. (1978). *Appl. Phys. Lett.* **33,** 1009.

Dexter, D. L., and Seitz, F. (1952). *Phys. Rev.* **86,** 964.

DiStefano, T. H., and Cuomo, J. J. (1977). *Appl. Phys. Lett.* **30,** 351.

Dow, J. D., and Redfield, D. (1972). *Phys. Rev. B* **5,** 594.

Dresner, J., and Stringfellow, G. B. (1968). *J. Phys. Chem. Solids* **29,** 303.

Elbaum, C. (1974). *Phys. Rev. Lett.* **32,** 376.

Fagen, E. A. (1979). *J. Appl. Phys.* **50,** 6505.

Feldman, C., Blum, N. A., Charles, H. K., Jr., and Satkiewicz, F. G. (1978). *J. Electron. Mater.* **7,** 309.

Filinsky, I. (1972). *Phys. Status. Solidi B* **49,** 577.

Fischer, H., and Pschunder, W. (1976). *Proc. 12th IEEE Photovoltaic Specialists Conf.,* p. 86.

Fossum, J. G., and Lindholm, F. A. (1980a). *IEEE Trans. Electron Devices* **ED-27,** 692.

Fossum, J. G., and Lindholm, F. A. (1980b). *IEEE Electron Device Lett.* **ED-1,** 267.

Fossum, J. G., Neugroschel, A., Lindholm, F. A., and Mazer, J. A. (1980). *Proc. 14th IEEE Photovoltaic Specialists Conf.,* p. 184.

Fraas, L. M. (1978). *J. Appl. Phys.* **49,** 871.

Fraas, L. M. (1980). *Proc. 14th IEEE Photovoltaic Specialists Conf.,* p. 1316.

Fraas, L. M., and Zonio, K. (1980). The electronic structure of grain boundaries in polycrystalline semiconductor thin films. *In* "Polycrystalline and Amorphous Thin Films and Devices" (L. L. Kazmerski, ed.), p. 153. Academic Press, New York.

Fraas, L. M., Bleha, W,. P., and Braatz, P. (1975). *J. Appl. Phys.* **46,** 491.

Fripp, A. L., Jr. (1975a). *IEEE Trans. Parts, Hybrids Packag.* **PHP-11,** 239.

Fripp, A. L., Jr. (1975b). *J. Appl. Phys.* **46,** 1240.

Ghandi, S. K., Borrego, J. M., Reep, D., Hsu, Y.-S., and Pande, K. P. (1979). *Appl. Phys. Lett.* **34,** 699.

Ghosh, A. K., Rose, A., Maruska, H. P., Eustace, D., and Feng, T. (1980a). *Appl. Phys. Lett.* **37,** 544.

Ghosh, A. K., Fishman, C., and Feng, T. (1980b). *J. Appl. Phys.* **51,** 446.

Gilles, J., and Van Cakenberghe, J. (1958). *Nature (London)* **182,** 862.

Gold, R. B. (1980). Ph. D. Thesis, Stanford Univ., Stanford Calif.

Graef, M. W. M., Giling, L. J., and Bloem, J. (1977). *J. Appl. Phys.* **48,** 3937.

Graef, M. W. M., Bloem, J., Giling, L. J., Monkowski, J. R., and Maes, J. W. C. (1979). *Proc. EC Photovoltaic Sol. Energy Conf., 2nd, Berlin,* p. 65. D. Reidel Publ., Dordrecht, Netherlands.

Green, M. A. (1978). *Solid-State Electron.* **21,** 1139.

Guenther, R. G. (1966). "The Use of Thin Films in Physical Investigations" (J. C. Anderson, ed.), p. 213. Academic Press, New York.

Hall, R. N. (1952). *Phys. Rev.* **87,** 387.

Hanoka, J. I. (1980). *Solar Cells* **1,** 123.

Hasson, G., Boos, J.-Y., Herbeuval, I., Biscondi, M., and Goux, C. (1972). *Surf. Sci.* **31,** 115.

Heavens, O. S. (1965). "Optical Properties of Thin Solid Films." Dover, New York.

Helmreich, D., and Seiter, H. (1979). *Proc. EC Photovoltaic Sol. Energy Conf., 2nd, Berlin,* p. 742. D. Reidel Publ., Dordrecht, Netherlands.

Henry, C. H., Logan, R. A., and Merrit, F. R. (1977). *Appl. Phys. Lett.* **31,** 454.

Henry, C. H., Logan, R. A., and Merrit, F. R. (1978). *J. Appl. Phys.* **49,** 3530.

Hilborn, R. B., Jr., and Lin, T. (1976). *Proc. National Workshop on Low Cost Solar Cells.* Unpublished.

Holt, D. B. (1974). The growth and structure of epitaxial films and heterojunctions of II–VI compounds, *Thin Solid Films* **24,** 1.

Hwang, W., Panayotatos, P., Poon, E., and Card, H. C. (1981). *Proc. 15th IEEE Photovoltaic Specialists Conf.,* p. 871.

Inoue, N., Wilmsen, C. W., and Jones, K. A. (1981). *Solar Cells* **3,** 35.

Itoh, H., Saitoh, T., Nakamura, N., Matsubara, S., Warabisako, T., and Tokuyama, T. (1978). *J. Cryst. Growth* **45,** 446.

Juruzik, J., and Zdanowicz, L. (1980). *Thin Solid Films* **67,** 285.

Kahle, W., and Berger, H. (1970). *Phys. Status Solidi A* **2,** 717.

Kamieniecki, E. (1976). *J. Phys. C.* **9,** 1211.

Kamins, T. I. (1971). *J. Appl. Phys.* **42,** 4357.

Kamins, T. I., Manoliu, J., and Tucker, R. N. (1972). Diffusion of impurities in polycrystalline Si. *J. Appl. Phys.* **43,** 83.

Kazmerski, L. L. (1978). *Solid-State Electron.* **21,** 1545.

Kazmerski, L. L. (1980a). *Proc. 14th IEEE Photovoltaic Specialists Conf.,* p. 281.

Kazmerski, L. L. (1980b). Electrical properties of polycrystalline semiconductor thin films. *In* "Polycrystalline and Amorphous Thin Films and Devices" (L. L. Kazmerski, ed.), p. 59. Academic Press, New York.

Kazmerski, L. L., and Juang, Y. I. (1977). *J. Vac. Sci. Technol.* **14,** 769.

Kazmerski, L. L., Berry, W. B., and Allen, C. W. (1972). *J. Appl. Phys.* **43,** 3515.

Kazmerski, L. L., Ireland, P. J., and Sheldon, P. (1980). *Proc. 14th IEEE Photovoltaic Specialists Conf.,* p. 1311.

Kim, K. M., Cullen, G. W., Berkman, S., and Bell, A. E. (1977). "Silicon Ribbon Growth by the Inverted Stepanov Technique." Final Report, DOE/JPL-954465. (Available from National Technical Information Service, Springfield, Virginia.)

Koliwad, K. M., and Daud, T. (1980). *Proc. 14th IEEE Photovoltaic Specialists Conf.,* p. 1204.

Koliwad, K. M., Daud, T., and Liu, J. K. (1979). *Proc. EC Photovoltaic Sol. Energy Conf., 2nd, Berlin,* p. 710. D. Reidel Publ., Dordrecht, Netherlands.

Korsh, G. J., and Muller, R. S. (1978). *Solid-State Electron.* **21,** 1045.

Krajenbrink, F., Hoberg, P., Montano, H., and Zonio, K. (1981). *J. Appl. Phys.* **52,** 1524.

Kumagawa, M., Sinami, H., Terasaki, T., and Nishizawa, J. (1968). *Jpn. J. Appl. Phys.* **7,** 1332.

Lanza, C., and Hovel, H. J. (1977). *IEEE Trans. Electron Devices* **ED-24,** 392.

Lanza, C., and Hovel, H. J. (1980). *IEEE Trans. Electron Devices* **ED-27,** 2085.

Leamy, H. J., Pike, G. E., and Seager, C. H., eds. (1982). "Grain Boundaries in Semiconductors" (*Proc. Mater. Res. Soc. Meet., Boston, 1981*). North-Holland Publ., Amsterdam.

Lesk, I. A., Baghdadi, A., Gurtler, R. W., Ellis, R. J., Wise, J. A., and Coleman, M. G. (1976). *Proc. 12th IEEE Photovoltaic Specialists Conf.,* p. 173.

Lindholm, F. A., and Fossum, J. G. (1981). *Proc. 15th IEEE Photovoltaic Specialists Conf.,* p. 422.

Lopez-Otero, A. (1977). *J. Cryst. Growth* **42,** 157.

Lopez-Otero, A. (1978). *Thin Solid Films* **49,** 3.

Maissel, L. and Glang, R. (1970). "Handbook of Thin Film Technology." McGraw-Hill, New York.

Makino, T., Nakamura, H., and Nakashita, T. (1980). *J. Appl. Phys.* **51,** 5868.

Makram-Ebeid, S. M. (1979). *Proc. EC Photovoltaic Sol. Energy Conf., 2nd. Berlin,* p. 792. D. Reidel Publ., Dordrecht, Netherlands.

Manasevit, H. M. (1974). A survey of the heteroepitaxial growth of semiconductor films on insulating substrates. *J. Cryst. Growth* **22,** 125.

Martinez, J., and Piqueras, J. (1980). *Solid-State Electron.* **23,** 297.

Maruska, H. P., Ghosh, A. K., Rose, A., and Feng, T. (1980). *Appl. Phys. Lett.* **36,** 381.

Mataré, H. F. (1971). "Defect Electronics in Semiconductors." Wiley (Interscience), New York.

McKelvey, J. P. (1957). *Phys. Rev.* **106,** 910.
McKelvey, J. P. (1961). *J. Appl. Phys.* **32,** 442.
Merchant, P., and Elbaum, C. (1976). *Solid State Commun.* **20,** 775.
Merchant, P., and Elbaum, C. (1979). *Phys. Rev. B* **19,** 2992.
Miller, G. L., and Orr, W. A. (1980). *Appl. Phys. Lett.* **37,** 1100.
Milnes, A. G., and Feucht, D. L. (1976). *Proc. 12th IEEE Photovoltaic Specialists Conf.*, p. 997.
Morrison, S. R. (1956). *Phys. Rev.* **104,** 619.
Movchan, B. A., and Demchishin, A. V. (1969). *Fiz. Metal. Metalloved.* **28,** 653. (In Russian.)
Mueller, R. K. (1961). *J. Appl. Phys.* **32,** 635, 640.
Nelson, R. J., Williams, J. S., Leamy, H. J., Miller, B., Casey, H. C., Jr., Parkinson, B. A., and Heller, A. (1980). *Appl. Phys. Lett.* **36,** 76.
Pearson, G. L. (1949). *Phys. Rev.* **76,** 459.
Petritz, R. L. (1956). *Phys. Rev.* **104,** 1508.
Petritz, R. L. (1958). *Phys. Rev.* **110,** 1254.
Pike, G. E., and Seager, C. H. (1979). *J. Appl. Phys.* **50,** 3414.
Raghuran, A. C., and Bunshah, R. F. (1972). *J. Vac. Sci. Technol.* **9,** 1385.
Rai-Choudhury, P., and Hower, P. L. (1973). *J. Electrochem. Soc.* **120,** 1761.
Rasmanis, E. (1963). *Semicond. Prod. Solid State Technol.* June, p. 30.
Rasminis, E. (1964). U.S. Patent No. 3,139,361.
Redfield, D. (1981). *Proc. 15th IEEE Photovoltaic Specialists Conf.*, p. 1179.
Robinson, P. H., and D'Aiello, R. V. (1981). *Appl. Phys. Lett.* **39,** 63.
Romeo, N. (1981). *J. Cryst. Growth* **52,** 692.
Rothwarf, A. (1976). *Proc. 12th IEEE Photovoltaic Specialists Conf.*, p. 488.
Sato, H. (1972). Film growth. *Annu. Rev. Mater. Sci.* **2,** 217.
Schins, W. J. H., Bezemer, J., and Radelaar, S. (1979). *Proc. EC Photovoltaic Sol. Energy Conf. 2nd, Berlin,* p. 734. D. Reidel Publ., Dordrecht, Netherlands.
Scholz, H., and Kluckhow, R. (1967). *In* "Crystal Growth" (H. S. Peiser, ed.), Proc. Inter. Conf. Cryst. Growth, Boston, p. 475. Pergamon, Oxford.
Schwuttke, G. H. (1979). *Proc. EC Photovoltaic Sol. Energy Conf., 2nd, Berlin,* pp. 123, 130. D. Reidel Publ., Dordrecht, Netherlands.
Seager, C. H. (1981). *J. Appl. Phys.* **52,** 3960.
Seager, C. H., and Castner, T. G. (1978). *J. Appl. Phys.* **49,** 3879.
Seager, C. H., and Ginley, D. S. (1979). *Appl. Phys. Lett.* 34, 337.
Seager, C. H., and Pike, G. E. (1979). *Appl. Phys. Lett.* **35,** 709; errata, *Appl. Phys. Lett.* **37,** 251 (1980).
Seager, C. H., Ginley, D. S., and Zook, J. D. (1980). *Appl. Phys. Lett.* **36,** 831.
Seto, J. Y. (1975). *J. Appl. Phys.* **46,** 5247.
Shigeta, J., Kotera, N., and Oi, T. (1976). *J. Appl. Phys.* **47,** 621.
Shockley, W. (1949). *Bell Syst. Tech. J.*, **28,** 435.
Shockley, W. (1950). "Electrons and Holes in Semiconductors." Van Nostrand, Princeton, New Jersey.
Shockley, W., and Read, W. T. (1952). *Phys. Rev.* **87,** 835.
Siekhaus, W. (1976). Personal communication.
Soclof, S. I., and Iles, P. A. (1975). *Proc. 11th IEEE Photovoltaic Specialists Conf.*, p. 56.
Sosnowski, L. (1959). *J. Phys. Chem. Solids* **8,** 142.
Stratton, R. (1956). *Proc. Phys. Soc. London, Sect. B* **69,** 513.
Szczyrbowski, J., and Czapla, A. (1977). *Thin Solid Films* **46,** 127.
Tarng, M. L. (1978). *J. Appl. Phys.* **49,** 4069.

Taylor, W. E., Odell, N. H., and Fan, H. Y. (1952). *Phys. Rev.* **88,** 867.

Thornton, J. A. (1974). *J. Vac. Sci. Technol.* **11,** 666.

Tsai, M.-J., and Bube, R. H. (1978). *J. Appl. Phys.* **49,** 3397.

Vogel, F. L., Read, W. T., and Lovell, L. C. (1954). *Phys. Rev.* **94,** 1791.

Volger, J. (1950). *Phys. Rev.* **79,** 1023.

Wertheim, G. K., and Pearson, G. L. (1957). *Phys. Rev.* **107,** 694.

Wu, C., and Bube, R. H. (1974). *J. Appl. Phys.* **45,** 648.

Yang, E. S., Poon, E. K., Wu, C. M., Hwang, W., and Card, H. C. (1981). *IEEE Trans. Electron Devices,* **ED-28,** 1131.

Yang, J. J. J., Dapkus, P. D., Dupuis, R. D., and Yingling, R. D. (1980). *J. Appl. Phys.* **51,** 3795.

Yeh, Y. C. M., Ernest, F. P., and Stirn, R. J. (1978). *Proc. 13th IEEE Photovoltaic Specialists Conf.,* p. 966.

Young, R. T., Lu, M. C., Westbrook, R. D., and Jellison, G. E., Jr. (1981). *Appl. Phys. Lett.* **38,** 628.

Zook, J. D., Schuldt, S. B., Maciolek, R. B., and Heaps, J. D. (1977). "Silicon on Ceramic Process." Annual Report No. 2, ERDA/JPL/954356-77/3. Unpublished

Zook, J. D., Schuldt, S. B., Maciolek, R. B., and Heaps, J. D. (1978). *Proc. 13th IEEE Photovoltaic Specialists Conf.,* p. 472.

CHAPTER 10

THE Cu_xS/CdS SOLAR CELL

A Case History of an All-Thin-Film Cell

From the late 1950s, for a period of almost 20 years, the only all thin-film photovoltaic cell available was the heterojunction between *p*-type Cu_xS and *n*-type CdS. In spite of its complexities this system is one of the prominent contenders for terrestrial photovoltaic solar energy conversion. Curiously, it is commonly referred to as the "CdS cell" (although CdS plays no role in light absorption or carrier generation), or more exactly as the "Cu_2S/CdS cell" (although the value of x deviates from exactly 2.000 in almost all cells).

Some of the reasons for the complex nature of the Cu_xS/CdS cell are the following. (1) The heterojunction structure is inherently complex because of the interface between two materials with different electron affinities, band gaps, and crystal structures. (2) The lattice mismatch and interdiffusion of components causes defect states at or near the interface that strongly affect the junction properties. In particular there is Cu diffusion into the CdS adjacent to the interface. The presence of this Cu-doped CdS layer gives rise to complex photoeffects that may control the junction transport. (3) A variety of Cu_xS phases exist at room temperature, each with quite different photovoltaic properties. Changes from one to another may occur during cell fabrication and extended operation because of loss of Cu by interaction with the atmosphere or by diffusion into the CdS. (4) In cells made with thin-film CdS, there are the complex effects of interface roughness, grain boundaries, and random crystal orientation, the last two of which are largely undetermined.

In spite of these difficulties, the cell has a fabrication procedure that is almost irresistible in its simplicity. A thin film of CdS, deposited by some method such as vacuum evaporation or spray pyrolysis, is dipped for a brief period into a warm aqueous solution containing cuprous ions to form a topotaxial layer of Cu_xS by a replacement reaction. After a heat treatment the cell is essentially completed. Solar efficiencies of 4% are readily obtained, and, more recently, values higher than 9.1% have been cited for the Cu_xS/CdS cell (Bragagnolo *et al.*, 1980) and 10.2% for a $Cu_xS/Zn_yCd_{1-y}S$ variation (Hall *et al.*, 1981).

10.1 HISTORICAL INTRODUCTION

The historical development of the Cu_xS/CdS cell is rather colorful, filled with twists and turns as researchers attempted to ferret out the basic mechanisms involved. The first reported observation of a photovoltaic effect in CdS with Cu, Ag, or Au contacts was by Reynolds *et al.* (1954; Reynolds and Czyzak, 1954). Because the spectral response extended to photon energies appreciably smaller than the band gap of CdS, they attributed the effect to a two-step photoexcitation process through impurity levels in the CdS. They reported $V_{oc} = 0.45$ V and $J_{sc} = 15$ mA cm^{-2} under solar illumination, quite respectable values for this type of cell. Apparently the first all-thin-film cells were made by Nadjakov *et al.* (1954) using Al and Au electrodes.

Williams and Bube (1960) compared electroplated junctions of several metals on CdS and found that Cu gave the largest response. Because their junctions were not heat treated, they assumed that very little Cu had diffused into or reacted with the CdS and offered a model in terms of photoemission from the Cu metal into the CdS. They therefore contested the previous model that placed the principal source of absorption and carrier generation in the CdS. However, they did not realize that Cu_xS is formed even when Cu is plated on CdS at room temperature.

Fabricius (1962) and Grimmeiss and Memming (1962) joined with the earlier suggestion of Woods and Champion (1960) to explain the operation of the Cu/CdS cell by postulating p/n junction formation due either to impurity band hole conduction or to actual p-type CdS. The fact that the low photon energy response persisted even after the Cu contact was completely removed in a cyanide solution following heat treatment led Grimmeiss and Memming to the conclusion that the photoexcitation had to be taking place in the CdS. Once again the formation of Cu_xS was overlooked, leading to the mistaken hypothesis that CdS was the active material.

The quandary was partially resolved in 1963 when Cusano (1963) and Keating (1963) suggested a model in which the Cu_xS forms a heterojunction structure with the CdS; Cusano showed that similar results could be obtained for a Cu_xTe heterojunction structure with CdTe. Cusano made the correct assertion that previous workers had in reality formed thin layers of Cu_xS on CdS. However, Cusano continued to argue that the major active photoabsorption occurred in the CdS or the CdTe. As late as 1966 Duc Cuong and Blair (1966) thought that the important response might be due to excitation from Cu impurity states in the CdS.

By 1967, with the work of Hill and Kermidas (1967), Selle *et al.* (1967), and Potter and Schalla (1967), the final step was taken in assigning principal photoabsorption to the Cu_xS itself and not to the CdS. Whereas it is true that the spectral response of Cu impurity in CdS is quite similar to that of Cu_xS on CdS, the impurity absorption is far too small to account for observed quantum efficiencies that approached unity. Egorova (1968) and Mytton (1968) showed that heat treatment was an important step in making an efficient cell and proposed that photoexcited electrons were injected from the Cu_xS into the CdS.

Research on all-thin-film cells paralleled that of the general research just described. Cabannes (1958) reported photovoltaic effects using Cu on CdS films. Moss (1960) reported a Cu/CdS thin-film solar cell with solar efficiency of about 1%. Several laboratories involved in the space program investigated these thin-film cells because of their potential for radiation-resistant, large-area, light weight cells. This occurred several years before they became of interest for terrestrial applications. Shirland and Hietanen (1965; see also Shirland, 1966) and Shiozawa *et al.* (1968) played major roles in these investigations.

As a unified model for the Cu_xS/CdS heterojunction became generally accepted, it was agreed that (1) the optically active region for photogeneration of carriers was the Cu_xS, (2) electrons photoexcited in the Cu_xS are injected into the CdS for collection, (3) under operating conditions the forward-bias current flows through the junction by recombination and/or tunneling via interface states, rather than over the barrier into the Cu_xS conduction band, (4) diffusion of Cu into the CdS occurs during heat treatment and widens the depletion layer on the CdS side by compensation of CdS donors, (5) low-energy photon response is enhanced by high-energy photon illumination because of the effects on the junction width, (6) dark and light forward-bias J–V curves commonly cross in the heat-treated cell, and (7) long-term loss of sensitivity is caused by changes in the Cu_xS composition or, in some cases, by additional diffusion of Cu into the CdS.

The final area of lack of agreement centered on the mechanism controlling the injection of photoexcited electrons from the Cu_xS into the CdS and the injection of carriers from the CdS to the Cu_xS for forward bias.

Although a variety of mechanisms were proposed, three basic possibilities seemed most likely: (1) control by a series photoconducting layer of insulating Cu-doped CdS (Shiozawa *et al.*, 1968), (2) control by a small conduction band spike between the Cu_xS and the CdS (Lindquist and Bube, 1972a; Schock *et al.*, 1977), and (3) control by recombination (Rothwarf, 1979) or tunneling/recombination through interface states (Fahrenbruch and Bube, 1974). Lack of supporting evidence for the first two of these possibilities in subsequent research led to acceptance of some form of the third possibility.

Extensive development of the Cu_xS/CdS cell has been carried out at the Institute for Energy Conversion of the University of Delaware (Rothwarf and Barnett, 1977; Bragagnolo *et al.*, 1980). Figure 10.1a shows a scale view of a typical cell. Approximately 1 μm of Zn is electroplated onto a substrate of electroformed Cu about 25 μm thick. The Zn provides isolation of the CdS from the Cu substrate, makes an ohmic contact, and, after alloying with the Cu substrate, provides a surface with good optical reflection. CdS is vacuum evaporated from the compound contained in a graphite source at about 1100°C and is deposited on the substrate (held at about 220°C). Typical values for a CdS film are a thickness of 20 μm, a resistivity of 1–10 Ω cm (measured through the film), and grain sizes in the range of 1–5 μm. The CdS layer may then be etched for about 5 sec in HCl to provide a textured, low-reflectance front surface. It is then dipped into an aqueous cuprous ion solution at about 90°C for 5–10 sec. The Cu_xS layer thus formed is 1000–3000 Å thick, penetrating down the grain boundaries into the CdS up to a distance of 1–2 μm, as shown in Fig. 10.1b. The resistivity of the Cu_xS is generally 0.01–0.1 Ω cm. The contact to the Cu_xS is provided by a gold-plated copper grid. This grid, together with a protective plastic cover material is attached to the Cu_xS with epoxy, using a lamination process that involves heat treatment of the cell at about 200°C for 15 to 60 min. To bring the cell efficiency to its optimum requires subsequent additional heat treatment in vacuum at 170°C for about 16 hr, and in reducing (hydrogen or carbon monoxide) ambients in some cases.

The highly textured surface and grain boundary penetration produced by HCl etching and the wet dip process may partially explain the remarkably high quantum efficiency of the Cu_xS layer, considering its small local thickness. A large part of the light travels through the Cu_xS layer oblique to the local junction plane so that the light is presented with a large absorption thickness but electron–hole pairs are generated close to the electrical junction (Rothwarf and Barnett, 1977).†

Recently, the "dry" or planar process of junction formation has been

† Also see Section 7.4.4.

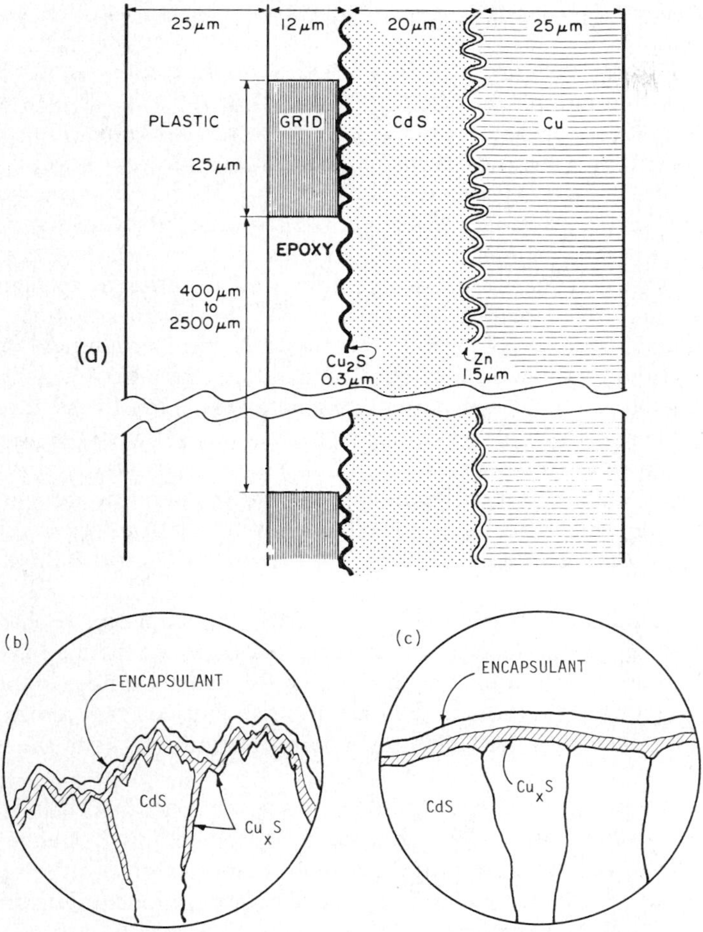

Fig. 10.1. (a) Schematic view of typical structure of the Cu_xS/CdS thin-film cell. [Redrawn from A. Rothwarf and A. M. Barnett, *IEEE Trans. Electron Devices,* **ED-24,** 381 (1977) © 1977 IEEE.] (b) Schematic view of Cu_xS film on texture (HCl) etched CdS. [Redrawn from R. B. Hall and J. D. Meakin, *Thin Solid Films* **63,** 203 (1979).] (c) Schematic view of Cu_xS film deposited by the planar (dry) process. [Redrawn from R. B. Hall and J. D. Meakin, Tech. Report IEC/PV/TR/79/6, Inst. Energy Convers., Univ. of Delaware, Newark (May 1979).]

used, in which CuCl is vacuum evaporated onto the CdS and subsequent mild heating causes the formation of the Cu_xS. The resulting interface, shown in Fig. 10.1c, is considerably more planar, resulting in less junction area and consequently a higher V_{oc}. However, J_{sc} values are lower than with the "wet" process, presumably because of lack of texturing.

10.2 PROPERTIES OF Cu_xS AND CdS

It is unikely that an inherently more complicated system could be chosen for thin-film photovoltaic cells than that of Cu_xS. The stable form of Cu_xS at room temperature is quite complex, containing 96 molecules per unit cell in the ideal Cu_2S form. At room temperature several phases are possible, some of which are shown in Fig. 10.2: (1) *chalcocite* ($x \simeq 1.995$–2.000) with an orthorhombic structure at room temperature transforming to hexagonal at 104°C; (2) *djurleite* ($x \simeq 1.96$) with an orthorhombic structure at room temperature transforming to tetragonal at 86–93°C and to cubic at 100°C; and (3) *digenite* ($x \simeq 1.8$) with a pseudo-cubic structure at room temperature transforming to cubic at 78°C. The photovoltaic properties of each of these forms of Cu_xS are quite different. These results are shown dramatically in a motion picture made by TeVelde and Dieleman (1973) using high-contrast television microphotography, in which reversible phase change patterns can be seen sweeping across the Cu_xS crystallites following minute temperature changes. The phase changes can also be observed by monitoring J_{sc}, V_{oc}, or junction capacitance during thermal cycling (Palz *et al.*, 1972; and Fahrenbruch, 1973).

In the topotaxial growth of Cu_xS on CdS, two Cu atoms replace a Cd atom, with the sulfur sublattice remaining relatively unchanged (TeVelde and Dieleman, 1973). Similarly, throughout the phase changes in the Cu_xS discussed earlier, it is thought that the sulfur sublattice again remains relatively stationary with the Cu atoms moving around it to form the various phases.

The kinetics of topotaxial Cu_xS growth on CdS layers have been examined by *in situ* optical absorbance measurements by Donaghey *et al.* (1980). These results were correlated with composition profiles and film structure as determined by galvanic electrochemical decomposition.

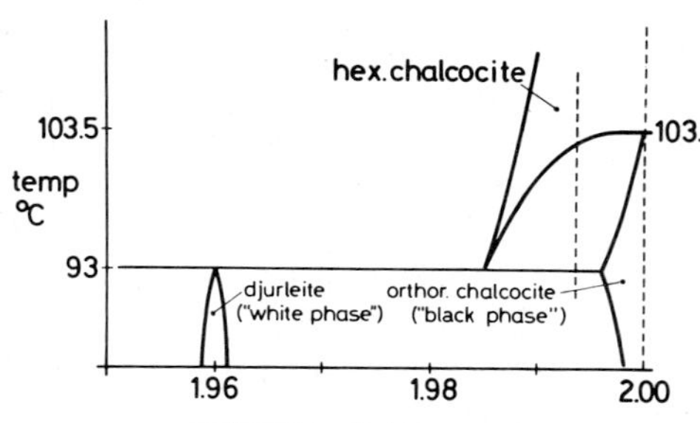

Fig. 10.2. Phase diagram for $Cu_{2-x}S$ near $x = 0$. [From J. Dieleman, *Int. NSF Workshop CdS Sol. Cells Other Abrupt Heterojunctions, Univ. of Delaware, Newark* p. 92 (1975).] Another, expanded phase diagram is given by R. W. Potter, II [*Econ. Geol.* **72,** 1524 (1977).]

Growth methods for Cu_xS include (1) the "wet" process mentioned earlier, where Cu ions from an aqueous solution replace the Cd atoms of CdS; (2) the "dry" method apparently first used by TeVelde (1975), in which CuCl is vacuum evaporated onto CdS, mild heating causes the solid-state ion exchange to occur, and finally the reaction product $CdCl_2$ is washed from the surface with alcohol; (3) deposition by reactive sputtering of Cu in H_2S and Ar atmosphere (Armantrout *et al.*, 1979; Jonath *et al.*, 1979); (4) an electrolytic method in which ion exchange from an aqueous Cu^+ ion solution is aided by the application of a small electric potential (Nakayama, 1969); (5) sulfurization of Cu in an H_2S and H_2 atmosphere at 200–400°C; (Shewchun *et al.*, 1975); and (6) spray pyrolysis, in which copper- and sulfur-containing solutions are sprayed on a heated CdS layer (Chamberlin and Skarman, 1966).

The stoichiometry and structural properties of Cu_xS films grown by the "wet" and "dry" processes are compared by Baron *et al.* (1978).

Chalcocite, Cu_2S, is a *p*-type, defect semiconductor with shallow Cu vacancy acceptors. Zn and Cd are donors in this material. Carrier densities observed usually range from 10^{18} to greater than 10^{21} cm^{-3}. Hall mobilities reported are usually low: 3–30 cm^2 V^{-1} sec^{-1} for single crystals and 2–7 cm^2 V^{-1} sec^{-1} for thin films (Stanley, 1975; Savelli and Bougnot, 1979; Bougnot *et al.*, 1975). Thus resistivities commonly range from 10^{-3} to 6×10^{-1} Ω cm for undoped material and to over 10^3 Ω cm for Cd-compensated material. Minority carrier diffusion lengths range from less than 0.1 μm to 0.6 μm in thin films (Gill and Bube, 1970a; Partain *et al.*, 1975). In addition, the material has a high ionic conductivity.

Figure 10.3 shows optical absorption spectra for three of the phases of Cu_xS, plus an additional $Cu_{1.9}S$ composition, in which the measured spectra have been corrected to remove contributions from free hole absorption in these degenerate *p*-type materials. It is evident that chalcocite is the most effective in light absorption. Related measurements on these materials indicate that, of all these possible Cu_xS phases, only chalcocite has an electron diffusion length greater than 50 Å; specific results for $Cu_{1.995}S$ were an electron diffusion length of 350 Å perpendicular to the *c*-axis and 900 Å at an angle of 35° to the *c*-axis.† It therefore appears that the most efficient Cu_xS phase for a photovoltaic cell is a chalcocite structure.

In order to check these considerations directly, TeVelde and Dieleman (1973) prepared a cell using the solid-state reaction between CuCl and CdS (the dry process) in such a way that regions with a particular local value of *x* could be identified by microabsorption spectral measure-

† A series of papers by B. J. Mulder, T. S. TeVelde, and J. Deileman is summarized in Dieleman (1975).

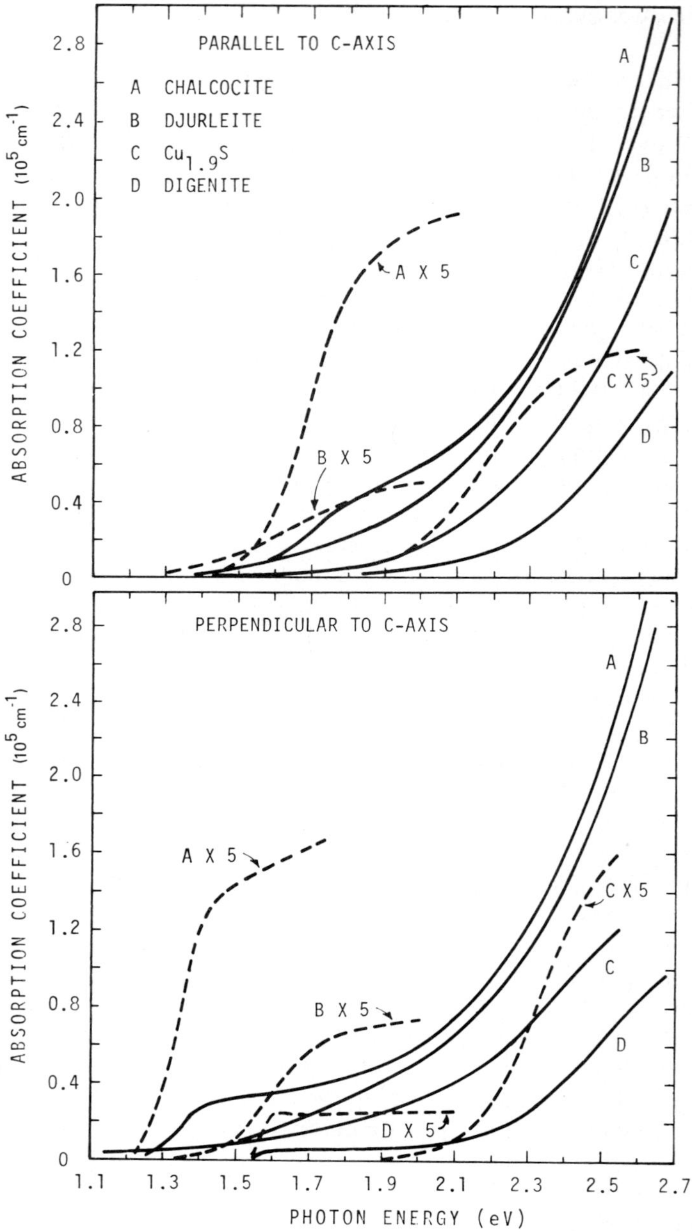

Fig. 10.3. Optical absorption spectra for four phases of Cu_xS for light polarized parallel and perpendicular to the *c*-axis respectively. The phases of Cu_xS are indicated in the figure. The solid curves show the sum of direct and indirect absorption, whereas the dashed curves show only the contribution to the absorption due to indirect transitions on an enlarged scale. [Redrawn from J. Dieleman, *Int. Workshop CdS Sol. Cells Other Abrupt Heterojunctions, Univ. of Delaware, Newark,* (K. W. Böer and J. D. Meakin, eds.), p. 92 (1975).] Note that Dieleman's data are slightly revised from those given by B. J. Mulder [*Phys. Stat. Sol. A* **13,** 79 (1972).]

ments, and then microphotovoltaic response measurements could be compared for different values of x. It was shown that the efficiency of a region corresponding to chalcocite was about an order of magnitude higher than that corresponding to djurleite; still poorer response was obtained for $Cu_{1.86}S$. Palz *et al.* (1972) found a similar relationship between stoichiometry and photovoltaic response. Changes in stoichiometry undoubtedly affect carrier mobility and recombination kinetics as well.

Thus the maximum efficiency in the Cu_xS/CdS system occurs for x very close to 2.00 (Rothwarf, 1979). Deviations from this value which may occur during the operating lifetime of a cell constitute degradation processes. Loss of Cu occurs primarily at the free surface of the Cu_xS where reaction with the air is possible, to form Cu_2O or $CuCO_3$. Loss of Cu can also occur at the CdS interface by diffusion of Cu into CdS under a concentration gradient. These can be reasonably rapid processes, compared to the desired 20-year operating lifetime of solar cells. Efforts to stabilize the Cu_xS/CdS cell against these degradation modes therefore require protection of the free surface against oxidation. Excess Cu diffusion into the CdS should also be minimized.

CdS has a direct band gap of 2.42 eV, giving it a transparent, sunny yellow appearance. It usually has the wurtzite hexagonal structure, although the zincblende cubic structure is found for preparation near room temperature. The material is polar with respect to the hexagonal *c*-axis and shows markedly greater chemical reactivity on the Cd-rich *a*-face than on the S-rich *b*-face (Miya, 1970).†

CdS is easily doped *n*-type with In, Sn, Al, Cl, or Br, all of which form shallow donors. Sulfur vacancies, induced by the presence of excess Cd during film or crystal growth, act as donors, and resistivities below 0.1 Ω cm can be achieved with no intentional impurity introduction. Deep acceptor states are formed by Cu, Ag, and Au impurities that readily compensate the native donors. Although many workers have attempted to prepare *p*-CdS, its existence has never been established beyond question; the tendency toward self-compensation is very strong in CdS.

Films of CdS have been deposited by perhaps a wider variety of methods than any other semiconductor material: vacuum evaporation, sputtering, reactive sputtering of Cd in H_2S, spray pyrolysis, screen printing, anodization (Moore *et al.*, 1978), vapor deposition, sedimentation followed by sintering, chemplating, electrolysis, and doctor blade, to name a few. CdS can be deposited by vacuum evaporation using the "three-temperature" method in which S and Cd evaporation sources are held at a

† For example, for the wet dipping process, the rate of Cu_xS growth is three times more rapid on the *a*-face (Fahrenbruch, 1973, p. 50).

suitable temperature (e.g., 110°C for S and 265°C for Cd) while the substrate is held at a third temperature (180–350°C). The vapor pressures of both Cd and S are many orders of magnitude lower over the compound CdS than over the elements, thus automatically controlling the stoichiometry of the deposited layer. Evaporation from the compound is usually used, since CdS evaporates congruently and the same automatic stoichiometry control obtains. In vacuum deposition, CdS usually prefers to grow with the hexagonal *c*-axis (1000) perpendicular to the substrate.

The electron mobility for CdS ranges from 100 to 400 $cm^2\ V^{-1}\ sec^{-1}$ at 300°K for single crystals. In thin films the mobility is commonly dominated by grain boundary potential barriers, and mobilities from well below 1 up to over 100 $cm^2\ V^{-1}\ sec^{-1}$ are commonly seen, depending strongly on the carrier density (see Chapter 9). Resistivities ranging from 0.001 to over 10^8 Ω cm can be achieved for films deposited on amorphous substrates.

Whereas vacuum evaporation of CdS followed by dipping to form the Cu_xS has been the popular method for fabrication of these solar cells, a number of other methods show promising simplicity and low-cost scale-up potential. A few of the many are noted here. Both CdS and Cu_xS films were vacuum evaporated by Vanderwel *et al.* (1980), giving an opportunity to study the properties of the Cu_xS layer separately. Cells with $\eta_s \simeq 6\%$ were obtained in preliminary experiments. Bordure *et al.* (1980) have produced cells by spray pyrolysis of CdS on ITO-coated Pyrex followed by dipping to produce the Cu_xS layer. These backwall cells were coated with ~100 Å of Cu and then annealed (Bogus and Mattes, 1972) to improve stoichiometry. Maximum η_s was 7.4%. Roderick (1980) used a similar backwall configuration with SnO_x coated float glass as a substrate for spray pyrolized CdS. Cu_xS was deposited by the "dry" process to yield $\eta_s = 7.3\%$ in research cells. Their pilot plant operation, geared to produce 2×10^6 ft^2 of panels per year, demonstrated a panel (3096 cm^2) efficiency of 2.6% in preliminary runs. Matsumoto *et al.* (1980) used screen printing to deposit the CdS layer with subsequent sintering. The Cu_xS layer was deposited electrochemically from a $CuSO_4$ solution. A maximum η_s of 8.9% (active area basis) was achieved.

10.3 PHOTOVOLTAIC PROCESSES AND THEORETICAL EFFICIENCY

An idealized band diagram of the Cu_xS/CdS heterojunction is shown in Fig. 10.4. Consideration of the best values available for electron affinities and band gaps yields a conduction band discontinuity of 0.2 eV, a valence band discontinuity of 1.40 eV, and a diffusion potential of $V_d =$

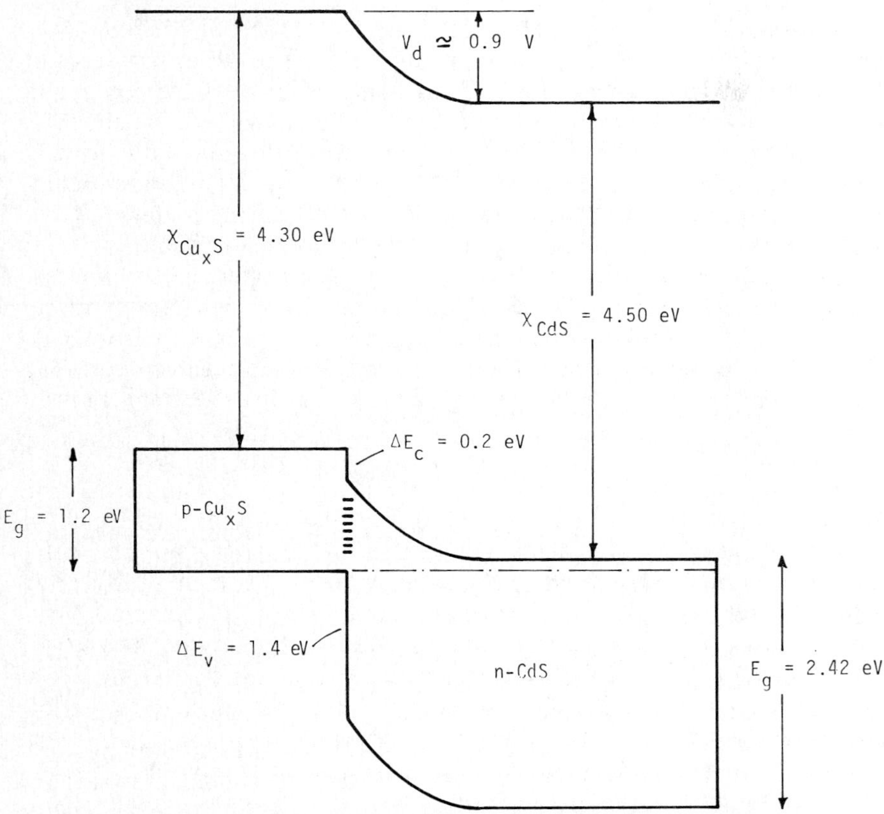

Fig. 10.4. Approximate energy band diagram of the Cu_xS/CdS heterojunction.

0.90 V. This assumes that the Fermi level lies at the valence band edge in the degenerate *p*-type Cu_xS and 0.1 eV below the conduction band edge in the *n*-type CdS.

In principle the cell could be operated in either a frontwall (light incident on the Cu_xS) or a backwall configuration (light incident on the CdS). With the structure shown in Fig. 10.1, of course, only the frontwall mode is possible. Rothwarf *et al.* (1975) have extensively modeled the photogenerated carrier collection for frontwall and backwall modes. In addition, they have studied the effect of an optical reflector placed on the rear surface of the cell, giving multiple passes for the illumination.

As with other heterojunctions, the quantum efficiency at each wavelength can be expressed as the product of two terms, representing, first, the efficiency for photogenerated carriers to reach the junction without re-

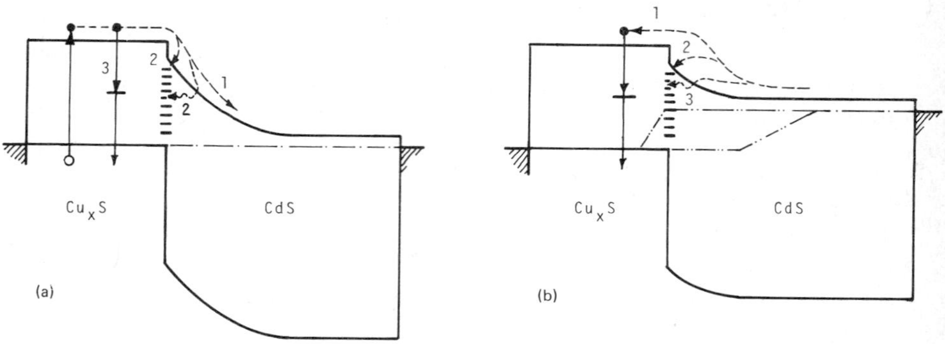

Fig. 10.5. (a) Mechanisms controlling the short-circuit current: 1, collection by the junction field; 2, recombination via interface states, possibly involving tunneling steps; and 3, bulk (and surface) recombination. (b) Mechanisms controlling the forward-bias diode current: 1, thermal excitation over the total barrier height; 2, thermally excited recombination through interface states; and 3, tunneling/recombination through interface states.

combination in the bulk, $g(\lambda)$, and, second, the probability for carriers to cross the junction and be collected without recombination through interface states, $h(V)$ (see Section 6.3.1). These various processes are indicated in Fig. 10.5a. Incident light with photon energy greater than 1.2 eV is absorbed in the Cu_xS, creating photoexcited electrons that diffuse to the junction interface. The electrons that reach the interface without recombining in the bulk have two possibilities: either (1) they cross into the CdS to be collected by the junction field or (2) they are captured by interface states and then recombine with holes in the Cu_xS at the interface, and hence are not collected. This latter process, represented by $h(V)$, can be expected to depend on bias voltage as well as the wavelength and intensity of the incident light which also affects the junction field.

For frontwall operation, most of the photons with energy greater than 1.2 eV in the solar spectrum are absorbed in the Cu_xS, and J_{sc} = 35 mA cm^{-2} is predicted, assuming no losses. In 1979 the best cells yielded about 22 mA cm^{-2} (total area basis, AM1.5, 87.9 mW cm^{-2}) (Hall and Meakin, 1979). For backwall operation only photons with energy between the band gap of CdS and that of Cu_xS are available, and J_{sc} is predicted to be 29 mA cm^{-2}, again assuming no losses.

An extensive loss analysis has guided the work at the Institute of Energy Conversion in optimizing the Cu_xS/CdS cell. Rothwarf and Barnett (1977) estimate that the total losses in J_{sc} for the frontwall cell (29% in this case) have the following fractional components†: 15% from recombina-

† The total and fractional losses are related by

$$\text{Total Loss} = 1 - \prod_n [1 - (\text{Fractional Loss})_n].$$

tion in the bulk Cu_xS, 5% from recombination at the interface between the Cu_xS and the CdS, 4% from grid shadowing, 5% from light absorption and reflection losses in the encapsulating layer, and 1% each for recombination losses at the front surface and at grain boundaries. The major area for improvement appears to be in the recombination loss in the bulk Cu_xS and future projections indicate that, for a frontwall cell, $J_{sc} = 26$ mA cm^{-2} is reasonable.

An important contribution ($\simeq 15\%$) to the total quantum efficiency of the frontwall cell is from light with 1.2 eV $< h\nu <$ 2.4 eV that passes through both the Cu_xS and the CdS and is reflected at the rear contact to obtain a second pass through the Cu_xS layer.

The three major mechanisms contributing to the diode current J_0 are shown in Fig. 10.5b: (1) thermally excited electrons in the CdS diffuse into the quasi-neutral region of the Cu_xS and recombine with holes, (2) thermally excited electrons in the CdS flow through interface states to recombine with holes at the interface in the Cu_xS, and (3) electrons in the CdS, with or without thermal assistance, tunnel into interface states with subsequent recombination with holes in the Cu_xS at the interface. An estimate of the first process, which is the standard Shockley diode current, gives a value of $J_{01} = 10^{-16}$ A cm^{-2} at 300°K, much smaller than any values observed. The second process, involving interfacial recombination, appears to dominate in good cells, with the diode curent I_{02} given by†

$$I_{02} = \mathscr{A}_j J_{02} = qS_i N_c^{CdS} \mathscr{A}_j \exp[-(E_g^{Cu_xS} - \Delta E_c)/kT]. \qquad (10.1)$$

Here S_i is the interface recombination velocity (with values between 3×10^5 and 5×10^6 cm sec^{-1}), N_c^{CdS} is the density of states in the conduction band of CdS (2×10^{18} cm^{-3} at 300°K), $E_g^{Cu_xS} = 1.2$ eV, and ΔE_c, the conduction band discontinuity, is about 0.2 eV. The diode junction area is $\mathscr{A}_j$.

The third process, tunneling through the junction, is dominant in cells before heat treatment, giving poor diode characteristics. The heat treatment step, which diffuses Cu acceptors into the CdS and widens the depletion layer, drastically reduces this tunneling current. Contributions to J_0 from the tunneling mechanism do not appear to be important in the normal operating range of high-efficiency Cu_xS/CdS thin-film cells under solar illumination. However, evidence for tunneling as an important transport mechanism can be found at low voltages and/or low temperatures, and in degraded cells (Haines and Bube, 1980). The tunneling mechanism is important over a somewhat wider range of voltages and temperatures in single-crystal CdS-based cells.

† The formulation of Eq. (10.1) is due to A. Rothwarf. See, for example, Rothwarf *et al.* (1979).

The actual junction area $\mathscr{A}_j$ may be considerably larger than the geometric area in the thin-film cell because of Cu_xS growth down CdS grain boundaries as suggested by Fig. 10.1b. Values of J_{02} of the order of 10^{-10} A cm^{-2} are predicted by Eq. (10.1), in reasonable agreement with measured values under AM1 illumination. Measured values of the thermal activation energy for J_0 are about 0.9 eV, also in good agreement with the formulation of Eq. (10.1).

For a cell in which the reverse saturation current is dominated by I_{02}, as given by Eq. (10.1), the total cell current under illumination is given by

$$I = \mathscr{A}_j J_{02}\{\exp[q(V - IR_s)/kT] - 1\} - \mathscr{A}_A J_L h(V), \qquad (10.2)$$

where R_s is the series resistance; $\mathscr{A}_j$ is the total junction diode area of the cell; $\mathscr{A}_A$ is the active, illuminated area; and it has been assumed that shunt resistances are large enough to be neglected. The total light generated current J_L from the Cu_xS is reduced by the interface collection factor $h(V)$ (Section 6.3.1). Note that the value of the diode quality factor A is unity, as is observed under solar illumination for thin-film cells. At smaller illumination levels and for single-crystal cells, the A factors observed are generally larger than 1. Calculation of the open-circuit voltage from Eq. (10.2), with substitution for J_{02} from Eq. (10.1) gives

$$qV_{oc} = E_g^{Cu_xS} - \Delta E_c + kT \ln[J_L h(V_{oc})/qS_i N_c^{CdS}] - kT \ln(\mathscr{A}_j/\mathscr{A}_A). \qquad (10.3)$$

In this expression the quantities $E_g^{Cu_xS}$, N_c^{CdS}, and ΔE_c are directly related to basic material properties and hence cannot be changed within the context of the Cu_xS/CdS cell. One area for improvement is in making $\mathscr{A}_j \simeq \mathscr{A}_A$, by using a different process for deposition of the Cu_xS that does not allow growth down the grain boundaries. With this improvement it is expected that V_{oc} could be increased to 0.54 V.

The value of S_i assumed earlier appears to be consistent with the 4% lattice mismatch between Cu_xS and CdS. The value of S_i has been measured using the variation in the interfacial collection efficiency $h(V)$ caused by changes in the electric field at the interface. The electric field is measured by junction capacitance. Fahrenbruch and Bube (1974) obtained $S_i \simeq 10^6$ cm sec^{-1} for single-crystal cells and Wyeth and Rothwarf (1979) and Rothwarf (1978) obtained values of approximately 3×10^5 cm sec^{-1} for thin-film cells. The effective value of S_i can presumably be changed by variation in the junction profile (e.g., variation of the CdS donor density profile) or by interfacial doping, although neither alternative is sufficiently understood to attempt at present.

Departures from the ideal fill factor for the Cu_xS/CdS cell arise from series resistance in the thin Cu_xS layer (and perhaps to a small degree in the compensated CdS:Cu layer near the interface) and from the voltage dependence of the interface collection function $h(V)$. The fill factor can also be reduced by too small a value for the shunt resistance, but in good cells this effect appears to be small. Taking into account various effects, a practical fill factor of about 0.74 is predicted.

These predictions for cell performance can be combined to obtain a value for the solar efficiency. For an ideal frontwall cell without losses the values obtained are $J_{sc} = 35$ mA cm^{-2}, $V_{oc} = 0.57$ V, fill factor $\simeq 0.81$, and solar efficiency $\simeq 16.1\%$, at 300°K. Substitution of lower, or more achievable, values for these three parameters leads us to expect a solar efficiency of 10.4%. These predictions can be compared with the data reported by Barnett *et al.* (1978): $J_{sc} = 21.8$ mA cm^{-2}, ff = 0.714, $V_{oc} = 0.516$, and $\eta_s = 9.14\%$ at $P_s = 87.9$ mW cm^{-2}, AM1.

10.4 THE RELATION OF Cu_xS LAYER PROPERTIES TO CELL EFFICIENCY AND STABILITY

The structure and stoichiometry of the Cu_xS layer is the major factor determining the number of photogenerated carriers available for collection and the stability of the finished Cu_xS/CdS cell. In addition, the stoichiometry influences the resistivity of the Cu_xS layer that is a large contributor to the series resistance of the device. The stoichiometry is in turn influenced both by the details of fabrication and encapsulation and by environmental conditions after deployment. The heterojunction interface and the CdS adjacent to it primarily determine the forward-bias diode current and cell voltage. These are discussed in the following sections.

10.4.1 Surface Recombination on Cu_xS

It is still not completely understood why front-surface recombination losses for the Cu_xS layer are so low. A rough calculation using published values of α and L_n and assuming $D_n = 10$ cm^2 sec^{-1} and $S_i = 10^6$ cm sec^{-1} indicates a loss of greater than 50% of J_{sc} at the light-incident surface. Using EBIC techniques, Partain *et al.* (1980) found a value of $S_i/v_d \simeq 1$, where $v_d = D_n/L_n$ is the diffusion velocity. The ratio did not appear to depend on heat treatment in air or H_2. Assuming a mobility of 4 cm^2 V^{-1} sec^{-1} and a diffusion length $L_n = 0.25$ μm (the value of L_n in the work cited) gives $S_i \simeq v_d = 4 \times 10^3$ cm sec^{-1}.

At least three possibilities exist: (1) Cu_xS has an unusually low surface recombination velocity inherently, (2) the Cu_xS layer is inhomogeneous in the thickness direction with a resulting electric field that drives photogenerated carriers away from the surface (e.g., a p^+/p structure caused by a Cd doping profile), and/or, (3) the processing of the cell passivates the surface recombination velocity (e.g., formation of a thin Cu_2O layer at the surface) (Pfisterer *et al.* 1975, 1976, 1980).

10.4.2 Cu_xS Layer Topography

The microtopography of the Cu_xS/CdS cell has been explored by Norian and Edington (1980), Donaghey *et al.* (1980), and Mukherjee *et al.* (1977). The thickness of the Cu_xS layer is found to be highly nonuniform with substantial intrusions down the grain boundaries of the CdS (as suggested by Fig. 10.1b), particularly in cells fabricated by the "wet" process. In cells prepared by other methods the layer is more planar (Fig. 10.1c). The topography of the Cu_xS film determines the optical absorption and photogenerated carrier collection geometry. As in the case of textured Si cells (Section 7.4.4), for light traveling in the Cu_xS at an oblique angle to the junction plane, the optical absorption coefficient is effectively increased. Peripheral effects of the topography are a decrease in optical reflection and an increase in junction (diode) area.

The three-dimensional aspect of these Cu_xS films has been included in a theoretical current transport model by Mukherjee *et al.* (1978), who predicted increases in long wavelength response (compared to the standard one-dimensional theory) and a substantial increase in diode current. Another theoretical analysis is given by Jacobs and De Mey (1981).

Experimentally, higher J_{sc} and lower V_{oc} values are observed with the highly textured "wet" process cells as compared with the more planar "dry" process cells.

10.4.3 Stoichiometry of the Cu_xS Layer

As was pointed out in Section 10.2, the optical and electrical properties of Cu_xS vary considerably with the value of x. The optical absorption increases strongly with x in the range $1.9 < x < 2.00$. Since Cu_xS is a defect semiconductor with Cu vacancies acting as acceptors, the carrier density decreases as x closely approaches 2.00. The carrier density, in turn, controls both the conductivity and the minority carrier lifetime (by way of Auger recombination). The optimum value of x for a solar cell is estimated at $\lesssim 1.995$, i.e., totally chalcocite but with a fairly large carrier density.

Control of the stoichiometry during the formation of the Cu_xS layer is accomplished by the following methods: (1) for the wet process—by addition of reducing agents to the dipping solution (e.g., hydrazine); (2) for spray pyrolysis, chemical vapor deposition, vacuum evaporation, or sputtering—by addition of excess Cu; and (3) for electrochemical deposition—by control of current density and temperature (Nakayama *et al.*, 1971).

In most cases adjustment of x after layer formation is required. This can be accomplished by heat treatment in various ambients (vacuum, H_2, or CO) (Bragagnolo *et al.*, 1980). In the range $1.994 < x < 2.000$ the stoichiometry and sheet resistivity can be cycled reversibly with sequential heat treatments in oxidizing and reducing ambients. Another method of stoichiometry control, developed by Bogus and Mattes (1972) involves the formation of a thin film of metallic Cu on the Cu_xS surface and subsequent annealing to level the Cu density across the layer. Formation of the Cu film is accomplished either by reduction of a portion of the Cu_xS with a glow discharge in atomic hyrogen or by direct vacuum evaporation of Cu (~100 Å). The latter method is used by Pfisterer *et al.* (1975, 1976) to create, in addition, a Cu_2O minority carrier mirror by subsequent heat treatment in air.

Evidence exists that there is a fairly large concentration of Cd (10^{17}–10^{19} cm^{-3}) in the Cu_xS layer (Burton *et al.*, 1981; Florio *et al.*, 1981; Martinuzzi *et al.*, 1972) with concentration maxima at the Cu_xS surface and at the Cu_xS/CdS interface. Since Cd is a donor in Cu_xS, the Cd concentration may affect the carrier density in the Cu_xS layer and hence influence both the minority carrier lifetime and the series resistance.

It is generally agreed that changes in stoichiometry can profoundly affect both the quantum efficiency of Cu_xS/CdS cells and the sheet resistivity of Cu_xS films. The exact mechanisms are uncertain and many theories have been advanced, several of which are outlined here.†

Palz *et al.* (1972) suggested that the primary effect of varying x was to change the proportion of the Cu_xS phases present in the layer (i.e., chalcocite, $x = 2.00$; djurleite, $x = 1.96$; etc.). They based this on their observations that J_{sc} for chalcocite was about a factor of 10 higher than that for djurleite, with an almost linear variation with x in the range $1.95 \leq x \leq 2.00$. Hence the stoichiometry has a large effect on cell output. They also observed an increase in sheet resistivity of the Cu_xS by a factor of 20 in going from $x = 1.95$ to $x = 2.00$. Pfisterer *et al.* (1979) added the possibility of changes in surface recombination velocity from the presence or absence of the Cu_2O layer and the improvement of the minority carrier diffusion length with increasing x. Rothwarf and Windawi (1980, 1981) focused on stoichiometry induced changes in the optical absorption coefficient α within the single phase of chalcocite ($1.995 < x < 2.000$) from which the most efficient solar cells are made. They attributed the change in α to the Burstein–Moss shift (Section 3.4) with changes in carrier den-

† A more extensive listing of mechanisms is given by Rothwarf and Windawi (1980).

sity. Thus raising x would decrease carrier density and raise the effective α, tending to increase J_{sc}. Based on these considerations, the authors derived a relation between J_{sc} (as an indicator of α) and the sheet resistivity (indicating carrier density) and found good agreement with experimental data. Later experimental work by Eser and Cambridge (1981) showed that the changes in α were produced by oxidizing or reducing heat treatments but that these could account only for changes of ~10% in J_{sc} whereas changes of 20 to 40% were commonly observed. No evidence for a Burstein–Moss shift was observed. They concluded that oxidizing and reducing heat treatments must also be changing the minority carrier lifetime and/or the interface transport $h(V)$. Rothwarf (1979) suggested changes in cell output could also occur because of increased band bending in the Cu_xS at the Cu_xS/CdS interface as the carrier density is decreased by variation of stoichiometry.

To summarize, changes in stoichiometry can modify (1) the absorption coefficient α (through a change in Cu_xS phase for large changes in x, or by more subtle changes for values of x close to 2.000), (2) the carrier density, (3) the lifetime, (4) the mobility, (5) the surface recombination, and (6) the band bending at the Cu_xS/CdS interface. All these modifications have a strong effect on cell output. Since modification of stoichiometry is such a significant means of optimizing the output, it is also a pervasive cause of cell degradation.

10.4.4 Stability of Cu_xS/CdS Cells

The most-asked question about the Cu_xS/CdS cell is "Is it stable?" The experimental results are ambiguous and will probably continue to be so until cells have had significant use on consumer's rooftops. While some cells degrade rapidly, other, properly encapsulated cells have remained stable for periods of 10 years or more in the laboratory.

It is generally agreed that degradation of cell output, when it occurs, is most rapid under load (or at forward bias), at elevated temperatures, and in humid, oxidizing atmospheres (Castel and Soubeyrand, 1981). The degradation is accelerated by the high ionic mobility in Cu_xS. Three of the possible degradation modes are discussed here. Not surprisingly, two of these modes are concerned with changes in stoichiometry of the Cu_xS.

(1) *Atmospheric decomposition of Cu_xS.* Decomposition of the Cu_xS into products such as CuO, Cu_2O, and $CuCO_3$ results in reducing the value of x, with the possibility of formation of nonoptimal phases if enough Cu is lost. Baron *et al.* (1978) have pointed out that the growth of only two monolayers of Cu_2O is sufficient to reduce x for a 0.4-μm-thick Cu_xS film from 2.000 to 1.995. Fortunately the oxide growth appears to be

self-limiting. Florio *et al.* (1981) have pointed out the involvement of Cd as a requirement for the atmospheric oxidation of Cu_xS. The application of a thin Cu layer and subsequent anneal (Bogus and Mattes, 1972) strongly increases the stability of Cu_xS/CdS cells, suggesting that cells with optimal stoichiometry are quite resistant to degradation whereas those with marginal stoichiometry degrade more rapidly.

(2) *Electrolytic Decomposition of* Cu_xS. During operation electric fields may be set up across the portion of the Cu_xS layer between grid lines and particularly adjacent to macroscopic defects. These may cause changes in stoichiometry by field-aided diffusion, and in extreme cases create free Cu by electrolytic decomposition (Mathieu *et al.,* 1973). The nodules of Cu thus formed can short out the cell. Zwicker *et al.* (1981) showed that lateral migration of Cu toward the grid lines could occur under load, resulting in stoichiometry changes and local J_{sc} loss, with V_{oc} and ff remaining unchanged. The resulting field-assisted diffusion pattern, as represented by an EBIC J_{sc} image, could be completely removed by heat treatment of the hermetically sealed cells in the dark at 105°C for 1 hr.

(3) *Over-Optimal Diffusion of Cu into the CdS.* A certain degree of diffusion is required to "form" the junction barrier and optimize both current collection and cell voltage. It is generally agreed that *excessive* heat treatment further widens the depletion layer and eventually decreases J_{sc} and η_S. It is doubtful whether this can happen at cell operating temperatures within the lifetime of the cell, but the question is still under debate. Partain and Birchenall (1975) outline the various diffusion processes involved. They estimate the rate of degradation on the basis of the loss of Cu to the CdS that would result in a Cu_xS stoichiometry deficit and consequent loss of J_{sc}.

A good overview of the the reliability problem is given by Meakin and Phillips (1979).

10.5 FUNDAMENTAL HETEROJUNCTION TRANSPORT PHENOMENA IN Cu_xS CELLS

Despite over 10 years of existence of the Cu_xS cell, by the mid-1960s important questions still persisted about the basic transport mechanisms active in the cell. We have already mentioned the indecision about the site of optical absorption and how this question was finally resolved in favor of a heterojunction, with optical absorption taking place primarily in the Cu_xS. Other still unexplained phenomena included (1) the crossing of the light and dark $J-V$ curves; (2) different spectral response of the cell, depending on its optical, thermal, and bias voltage history and/or the presence of a bias illumination during the actual measurement of spectral response; and (3) the existence of both slow and rapid degradation processes, particularly in heat-treated cells. In order to investigate these phenomena, considerable research was done on Cu_xS/CdS cells made with both polycrystalline thin-film and single-crystal CdS, resulting in a number of theoretical models.

10.5.1 Summary of Proposed Models

By 1974, various models had been advanced to explain the Cu_xS/CdS junction transport:

(1) *The "Clevite" or photoconductor layer model* was proposed by Shiozawa *et al.* (1969; see also Shiozawa *et al.*, 1968). In this model carrier transport is controlled by the series resistance of a photoconductive Cu-doped CdS layer adjacent to the junction, formed by diffusion of Cu during processing.

(2) *The conduction band spike moel* was suggested by Gill and Bube (1970b) and Lindquist and Bube (1972a). A spike resulting from electron affinity mismatch in the conduction band at the Cu_xS/CdS interface was postulated. The thickness of the spike was controlled by deep, photoactive states in the CdS : Cu layer and at the interface. The transport of photogenerated carriers was controlled by tunneling through the spike or by thermal excitation over it. As a possible alternative mechanism, Gill suggested recombination through interface states.

(3) *TeVelde's model* invoked a variation of the conduction band discontinuity ΔE_c and the CdS band bending caused by the reversible absorption of oxygen at the Cu_xS/CdS interface (TeVelde, 1973). TeVelde's values of ΔE_c varied from positive, corresponding to a drop in the conduction band going from the Cu_xS to the CdS, to negative, corresponding to a conduction band spike that limits carrier flow. TeVelde also introduced interfacial recombination as a loss mechanism for $\Delta E_c > 0$ case.

(4) In the *interfacial tunneling/recombination model* of Fahrenbruch and Bube (1974), the electric field at the Cu_xS/CdS interface modulates the photogenerated carrier loss there due to a tunneling/recombination pathway through states in the interface region. The electric field at the interface is, in turn, controlled by the occupation of deep, photoactive states in the CdS : Cu layer and at the interface.

(5) The *high-field domain model*, proposed by Böer (1976), attributed at least part of the control of photogenerated carrier transport to the formation of high-field domains in the CdS : Cu layer. This model includes interfacial recombination also, which reduces V_{oc} but has less influence on J_{sc}.

(6) *The Institute of Energy Conversion Model,* proposed by Rothwarf (1976; see also Rothwarf *et al.*, 1978; Rothwarf, 1979)† describes transport in terms of electric field control of interfacial recombination, which is considered to be lumped into an effective interface recombination velocity, yielding the diode current expressions of Eq. (10.1) and (10.2). In this model the depletion layer is divided into two parts, a photoactive layer of about 0.3 μm thickness adjacent to the interface, and another considerably wider section extending as much as 3 μm into the CdS bulk that dominates the capacitance in the dark (Hall and Singh, 1979). Tunneling is not included in this model.

(7) *The space-charge-limited current model* of Moorthy *et al.* (1979) invokes an insulating layer of CdS : Cu adjacent to the interface for transport control. Control of forward-bias current in both dark and light is dominated by space-charge-limited flow and the $J-V$ curves are represented by power law relations $J = J_0 V^m$, where $m \geqslant 2$.

A common feature of all the models, except TeVelde's, is the presence of a photoactive Cu-doped CdS layer at the interface that accounts for most of the illumination wavelength-dependent effects. In addition, the later models include some form of interfacial recombination as a loss mechanism for photogenerated carriers. The remaining areas of disagreement center around the extent of the influence of the interfacial recombination on transport, the details of the interfacial recombination process itself, and the mechanism(s) by which transport is controlled (i.e., electric field control of interfacial recombination, high field domains, or space-charge-limited current effects).‡

10.5.2 Effects of Heat Treatment: Thermal and Optical Quenching Phenomena and Photocapacitance

The results described in the remainder of Section 10.5 provide a unified picture of the diverse phenomena that are characteristic of the Cu_xS/CdS cell. They also show how complicated the detailed description of these cells is. To eliminate the complications of grain boundaries, many of the experiments were done using single-crystal CdS rather than polycrystalline thin films. Most of the photoelectronic characteristics of thin-film cells and single-crystal cells are qualitatively similar, but the optical and thermal properties are much exaggerated in the latter case. For example, the spectral response curves for non-heat-treated single-crystal cells, shown in Fig. 10.6, are qualitatively identical to those for thin-film

† This model is also discussed briefly in Section 5.5.4.

‡ The authors now favor a model combining (4) and (6) with thermally assisted tunneling/recombination and direct thermal activation to interfacial recombination acting in parallel.

(a)

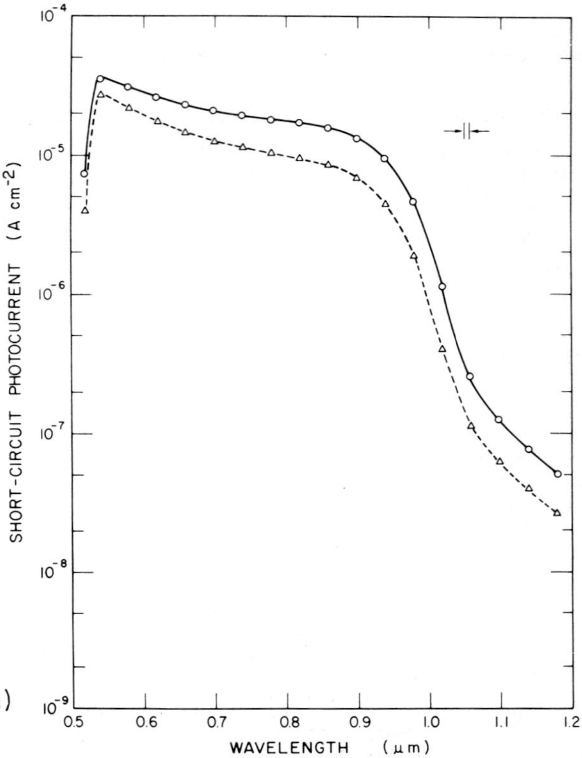

SHORT-CIRCUIT PHOTOCURRENT (A cm^{-2})
10^{-4}
10^{-5}
10^{-6}
10^{-7}
10^{-8}
10^{-9}
0.5
0.6
0.7
0.8
0.9
1.0
1.1
1.2
WAVELENGTH (μm)

(b)

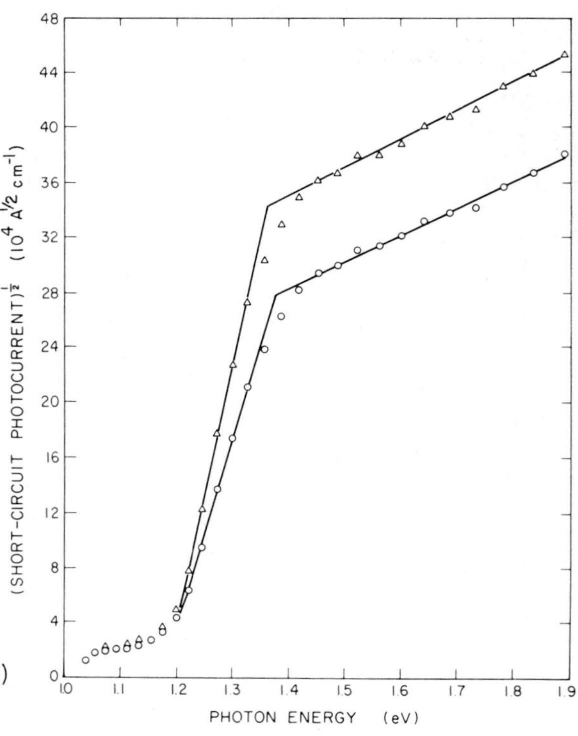

(SHORT-CIRCUIT PHOTOCURRENT)$^{\frac{1}{2}}$ (10^4 A$^{1/2}$ cm^{-1})
48
44
40
36
32
28
24
20
16
12
8
4
0
1.0
1.1
1.2
1.3
1.4
1.5
1.6
1.7
1.8
1.9
PHOTON ENERGY (eV)

cells, with the single-crystal cell showing a quantum efficiency of 0.1–0.2, compared to 0.5–0.8 for thin-film cells, and both showing the long wavelength onset of J_{sc} at $h\nu = 1.2$ eV. Thermal and optical data for thin-film cells are given by Lindmayer and Revesz (1971), for example.

Despite the qualitative similarities between cells based on single-crystal and polycrystalline thin-film CdS, it must be emphasized that there are large quantitative differences between them. For example, infrared quenching is normally a small effect in thin-film cells, whereas for single-crystal-based cells infrared quenching can produce reductions in J_{sc} by factors of 10^3 or larger. Another difference is that the diode A factors for single-crystal cells are quite variable, depending on preparation, temperature, and bias-voltage histories, and reaching values of 2 to 3 or more. For thin-film cells under high forward bias and in AM1 illumination, $A = 1$ is commonly seen. Thus it is important to realize that one of the purposes of the remainder of Section 10.5, which outlines the complexities of behavior for single-crystal cells, is to guide and aid in interpreting detailed laboratory measurements on *both* single-crystal and thin-film cells.

Although thin-film Cu_xS/CdS cells require heat treatment in order to realize desirable diode and photovoltaic behavior, cells made with single-crystal CdS have excellent diode characteristics and considerable photovoltaic response without heat treatment (beyond that intrinsic in the formation of the Cu_xS layer by dipping in 70–90°C solution of cuprous ions). A variety of non-heat-treated cells made with Cu_xS thicknesses from 10 to 200 μm (and hence operated in the backwall mode through the CdS crystal) showed open-circuit voltages of about 0.45 V, quantum efficiencies of about 0.1, fill factors of about 0.7, and light $J-V$ curves that did not cross the dark $J-V$ curves (Lindquist and Bube, 1972a). The low quantum efficiency may be caused at least partially by optical losses from reflection and by transmission through the CdS bulk crystal as well as poor stoichiometry in the Cu_xS layer. These results suggest that the necessity of the heat treatment for the thin-film cells is connected with the details of the polycrystalline structure involving Cu_xS growth down CdS grain boundaries. Any heat treatment of the single-crystal CdS cells causes the short-circuit current to decrease.

Nearly all the complexities of the Cu_xS/CdS single-crystal cell appear

---

Fig. 10.6. Spectral response of short-circuit current for several different non-heat-treated Cu_xS/CdS cells prepared with single-crystal CdS, plotted in two different ways. The cells were made on polished (0001) CdS faces. [After P. F. Lindquist and R. H. Bube, *J. Electrochem. Soc.* **119,** 936 (1972). Reprinted by permission of the publisher, The Electrochemical Society, Inc.]

to be generated by heat treatment. Before heat treatment the light and dark $J-V$ curves do not cross, spectral response does not depend on bias irradiation, slow transients do not occur in the spectral response, there is very little effect of illumination on the junction capacitance, and rapid degradation phenomena are absent. For these non-heat-treated single-crystal cells, the forward current in the dark appears to be totally controlled by tunneling, as shown in Fig. 10.7. These data can be described by the relation $J(V, T) = J_0(T) \exp(\alpha V)$, where $J_0(T)$ appears to be thermally activated near 300°K with an activation energy of 0.26 eV, somewhat smaller than that seen in heat-treated cells, and α does not depend on temperature.

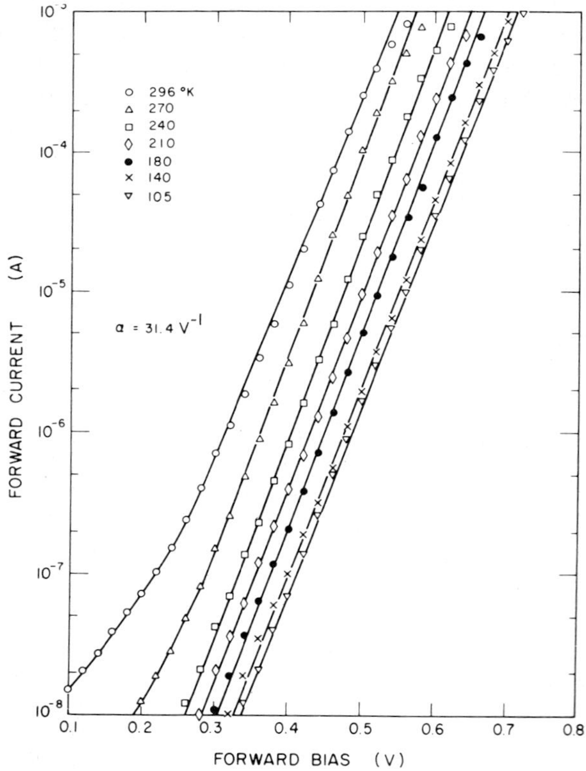

Fig. 10.7. Forward-bias $I-V$ curves at different temperatures for a non-heat-treated Cu_xS/CdS cell prepared with single-cystal CdS. The cell was made on a polished and etched (0001) CdS face. The donor density in the CdS was $\simeq 10^{16}$ cm^{-3}. [After P. F. Lindquist and R. H. Bube, *J. Electrochem. Soc.* **119**, 936 (1972). Reprinted by permission of the publisher, The Electrochemical Society, Inc.]

One of the major effects of heat treatment is to cause diffusion of Cu from the Cu_xS layer into the CdS crystal, thus broadening the depletion layer in the CdS near the junction interface. It is in this Cu-doped CdS depletion layer that the thermally and optically induced changes occur that ultimately control junction transport.

Figure 10.8 shows the light and dark $J-V$ curves for a *typical* Cu_xS/CdS cell on single-crystal CdS before and after heat treatment. Effects induced by heat treatment are (1) a large decrease in dark forward-bias current, (2) crossing of the light and dark $J-V$ curves, (3) a slight increase in V_{oc}, and (4) a marked decrease in J_{sc}. A brief explanation of these effects follows. Heat treatment decreases the dark forward-bias current by increasing the depletion layer width W_d and markedly decreasing the tunneling forward current; the light and dark $J-V$ curves cross after heat treatment because illumination shrinks the depletion layer and allows an increased tunneling current to flow. The open-circuit voltage is increased because of a decrease in J_0 due to the decrease in for-

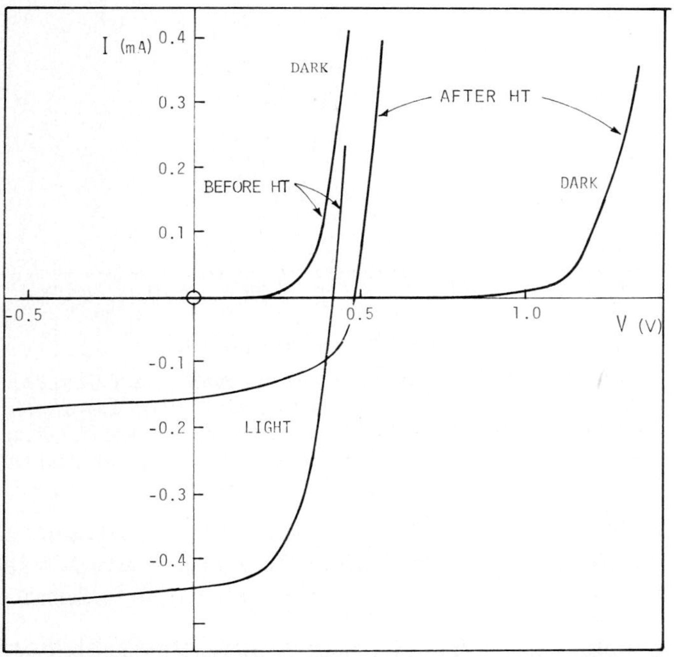

Fig. 10.8. Effect of heat treatment (1 min in air at 250°C) on the $J-V$ curves in light and dark for a backwall Cu_xS/CdS cell prepared on single-crystal CdS. Although reduction of J_{sc} with heat treatment is typical, there are examples of *both* V_{oc} and J_{sc} increasing with heat treatment. [After W. D. Gill and R. H. Bube, *J. Appl. Phys.* **41,** 3731 (1970).]

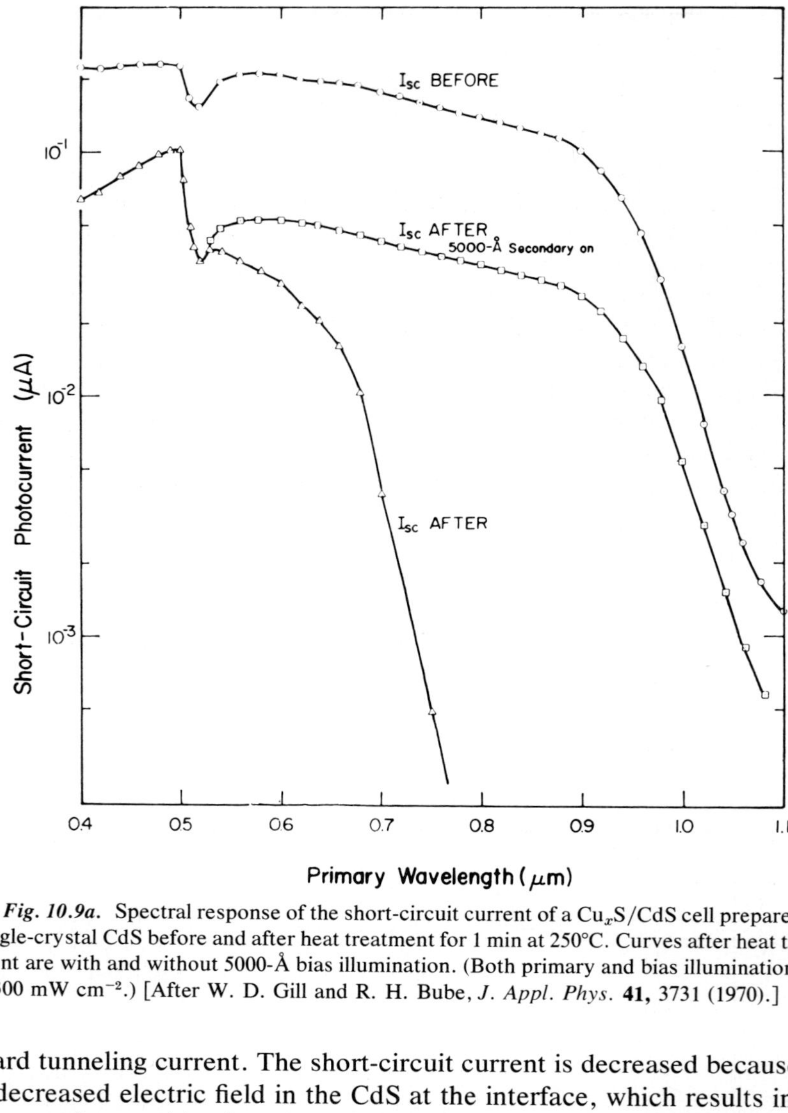

Fig. 10.9a. Spectral response of the short-circuit current of a Cu_xS/CdS cell prepared on single-crystal CdS before and after heat treatment for 1 min at 250°C. Curves after heat treatment are with and without 5000-Å bias illumination. (Both primary and bias illumination are $\simeq$300 mW cm^{-2}.) [After W. D. Gill and R. H. Bube, *J. Appl. Phys.* **41,** 3731 (1970).]

ward tunneling current. The short-circuit current is decreased because of a decreased electric field in the CdS at the interface, which results in an increased recombination through interface states, as is described more fully in Section 10.5.4.†

Figure 10.9a shows the effect of heat treatment on the spectral response

† Note that changes in the stoichiometry of the Cu_xS layer during heat treatment can also induce changes in J_{sc}, particularly for thin Cu_xS layers.

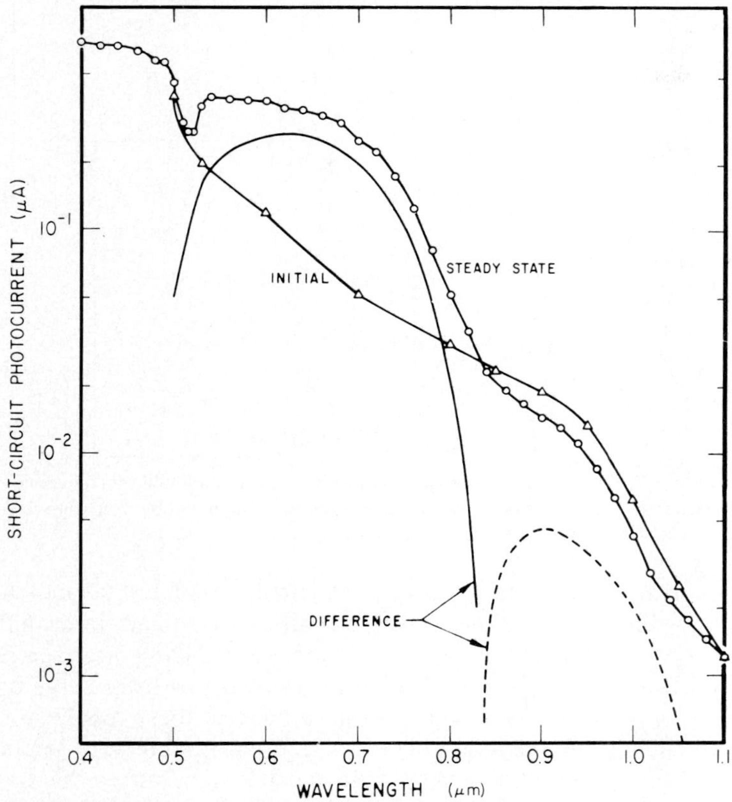

Fig. 10.9b. Spectral response of the initial and steady-state short-circuit current for a heat-treated Cu_xS/CdS cell. The difference curves represent the slow component of response; solid curve, enhancement; dashed curve, quenching. [After W. D. Gill and R. H. Bube, *J. Appl. Phys.* **41,** 3731 (1970).]

of the Cu_xS/CdS cell prepared on single-crystal CdS. Heat treatment causes a drastic reduction in the short-circuit current for the longer wavelength range. However, this reduction is removed if the spectral response is measured in the presence of a 0.5-μm-bias illumination. Heat treatment also causes slow transients in the spectral response, as shown in Fig. 10.9b. For wavelengths greater than 0.83 μm, the short-circuit current is *decreased* from its initial value (quenching), whereas at shorter wavelengths J_{sc} is *increased* from its initial value (enhancement). These variations of J_{sc} can be associated with the width of the depletion layer. Intrinsic photoexcitation across the band gap or extrinsic photoexcitation

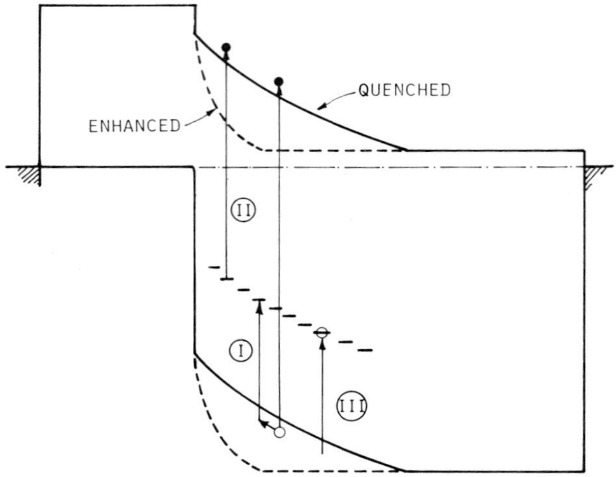

Fig. 10.10. Variation of depletion layer width for a Cu_xS/CdS cell by (I) intrinsic excitation and (II) extrinsic excitation. Quenching of short-circuit current by transition (III), representing either an optical or a thermal excitation.

from levels in the depletion layer to the conduction band (as indicated in Fig. 10.10) deposits additional positive charge in localized states in the depletion layer, shrinking the depletion layer and reducing interface recombination. Longer-wavelength irradiation removes positive charge by photoexcitation of electrons from the valence band to these localized states, thus broadening the depletion layer and increasing interface recombination. If intrinsic illumination is present as a bias light during the measurement of spectral response, positive charge is kept in the depletion layer and it remains narrow. Time constants for building up or removing this positive charge correspond to the time constants observed in the transient behavior. An enhanced state persists for some time at temperatures near or below 300°K if the cell is irradiated with short-wavelength light before measuring the spectral response. Measurements of rate of decay of this enhancement at several different temperatures indicate a thermal activation energy of 0.95 eV for the process. This corresponds to the thermal release of positive charge from localized states in the depletion layer to the valence band.

The reaction of the Cu_xS/CdS cell to optical and thermal inputs is summarized in Table 10.I.

The spectral response of the quenching process can be investigated further by establishing a short-circuit current with primary illumination (bias light) and then measuring the changes produced by varying the photon energy of a secondary illumination. The effect can be observed

Table 10.1
Summary of Cu_xS/CdS Cell States

State	Origin	Consequence
Enhanced	Optical excitation, short wavelengths	Larger net positive charge[a] Thinner depletion layer Decreased V_{oc} Increased J_{sc} Large junction capacitance
Quenched	Thermal and/or optical excitation, long wavelengths (thermal quenching only above $\simeq$200°K)	Smaller net positive charge[a] Thicker depletion layer Increased V_{oc} Decreased J_{sc} Small junction capacitance

[a] Within the CdS depletion layer, adjacent to the metallurgical interface.

either by measuring the short-circuit current directly, as in Fig. 10.11a, or by measuring the decrease in the junction capacitance, as shown in Fig. 10.11b. In the latter figure, the junction capacitance has been increased previously by intrinsic radiation and is stable against thermal decay at sufficiently low temperatures. One of the two quenching bands shown in Fig. 10.11 corresponds to the direct transitions (III) shown in Fig. 10.10, and the other corresponds to a transition involving an excited state of an imperfection. These two quenching bands are well known from the quenching of photoconductivity and the transitions involving Cu as an impurity in CdS (see, e.g., Bube, 1976).

Measurements of enhancement and quenching effects on thin-film Cu_xS/CdS cells show that qualitatively the same phenomena are observed except that the second long-wavelength optical quenching band at 1.5 μm cannot be detected. In the thin-film case the effect of quenching on the cell parameters is quantitatively much smaller. For example, optical quenching reduces J_{sc} to between 10 and 80% of its unquenched value in the thin-film case, whereas it may be reduced to 0.1% or less in the single-crystal cell.

The presence of an enhanced state due to trapped positive charge in the depletion layer of a Cu_xS/CdS cell can be correlated in a striking way with the short-circuit current in both non-heat-treated and heat-treated cells. Figure 10.12a shows the relationships between short-circuit current and photocapacitance as both are varied by excitation with 0.7-μm light. Normalizing the two curves to the same intensity makes them essentially identical. Even in the non-heat-treated cell a detectable variation in J_{sc} with enhancement versus quenching conditions can be detected, but the

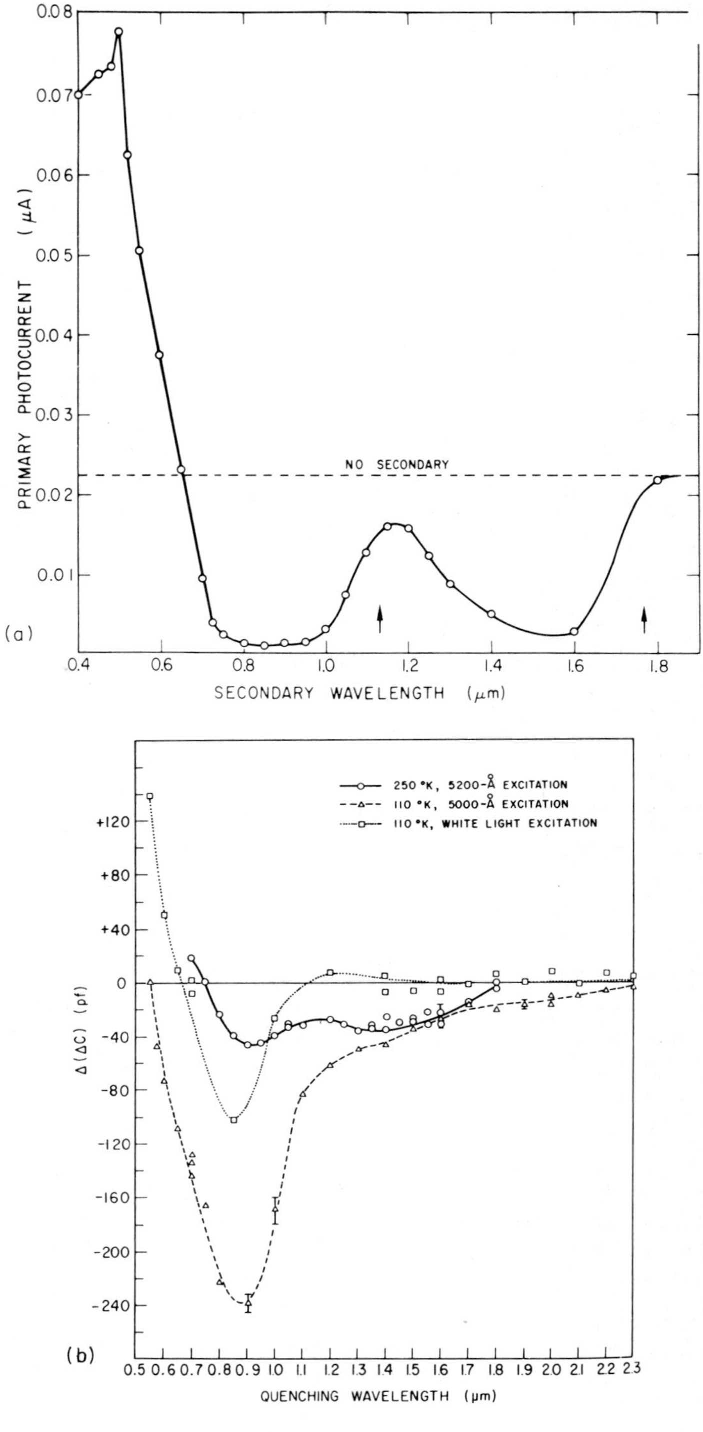
PRIMARY PHOTOCURRENT (μA)
0.08
0.07
0.06
0.05
0.04
0.03
0.02
0.01
NO SECONDARY
(a)
0.4
0.6
0.8
1.0
1.2
1.4
1.6
1.8
SECONDARY WAVELENGTH (μm)
250 °K, 5200-Å EXCITATION
110 °K, 5000-Å EXCITATION
110 °K, WHITE LIGHT EXCITATION
Δ(ΔC) (pf)
+120
+80
+40
0
-40
-80
-120
-160
-200
-240
(b)
0.5 0.6 0.7 0.8 0.9 1.0 1.1 1.2 1.3 1.4 1.5 1.6 1.7 1.8 1.9 2.0 2.1 2.2 2.3
QUENCHING WAVELENGTH (μm)

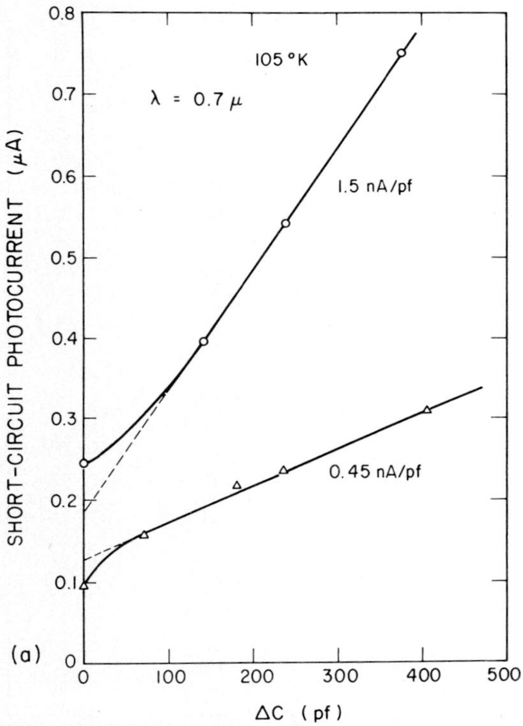

Fig. 10.12a. Short-circuit current of a Cu_xS/CdS cell prepared on single-crystal CdS without subsequent heat treatment for 0.7-μm illumination, as a function of the magnitude of the photocapacitance of the junction (for two illumination intensities). Illumination is back-wall and cell area is 0.02 cm². All data at 150°K. [After P. F. Lindquist and R. H. Bube, *J. Appl. Phys.* **43,** 2839 (1972).]

effect is relatively small and the current varies approximately linearly with the photocapacitance. After 2 min of heat treatment at 200°C, however, the short-circuit current is strongly dependent on the photocapacitance, now varying approximately exponentially over several orders of

---

Fig. 10.11. (a) Enhancement and quenching of short-circuit current of a Cu_xS/CdS cell prepared on single-crystal CdS by a secondary light. Primary excitation was by 6550-Å light. Arrows indicate where the onset of optical quenching of photoconductivity occurs in sensitive CdS crystals. [After W. D. Gill and R. H. Bube, *J. Appl. Phys.* **41,** 3731 (1970).] (b) Optical quenching spectra of photocapacitance of 110 and 250°K for a Cu_xS/CdS cell prepared on single-crystal CdS. The photocapacitance ΔC had previously been generated by intrinsic excitation; the temperature was sufficiently low that positive charge induced into the depletion layer by this excitation was thermally stable. [After P. F. Lindquist and R. H. Bube, *J. Appl. Phys.* **43,** 2839 (1972). Also see T. Suda and R. H. Bube, *J. Appl. Phys.* **52,** 6218 (1981).]

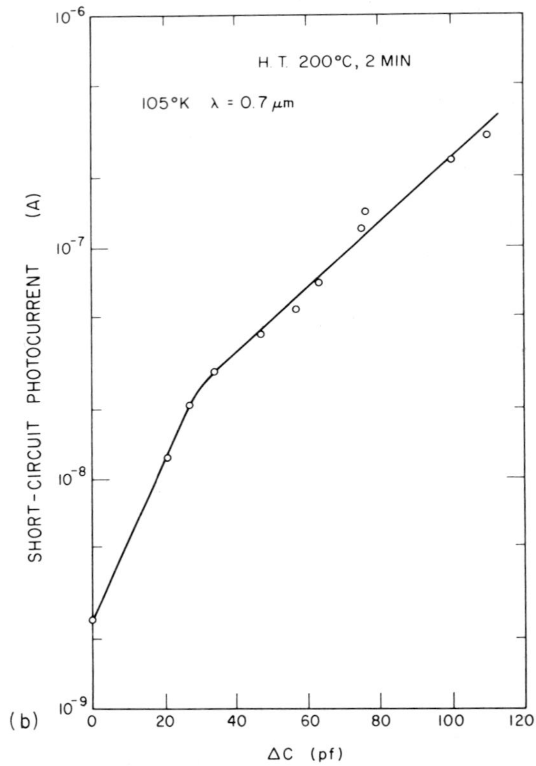

Fig. 10.12b. Short-circuit current of the cell of Fig. 10.12a as a function of photocapacitance after heat treatment for 2 min at 200°C. [After P. F. Lindquist and R. H. Bube, *J. Appl. Phys.* **43,** 2839 (1972).]

magnitude, as shown in Fig. 10.12b. Qualitatively similar optical quenching phenomena have been observed in Schottky barrier solar cells made on *p*-Si (Kul'sha *et al.,* 1976).

10.5.3 Thermally Restorable Optical Degradation

Another phenomenon associated with the properties of the depletion layer in heat-treated Cu_xS/CdS cells is a large and rapid degradation process introduced by optical excitation *after* heat treatment. The degradation can be reversed by subsequent heating at about 100°C in the dark. This phenomenon has been given the name of *thermally restorable optical degradation* (TROD) (Kanev *et al.,* 1971; Fahrenbruch and Bube, 1974). It is strikingly similar to a decrease in electron lifetime on optical illumination that has been observed in single-crystal CdS photoconductors that have a density of Cu impurity in excess of the density of excess Cd donors. Apparently similar optical degradation phenomena occur in Si solar cells that have been exposed to high-energy radiation as reported by Weizer *et al.* (1978) and Cheng *et al.* (1978). In Weizer's results the effect appears to depend on the presence of Ag impurity energy levels close to the junction interface.

Figure 10.13 shows the variation with time of the short-circuit current in single-crystal Cu_xS/CdS cells after 100°C heating in the dark, and also of the photocurrent generated in a CdS:Cd:Cu single crystal. Both currents may decrease by several orders of magnitude within minutes in room light. Thus, unless the cell is heat treated in the dark and kept there, only the optically degraded state can be observed. Optical degradation is a thermally activated process that becomes increasingly slow at reduced temperatures. For both the Cu_xS/CdS cell and the CdS:Cd:Cu photoconductor, the activation energy for the process is about 0.4 eV. The decrease in current caused by this optical degradation can be completely reversed by heating above 100°C in darkness, called here "restoration."† This temperature is low compared to the original heat-treatment temperature. For both the Cu_xS/CdS cell and the CdS:Cd:Cu photoconductor, the restoration process is also thermally activated, with an activation energy of about 1.6 eV. The process is fully reversible; a particular cell can pass through many degradation/restoration cycles without change in the magnitudes of the currents observed. The degree of degradation/restoration is apparently independent of the degree of enhancement/quenching discussed earlier. The consequence is that four distinct states of the cell can be defined: restored-enhanced (RE), optically degraded-enhanced (DE), restored-quenched (RQ), and degraded-quenched (DQ).

A complete cycle of these states can be traced with the short-circuit current of a single Cu_xS/CdS cell, as shown in Fig. 10.14. The cycle starts at 300°K with the cell in the DQ state. As the temperature is decreased, the cell goes to the DE state because of the removal of the thermal quenching component. Little hysteresis occurs on returning to 300°K. From 325 to 375°K, J_{sc} drops to immeasurably small values as the state changes from DE to DQ and then increases as the temperature is raised and thermal restoration produces the RQ state. As the temperature is decreased from 375°K, the enhancement/quenching ratio increases and the

† Although the cells discussed here were "restored" by heat treatment above 100°C in vacuum, similar results are obtained in air, H_2, or Ar. The Cu_xS layer thickness is $\approx 2\ \mu m$, which may explain their insensitivity to oxidation or reduction at these temperatures.

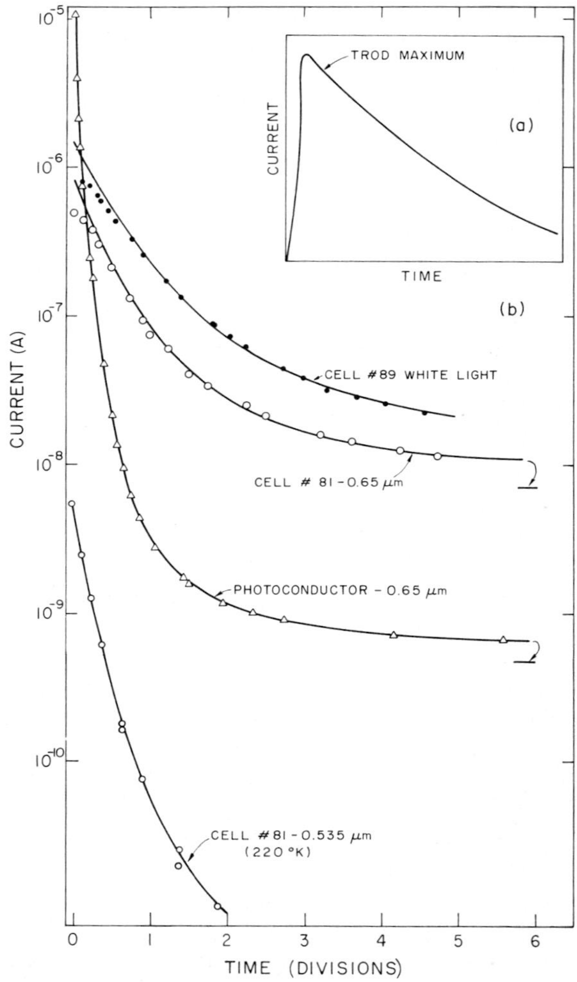

Fig. 10.13. (a) Schematic TROD curve showing the variation of J_{sc} with time after beginning photoexcitation. The rising part of the curve corresponds generally to enhancement, and the falling part to degradation. (b) Short-circuit current for Cu_xS/CdS cells and photocurrent for a CdS:Cd:Cu crystal as a function of time during degradation. Time scales for the four curves from top to bottom are 200, 4000, 4000, and 100 sec/div, respectively. The upper three curves are measured at 300°K. Fully degraded values of the current for two of the curves are indicated on the right-hand side. All samples have an illuminated area of ~0.04 cm^2. [After A. L. Fahrenbruch and R. H. Bube, *J. Appl. Phys.* **45,** 1264 (1974).] (Cells are in backwall mode.)

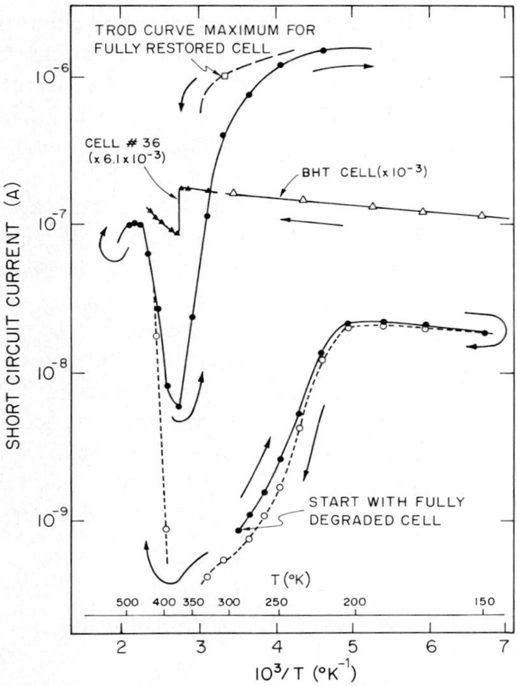

Fig. 10.14. A complete thermal cycle of short-circuit current for a cell of Cu_xS/CdS previously subjected to long heat treatment. Exciting light was white with an intensity of 1.4 mW cm^{-2}. Other curves shown include: (1) a before-heat-treatment (BHT) curve showing the typical small dependence on temperature, (2) data for cell 36 obtained after a short heat treatment at 150°C (normalized to the BHT data) showing the abrupt change in J_{sc} with a phase transformation in the Cu_xS at 80°C, and (3) an indication of the maximum current for the fully restored cell. [After A. L. Fahrenbruch and R. H. Bube, *J. Appl. Phys.* **45,** 1264 (1974).]

cell enters the RE state with also a small effect due to optical degradation. The cell in the restored state also shows little hysteresis between 150 and 300°K. Similar representations of the TROD cycle can be obtained with measurements of the temperature variation of V_{oc}, I_{dark} at constant forward bias, or junction capacitance.

Phenomena associated with the TROD effect have also been observed in Clevite all-thin-film Cu_xS/CdS cells but with greatly reduced magnitude. Typically, optical degradation corresponds to about 15% decrease in J_{sc} for white-light excitation, although spectral measurements indicate that the long-wavelength response is decreased by a factor of 4.

The mechanism of the optical degradation process is not fully known.

An explanation has been proposed from research on the process in photoconducting CdS:Cd:Cu (Kanev *et al.*, 1969). They suggested that optical excitation shifts an electron from a low-recombination-rate center (possibly a neutral Cu–Cd complex) to a unique trap surrounded by a repulsive barrier. When an electron is thermally activated to surmount this barrier and is trapped, the degradation process can occur. After the electron is trapped, capture of a hole by the initially low-recombination-rate center converts it into a much larger recombination-rate center. Thermal restoration involves freeing the trapped electron from the deep repulsive trap in which optical excitation had placed it.

The necessity for both optical and thermal inputs suggests the formation of a complex center with both electronic and ionic motion; this complex center then behaves as a rapid recombination center. The TROD processes appear to be similar to the "photochemical reactions" described by Tscholl (1968) and Korsunskaya *et al.* (1966) in CdS crystals doped with various metals such as Cu, Ag, or Ni. Im *et al.* (1970) observed similar effects in CdS_ySe_{1-y} and in electron-radiation-damaged CdS crystals.

10.5.4 Electric Field Control Model for Junction Transport

As mentioned in our discussion of enhancement/quenching effects, loss of photoexcited electrons via recombination through interface states is greater when the depletion layer is wider. This is dramatized by the relation observed between the short-circuit current and the dark forward-bias current for a constant bias voltage. A cell with about 1 hr of heat treatment was restored and then observed during the process of optical degradation at 300°K, corresponding to tracing out the TROD curve of Fig. 10.13a. Measurements of dark forward-bias current and stable short-circuit current for white-light photoexcitation were made at a number of points along the curve, both before and after the maximum, by interrupting the degradation radiation. Results for a forward bias of 0.1 V given in Fig. 10.15 show an almost linear dependence over three orders of magnitude below the saturation of J_{sc}. Similar data are also shown in Fig. 10.15 for other conditions.

Since the single curve just described represents a variety of conditions from restored to degraded, and from partially enhanced to partially quenched, it seems likely that the same mechanism controls both J_{sc} (a "reverse" current, under photoexcitation) and J_d (a "forward" current under bias in the dark). Fahrenbruch and Bube (1974) suggested that the high electric field at the interface increases J_d tunneling through interface states. This also increases J_{sc} by reducing the probability for the loss of in-

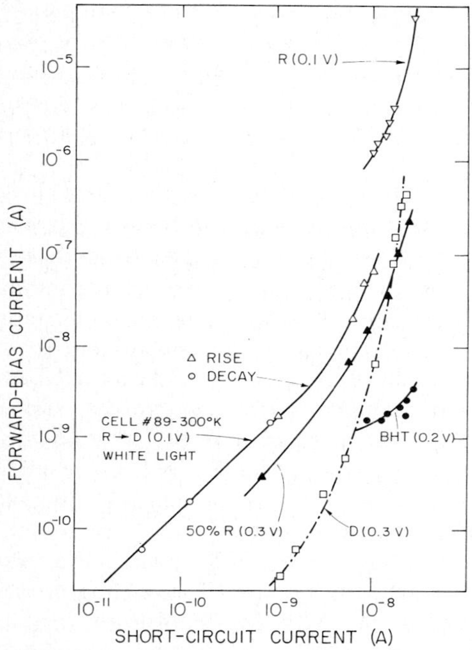

Fig. 10.15. Short-circuit current versus dark forward-bias current for the states of a Cu_xS/CdS heat-treated cell and for the bias voltages indicated. The curve marked BHT represents data obtained before heat treatment. All curves are for cell 61 at 150°K with 0.7 μm light, except for the single curve for cell 89 at 300°K as marked. [After A. L. Fahrenbruch and R. H. Bube, *J. Appl. Phys.* **45,** 1264 (1974).]

jected electrons via the tunneling/recombination pathway to the Cu_2S valence band (process 2 in Fig. 10.5a). The high electric field at the interface is correlated with a small depletion layer width. A simple calculation based on this tunneling/recombination model for control of the short-circuit current shows that

$$J_{sc} = J_{sco} \exp(-BW_d) = J_{sco} \exp(-B'/C), \tag{10.4}$$

where A and A' are constants, W_d is the effective depletion layer width, and C is the corresponding effective junction capacitance.† This behavior

† Equation (10.4) can also be expressed as

$$J_{sc} = J_{sco} \exp(-L/\tau\mathscr{E}\mu) = J_{sco} \exp(-S_i/\mathscr{E}\mu) \simeq J_{sco}/[1 + (S_i/\mathscr{E}\mu)], \tag{10.4'}$$

where L, τ, $\mathscr{E}$, and μ are the minority carrier diffusion length, effective lifetime, electric field, and mobility in the vicinity of the interface and S_i is an effective interfacial recombination velocity. In this form the equivalence of the exponential term to an interfacial collection function $h(V)$ is more obvious.

agrees well with what is observed (recall, for example, Fig. 10.12b). The same model can be used to explain the shape of the optical degradation curves in Fig. 10.13 by assuming that the optical degradation process creates states that participate in the tunneling/recombination process; the curves drawn for the cell in Fig. 10.13 have the form predicted by such a calculation.

An interesting feature of this model is the apparent inconsistency in the claim that a decrease in tunneling width at the interface leads to both an increase in dark forward-bias current and an increase in short-circuit, photogenerated current, even though tunneling/recombination is the primary mechanism for loss of J_{sc}. The resolution of this apparent inconsistency lies in the realization that J_{sc} is determined by two *competing* mechanisms. First is the effect of the electric field on the carrier density available for recombination; higher fields give smaller densities and hence smaller recombination rates. Second is the relationship of the electric field to the depletion layer width; higher fields correspond to smaller depletion layer widths and hence larger tunneling/recombination rates. The former effect dominates in the control of J_{sc}. For the dark forward-bias current, however, the predominant mechanism is the reduction of barrier thickness and the consequent increase in tunneling probability. The density of carriers available for forward-bias current transport is not affected by changes in the electric field.†

The model just described predicts the relationship between W_d and maximum power shown schematically in Fig. 10.16. For small W_d, J_0 is large and V_{oc} small; the junction is "leaky." But J_{sc} is large because of the large electric field at the interface. For large W_d, the value of J_0 is small and V_{oc} large, but many of the photogenerated carriers are lost to interfacial recombination as the electric field is reduced. An optimum lies between. In thin-film Cu_xS/CdS cells, a heat treatment is required to widen the depletion layer and achieve the maximum power output. In single-crystal cells the situation appears to be confined to the right of the maximum in Fig. 10.16.

Including the variation of W_d with illumination intensity, wavelength, and temperature, the model described is sufficient to explain, at this level, all the observed phenomena described in this section.

† A second aiding mechanism that may take place in junctions with lower carrier density in the Cu_xS (due, for example, to Cd compensation of Cu defect acceptors) is given here. Smaller depletion layer widths imply a greater unbalanced charge density on the CdS side of the junction. This gives a greater proportion of the diffusion potential borne by the Cu_xS side of the junction, effectively giving a lower barrier on the CdS side. Since the effective carrier density available for tunneling/recombination depends only on the tunneling energy, the forward-bias current is increased.

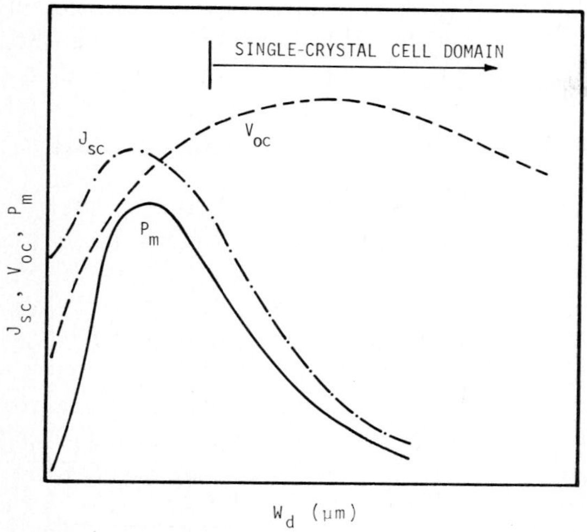

Fig. 10.16. Schematic plot of Cu_xS/CdS cell parameters versus depletion layer width.

Research by Haines and Bube (1980) on Cu_xS/CdS cells prepared on single-crystal and thin-film CdS by both the dipping and the dry processes† supports and extends the conclusions drawn earlier. According to this work, the behavior of the photovoltaic parameters can be described in terms of a model that involves a deep donor/like energy level in the CdS adjacent to the metallurgical interface. This center, with an ionization energy of 0.45 eV, acts as a tunneling/recombination site. The effects of temperature and spectral content on the junction transport are due to changes in the net positive charge near the interface, as described in Section 10.5.2. These changes are, in turn, determined by the joint action of this deep donor and the familiar Cu acceptor states in the CdS. Bias light techniques were used to investigate the phenomena, which allowed the net positive charge to be varied while maintaining constant J_{sc}. The deep donor acts as a recombination center controlling J_{sc}; for narrow barrier widths fewer of the deep donors lie above the Fermi level and the electric field is large so that the interfacial recombination is small. As the barrier width is increased, the number of deep donors available to act as recombination centers increases and the electric field also decreases. Thus a larger loss in J_{sc} occurs. This indicates that the interfacial recombination loss changes with variations in the spectral content of the incident light.

† These cells are as-prepared or with little heat treatment.

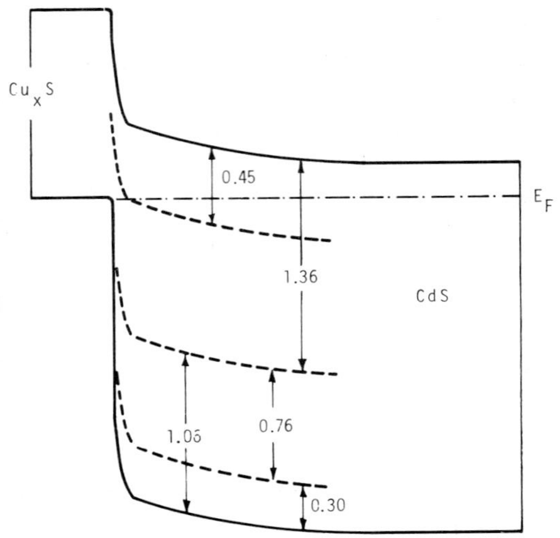

Fig. 10.17. Energies of the transitions between levels in the CdS layer that control the photovoltaic properties (in eV). [Redrawn from W. G. Haines and R. H. Bube, *IEEE Trans. Electron Devices*, **ED-27**, 2133 (1980). © 1980 IEEE.]

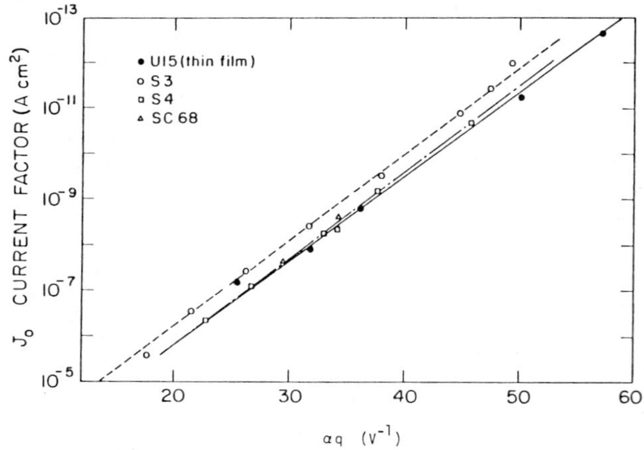

Fig. 10.18. Diode pre-exponential factor J_0 as a function of αq [as in $J = J_0[\exp(\alpha qV) - 1]$] measured by varying the intensity of bias light (cf., Fig. 5.29). Samples: U15, polycrystalline thin film; S3, S4, and SC 68, on single-crystal CdS substrates. Slope of curves is ~0.44 eV. [After W. G. Haines and R. H. Bube, *IEEE Trans. Electron Devices* **ED-27**, 2133 (1980). © 1980 IEEE.]

The combined effect is for large J_{sc} to correspond to small V_{oc}. These results are consistent with a density of deep donors of greater than 10^{19} cm^{-3} and the band diagram shown in Fig. 10.17.

A particularly interesting facet of the work described above is the way in which the diode parameters J_0 and A are governed by the magnitude of the net positive charge adjacent to the interface. This is illustrated by the plot of ln J_0 versus αq for a variety of charge densities, shown in Fig. 10.18 ($\alpha q = q/AkT$).

10.6 $Cu_xS/Zn_yCd_{1-y}S$ CELLS

Efforts to extend and improve the performance of Cu_xS/CdS cells have turned naturally to the substitution of $Zn_yCd_{1-y}S$ solid solutions for the CdS. The increase in band-gap achieved by partial substitution of Zn for Cd increases the band-gap window and the short-circuit current for a backwall cell. But the largest improvement from the use of $Zn_yCd_{1-y}S$ is a decrease in the conduction band discontinuity ΔE_c and hence an increase in expected open-circuit voltage. It is anticipated that $V_{oc} = 0.74$ V can be achieved in a $Cu_xS/Zn_yCd_{1-y}S$ cell (Rothwarf and Barnett, 1977), using a junction area equal to the geometric cell area [Eq. (10.3)]. Coupling this open-circuit voltage with a short-circuit current of 26 mA cm^{-2} and a fill factor of 0.78 leads to a solar efficiency of about 15% as a reasonable goal. A calculation of the ideal efficiency of such a $Cu_xS/Zn_yCd_{1-y}S$ cell gives a value of 26% without losses.

The maximum benefit of the $Zn_yCd_{1-y}S$ would be obtained for a value of y sufficiently large to make $\Delta E_c = 0$; that is, $y \simeq 0.33$. However, the bulk resistance of $Zn_yCd_{1-y}S$ increases rapidly with increasing y, and it is unlikely that values of y greater than about 0.20 can be effectively used in solar cells. The dependence of resistivity on y is given in Fig. 10.19 for both single-crystal and thin-film $Zn_yCd_{1-y}S$. In single-crystal samples there is an abrupt increase in resistivity between 10% and 20% Zn by some nine orders of magnitude; apparently this is correlated with an increase in the ionization energy of dominant donors from 0.03 eV in CdS to 0.30 eV in $Zn_{0.2}Cd_{0.8}S$. In thin films, apparently independent of the method of deposition, the resistivity increases more rapidly between $y = 0$ and $y = 0.2$ than for higher values of y but exhibits a continuous increase with y over the whole range. Hench and Hall (1979) have prepared films with $y = 0.18$ that have through-the-film resistivities of about 10 Ω cm.

Measurements on dipped $Cu_xS/Zn_{0.16}Cd_{0.84}S$ cells by Hall *et al.* (1981)

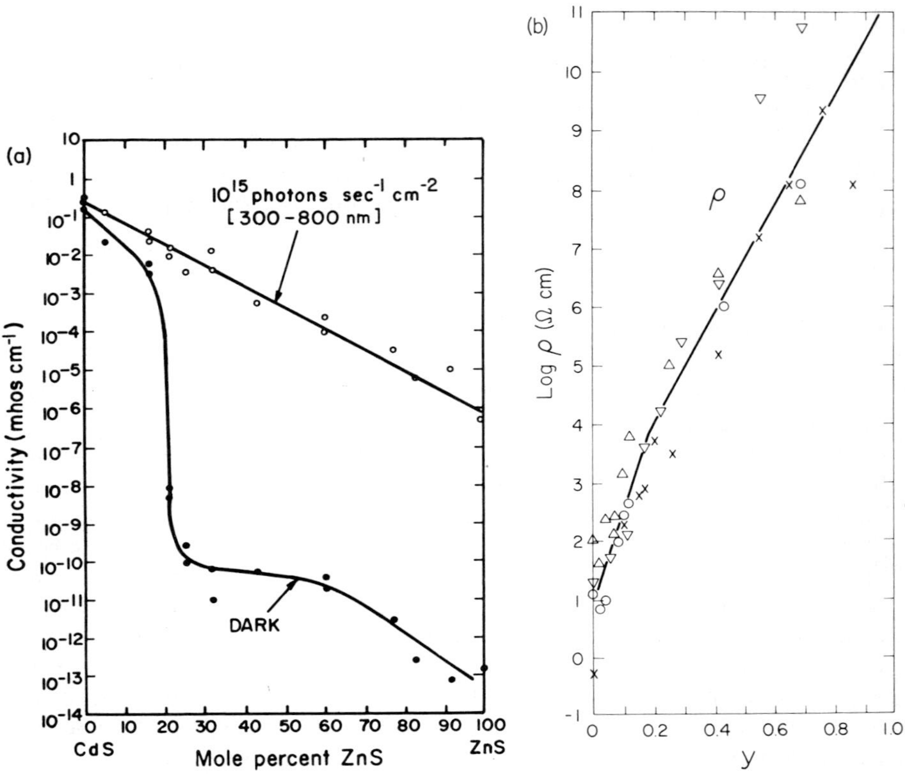

Fig. 10.19. (a) Variation of conductivity with y in $Zn_yCd_{1-y}S$ single crystals in dark and under photoexcitation. [After E. A. Davis and E. L. Lind, *J. Phys. Chem. Solids* **29,** 79 (1968).] (b) Variation of resistivity with y in $Zn_yCd_{1-y}S$ thin films in the dark. Circles and triangles are for films deposited under different conditions by spray pyrolysis. [After R. S. Feigelson, A. N'Diaye, S.-Y. Yin, and R. H. Bube, *J. Appl. Phys.* **48,** 3162 (1977).] Crosses are for films deposited by vacuum evaporation. [After L. C. Burton, B. Baron, W. Devaney, T. L. Hench, S. Lorenz, and J. D. Meakin, *Proc. 12th IEEE Photovoltaic Specialists Conf.* (1976), p. 526. © 1976 IEEE.]

do indeed give increased V_{oc}, but with slightly decreased J_{sc} (V_{oc} = 0.599 V, J_{sc} = 18.5 mA cm^{-2} in 81 mW cm^{-2} direct sunlight, giving η_s = 10.2%).

10.7 SUMMARY

The Cu_xS/CdS cell occupies an almost unique status in photovoltaics. On the one hand, the ease of deposition of CdS by a variety of methods and the apparent ease of fabrication of cells give it a veneer of

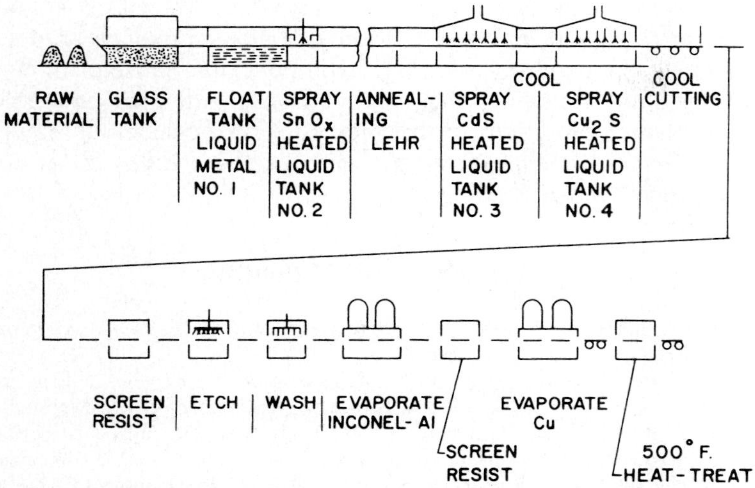

Fig. 10.20. Schematic of float glass plant with additional section for automatic production of solar cells. [From J. F. Jordan, *Proc. 11th IEEE Photovoltaic Specialists Conf.* (1975), p. 508. © 1975 IEEE.]

simplicity and promise inexpensive, large-scale production. On the other hand, it is certainly one of the more complex photovoltaic systems. Almost all the phenomena observed in the Cu_xS/CdS cell have been seen to a lesser degree in other systems such as Si cells, but in Cu_xS/CdS they are magnified to an extent that would seem almost pathological in the other cells. The Cu_xS/CdS cell continues to occupy a prominent place in the world of photovoltaics by virtue of having been the most developed, and the most efficient polycrystalline, thin-film cell for more than 25 years.†

Fabrication of the cell continues to require a fortuitous combination of alchemy and science. Both the "wet" and "dry" processes of junction formation are still not completely understood. Junction transport is on the verge of being understood, but knowledge about the recombination centers in the vicinity of the interface and about the effects of grain boundaries on junction transport is so far insufficient to manipulate these quantities. The question of stability is also still unresolved.

The commercial scale-up possibilities of the Cu_xS/CdS cell are illustrated by an interesting concept due to Jordan (1975). In this assembly-line process, the output of a float-glass manufacturing plant is fed directly

† The authors feel that they and the rest of the photovoltaic community owe a debt of gratitude to Karl Böer. Böer's scientific contributions and his enthusiasm kept interest and research funds up through the lean years when 6% efficiency was seen only in the most exceptional of Cu_xS/CdS cells.

into a photovoltaic cell fabrication system while the glass is still at high temperature, Fig. 10.20. A layer of SnO_x, followed by a layer of CdS, is deposited by spray pyrolysis onto the still-hot glass. Subsequent high-rate processes apply contacts and protective layers. Although cells produced by pilot plant explorations of this method show considerably lower efficiency at this time, high throughput may be sufficient to offset a somewhat lower efficiency figure.

GENERAL REFERENCES

Böer, K. W., and Meakin, J. D., eds. (1975). *Int. NSF Workshop CdS Sol. Cells Other Abrupt Heterojunctions, Univ. of Delaware, Newark.*

DeMeo, E. (1976). "Assessment of CdS Photovoltaic Arrays for Large Scale Electric Utility Applications." EPRI Rep. EER-188. Brown Univ., Providence, Rhode Island.

Shiozawa, L. R., Augustine, F., Sullivan, G. A., Smith, J. M., III, and Cook, W. R. (1969). "Research on the Mechanism of the Photovoltaic Effect in High-Efficiency CdS Thin-Film Solar Cells." Clevite Corp. Final Rep. (Air Force Contract AF 33 (615)-5224, 1966-1969). Aerosp. Res. Lab., Office of Aerosp. Res., Wright-Patterson Air Force Base, Ohio.

Stanley, A. G. (1975). Cadmium sulfide solar cells. *Appl. Solid State Sci.* **5,** 251.

Stanley, A. G. (1978). "Cadmium Sulfide," Vols. I and II, DOE/JPL Tech. Rep., JPL 78-77 **1** and **2.** Pasadena, California.

REFERENCES

Armantrout, G. A., Miller, D. E., Vindelov, K. E., and Brown, T. G. (1979). *J. Vac. Sci. Technol.* **16,** 212.

Barnett, A. M., Bragagnolo, J. A., Hall, R. B., Phillips, J. E., and Meakin, J. D. (1978). *Proc. 13th IEEE Photovoltaic Specialists Conf.,* p. 419.

Baron, B., Catalano, A. W., and Fagen, E. A. (1978). *Proc. 13th IEEE Photovoltaic Specialists Conf.,* p. 406.

Böer, K. W. (1976). *Phys. Rev. B* **13,** 5373.

Bogus, K., and Mattes, S. (1972). *Proc. 9th IEEE Photovoltaic Specialists Conf.,* p. 106.

Bordure, G., Llinares, C., Bougnot, J., Perotin, M., Luqet, H., Belgacem, S., and Saveli, M. (1980). *Proc. 3rd EC Photovoltaic Sol. Energy Conf., Cannes,* p. 335. D. Reidel Publ., Dordrecht, Holland.

Bougnot, J., Guastavino, F., Mousalli, G. M., and Savelli, M. (1975). *Int. NSF Workshop CdS Sol. Cells Other Abrupt Heterojunctions,* Univ. of Delaware, Newark (K. W. Böer and J. D. Meakin, eds.), p. 337.

Bragagnolo, J. A., Barnett, A. M., Phillips, J. E., Hall, R. B., Rothwarf, A., and Meakin, J. (1980). *IEEE Trans. Electron Devices* **ED-27,** 645.

Bube, R. H. (1976). Covalent semiconductors. *In* "Photoconductivity and Related Phenomena" (J. Mort and D. M. Pai, eds.), p. 117. Elsevier, Amsterdam.

Burton, L. C., Uppal, P. N., and Dwight, D. W. (1981). *Proc. 15th IEEE Photovoltaic Specialists Conf.,* p. 780.

Cabannes, F. (1958). *C. R. Acad. Sci.* **246,** 257.

Castel, E. D., and Soubeyrand, M. J. (1981). *Proc. 15th IEEE Photovoltaic Specialists Conf.*, p. 772.
Chamberlin, R. R., and Skarman, J. S. (1966). *J. Electrochem. Soc.* **113,** 86 [and *Solid-State Electron.* **9,** 819].
Cheng, L. J., Turner, G. B., Downing, R. G., Wang, E. Y., and Seaman, C. H. (1978). *Proc. 13th IEEE Photovoltaic Specialists Conf.*, p. 1333.
Cusano, D. A. (1963). *Solid-State Electron.* **6,** 217.
Dieleman, J. (1975). *Int. NSF Workshop CdS Sol. Cells Other Abrupt Heterojunctions, Univ. of Delaware, Newark* (K. W. Böer and J. D. Meakin, eds.), p. 92.
Donaghey, L. F., Duisman, J. A., Peterson, T. M., and Ryason, P. R. (1980). *Proc. 14th IEEE Photovoltaic Specialits Conf.*, p. 728.
Duc Cuong, N., and Blair, J. (1966). *J. Appl. Phys.* **37,** 1660.
Egorova, I. V. (1968). *Sov. Phys.—Semicond.* **2,** 266.
Eser, E. and Cambridge, J. A. (1981). In "CdS/Cu_2S Heterojunction Cell Research." SERI Final Report No. XS-9-8310-1-06, June, 1981, Appendix C. Inst. Energy Conver., Univ. of Delaware, Newark.
Fabricius, E. D. (1962). *J. Appl. Phys.* **33,** 1597.
Fahrenbruch, A. L. (1973). Ph.D. Thesis, Stanford Univ., Stanford, California.
Fahrenbruch, A. L., and Bube, R. H. (1974). *J. Appl. Phys.* **45,** 1264.
Florio, J. V., Matysik, K. J., and Ramos, F. G. (1981). *Proc. 15th IEEE Photovoltaic Specialists Conf.*, p. 793.
Gill, W. D., and Bube, R. H. (1970a). *J. Appl. Phys.* **41,** 1694.
Gill, W. D., and Bube, R. H. (1970b). *J. Appl. Phys.* **41,** 3731.
Grimmeiss, H. G., and Memming, R. (1962). *J. Appl. Phys.* **33,** 2217, 3596.
Haines, W. G., and Bube, R. H. (1980). *IEEE Trans. Electron Devices* **ED-27,** 2133.
Hall, R. B., and Meakin, J. D. (1979). Tech. Rep. IEC/PV/TR/79/6. Inst. Energy Convers., Univ. of Delaware, Newark.
Hall, R. B., and Singh, V. P. (1979). *J. Appl. Phys.* **50,** 6406.
Hall, R. B., Birkmire, R. W., Phillips, J. E., and Meakin, J. D. (1981). *Proc. 15th IEEE Photovoltaic Specialists Conf.*, p. 777.
Hench, T. L., and Hall, R. B., (1979). *Proc. 2nd EC Photovoltaic Sol. Energy Conf., Berlin*, p. 379. D. Reidel Publ., Dordrecht, Holland.
Hill, E. R., and Kermidas, B. G. (1967). *IEEE Trans. Electron. Devices* **ED-14,** 22.
Im, H.-B., Matthews, H. E., and Bube, R. H. (1970). *J. Appl. Phys.* **41,** 2581.
Jacobs, B., and De Mey, G. (1981). *IEEE Trans. Electron Devices* **ED-28,** 289.
Jonath, A. D., Anderson, W. W., Thornton, J. A., and Cornog, D. G. (1979). *J. Vac. Sci. Technol.* **16,** 200.
Jordon, J. (1975). *Proc. 11th IEEE Photovoltaic Specialists Conf.* p. 508.
Kanev, S. K., Fahrenbruch, A. L., and Bube, R. H. (1971). *Appl. Phys. Lett.* **19,** 459.
Kanev, S., Stoyanov, V., and Lakova, M. (1969). *Dokl. Bolg. Akad. Nauk.* **22,** 863.
Keating, P. N. (1963). *J. Phys. Chem. Solids* **24,** 1101.
Korsunskaya, N. E., Markevich, I. V., and Sheinkman, M. K. (1966). *Phys. Status Solidi* **13,** 25.
Kul'sha, A. M., Sukhotskii, V. P., and Yarzhembitskii, V. B. (1976). *Sov. Phys.—Semicond.* **10,** 1180.
Lindmayer, J., and Revesz, A. G. (1971). *Solid-State Electron.* **14,** 647.
Lindquist, P. F., and Bube, R. H. (1972a). *J. Appl. Phys.* **43,** 2839.
Lindquist, P. F., and Bube, R. H. (1972b). *J. Electrochem. Soc.* **119,** 936.
Martinuzzi, S., Cabane-Brouty, F., and Bretzner, J. F. (1972). *Proc. 9th IEEE Photovoltaic Specialists Conf.*, p. 111.

Mathieu, H. J., Reinhartz, K. K., and Rickert, H. (1973). *Proc. 10th IEEE Photovoltaic Specialists Conf.*, p. 93.
Matsumoto, H., Nakayama, N., and Ikegami, S. (1980). *Jpn. J. Appl. Phys.* **19,** 129.
Meakin, J. D., and Phillips, J. E. (1979). In "Cadmium Sulfide/Copper Sulfide Heterojunction Cell Research" (J. D. Meakin, ed.), App. F. Dept. of Energy Final Rep. No. EG-77-C-03-1576-FR, May 1979. Inst. Energy Conver., Univ. of Delaware, Newark.
Miya, N. (1970). *Jpn. J. Appl. Phys.* **9,** 768.
Moore, K., Casperd, A., Casselton, R. E. W., and Moodie, W. (1977). *Proc. 1st EC Photovoltaic Sol. Energy Conf., Luxembourg,* p. 1278. D. Reidel Publ. Dordrecht, Holland.
Moorthy, R., Partain, L., Okubo, K., and Henderson, D. (1979). *Tech. Dig.—Int. Electron Devices Meet., Washington, D.C.*, p. 442.
Moss, H. (1960). *Proc. Natl. Aeronaut. Electron. Conf.*, p. 47.
Mukherjee, M. K., Pfisterer, F., Hewig, G. H., Schock, H. W., and Bloss, W. H. (1977). *J. Appl. Phys.* **48,** 1538.
Mukherjee, M. K., Saha, A. R., and Das, S. N. (1978). *IEEE Trans. Electron Devices* **ED-25,** 285.
Mytton, R. J. (1968). *Br. J. Appl. Phys.* **1,** 721.
Nadjakov, G., *et al.* (1954). *Izv. Bulg. Akad. Nauk* **4,** 10.
Nakayama, N. (1969). *Jpn. J. Appl. Phys.* **8,** 450.
Nakayama, N., Gyobu, A., and Morimoto, N. (1971). *Jpn. J. Appl. Phys.* **10,** 1415.
Norian, K. H., and Edington, J. W. (1980). *Proc. 14th IEEE Photovoltaic Specialists Conf.*, p. 700.
Palz, W., Besson, J., Nguyen Duy, T., and Vedel, J. (1972). *Proc. 9th IEEE Photovoltaic Specialists Conf.*, p. 91.
Partain, L. D., and Birchenall, C. E. (1975). *Int. NSF Workshop CdS Solar Cells Other Abrupt Heterojunctions, Univ. of Delaware, Newark* (K. W. Böer and J. D. Meakin, eds.), p. 355.
Partain, L., Oakes, J. H., and Greenfield, I. G. (1975). *Int. NSF Workshop CdS Sol. Cells Other Abrupt Heterojunctions* Univ. of Delaware, Newark. (K. W. Böer and J. D. Meakin, eds.), p. 346.
Partain, L., Armantrout, G., and Okubo, D. (1980). *IEEE Trans. Electron Devices* **ED-27,** 2127.
Pfisterer, F., Hewig, G. H., and Bloss, W. H. (1975). *Proc. 11th IEEE Photovoltaic Specialists Conf.*, p. 460.
Pfisterer, F., Schock, H. W., and Bloss, W. H. (1976). *Proc. 12th IEEE Photovoltaic Specialists Conf.*, p. 502.
Pfisterer, F., Schock, H. W., and Hewig, G. H. (1979). *Proc. 2nd EC Photovoltaic Sol. Energy Conf., Berlin,* p. 352. D. Reidel Publ., Dordrecht, Holland.
Pfisterer, F., Schock, H. W., and Wörner, J. (1980). *Proc. 3rd EC Photovoltaic Sol. Energy Conf., Cannes,* p. 762. D. Reidel Publ., Dordrecht, Holland.
Potter, A. E., Jr., and Schalla, R. L. (1967). "Mechanism of CdS Film Cell," NASA Tech. Note, NASA TN D-3849. Washington, D.C.
Reynolds, D. C., and Czyzak, S. J. (1954). *Phys. Rev.* **96,** 1705.
Reynolds, D. C., Leies, G., Antes, L. L., and Marburger, R. E. (1954). *Phys. Rev.* **96,** 533.
Roderick, G. A. (1980). *Proc. 3rd EC Photovoltaic Sol. Energy Conf., Cannes,* p. 327. D. Reidel Publ., Dordrecht, Holland.
Rothwarf, A. (1976). "Theoretical Prospects for the CdS-Cu_2S Cell." Tech. Rep. NSF/RANN/AER-72-03478, A04/TR 76/1. Inst. Energy Convers., Univ. of Delaware, Newark.
Rothwarf, A. (1979). *Proc. 2nd EC Photovoltaic Sol. Energy Conf., Berlin,* p. 370. D. Reidel Publ., Dordrecht, Holland.

Rothwarf, A., and Barnett, A. M. (1977). *IEEE Trans. Electron Devices* **ED-24,** 381.
Rothwarf, A., and Windawi, H. (1980). *Proc. 14th IEEE Photovoltaic Specialists Conf.,* p. 722.
Rothwarf, A., and Windawi, H. (1981). *IEEE Trans. Electron Devices* **ED-28,** 64.
Rothwarf, A., Burton, L. C., Hadley, H. C., Jr., and Storti, G. M. (1975). *Proc. 11th IEEE Photovoltaic Specialists Conf.,* p. 476.
Rothwarf, A., Phillips, J., and Wyeth, N. C. (1978). *Proc. 13th IEEE Photovoltaics Specialists Conf.,* p. 399.
Savelli, M., and Bougnot, J. (1979). Problems of the Cu_2S/CdS cell. *In* "Solar Energy Conversion—Solid State Physics Aspects" (B. O. Seraphin, ed.), Topics in Applied Physics, Vol. 31, p. 213. Springer-Verlag, Berlin and New York.
Schock, H. W., Gilger, G., Bloss, W. H., and Pfisterer, F. (1977). *Vacuum* **27,** 281.
Selle, B., Ludwig, W., and Mach, R. (1967). *Phys. Status Solidi* **24,** K145.
Shewchun, J., Loferski, J. J., Wold, A., Arnott, R., DeMeo, E. A., Beaulieu, R., Wu, C.-C., and Hwang, H. L. (1975). *Proc. 11th IEEE Photovoltaic Specialists Conf.* p. 482.
Shiozawa, L. R., Sullivan, G. A., and Augustine, F. (1968). *Proc. 7th IEEE Photovoltaic Specialists Conference,* p. 39.
Shiozawa, L. R., Augustine, F., Sullivan, G. A., Smith, J. M., III, and Cook, W. R., (1969). "Research on the Mechanism of the Photovoltaic Effect in High-Efficiency CdS Thin-Film Solar Cells." Clevite Corp. Final Rep. (Air Force Contract AF 33(615)-5224, 1966-1969). Aerosp. Res. Lab., Office of Aerosp. Res., Wright-Patterson Air Force Base, Ohio.
Shirland, F. A. (1966). *Advanced Energy Convers.* **6,** 201. [Reprinted in "Solar Cells" (C. E. Backus, ed.), p. 43. IEEE Press, New York, 1976.]
Shirland, F. A., and Hietanen, J. R. (1965). *Proc. 19th Annu. Power Sources Conf.,* p. 177.
Stanley, A. G. (1975). Cadmium sulfide solar cells. *Appl. Solid State Sci.* **5,** 251.
TeVelde, T. S. (1973). *Solid-State Electron.* **16,** 1305.
TeVelde, T. S. (1975). *Energy Convers.* **15,** 111.
TeVelde, T. S., and Dieleman, J., (1973). *Philips Res. Rep.,* **28,** 573.
Tscholl, E. (1968). The photochemical interpretation of slow phenomena in CdS. *Philips Res. Rep. Suppl.* **6.**
Vanderwel, T., Clarke, R., Scholz, F., Burk, D., Dalacu, N., and Shewchun, J. (1980). *Proc. 14th IEEE Photovoltaics Specialists Conf.,* p. 694.
Weizer, V. G., Brandhorst, H. W., Broder, J. D., Hart, R. E., and Lamneck, J. H. (1978). *Proc. 13th IEEE Photovoltaic Specialists Conf.,* p. 1327.
Williams, R., and Bube, R. H. (1960). *J. Appl. Phys.* **31,** 968.
Woods, J., and Champion, J. A. (1960). *J. Electron. Control* **7,** 243.
Wyeth, N. C., and Rothwarf, A. (1979). *J. Vac. Sci. Technol.* **16,** 1402.
Zwicker, H. R., Brickman, L. A., Hadley, H. C., and Matysik, K. J. (1981). *Proc. 15th IEEE Photovoltaic Specialists Conf.,* p. 787.

CHAPTER 11

OTHER PHOTOVOLTAIC CELLS

So far in this book we have considered two major materials systems based on single-crystal technology: Si and GaAs. With their variations these promise to make large contributions to terrestrial solar concentrator systems and to space power systems. The all-thin-film system Cu_xS/CdS is, at least, a prototype of cells that will be used for large thin-film terrestrial applications. In this chapter we briefly describe a sampling of other photovoltaic cells that have demonstrated both scientific and practical interest. Since the number of possible photovoltaic converters is so large that a survey of them rapidly becomes obsolete, these descriptions are not meant to be exhaustive but rather illustrative.

We begin with a discussion of Schottky barrier, metal/insulator/semiconductor (MIS), and semiconductor/insulator/semiconductor (SIS) cells. As discussed in Section 5.7.4, the distinction between a Schottky barrier and a junction with an intervening insulating layer appears to be a matter of degree, especially since it is extremely difficult to avoid such a layer in the fabrication of most junctions. Perhaps equally important is the chemical reactivity of the semiconductor junction components on an atomic scale at the interface.†

Highly conductive metal oxides form the window layers in a number of solar cells. Strictly speaking these are SIS structures, but since the

† See the papers in the Proceedings of the Sixth Conference on the Physics of Compound Semiconductor Interfaces, edited by P. Mark, published as *J. Vac. Sci. Technol.* **16**(5), (1979). In particular, L. Brillson, *J. Vac. Sci. Technol.* **16**(5), 1137 (1979).

metal-oxide layers are usually degenerate semiconductors, the cells often behave like MIS diodes. Thus the distinction between these cells and those with pure metals as window layers is also indistinct.

In Sections 11.2–11.5 we discuss solar cells based on amorphous Si, InP, CdTe, and the I–III–VI_2 chalcogenides, all of which are candidates for thin-film devices.

In Section 11.6 we consider some exploratory materials, such as Zn_3P_2, WSe_2, quaternary compounds, and the polymers $(SN)_x$ and $(CH)_x$.

11.1 SCHOTTKY, MIS, AND SIS JUNCTION SOLAR CELLS

Schottky barrier junctions involving Si, GaAs, and InP have been considered to simplify and to reduce the cost of photovoltaic cells using these high-efficiency materials.† Many advantages can be derived from going to a Schottky barrier structure, including reduced fabrication temperatures, which avoid degradation of minority carrier lifetime; lower cost of barrier formation; reduction of the losses associated with shallow homojunctions and surface recombination; increased short-wavelength response; and easier large-scale fabrication. Last, but hardly least, is the inherent adaptability of the Schottky barrier to polycrystalline thin-film systems, where rapid grain-boundary diffusion makes formation of p/n junctions by diffusion difficult.

Of course there are also disadvantages to the Schottky barrier device. Because the diode preexponential factor J_0 is governed by thermionic emission across the Schottky barrier, its magnitude is relatively large, and the measured values of V_{oc} are usually smaller than those of homojunction or heterojunction cells. This basic handicap of a Schottky barrier cell can be overcome by introducing a thin, insulating layer between the barrier metal and the semiconductor, as shown in Fig. 11.1. The result is called an MIS structure (or MOS structure, for the special case when the insulator is an oxide). If the metal is replaced by a semiconductor, then an SIS device is formed. Even when the insulating interfacial layer is thin enough not to hinder the collection of photogenerated carriers, it can still improve cell performance by decreasing the dark diode current J_0, by increasing the effective barrier at the semiconductor surface, or by increasing the value of the diode factor.

The specific mechanisms by which the oxide interface layer improves

† Of course, Schottky barrier cells have been made with most of the other common semiconductors as well. For a listing, see Bucher (1978).

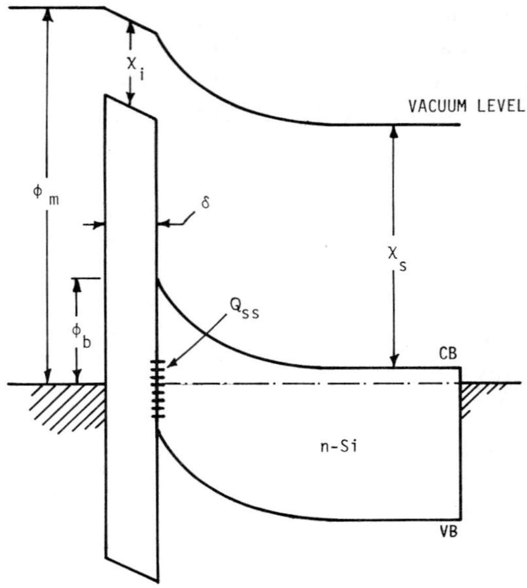

Fig. 11.1 Schematic band diagram of *n*-Si-based MIS junction. The insulator thickness is exaggerated for clarity.

the performance of MIS devices have received considerable attention, but without complete resolution to date. The theories of Schottky, MIS, and SIS junctions are discussed in Section 5.7 and MIS junctions are the subject of an excellent review by Pulfrey (1978).

The composition and structure of the insulator layer, as well as its thickness, are critical. Both Si and GaAs form thin insulating oxides simply by being exposed to air at room temperature, but elevated temperatures (200–900°C) are usually necessary to form the 20–30-Å thickness required for an MIS device in a reasonable time (Pulfrey, 1978). SiO_x is the most widely researched insulator for Si-based MIS solar cells. Many different procedures exist for the formation of such layers, including heating in wet or dry O_2, wet N_2, steam, and other gas mixtures; oxidation in acids such as HNO_3 or HF + H_2O; anodization; and vacuum evaporation or chemical vapor deposition of the layer, to name a few.

The electronic properties of oxides are strongly dependent on the details of their formation. The photovoltaic parameters of the resulting MIS cells, especially the value of V_{oc}, depend critically on which method is used and how it is applied. For example, procedures that leave an incompletely oxidized layer appear to yield lower V_{oc} for M/SiO_x/*n*-Si cells (Pulfrey, 1978), whereas methods leaving traces of OH or F in the layer are reported to give higher V_{oc} for M/SiO_x/*n*-Si devices (Kipperman *et al.*, 1978a, 1978b). In contrast, for *p*-Si-based devices the presence of OH appears detrimental and "dry" oxides give higher V_{oc}.

The importance of the chemical processes inherent in the fabrication of MIS junctions is illustrated dramatically by the results of Ponpon and Siffert (1978). They fabricated Au/*n*-Si devices by vacuum evaporation of Au and measured the electrical parameters of the resulting devices while they were still *in vacuo* ($p \simeq 10^{-6}$ Torr). The *n*-Si substrates had a very thin native oxide layer before deposition of the Au. They found these junctions were *not* rectifying and showed negligible photovoltaic response before exposure to air. When exposed to air the devices then developed the electrical properties generally seen by other researchers (e.g., barrier heights of $\simeq$0.8 eV) after periods of $\simeq$3000 min in dry air or $\simeq$30 min in moist air (for 100 Å of Au). The time rates for evolution of the electrical properties ("aging") were consistent with rates for the diffusion of gases through thin metal films. Ponpon and Siffert concluded that the presence of an Au overlayer increased the driving force for the oxidation of Si so that the initial native oxide began renewed growth after the Au deposition. In addition, this secondary oxide layer growth was a requirement for rectification in this system. Similar "aging" processes in Schottky and MIS devices have been noted by other investigators (Turner and Rhoderick, 1968).

In the case of GaAs, the oxide formation process is more complex (Schwartz *et al.*, 1979), involving Ga_2O_3, As_2O_3, and $GaAsO_4$, but apparently similar sensitivity of the electrical properties to formation details is observed.

Good results have been obtained by deposition of foreign insulating materials by means of vacuum evaporation or CVD. Of particular interest is the use of vacuum-evaporated, insulating Sb_2O_3 films for *p*-GaAs MIS cells, giving comparable or higher η_s than is available by other methods (Stirn *et al.*, 1977). As another example, the electrical properties and stability of silicon nitride have been shown to be excellent for integrated-circuit applications and Al/Si_3N_4/*p*-Si solar cells have been fabricated with $\eta_s = 10.2\%$ (Charlson and Richardson, 1978).

The intricacies of the formation of the insulator layer just discussed [e.g., the dependence on preparation variables and the need for postfabrication stabilization (aging) for some systems] are undoubtedly the cause of much of the scatter in the data for barrier heights and diode parameters reported in the literature.

The theory and current status of MIS and SIS cells are reviewed by Shewchun *et al.* (1980) and by Ng and Card (1980). A review of "CIS" solar cells (i.e., conductor/insulator/semiconductor), including MIS, SIS, and semiconductor/insulator junctions with electrolytes and metal-like polymers such as SN_x, is given by Singh *et al.* (1981).

Because the insulating layers of these cells are only 15 to 20 Å thick there is some question about their long-term stability. These questions are discussed by Kleta and Pulfrey (1980) and Godfrey and Green (1979).

11.1.1 Si-Based Junctions with Metal Barriers

MIS junctions on Si have been investigated extensively and the solar efficiencies of the MIS cells prepared to date are comparable to the best *p/n* Si homojunction cells. Theoretical estimates for the maximum efficiencies of the devices are 16–21%.

One example of an early MIS cell is the Cr/*i*/*p*-Si device with the layered structure:

Al, ohmic contact;
p-Si, 2 Ω cm, ⟨100⟩;
SiO_2, 20 Å, by oxidation in air at ≃620°C;
Cr, 30 Å;
Cu, 50 Å to increase conductivity;
Cr, 10 Å, as an oxidation barrier;
Al, grid fingers, 1 μm thick; and
SiO, 700 Å, as an antireflection coating.

For this cell, values of $V_{oc} = 0.58$ V, $J_{sc} = 20$ mA cm^{-2}, ff = 0.72, and $\eta_s = 12\%$ are reported (area = 1.5 cm^2) (Anderson *et al.*, 1974, 1977; Anderson, 1977). Another example is the Al/*i*/*p*-Si junction using

p-Si, 3.5–15 Ω cm;
SiO_2, 20–30 Å, by oxidation in N_2/O_2 at 480–500°C;
Al, 100–150 Å;

that gives $V_{oc} = 0.45$ V, $J_{sc} = 26.5$ mA cm^{-2}, and $\eta_s = 8\%$ (Charlson and Lien, 1975). Another variation is the Au/*i*/*n*-S; junction with the structure

n-Si, 1–10 Ω cm;
SiO_2, 10–40 Å, by vacuum evaporation;
Au, 100 Å;

giving $V_{oc} = 0.54$, $J_{sc} = 23$ mA cm^{-2}, ff = 0.75, and $\eta_s = 9\%$ (Ponpon and Siffert, 1976).

One striking effect of the oxide interlayer in the Al/*i*/*p*-Si device is a large increase in barrier height. The barrier height for Al on a "clean" Si surface is only about 0.4 eV, but with an oxide interface layer, the value increases to about 0.85 eV (Charlson and Lien, 1975). From the Cr/*i*/*p*-Si MIS device work, there are indications that the effective work function of Cr is reduced from its normal bulk value of 4.55 eV to 4.12 eV

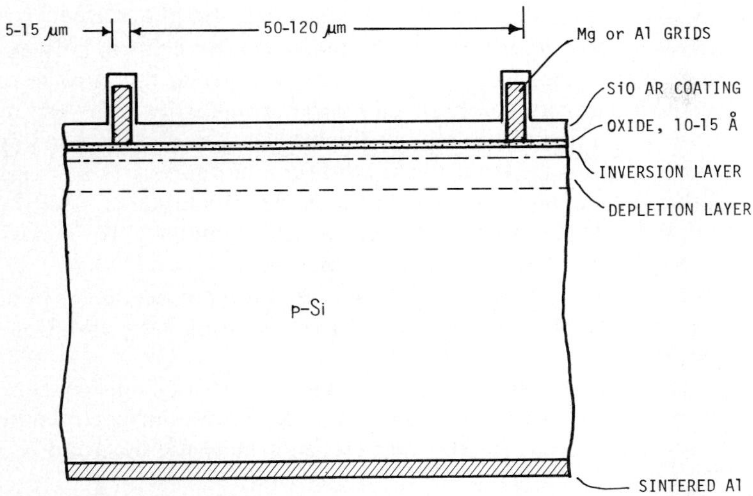

Fig. 11.2. MIS-grating-type inversion layer cell. [Redrawn from R. B. Godfrey and M. A. Green, *Appl. Phys. Lett.* **34,** 790 (1979).]

when the Cr is deposited slowly onto the oxide layer (Anderson *et al.*, 1974, 1977).

Successful application of several MIS concepts is illustrated by the "grating"† inversion layer MIS cell (Van Halen *et al.*, 1976). The charge in the insulator layer adjacent to the interface in this cell not only is sufficient to form a barrier without the presence of the metal contact, but is strong enough to drive the semiconductor surface to inversion (Salter and Thomas, 1975). To take advantage of this effect the metal contact is applied only at grid lines and lateral current collection is through the inversion layer by *both* drift and diffusion (Fig. 11.2). Grid line separation must be small, 50–150 μm typically, and so photolithographic fabrication is required. In a sense this device is a *n*/*p* homojunction in which the front, *n*-type layer is formed by inversion rather than by doping; this approach eliminates heavy doping effects for the *n*-layer, such as band-gap narrowing and mobility and lifetime degradation.

Godfrey and Green (1979) have reported $\eta_s = 17.6\%$ (*active area basis,* AM1, simulated) for a 3 cm^2, "grating" inversion layer cell in which the metal (Mg or Al) forming the barrier covers only a small fraction of the light-incident area.‡ The float-zone Si base is *p*-type with a

† The term "grating" describes the use of grid lines with spacing on the order of the minority carrier diffusion length so that carriers are collected by diffusion (Loferski *et al.*, 1973).

‡ Also see Thomas *et al.* (1980).

resistivity of 0.1 Ω cm. The electric charge within the thin (10–15 Å) SiO_2 layer and in the SiO layer forming the antireflection coating between the metal grids is sufficient to form the inversion layer at the surface of the *p*-Si. The diode current is dominated by minority carrier transport in this device and so V_{oc} increases almost linearly with a decrease in the log resistivity for the Si base from 10 to 0.10 Ω cm, as shown in Fig. 11.3. The largest values of V_{oc} obtained are 0.651–0.655 V (simulated AM0), compared with 0.620 V (Fossum and Burgess, 1978) and 0.637 V (D'Aiello *et al.*, 1976) for the best Si *n*/*p* homojunction cells.

Tarr and Pulfrey (1979) have substantiated the dominance of minority carrier transport in $Al/SiO_x/p$-Si devices by examining the variation of J_0 with temperature.

Charged oxide layers can also be used to enhance the emitter efficiency of *p*/*n* homojunction cells by suppressing front-surface recombination (Neugroschel *et al.*, 1978). The best features of the MIS and *n*/*p* ho-

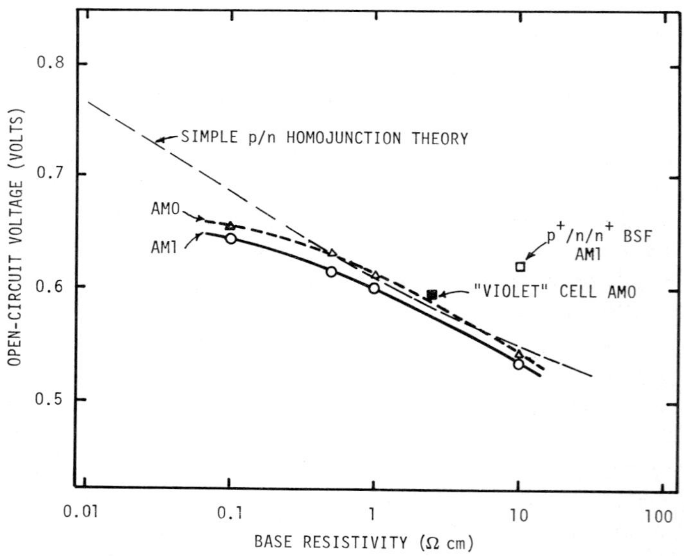

Fig. 11.3. Measured open-circuit voltage versus base resistivity for *p*-Si based MIS solar cells (△, ○). [Drawn using data from R. B. Godfrey and M. A. Green, *Appl. Phys. Lett.* **34,** 790 (1979).] Also shown is a curve (thin broken line) for *p*-Si based *n*/*p* homojunctions from a "simple" theory with no heavy doping effects and no back-surface field. [Data from M. P. Godlewski, H. W. Brandhorst, and C. R. Baraona, *Proc. 11th IEEE Photovoltaic Specialists Conf.* (1976), p. 32.] Isolated values are also shown for a "violet" type Si *n*/*p* homojunction cell [■] and a high-efficiency, back-surface field (BSF) $p^+/n/n^+$ homojunction cell [□]. [Data for the BSF cell from J. G. Fossum and E. L. Burgess, *Appl. Phys. Lett.* **33,** 238 (1978).] In the latter case the resistivity is that of the *n*-type base layer.

mojunction designs have been combined in a metal/insulator/n-Si/p-Si (MINP) cell by Blakers and Green (1981). In this design a thin (0.05 to 0.10 μm) n-layer is formed on a high-quality p-Si base by either diffusion or ion implantation. The n-Si layer is relatively lightly doped (300 to 1000 $\Omega\,\square^{-1}$) with respect to conventional Si cell technology. A metal/insulator grid contact and a SiO antireflection coating are applied using technology similar to that just described for the grating cell (e.g., see Green *et al.*, 1980), but with wider grid spacing. The performance advantages of the MINP devices with respect to conventional n/p homojunction devices are attributed to a major reduction in the effective recombination velocity under the top grid contact and reduced bulk recombination in the emitter region with its lower doping density. Using a 0.1-Ω-cm float-zone p-Si base with a back-surface field and the MINP configuration, Blakers and Green have obtained a V_{oc} of 0.678 V (simulated AM0).

11.1.2 GaAs- and InP-Based MIS Devices

The effects of an interfacial insulator layer are also critical in improving Schottky barrier performance on GaAs. Figure 11.4 shows the dramatic effect on the open-circuit voltage of various stages of oxidation of an n-GaAs surface before application of a Au barrier metal contact. The magnitude of the diode current J_0 was actually found to increase with increasing "oxide" thickness in some cases, corresponding to new routes of conduction for low bias voltage. However, the increase in the diode factor A caused by the oxide layer more than compensated for the increase in J_0, so that the value of the open-circuit voltage is increased (Stirn and Yeh, 1975). Solar efficiencies up to 15% were measured corresponding to the data of Fig. 11.4. These were called AMOS cells: antireflection-coated metal oxide semiconductor.

A variety of techniques have been tested involving different surface treatments of the GaAs and different methods of producing the "oxide" layer (Stirn and Yeh, 1977; Stirn *et al.*, 1977). The bests results were obtained by first etching the surface with $NH_4OH : H_2O_2 : H_2O = 1:1:10$ and then with $H_2SO_4 : H_2O_2 : H_2O = 10:1:1$ before oxidation. Development of the oxide layer either by thermal oxidation in water vapor or by vacuum deposition of Sb_2O_3 yields similar results: $V_{oc} = 0.73$–0.83, $J_{sc} =$ 26 mA cm^{-2}, ff $= 0.79$–0.83, and $\eta_s = 15.5$–17%. These cells have diode factors in the range of 1.02–1.15 and low diode currents of 10^{-12}–10^{-14} A cm^{-2}. Use of an antireflection coating is essential because of the average reflectivity (45%) of the Au/GaAs surface; such a coating increases the short-circuit current by a factor of 1.67.

The mechanisms by which "oxide" layers affect the properties of the

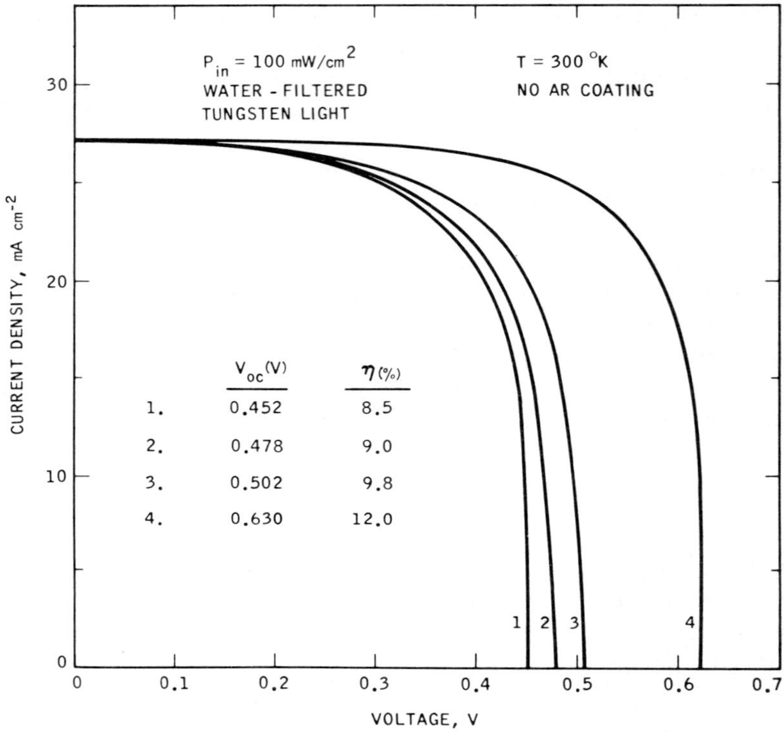

Fig. 11.4. Light *J–V* curves for Au/GaAs Schottky barriers for various treatments of the GaAs surface before application of the metal barrier contact. No antireflection coating is present. (1) "Clean" interface, (2) GaAs surface exposed to air at 300°K for 4 hr, (3) exposed to air at 300°K for 95 hr, and (4) exposed to air at 403°K for 70 hr. [After R. J. Stirn and Y. C. M. Yeh, *Appl. Phys. Lett.* **27,** 95 (1975).]

Au/GaAs junctions fall into two categories: increases in barrier height and increases in diode factor. An "unoxidized" GaAs surface has an oxide layer 10–20 Å thick and a barrier height of 0.90–0.98 eV. Oxidation of such a surface increases the barrier height to 1.03 eV, either through a modification of the interface states that control the barrier height or by the presence of negative charge within the oxide layer. Crystallographic orientation also plays a significant role in determining the processes affecting the diode current and hence the behavior of the device as a photovoltaic cell. For AMOS cells prepared on (100) GaAs surfaces, oxidation causes the diode factor to increase from 1.00 to about 1.2, *without* increase in J_0. In contrast, on (111) GaAs surfaces, J_0 does increase on oxidation by about an order of magnitude, as described previously for the earlier

AMOS work, and diode factors of 1.4–1.6 are found. On $(\overline{111})$ GaAs surfaces, J_0 increases by three orders of magnitude on oxidation. In this case, even though the diode factor is increased to even larger values, the resulting V_{oc} is about 0.10 V smaller than those for cells prepared on (111) GaAs.

These MIS results have similarities to those for the Au/*n*-AlGaAs/*n*-GaAs structure discussed in Section 8.2.6. In that case the Schottky barrier voltage is increased by means of a thin (≃500 Å) intervening layer of a large-band-gap semiconductor made insulating by depletion, rather than by an intrinsic insulator.

High-efficiency MIS structures have also been fabricated using InP. In contrast to the case for GaAs, the barrier heights for *p*-InP are higher than those for *n*-InP. Kamimura *et al.* (1981) fabricated Al/*i*/*p*-InP devices for which the insulator layers were grown by immersion in a Br : H_2O solution and subsequent annealing in dry O_2 at 200°C for 1 hr. Cells with antireflection coatings produced $V_{oc} = 0.775$ V and $\eta_s = 14.5\%$ under 72-mW cm^{-2} illumination.

11.1.3 Si-Based Junctions with Conductive Metal-Oxide Window Layers

Efficient solar cells on Si have been prepared using metal-oxide semiconductors such as SnO_2, In_2O_3, or solid solutions of indium tin oxides (ITO) as the *n*-type semiconductor window layers. The oxide semiconductors (OS) belong to an interesting class of materials that behave as transparent conductors. Although they have large band gaps (~3 eV) they can be made highly degenerate, reaching values of resistivity as low as 2×10^{-4} Ω cm that correspond to electron densities of 10^{21} cm^{-3}. Junctions of OS/Si are attractive compared to conventional Si solar cells for a number of reasons. The high electrical conductivity of the OS layers greatly simplified lateral current collection in the front layer and fabrication of front contacts. The index of refraction of these materials can provide an inherent antireflection coating. Finally, the large band gaps and high transparencies of the materials improve short-wavelength response by eliminating the surface "dead layer" found in conventional *p*/*n* homojunction cells.

The properties of OS/Si devices are extremely sensitive to fabrication variables. Efficiencies ranging from 1% to over 12%† were reported for

† Mizrah and Adler (1976) report $\eta_s = 1\%$ for ITO/*n*-Si. Lai *et al.* (1975) report $\eta_s = 6\%$ for In_2O_3/*p*-Si. Franz *et al.* (1977) report $\eta_s = 9.9\%$ for SnO_2/*n*-Si. Dubow *et al.* (1976) report $\eta_s = 12\%$ for ITO/*p*-Si, with $V_{oc} = 0.61$ V, $J_{sc} = 32$ mA cm^{-2}, and ff = 0.70 (AM1.2, sim.).

OS/*p*-Si devices in the period from 1975 to 1977. Surprisingly, almost equally good cells can be made on both *p*-Si (Dubow *et al.*, 1976) and *n*-Si.† This fact immediately rules out a *simple* heterojunction structure. Both the latter devices were fabricated at relatively high temperatures, the first by ion-beam deposition (Dubow *et al.*, 1976) and the second by spray pyrolysis (Manifacier and Szepessy, 1977). The issue is further complicated by the fact that ohmic contacts are produced by ion-beam deposition of ITO on *n*-Si (Dubow *et al.*, 1976) and by vacuum evaporation or spray pyrolysis of ITO on *p*-Si (Manifacier and Szepessy, 1977; Feng *et al.*, 1979a, b).

Indeed, one might wonder how such high efficiencies are possible for this system. The lattice mismatch between ITO and Si is considerable, and the reported values of electron affinity for ITO do not mesh well with a simple heterojunction model. The electron affinity value given by Thompson and Anderson (1978), 4.5 eV, appears unsuitable for making good cells with either *n*- or *p*-type Si, whereas the value reported by Shewchun *et al.* (1978a), 4.1 eV, indicates good diodes only for *p*-Si.

An explanation for the apparently anomalous differences between cells produced by ion-beam sputter deposition and by spray pyrolysis or vacuum evaporation is given by Ashok *et al.* (1981). They showed that the damage created by ion-beam etching prior to ITO deposition‡ causes the conduction and valence bands to bend downward at the surfaces of both *n*- and *p*-type Si. Thus sputter damage produces an accumulated, ohmic contact on *n*-Si, and a buried homojunction (*n*-ITO/*n*-Si/*p*-Si) on the *p*-Si base. In general, such damage might result from predeposition sputter-etch cleaning or during the initial moments of ITO deposition by sputtering. Such sputter-damage-induced type conversion has also been observed at the surfaces of *p*-InP (Tsai *et al.*, 1980; Bachmann *et al.*, 1979), *p*-GaAs (Bachmann *et al.*, 1979), and *p*-CdTe (Courreges *et al.*, 1980).

An insulating "oxide" interface layer also plays a decisive role in determining the properties of particular cells (Wager and Wilmsen, 1979). In the case of ITO/Si junctions, for example, the "oxide" insulator layer may be even more complex than SiO_2. The potential barrier height is partially determined by the pinning of the Fermi level at the Si/insulator and insulator/OS interfaces, rather than totally by the electron affinity difference between the Si and the OS layer. The behavior of such junctions should therefore be considered in terms of the kind of model used to de-

† Manifacier and Szepessy (1977) report $\eta_s = 10\%$ for ITO/*n*-Si, with $V_{oc} = 0.50$ V, $J_{sc} = 32$ mA cm^{-2}, and ff = 0.65 (AM1, sim.).

‡ Ion-beam etching is a preliminary to ITO deposition for the cells of Dubow *et al.* (1976).

scribe MIS Schottky barriers (Ghosh *et al.*, 1978a), or an equivalent SIS formulation (Shewchun *et al.*, 1978a, b). It appears, in addition, that the insulating layer acts as a buffer to accommodate the large mismatch between the OS and the Si.

The theory of SIS junctions for photovoltaics is quite similar to that for MIS structures; since the OS are degenerate, they behave in some sense as metals. If the thickness of the insulator layer is greater than 30 Å, then the minority carrier tunneling current through the insulator is very small, and the OS and the semiconductor are essentially in separate thermal equilibrium [i.e., a capacitor with an insulating dielectric (Green *et al.*, 1974)]. If the insulator layer is less than 10 Å thick, then the insulator can play a more subtle role. In this case we have essentially a Schottky barrier or heterojunction with the band profile distorted by fixed charge in the insulator and/or at the interfaces. However, for insulator layer thicknesses between 10 and 30 Å, the tunneling current is large enough to cause the semiconductors to depart from thermal equilibrium under bias and "tunnel SIS diodes" are formed. Over a certain range of bias voltage in such nonequilibrium devices, the majority carrier diode current is suppressed sufficiently so that the total diode current is controlled by generation/recombination mechanisms in the semiconductor depletion layer, or by injection/diffusion into the quasi-neutral bulk (both minority carrier transport) with the tunnel current simply providing an "ohmic contact." This results in a large decrease in the magnitude of the diode current for the device.

An essential difference between MIS and SIS devices is that in the SIS case, there is no large density of states within the forbidden gap of the bulk semiconductor (here the OS) to be involved in tunneling (Fig. 11.5). Thus in an *n*-OS/*p*-Si device, for example, holes from the *p*/Si must tunnel to states at the OS/insulator interface, then recombine with electrons from the conduction band of the OS via these interface states. An alternative route is for electrons from the OS conduction band to tunnel through the insulator layer and recombine with holes from the semiconductor at states at the insulator/semiconductor interface. If these tunneling/recombination steps are not a limiting factor in the transport route, then the two models are equivalent.

Ghosh *et al.* (1978a, b) used an MIS-type model of Card and Rhoderick (1971) for the thermionic, majority carrier transport in their consideration of an SnO_2/*n*-Si cell. To this they added a parallel, injection/diffusion component for minority carrier (hole) transport into the bulk *n*-Si, assuming a tunneling transmission close to unity for the insulating layer. By proper choice of insulator layer thickness, the majority carrier current component could be suppressed sufficiently so that minority carrier trans-

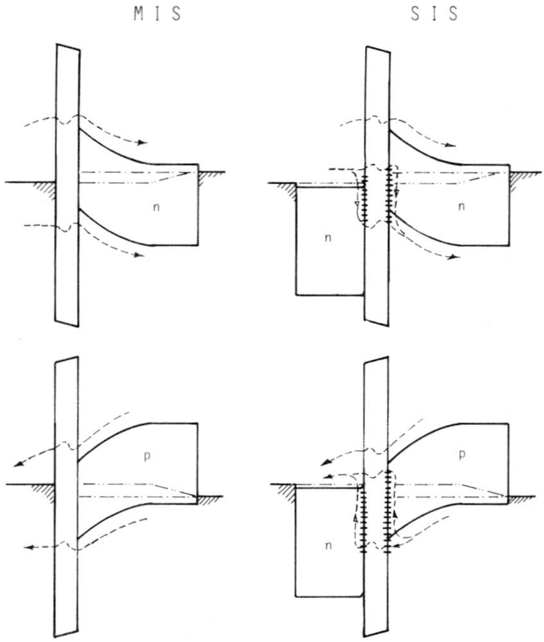

Fig. 11.5. Comparison of current pathways in MIS and SIS junctions.

port dominated at normal solar intensities. The SnO_2 layer was deposited by electron beam evaporation and the *i*-layer was formed by annealing in air at 300–350° C for 1–2 h, giving η_s = 10–12% (Fishman *et al.*, 1978). These authors estimated a maximum theoretical solar efficiency of 20%.

An unexpected aspect of this work is a strong dependence of η_s on the angle of incidence θ of the SnO_2 vapor stream onto the Si substrate during deposition (Feng *et al.*, 1979a). Deposition-angle-dependent η_s effects were also seen for similarly prepared ITO/*n*-Si cells (e.g., $\eta_s = 3\%$ for $\theta = 30°$, $\eta_s = 10\%$ for $\theta = 45°$) (Feng *et al.*, 1979b). Feng *et al.* (1979a) developed a model for the phenomenon in which the orientation and phase of the SnO_2 crystallites depends on θ. This, in turn, produces changes in the minority carrier tunneling transmission coefficient at the interfacial SiO_2 layer and hence changes the η_s.

Shewchun *et al.* (1978a) devised an SIS model to analyze the ITO/*p*-Si devices fabricated by Dubow *et al.* (1976). In this theory the *p*-Si is assumed to be in a nonequilibrium mode, i.e., transport is controlled by generation/recombination processes in the bulk semiconductor and $E_{Fn} \neq E_{Fp}$. The OS is assumed to be in equilibrium, i.e., no appreciable excess carrier density is present and $E_{Fn} = E_{Fp} = E_F$. Since they assume that interface states are absent at the OS/insulator interface, majority tunneling current is blocked by the forbidden gap of the OS (providing that tunneling/recombination at the insulator/*p*-Si interface for majority carriers is small). Thus the device should operate in the minority carrier mode and efficiencies of $\simeq 20\%$ are predicted. Shewchun *et al.* (1979) found that an ITO composition of 9 mole % of SnO_2 gave the best devices: $V_{oc} = 0.52$ V, $J_{sc} = 32$ mA cm^{-2}, ff $= 0.71$, and $\eta_s = 12.8\%$ under AM1.2 sunlight. The data for these cells are consistent with the theoretical predictions, but additional components of diode current, e.g., recombination/generation, are sufficiently large in these cells so that it is impossible to say unambiguously that minority carrier transport predominates under solar illumination.

DeVisschere and Pauwels (1978) have considered the photovoltaic efficiency of SIS heterojunctions in a more general way. Because of the complexity of the problem, several restrictive assumptions had to be made. However, they argue that an insulating layer is of advantage only when photogeneration occurs in the more lightly doped semiconductor and then only for one sign of the band discontinuity of interest.

A decrease in the output of SnO_2/*n*-Si cells (Nash and Anderson, 1977; Fishman *et al.*, 1979) and ITO/*p*-Si cells (Goodnik *et al.*, 1980) with time has been reported to occur both at room temperature, and at an accelerated rate at higher temperatures. This degradation is presumably due to the growth in width of the insulator layer with time. Its implication for the practical use of the devices involving MO/Si junctions has yet to be determined. Maruska *et al.* (1981) have shown the existence of two separate degradation mechanisms: one requiring optical excitation that increases J_0; and the other, a thermal mechanism that decreases the diode factor A and J_L.

Several other junctions involving OS have been made that may or may not include insulating layers, including the ITO/*p*-InP devices described in Section 8.3 (Sree Harsha *et al.*, 1977; Tsai *et al.*, 1980); SnO_2/*p*-GaAs, SnO_2/*n*-GaAs, and SnO_2/Ge devices (Wang and Legge, 1978; Hsu and Wang, 1978); ITO/*p*-CdTe junctions ($\eta_s = 8.0\%$, *active area basis,* AM1.5) (Courreges *et al.*, 1980); ZnO/*p*-CdTe junctions ($\eta_s = 9.2\%$, *active area basis,* AM1.5) (Aranovich *et al.*, 1980), ITO/*n*-$CuInSe_2$ cells ($\eta_s = 8.5\%$ on single crystal, $\eta_s = 2.1\%$ on polycrystalline thin films) (Kazmerski and Sheldon, 1978); and *n*-CdS/SiO_2/*p*-Si devices (Bhattacharyya *et al.*, 1979).

11.1.4 MIS Cells with Polycrystalline Semiconductors

After only a few years of concentrated development, the efficiencies of MIS cells on single-crystal Si rival those of *p*/*n* Si homojunction cells.

Similarly, the efficiencies of single-crystal GaAs-based MIS devices are close to those of the AlGaAs/GaAs heteroface and GaAs *p/n* homojunction devices. But the major potential of MIS devices, and much of the reason for their development, is for application to large-area, polycrystalline (PX) devices.

As discussed in Chapter 9, there are losses associated with the use of polycrystalline materials in photovoltaic cells. Because of enhanced recombination at the grain boundaries, the minority carrier diffusion length L is reduced, commonly being limited to the grain diameter or less. For fibrously oriented grains, with the fiber axis perpendicular to the junction plane, and with a sufficiently large grain size, large J_L values can still be observed. A larger loss is associated with increased J_0 (and possibly the effects of a bias-dependent light-generated carrier collection, Section 6.3.1) for PX cells, producing substantial decreases in V_{oc} and ff. The grain boundaries intersecting the junction plane are qualitatively similar to the edges of single-crystal diode structures, where increased recombination losses are commonly seen. Similar mechanisms may, in fact, be responsible.

A further difficulty arises because of the different orientation of the PX grains. If a native oxide is used for the insulating layer, it grows at different rates and has different interface state properties and effective barrier potentials on the differently oriented grains. Thus J_0 tends to be limited by the J_0 of the worst-case grains. This can be partially overcome by the use of deposited, nonnative insulator layers. From a fabrication viewpoint, the faceting of individual grains during growth produces a rather nonplanar surface that can make insulator deposition and metallization difficult.

A semiempirical theoretical maximum efficiency calculation for PX MIS cells was done by Lanza and Hovel (1976). They found that $\eta_s \simeq 14\%$ could be achieved for Au/*i*/*n*-GaAs with a grain size of 10 μm or greater, and $\eta_s \simeq 8\%$ could be obtained for Au/*i*/*p*-Si devices with greater than 50-μm grain size. The surface recombination velocity at the grain boundaries was taken as infinite to obtain the "worst-case" result. These results indicated that the efficiencies were decreased proportionately much more than were the short-circuit currents.

Meaningful comparison of experimental results requires care because of the large variety of PX structures in materials used to fabricate cells. Structures range from that of directionally solidified Wacker SILSO™ Si with millimeter-size fibrous grains (which is not unlike high-quality single-crystal material), to the almost randomly oriented, 2-μm grains found in some GaAs layers.

Green *et al.* (1978) fabricated an (Al or Mg)/SiO_x/*p*-Si grating type, inversion-layer cell on Wacker SILSO™ and obtained $\eta_s = 10.5\%$ at AM1. In these cells J_{sc} and V_{oc} were both 10–11% lower than for comparable single-crystal-based devices. MIS cells made on PX Si grown on carbon ribbon ("ribbon-against-drop" process in which a carbon ribbon is moved against a drop of molten Si) with grains several millimeters across yielded $\eta_s = 3.9\%$ with a considerably reduced V_{oc} ($V_{oc} = 0.38$ V, $J_{sc} = 21.1$ mA cm^{-2}) (Fabre *et al.*, 1978). A review of MIS and SIS cells on PX Si is given by Cheek and Mertens (1980).

For PX GaAs MIS cells, good results have been obtained by vacuum deposition of an Sb_2O_3 insulating layer (Yeh *et al.*, 1978). In this case, 2–3-μm-thick *n*-GaAs was grown on a laser-recrystallized Ge layer on a W-coated substrate. Table 11.I shows a comparison of three similar devices on bases with different grain size and a GaAs *p/n* single-crystal homojunction cells. Going to PX GaAs leaves J_{sc} essentially unchanged, but considerably reduces V_{oc} and ff.

S. S. Chu *et al.* (1978) used CVD to grow a somewhat thicker GaAs layer to form the following structure:

TiO_2, antireflection coating;
Au;
Oxide insulator layer, thermally grown;
n-GaAs, 10–20 μm;
n^+-GaAs, 5–10 μm;
W, film; and
Graphite, substrate;

Table 11.I
Photovoltaic Parameters for MIS GaAs Cells

Cell	Cell area, cm^2	Grain size, μm	J_{sc}, mA cm^{-2}	ff	V_{oc}, V	η_s, %
Single-crystal homojunction[a]	0.5	—	25	0.81	0.97	20
Single-crystal MIS[b]	1.0	—	25.7	0.80	0.80	16.2
Polycrystal slice MIS[c]	0.2	100	25	0.78	0.71	14
Grown polycrystal layer on laser-recrystallized-Ge MIS[c]	0.0314	?	14.7[d] (24.5)	0.60	0.53	4.8[d] (8)

[a] J. C.-C. Fan and C. O. Bozler, *Proc. 13th IEEE Photovoltaic Specialists Conf.* (1978), p. 953. Measured at 98 mW cm^{-2}, AM1, natural sunlight.

[b] R. J. Stirn, Y. C. M. Yeh, E. Y. Wang, F. P. Ernest, and C. J. Wu, *Tech. Dig.—Int. Electron Devices Meet.* p. 48 (1977). Measured at 100 mW cm^{-2}, AM1, simulated.

[c] Y. C. M. Yeh, F. P. Ernest, and R. J. Stirn, *Proc. 13th IEEE Photovoltaic Specialists Conf.* (1978), p. 966. Measured at 100 mW cm^{-2} AM1, simulated.

[d] All cells have antireflection coatings except for this one. Data in parentheses are calculated results for the addition of an antireflection coating.

yielding $\eta_s = 6.5\%$ for a large, 9-cm^2 cell. Blakeslee and Vernon (1979) discuss the relation of grain size and solar efficiency for GaAs-based devices.

11.2 AMORPHOUS SILICON

Although widely known for many years, amorphous materials were not included in modern solid-state physics by serious investigation until about 1960 (LeComber and Mort, 1973; Mott and Davis, 1971). One of the characteristics of most such amorphous semiconductors is a high density of localized states in the normally forbidden energy gap. With significant exceptions, they are relatively insensitive to low-density impurity doping effects. A hydrogenated amorphous silicon a-Si : H, produced by glow discharge decomposition of silane (SiH_4), was introduced by Chittick *et al.* (1969) who used a deposition method devised by Sterling and Swann (1965). Subsequent investigations have shown that a substantial proportion of hydrogen (up to 20–30%) is incorporated into the amorphous silicon and that, as a consequence, the properties of the material are markedly changed.

One of the most striking differences between crystalline Si and a-Si : H is the change in the optical absorption spectrum, shown in Fig. 11.6. Over most of the visible portion of the spectrum the absorption constant is appreciably increased, becoming much more like that of a direct-band-gap material. Just about a micrometer of a-Si : H is necessary to absorb most of the available AM1 photons with energies greater than the equivalent optical band gap (1.55 eV at room temperature). Other effects of hydrogenation are equally striking: a decrease in the density of localized states at the Fermi energy and an increase in resistivity (Lecomber and Spear, 1970; Spear and Lecomber, 1972; Molhotra and Neudeck, 1976), and the observation of photoluminescence emission from the a-Si : H that depends on the degree of hydrogenation (Engemann and Fischer, 1974; Pankove, 1978). Like crystalline silicon, a-Si : H can be doped both *n*- and *p*-type (Spear and LeComber, 1975; Spear *et al.*, 1976; Knights, 1976; Jan *et al.*, 1979, 1980). Since the a-Si : H has relatively high resistivity with a low carrier mobility, its behavior in solar cells is considerably different from that of normal single-crystal Si. A variety of different junction structures involving a-Si : H have been prepared, including heterojunctions, Schottky barriers, MIS junctions, and *p*/*i*/*n* junctions (Carlson and Wronski, 1976; Wronski *et al.*, 1976; Wronski, 1977; Carlson 1977, 1980). Homojunctions of a-Si : H have not proved to be efficient solar cells. Amorphous Si solar cells are among the most promising thin-film devices.

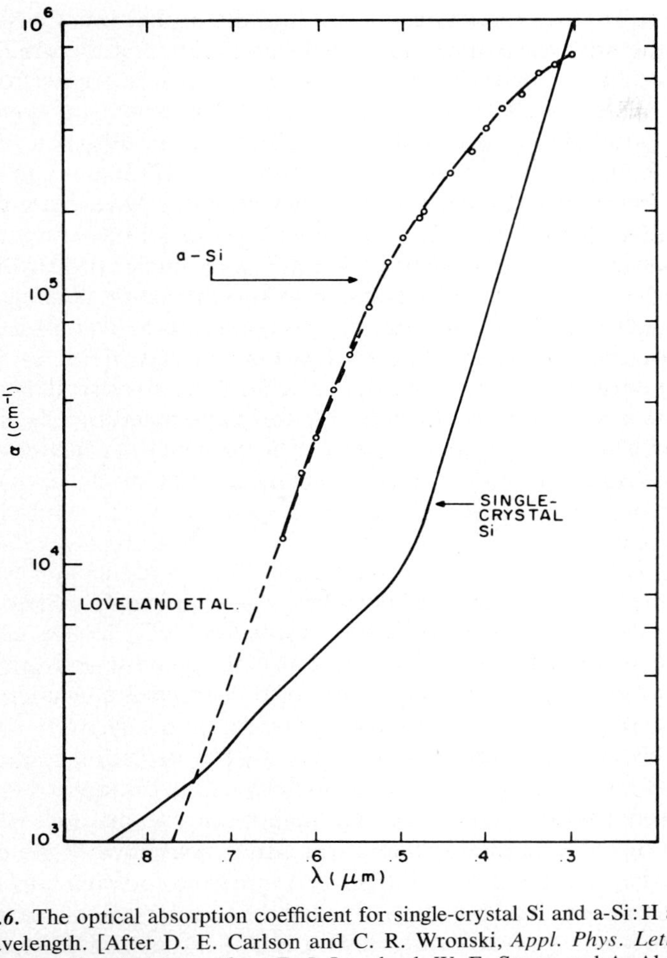

Fig. 11.6. The optical absorption coefficient for single-crystal Si and a-Si:H as a function of wavelength. [After D. E. Carlson and C. R. Wronski, *Appl. Phys. Lett.* **28,** 671 (1976). Long-wavelength data are from R. J. Loveland, W. E. Spear, and A. Al-Sharboty, *J. Non-Cryst. Solids* **13,** 55 (1973–1974).]

The a-Si:H material of interest for solar cells is usually deposited by glow discharge decomposition of silane SiH_4 or rf sputtering in $SiH_4 + H_2$ (capacitively coupled, planar). Substrate temperatures of 200–400°C are used and film deposition rates of 0.005–0.03 μm min^{-1} are obtained. Contamination from the walls and fixturing of the deposition chamber is alleviated by relatively high flow rates of SiH_4 and H_2. Despite this, materials used in high η_s cells usually contain large concentrations of O, N, and C ($>10^{18}$ cm^{-3}). Since the final properties of the material depend critically on the extent to which bonds are satisfied, the properties of a-Si : H vary extensively with the temperature of the substrate and the precise technique used for the decomposition. Resistivities that decrease from 10^8 Ω cm to 10^5 Ω cm with increasing substrate temperature are typical. The material is relatively stable, although effects of out-diffusion of H at T $\gtrsim$350°C (Carlson and Magee, 1979; Street *et al.*, 1978) and light-induced changes increasing the density of localized states have been observed (Staebler and Wronski, 1977, 1980; Staebler *et al.*, 1981).

Pawlewicz (1978) has reported that a-Si with similar properties can be deposited by rf sputtering. By using only Ar (eliminating the H_2) a-Si with resistivity as low as 110 Ω cm has been prepared, with even higher optical absorption for low-energy photons than the material of Fig. 11.6.

Since diffusion lengths of carriers in a-Si : H are so short (0.03–0.2 μm for *p*-type a-Si : H) it is highly desirable for photogeneration to occur in a region with a high drift field. The role of the minority carrier diffusion length is then replaced by the drift length L_{dr} given by

$$L_{dr} = \mu\mathscr{E}(V)\tau \simeq L_0\mathscr{E}(V)/\mathscr{E}_c, \tag{11.1}$$

where L_0 is the zero-field diffusion length, $\mathscr{E}(V)$ is the local electric field (depending on bias voltage and position), τ is the minority carrier lifetime, and $\mathscr{E}_c = kT/qL_0$ is the equivalent diffusion field (cf., Section 4.5.1). Because the drift length is bias dependent so is the quantum efficiency. Thus the ff may be appreciably reduced by the decreased collection of photogenerated carriers at forward bias (see Section 6.3.1).

It is therefore desirable to have as wide a depletion region in the a-Si : H as possible, so that the electric field is high throughout the region of photogeneration. This means that high-resistivity material is desired, but such material can contribute to the series resistance of the cell. The decrease in depletion layer width with increasing forward bias can also affect η_Q; a cell that may be fully depleted under short-circuit conditions can, under forward bias, have a relatively large field-free region, where transport is by diffusion with very short diffusion lengths (a "dead" region). A further subtlety arises from the fact that a-Si : H is photoconductive. Illumination increases the diffusion voltage of the junction, thus increasing the depletion layer width and at least partially compensating for the decrease in width. These considerations favor the *p*/*i*/*n* structure.

Three representative types of a-Si : H based cells are shown in Fig. 11.7: (1) a heterojunction with SnO_2, (2) a *p*/*i*/*n* junction, and (3) a Schottky barrier cell using Pt as the barrier metal. The wavelength dependences of collection efficiency for typical cells of each of these

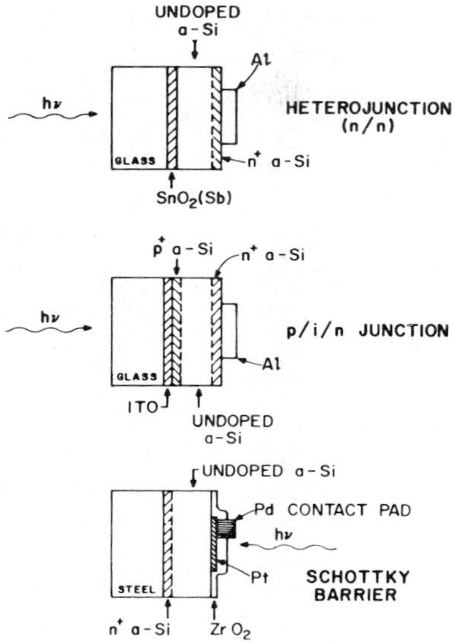

Fig. 11.7. Schematic diagrams of three types of junctions involving a-Si:H. [After D. E. Carlson, *IEEE Trans. Electron Devices* **ED-24,** 449 (1977). © 1977 IEEE.]

types are shown in Fig. 11.8. Since the total structure of each of these cells is less than 1 μm thick, these are truly thin-film cells.

The best of the group of heterojunction cells reported by Carlson (1977) were made by depositing undoped a-Si:H on ITO or SnO_2:Sb-coated glass substrates, using a thin P-doped a-Si:H layer to make ohmic contact to an evaporated Al contact. Values of $V_{oc} = 0.43$ V and $J_{sc} =$ 10.0 mA cm^{-2} were obtained, but the fill factor was only 0.28, giving a solar efficiency of 1.2%.

A useful heterojunction partner for a-Si is hydrogenated, amorphous silicon carbide a-SiC:H, first deposited by Anderson and Spear (1977) by glow discharge decomposition of SiH_4 and CH_4 mixtures. The optical band gap of a-SiC:H increases as the proportion of CH_4 is increased. Better doping efficiency, higher conductivity, and larger photoconductivity (suggesting longer lifetime) compared to a-Si:H are observed (Tawada *et al.*, 1981).

Tawada *et al.* (1981) fabricated n-a-Si:H/i-a-Si:H/p^+-a-SiC:H solar cells on ITO-coated glass (light was incident through the glass onto the p^+-a-SiC:H). Since the optical band gap is increased by addition of C, the loss in the p-layer was reduced, the blue response increased, and the diffusion voltage (and hence V_{oc}) increased. Small area cells (2 mm in diameter) produced $V_{oc} = 0.909$ V and $\eta_s = 7.55\%$, a record for any amorphous

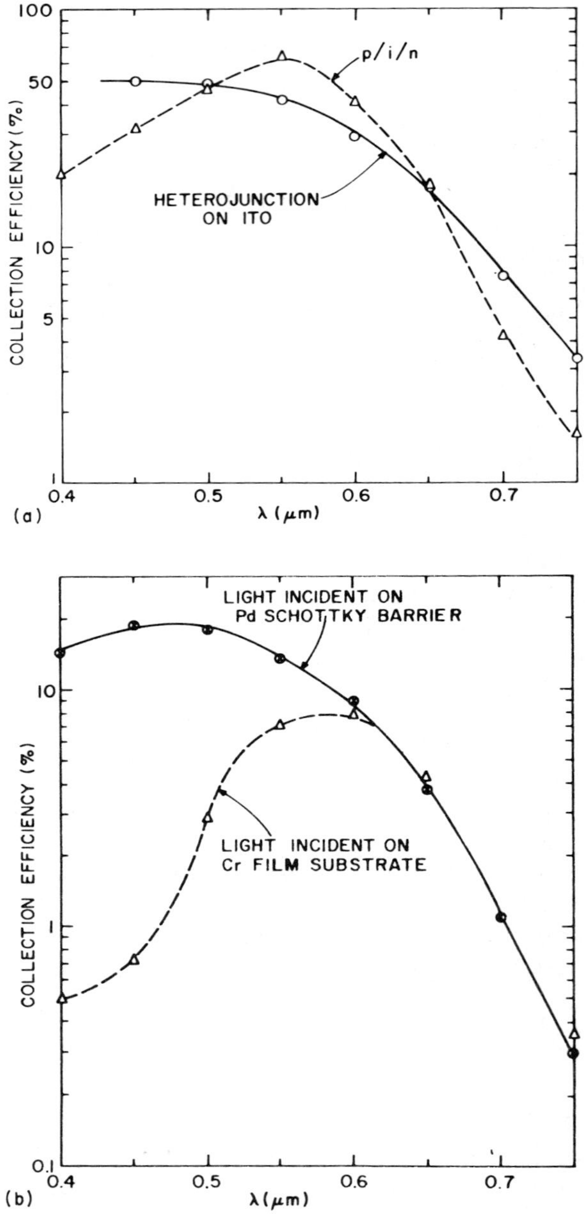

Fig. 11.8. Wavelength dependence of the collection efficiency for three different types of a-Si:H solar cells. (a) ITO/a-Si:H heterojunction and *p*/*i*/*n* a-Si:H junction; (b) Pd Schottky barrier on a-Si:H deposited on a thin Cr film substrate. [After D. E. Carlson, *IEEE Trans. Electron Devices* **ED-24,** 449 (1977). © 1977 IEEE.]

Si device at that time. Large area cells (1 cm^2) gave $\eta_s = 6.8\%$. Catalano *et al.* (1982) reported similar cells with $\eta_s = 10.1\%$ after refinement in cell design.

The small area Schottky barrier cells reported by Carlson (1977) (Figs. 11.7 and 11.8) were made by depositing on a stainless steel substrate successive layers of the following: a thin P-doped n^+-a-Si:H layer, ~1 μm of undoped a-Si:H, 50 Å of the Schottky barrier metal (Pt), a Pd contact pad about 1000 Å thick deposited on one edge of the Pt layer, and an antireflection coating of about 450 Å of ZrO_2. Values of $V_{oc} = 0.80$ V and $J_{sc} = 12.0$ mA cm^{-2} and a fill factor of 0.58 were obtained, yielding a solar efficiency of 5.5%. The open-circuit voltage apparently increases with the work function of the metal chosen for the Schottky barrier, indicating that the Fermi level is only partially pinned at the surface, as illustrated in Fig. 11.9.

The Schottky barrier structure has the following advantages: (1) with front illumination, most of the light absorption takes place in the high-field

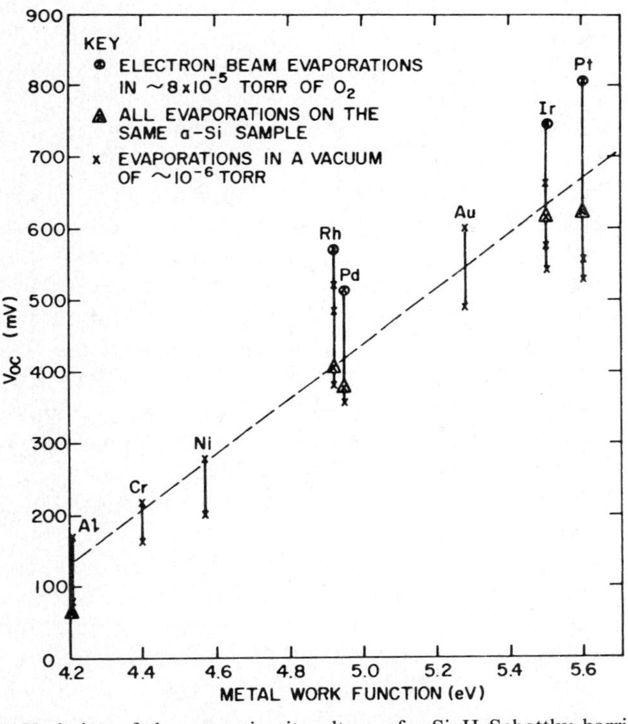

Fig. 11.9. Variation of the open-circuit voltage of a-Si:H Schottky barriers with the work function of the metal used to form the barrier. [After D. E. Carlson, *IEEE Trans. Electron Devices* **ED-24**, 449 (1977). © 1977 IEEE.]

region of the barrier; (2) only a single doped layer is required at the rear of the cell for an ohmic contact, thus preserving the carrier lifetimes of the undoped a-Si:H; (3) in the dark the Schottky barrier characteristics are almost ideal, with diode factors approaching unity; and (4) barrier heights of about 1.1 eV have been measured for Pt on a-Si:H, giving large values of V_{oc}.

The enhancement in V_{oc} realized by going to MIS structures can also be obtained with a-Si:H. Wilson *et al.* (1978) and McGill *et al.* (1979) fabricated Ni/TiO_x/a-Si cells with varying thicknesses of TiO_x. The 1-μm-thick a-Si:H layer was undoped. In this case the presence of the oxide layer increased both V_{oc} *and* J_{sc}, the latter apparently by providing a greater drift field for carrier collection. The solar efficiency was maximum for a TiO_x thickness of 20 Å, reaching $\eta_s = 4.8\%$, a value over four times that for similar cells without the layer.

The addition of F to produce a-Si:H:F lowers the density of gap states in the material (Madan and Ovshinsky, 1980), improves doping efficiency (Spear and Lecomber, 1976), and increases photoconductivity. This feature is used to advantage for MIS devices in the configuration ZnS/Au:Pd/*i*-Nb_2O_5/*i*-a-Si:H:F/n^+-a-Si:H:F that were fabricated by Madan *et al.* (1980) on Mo-coated glass. The ZnS serves as an antireflection coating. They obtained $\eta_s = 6.3\%$, with $V_{oc} = 0.80$ V, $ff = 0.61$, and $J_{sc} = 12.9$ mA cm^{-2} under simulated AM1 illumination (100 mW cm^{-2}, cell area 4.2 mm^2).

The *p/i/n* junction cells reported by Carlson (1977) (Figs. 11.7 and 11.8) were made by depositing the following successive layers on ITO-coated glass: B-doped p^+ a-Si:H a few hundred angstroms thick; ~1 μm undoped a-Si:H; a few hundred angstroms of P-doped n^+ a-Si:H; and finally Al or Cr metal electrodes. Values obtained for these small-area cells were $V_{oc} = 0.57$ V and $J_{sc} = 10.5$ mA cm^{-2}, with $ff = 0.40$, yielding $\eta_s = 2.4\%$. With a thicker p^+ layer, higher values of V_{oc} up to 0.80 V were reported, but J_{sc} was substantially reduced. The decrease in collection efficiency for short wavelengths shown in Fig. 11.8 is apparently related to a decrease in carrier lifetime in the doped layers. Later, large-area cells have demonstrated $\eta_s = 6.1\%$ and values of V_{oc} up to 0.91 V (not on the same cell, however) (Carlson, 1980).

These *p/i/n* structures are usually almost fully depleted and photoconductive effects are evident. Minority carrier lifetime, electric field, and photogenerated carrier collection are thus dependent on illumination intensity. In addition, a field-free "dead" region can be present within the *i*-layer. The width of this high-recombination region varies with light intensity. The *p/i/n* a-Si:H cell including these effects has been modeled by

Dalal (1980) and Rothwarf (1981). The "dead" layer is shown by them to account for the observed variations in photocapacitance and spectral response with changes in illumination and bias. The photoconductive effects also produce large differences in the diode characteristics of *p/i/n* cells in dark and light.

Doping increases the defect density in a-Si : H and reduces the carrier lifetime, resulting in *n*- and *p*-layers in the *p/i/n* cells that have high recombination. Thus these layers, especially the layer on which the light is incident, must be kept as thin as possible. Doping is generally not as efficient as in crystalline Si and E_F cannot be moved as close to the conduction (valence) band in *n*-type (*p*-type) materials as might be desired. This results in diffusion voltage values that are considerably smaller than would be indicated by the optical band gap.

Because of the influence of illumination on the quantum efficiency, mentioned previously, the measurement of spectral response usually requires bias illumination. The spectral response is peaked near 0.5 μm giving high η_s under fluorescent light or diffuse sunlight.

The *p/i/n* structure appears to have some advantages over the MIS structure with regard to stability and scale-up. Early MIS and Schottky devices degraded slowly on exposure to air, apparently as a result of the interaction of H_2O vapor at the metal/semiconductor interface; encapsulation is required for MIS devices. The *p/i/n* devices have much better stability even without encapsulation.

Small-area MIS devices have achieved efficiencies slightly greater that *p/i/n* cells (Carlson, 1980). However, problems with nonuniformity of the thin films of the metal (50–100 Å) and insulator layers (10–20 Å) have restricted the η_s of large-area (>1 cm^2) devices [e.g., to <3%, Carlson (1980)]. These problems appear to have been circumvented by the structure of the *p/i/n* cell. For example, the effect of pinhole shunting paths is small in *p/i/n* devices because the high sheet resistance of the layers effectively isolates the pinholes.

The interest in a-Si : H solar cells is sustained by the possibility of obtaining low-cost, large-area cells with a simple, readily available material, in spite of the rather low values of solar efficiency reported here. Problem areas that are undergoing intensive research are short diffusion lengths in the *i*-layer, extremely short diffusion lengths in the *n*- and *p*-layers (particularly on the illuminated side), and low V_d and high contact resistivity because of low doping efficiency. If it is assumed that limits on the values of V_{oc} and J_{sc} are about 0.80 V and 22 mA cm^{-2}, then a limiting fill factor in the range of 0.78 to 0.87 indicates a maximum feasible solar efficiency of 14 to 15%.

Reviews of the photoelectronic properties of a-Si : H material are given by Fritzche (1980) and Pankove and Carlson (1980). A SERI/DOE conference workshop on a-Si and a-Si based solar cells has also been published (SERI, 1980).

11.3 InP-BASED THIN-FILM CELLS

Since InP is an expensive material for large-area cells, it is desirable to use only the minimum thickness required. Layers of InP less than 2 μm thick are required to absorb almost all the solar photons. Single-crystal InP cells have been discussed in Section 8.3. Here we focus on their thin-film counterparts, in which the grain size in the polycrystalline layer is expected to be of the same order as its thickness.

Thin-film CdS/InP junctions have been prepared on polycrystalline InP deposited by chemical vapor deposition on carbon substrates (Bachmann *et al.*, 1976; Shay *et al.*, 1977). Deposition of InP directly on carbon resulted in a grain size of a few micrometers. Cells prepared by subsequent vacuum evaporation of CdS showed V_{oc} = 0.40 V, J_{sc} = 20 mA cm^{-2}, ff = 0.31, and η_s = 2.8%. The striking result was that the value of J_{sc} was nearly the same for polycrystalline InP as for the single-crystal material. However, V_{oc} was lower because of increased diode current because of the polycrystalline nature of the InP layer. The reduction in ff was apparently due to a poor InP/C contact. Subsequently, improved performance was achieved in cells for which the CdS was deposited by the H_2S/H_2 chemical vapor deposition method, and the contact to the InP was improved by depositing it on a 25-μm p^+-GaAs interlayer on carbon. The highest efficiency of 5.7% corresponded to V_{oc} = 0.46 V, J_{sc} = 13.5 mA cm^{-2}, and ff = 0.68 for AM2 illumination.

Figure 11.10 compares the properties of CdS/InP cells prepared on single-crystal InP to those prepared on polycrystalline InP. The quantum efficiency of the PX cells is only slightly lower than the single-crystal devices as shown in Fig. 11.10a. These data indicate that even in these PX cells, recombination of light-generated current at the heterojunction interface plays a very small role, at least under short-circuit conditions. The cells have a random orientation of the InP grains as shown by x-ray diffraction studies. Figure 11.10a also shows some response in the polycrystalline cells for wavelengths longer than the absorption edge of InP, indicating the presence of band tails in the polycrystalline InP grains. The open-circuit voltage increases as the degree of band-tailing decreases. Figure 11.10b shows that the appreciable reduction in V_{oc} in the PX cell, relative to the single-crystal cell, is caused both by a decrease in A factor (from $A \simeq 2.0$ to $A \simeq 1.7$) and by a hundredfold increase in J_0.

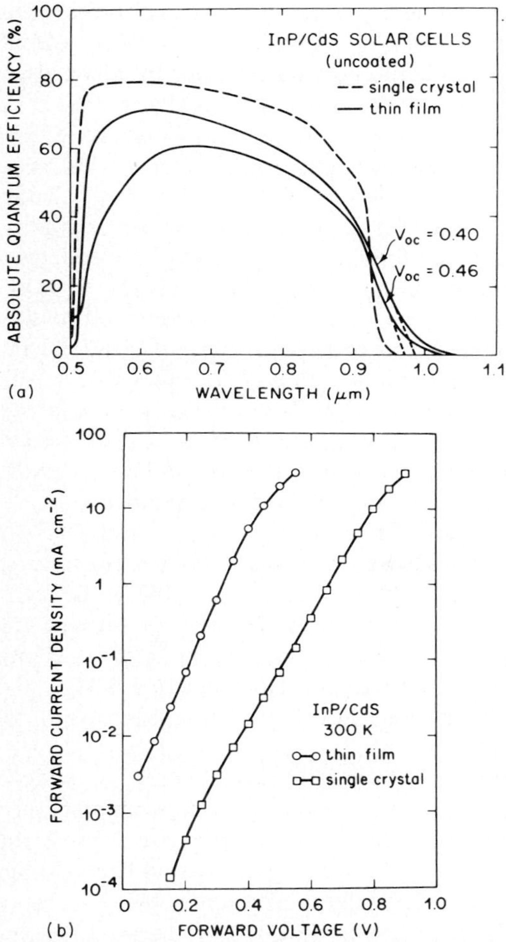

Fig. 11.10. Comparisons of CdS/InP photovoltaic junction properties for single-crystal InP and polycrystalline InP cells. (a) Wavelength dependence of the quantum efficiency. (b) Forward-bias J–V curves. [After J. L. Shay, S. Wagner, M. Bettini, K. J. Bachmann, and E. Buehler, *IEEE Trans Electron Devices* **ED-24,** 483 (1977). © 1977 IEEE.]

11.4 CdTe-BASED CELLS

Of the six chalcogenide compounds of Zn and Cd, only CdTe can be made highly conductive in both *n*- and *p*-type. CdTe is also the only material of this group with a sufficiently small band gap (E_g = 1.50 eV, direct) to be attractive as an absorber layer for solar cells.

There appears to be surprisingly little work done on the CdTe p/n homojunction (Gu *et al.*, 1975; Lincot *et al.*, 1979). Mimila-Arroyo *et al.* (1979) have fabricated homojunction cells by close-spaced vapor transport of thin p-CdTe:As layers on single-crystal n-CdTe substrates. Extremely low, dark, diode preexponential current in the range of $10^{-13} < J_0 < 10^{-10}$ A cm^{-2} were measured with A factors that were independent of temperature in the range 160–273°K: $1.3 < A < 1.7$. Spectral response measurements indicated a front-surface recombination velocity of $\simeq 10^6$ cm sec^{-1}, which limited J_{sc} to 12 mA cm^{-2} under solar illumination. Other photovoltaic parameters are $V_{oc} = 0.723$ V, ff = 0.63, and $\eta_s = 7.8\%$ after correction for reflection loss (*active-area basis*). (See item 1 of the note added in proof.) M. Chu *et al.* (1978) have made shallow homojunctions by ion implantation of As into n-CdTe, followed by annealing to restore the electrical activity of the implanted ions. Under AM1 illumination they obtained $V_{oc} = 0.84$ V, $J_{sc} = 6.5$ mA cm^2, and $\eta_s = 3\%$. In this case efficiency appears to be limited by a small diffusion length in the implanted layer and high series resistance.

Research on the Te analog of the Cu_xS/CdS junction i.e., p-Cu_xTe/n-CdTe, has shown quite high efficiencies, up to $\eta_s = 7.5\%$ (active area basis) (Cusano, 1963; Justi *et al.*, 1973). Schottky barrier cells, Pt/n-CdTe and Au/n-CdTe, have also been fabricated and show high values of V_{oc} (e.g., $V_{oc} = 0.68$ V for the Pt barrier) (Ponpon and Siffert, 1977). Finally, graded band-gap structures of $Cd_xHg_{1-x}Te$ have been fabricated by Marfaing and Chevallier (1971), but no photovoltaic parameters are reported.

The heterojunction configuration, using a wide-band-gap, high-conductivity window layer, appears most promising. To achieve greater long-term junction stability, it is desirable to use slowly diffusing elements for the window layer, preferably with a cation from Column II of the Periodic Table. These considerations have led to the investigation of CdS/CdTe, ITO/CdTe, and ZnO/CdTe heterojunctions (Bube *et al.*, 1977; Yamaguchi *et al.*, 1977; Aranovich *et al.*, 1980). Most of the work to date has been done with single-crystal CdTe, with the various window layers deposited by vacuum evaporation, close-spaced vapor transport, vapor phase deposition, rf sputtering, or spray pyrolysis.

One of the first CdS/CdTe heterojunction solar cells was made by Bonnet and Rabinhorst (1972) by CVD techniques. This all-thin-film cell yielded $\eta_s = 5$–6%.

Vacuum evaporation of CdS onto single-crystal CdTe was used by Mitchell *et al.* (1977) to prepare a cell that showed $\eta_s = 8\%$ with $V_{oc} = 0.63$ V, $J_{sc} = 16.1$ mA cm^{-2}, and ff = 0.66 for simulated solar illumination of 85 mW cm^{-2} (active-area basis). This cell appeared to have the characteristics of a genuine heterojunction. The second process, epitaxial vapor growth of CdS in H_2 onto single-crystal CdTe, was used by Yamaguchi *et al.* (1976, 1977) to fabricate a device having $\eta_s = 10.5\%$, total-area basis (in terms of active area $\eta_s = 11.7\%$), with $V_{oc} = 0.67$ V, $J_{sc} = 18.4$ mA cm^{-2}, and ff = 0.60 (solar illumination, 68 mW cm^{-2}). The latter cell is reported to be actually a buried homojunction with an *n*-CdS/*n*-CdTe/*p*-CdTe structure formed by diffusion of donors (In) from the CdS into the CdTe during fabrication. The electrical properties of the vacuum-evaporated and vapor-growth devices are quite similar: the evaporated CdS/CdTe junction has $A = 1.89$ and $J_0 = 1.7 \times 10^{-8}$ A cm^{-2}, whereas the vapor-deposited CdS/CdTe junction has $A = 1.80$ and $J_0 = 3.7 \times 10^{-9}$ A cm^{-2}.

The spectral response of the vacuum-evaporated CdS/CdTe cell described earlier is shown in Fig. 6.20, both for zero applied bias and for -1 V reverse bias. The short- and long-wavelength cutoffs are defined by the band gaps of CdS and CdTe as expected. The *bias-independent* increase in quantum efficiency is caused by the wavelength dependence of the optical absorption and a fairly small electron diffusion length of 0.4 μm in the CdTe. The *wavelength-independent* increase in quantum efficiency with reverse bias is caused by the effects of the increased electric field at the junction, which reduces the loss of photogenerated electrons by recombination at interface states, as described in Section 6.3.1. Solar efficiencies in the 10–15% range should be realizable by reasonable increases in electron diffusion length and hole density in the CdTe.

CdS/CdTe heterojunction cells have also been prepared by deposition of CdS by spray pyrolysis (Ma *et al.*, 1977) onto single-crystal CdTe [CdTe with the same properties as that used by Mitchell *et al.* (1977)]. The spectral response of three of these cells prepared under slightly different conditions is given in Fig. 11.11. The highest-efficiency cell has an almost panchromatic quantum efficiency, with $V_{oc} = 0.53$ V, $J_{sc} = 16.4$ mA cm^{-2}, and a solar efficiency of 6.5% without correction for reflection losses. However, cells with reduced quantum efficiency are also obtained with V_{oc} as large as 0.74 V. These latter cells show a spectral response more like that expected from a homojunction or may involve some kind of SIS structure. The inverse relationship between V_{oc} and J_{sc} indicated in Fig. 11.11 also suggests either of these structures. Substitution of 10% Zn for Cd to produce $Zn_{0.1}Cd_{0.9}S$/CdTe junctions by spray pyrolysis deposition of the ZnCdS yields cells with V_{oc} as high as 0.82 V, and solar efficiencies in the 6–8% range (Yin *et al.*, 1978).

Like all the direct band-gap materials, CdTe and CdS must be used in the form of PX thin films to meet solar cell cost goals (unless solar concentration is used). Such PX thin-film devices have been fabricated by CVD (Bonnet and Rabinhorst, 1972), CSVT (Tyan, 1980), and screen printing (Nakayama *et al.*, 1976), among others. Tyan (1980) fabricated all-PX-

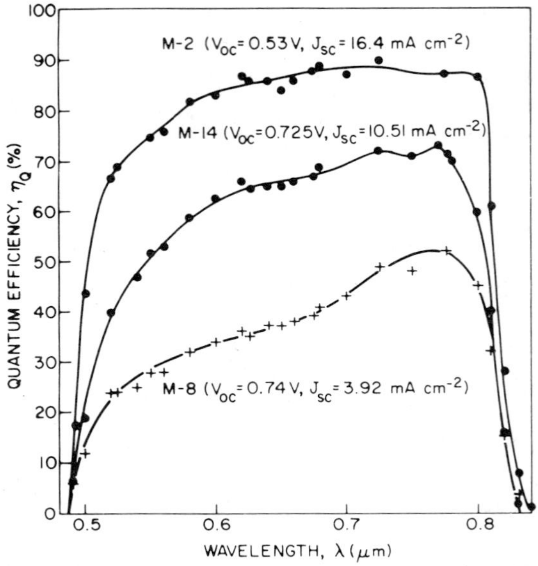

Fig. 11.11. Spectral dependence of the quantum efficiency for three CdS/CdTe junctions prepared by spray pyrolysis deposition of CdS onto single-crystal *p*-CdTe. Values of substrate temperature T_s and postdeposition heat-treatment temperature T_{HT} (for 5 min in H_2) are M-2 (T_s = 450°C, T_{HT} = 400°C), M-14 (T_s = 400°C, T_{HT} = 400°C), and M-8 (T_s = 400°C, T_{HT} = 420°C). [After Y. Y. Ma, A. L. Fahrenbruch, and R. H. Bube, *Appl. Phys. Lett.* **30,** 423 (1977).]

thin-film *p*-CdTe:Au/*n*-CdS solar cells on ITO coated glass by CSVT. The presence of a small partial pressure of O_2 and an optimal substrate temperature of 590–600°C during deposition were both reported to be necessary to obtain the maximum $\eta_s = 8.9\%$ ($V_{oc} = 0.705$ V, $J_{sc} =$ 16.3 mA cm^{-2}, ff = 0.58, simulated AM2). (See item 2 of the note added in proof.)

Success in producing reasonably efficient CdS/CdTe cells by simple screen printing techniques (See item 3 of the note added in proof.) has been reported by Nakayama *et al.* (1976), although again the evidence points toward a buried junction. The process involves the screen printing of a 20-μm layer of 0.2-Ω cm CdS onto an In_2O_3-coated glass substrate, followed by the screen printing of a 10-μm layer of *n*-type CdTe with resistivity of 0.1–1 Ω cm; a layer of Cu_xTe is formed on the *n*-CdTe by the dipping process and Ag-paint electrodes are applied to the Cu_xTe. The whole cell is embedded in resin after heat treatment for 10 min at 200°C. For 140 mW cm^{-2} of AM0 radiation, values of $V_{oc} = 0.69$ V, $J_{sc} = 31.1$ mA cm^{-2}, and efficiency = 8.1% are reported. The electron-beam-induced current plot

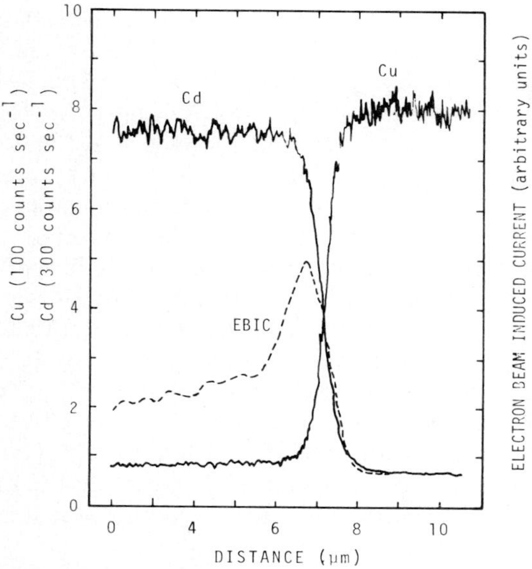

Fig. 11.12. Comparison of the position of the electrical junction (as indicated by electron-beam induced current EBIC) with the metallurgical junction (as indicated by microprobe analysis). For an *n*-CdTe single-crystal dipped to form a layer of Cu_xTe and then heat treated for 10 min at 200°C, to simulate similar phenomena in the screen printed CdS/CdTe/Cu_xTe solar cell. [Redrawn from N. Nakayama, H. Matsumoto, K. Yamaguchi, S. Ikegami, and Y. Hioki, *Jpn. J. Appl. Phys.* **15,** 2281 (1976).]

shown in Fig. 11.12 indicates that Cu has diffused into the *n*-CdTe from the Cu_xTe to produce a *n*-CdS/*n*-CdTe/*p*-CdTe:Cu/*p*-Cu_xTe buried junction.

In terms of their band gaps and expected diffusion voltages with CdTe, oxides such as ITO and ZnO are attractive as window materials in heterojunctions with CdTe. A typical degenerate n^+-ITO layer on CdTe is theoretically capable of producing a junction yielding 55% larger J_{sc} and 39% larger diffusion voltage than a CdS/CdTe junction; similarly, a n^+-ZnO layer on CdTe is capable of producing a junction yielding 52% larger J_{sc} and 39% larger diffusion voltage than CdS/CdTe. An unresolved question about such junctions is how adequately they can be characterized as abrupt heterojunctions (for which these estimates are derived) or whether their final character is really that of a buried homojunction or that of an SIS junction.

An ITO/CdTe junction prepared by rf sputtering a 4.5×10^{-3} Ω cm, 3000 Å thick n^+-ITO layer onto a *p*-CdTe single crystal showed V_{oc} = 0.82 V, J_{sc} = 14.5 mA cm^{-2}, ff = 0.55, and η_s = 8.0% (85 mW cm^{-2},

AM1.5; Bube, 1977). Unlike the CdS/CdTe heterojunctions, these ITO/CdTe cells lose their diode properties if heat treated even at moderate temperatures. The high value of V_{oc} and the spectral response obtained suggests homojunction formation during rf sputtering. This V_{oc} value can be compared with the high value of V_{oc} in Fig. 11.11 for a CdS/CdTe device showing homojunction-like spectral response, and with a value of $V_{oc} = 0.84$ V obtained from a CdTe homojunction prepared by implantation of As acceptors into *n*-CdTe (M. Chu *et al.*, 1978). Further research confirmed that a buried homojunction was formed by conversion of the CdTe surface to *n*-type by sputtering damage during the deposition of ITO (Courreges *et al.*, 1980). (See item 4 of the note added in proof.)

Junctions of ZnO/CdTe have been prepared by spray pyrolysis deposition of *n*-ZnO films onto single-crystal *p*-CdTe:P (Aranovich *et al.*, 1980). Genuine panchromatic, heterojunction-type spectral response is obtained for these cells with a high $J_{sc} = 18$ mA cm^{-2}. The value of V_{oc} is strongly dependent on the substrate temperature during spray pyrolysis, with an optimal value of 0.55 V, corresponding to a solar efficiency of 9.2% (active area, AM1.5) or 7.9% (total area).

11.5 CELLS BASED ON THE I–III–VI_2 CHALCOPYRITES

Among the more complex materials that are of interest for photovoltaic cells, the I–III–VI_2 materials, particularly $CuInS_2$ and $CuInSe_2$, stand out because of their demonstrated high solar efficiencies.

$CuInSe_2$ has a direct band gap of 1.04 eV. Its chalcopyrite structure with lattice constant $a = c/2 = 5.782$ Å makes a good match to wurtzite CdS with $2^{1/2}a = 5.850$ Å with only 1.2% lattice mismatch. CdS/$CuInSe_2$ junctions were prepared by the vacuum evaporation of CdS onto single-crystal *p*-type $CuInSe_2$ that had been Se-annealed to control its *p*-type conductivity by varying stoichiometry (Wagner *et al.*, 1974; Shay *et al.*, 1975). Microcracks in the $CuInSe_2$ appeared to limit the open-circuit voltage, but for small areas, the following values were obtained: $V_{oc} = 0.49$ V, $J_{sc} = 38$ mA cm^{-2}, ff = 0.60, and $\eta_s = 12\%$ (illumination 92 mW cm^{-2}, using an SiO antireflection coating). This was the sixth system, along with junctions based on Si, CdTe, GaAs, and InP and the Cu_xS/ZnCdS cell to show a solar efficiency in excess of 10%.

An extensive investigation of all-thin-film cells involving these chalcopyrites as heterojunctions with CdS has been carried out by Kazmerski and co-workers. (1975, 1976; Kazmerski, 1977). For frontwall operation, a 0.25–0.50 μm layer of $CuInSe_2$ was evaporated from a two-source system ($CuInSe_2$ + Se) onto 6–8 μm of evaporated CdS. For backwall operation, CdS was evaporated, without breaking vacuum, to a thickness of 6 μm onto 5–6 μm of $CuInSe_2$ on Au-coated glass. These all-thin-film CdS/$CuInSe_2$ cells show a panchromatic response from 0.58 to 1.3 μm, with V_{oc} = 0.49 V, J_{sc} = 25.1 mA cm^{-2}, ff = 0.54, and efficiency = 6.6% (illumination, 100 mW cm^{-2} from a tungsten–halogen lamp).

Using similar techniques, junctions were prepared from CdS and $CuInS_2$. Light J–V curves for this family of junctions are shown in Fig. 11.13 and typical spectral response of quantum efficiency curves are shown in Fig. 11.14, where they are compared with CdS/$CuInSe_2$ cells prepared with single-crystal $CuInSe_2$. The best efficiency was 3.25% for the CdS/$CuInS_2$ all-thin-film junction. Heterojunctions formed between CdS and $CuInTe_2$ were reported to show no photovoltaic sensitivity.

All-thin-film homojunctions of $CuInS_2$ and $CuInSe_2$ were also prepared by Kazmerski and co-workers by altering the flux of S or Se during vacuum evaporation of the compounds. Thus the p- and n-type layers

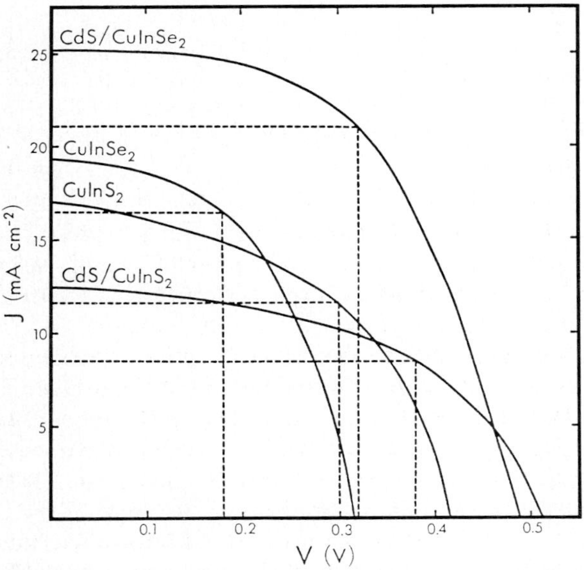

Fig. 11.13. Light J–V curves for all-thin-film heterojunctions of CdS/$CuInSe_2$ and CdS/$CuInS_2$, and for all-thin-film homojunctions of $CuInSe_2$ and $CuInS_2$. Illumination was 100 mW cm^{-2} by tungsten–halogen. [After L. L. Kazmerski, Univ. of Maine, Orono, Second Quarter Report to NSF-RANN and ERDA, NSF/AER 75-19576/PR/77/2 (1977), p. 13, Fig. 1.]

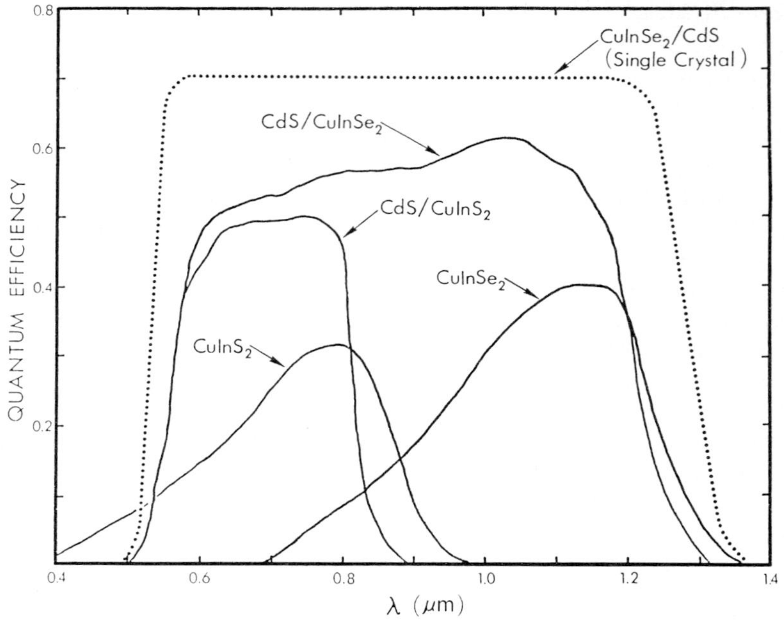

Fig. 11.14. Spectral dependence of quantum efficiency for all-thin-film heterojunctions of $CdS/CuInSe_2$ and $CdS/CuInS_2$, and for homojunctions of $CuInSe_2$ and $CuInS_2$. [After L. L. Kazmerski, Univ. of Maine, Orono, Second Quarter Report to NSF-RANN and ERDA, NSF/AER-75-19676/PR/77/2 (1977), p. 16, Fig. 3.]

were deposited without breaking vacuum. Solar efficiencies of 3.6% for the p/n $CuInS_2$ (Kazmerski and Sanborn, 1977) and 3.0% for the p/n $CuInSe_2$ cells were obtained with the $J-V$ curves and spectral responses shown in Figs. 11.13 and 11.14.

All-PX-thin-film $CdS/CuInSe_2$ heterojunction devices with $\eta_s = 9.5\%$ (total-area basis, SiO antireflection coating) have been fabricated by Mickelsen and Chen (1981). The active layers are sequentially deposited by vacuum evaporation onto Mo-coated alumina substrates and the total semiconductor thickness is 5 μm. Under simulated AM1 illumination (102 mW cm^{-2}) these cells produce $V_{oc} = 0.396$ V, $J_{sc} = 39$ mA cm^{-2}, and ff $= 0.63$. (See item 5 of the note added in proof.) These results are particularly encouraging in light of the best single-crystal $CuInSe_2$ cell data ($\eta_s = 12\%$). Despite a grain size of only ≈ 1 μm in the PX device, the quantum efficiency is high (0.9) and panchromatic, thus the grain boundary recombination loss in these $CuInSe_2$ layers appears to be minimal.

11.6 EXPLORATORY MATERIALS

As the search for efficient solar cells continues, so does the hope that perhaps some of the problems encountered in many of the present systems can be overcome by the discovery of some new materials system. Considerable research has already been done on the basic binary compounds with near-optimal direct band gaps: GaAs, InP, and CdTe. The only other III–V compound with similar band gap is AlSb, but this material is unstable in the presence of water vapor. No other II–VI materials exist with suitable band gap. The quest for binary compounds is therefore extended to other, less-explored combinations.

One binary compound that has acquired interest as the result of such a search is Zn_3P_2 (Catalano *et al.*, 1978). It has a band gap of close to 1.5 eV and can be readily prepared with *p*-type conductivity both in single crystals and thin films. A favorable consideration for the use of Zn_3P_2 in thin-film cells is the fact that Fe makes an excellent substrate for Zn_3P_2 films, having good lattice match and similar coefficient of thermal expansion as well as forming a low-resistance ohmic contact. Schottky barrier cells with Mg on single-crystal p-Zn_2P_2 have demonstrated $\eta_s = 6\%$ (Bhushan and Catalano, 1981). Thin films can be grown by CSVT in Ar with large grains ($\lesssim 10\ \mu m$) and long minority carrier diffusion lengths (0.5–4 μm). Schottky barrier cells based on this PX material have produced $\eta_s = 4.3\%$ (total area) (Bhushan, 1982).

Another binary compound that has shown early promise is WSe_2. Schottky barrier cells with the Al/p-WSe_2 structure have yielded $\eta_s = 5.3\%$ in preliminary experiments (Clemen and Bucher, 1978). The material has a direct band gap of 1.35 eV, an extremely high optical absorption coefficient ($> 10^5\ cm^{-1}$ over most of the range above its band-gap energy), excellent chemical stability, and can be made in either *n*- or *p*-type. Difficulties have been encountered with depositing films with the requisite structural properties however (Miller, 1981).

In addition to other possible binary compounds of a more exotic nature that may be developed, there are the ternary compounds such as $Al_{1-x}Ga_xAs$, $CuInSe_2$, $GaAs_{1-x}P_x$, $Ga_{1-x}In_xAs$, and $GaAs_{1-x}Sb_x$, which have already been used in solar cells. Undoubtedly there will be research on other ternary compounds of similar complexity, and research on quaternary and pentenary systems has begun (e.g., Loferski *et al.*, 1981; Antypas and Moon, 1974). Given sufficient freedom in the number of components of these solid solutions, it is possible (at least in principle) to design solar cell materials with both the desired band gap and the proper lattice constant to minimize mismatch in heterojunction structures. It is considered by some to be unlikely that such complex materials in general can meet the stringent demands of reproducibility and operating lifetime demanded for solar cells. However, it is possible that specific families of such materials may have the required properties. A limited but nevertheless suggestive example of success in spite of complexity is a $Al_{0.92}Ga_{0.08}As/Al_{0.14}Ga_{0.86}As$ solar cell that has yielded a solar efficiency of 15.5% and a maximum power output of 7.7 W at a V_{oc} just less than 1 V under illumination by 899 suns (Moon *et al.*, 1978). A representative listing of ternary tetrahedral compounds suitable for solar cells is given by Wagner and Bridenbaugh (1977). A summary is shown in Fig. 11.15. Pentenary compounds are discussed by Chapman *et al.* (1979) with $Cu_{1-y}Ag_yInS_{2(1-x)}Se_{2x}$ as a prototype. Schoijet (1979) gives a large listing of the more-univestigated binary and ternary compounds with their electronic and optical properties that might be suitable for solar cells.

Another set of hitherto unlikely candidates for photovoltaic solar energy conversion are the organic and nonorganic polymers.

Polymeric sulfur nitride $(SN)_x$, when properly grown, behaves as a highly anisotropic semimetal with a high dc conductivity, greater than 10^3 $(\Omega\ cm)^{-1}$, along the polymer chains at room temperature (Chiang *et al.*, 1976). Barrier height measurements with respect to various semiconductors show that $(SN)_x$ has a larger work function than any

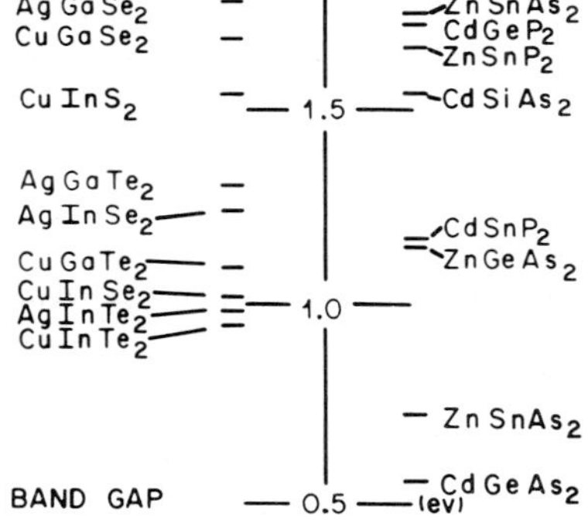

Fig. 11.15. Lowest energy gaps of I–III–VI_2 and II–IV–VI_2 semiconductors in the range of interest for solar photovoltaic conversion. [From S. Wagner and P. M. Bridenbaugh, *J. Cryst. Growth* **39**, 151 (1977).]

of the elemental metals (Scranton *et al.*, 1976; Best *et al.*, 1976). Thus the material is interesting for Schottky barrier analogs.

Cohen and Harris (1978) reported the first polymer/semiconductor solar cell, using a $(SN)_x/n$-GaAs structure with no intentional insulator interlayer. Values of V_{oc} greater than 0.70 V were obtained with AM1 illumination as compared to $V_{oc} = 0.49$ V for Au Schottky barriers on the same *p*-GaAs substrate material. Although the cells were not optimized and no antireflection coating was used, giving a low $J_{sc} = 10.3$ mA cm^{-2}, a solar efficiency of 6% was obtained (AM1, 70 mW cm^{-2}). Experimental cells on *n*-Si yielded similar enhancements in V_{oc}.

A similar metallic polymer is polyacetylene or $(CH)_x$. It has a band gap of ~1.6 eV and can be doped either *n*- or *p*-type. Homojunctions, heterojunctions with ZnS and CdS, and Schottky barriers with Ba have been made showing large built-in voltages, but they appear to have low quantum efficiencies for solar illumination because of small minority carrier diffusion lengths (Waldrop *et al.*, 1981).

Organic materials can also be used as the active, absorber layers of solar cells, but solar efficiencies reported to date are extremely low, probably because of the short minority carrier lifetimes and large series resistances characteristic of these materials. Although the monochromatic quantum efficiencies of these materials can approach 100%, the total quantum efficiencies for solar illumination are quite small. An example of such a device is the Cr/(chlorophyll-a)/Hg junction that gives $V_{oc} = 0.2$–0.5 V and a monochromatic conversion efficiency of 0.016% with illumination at 0.745 μm (Tang and Albrecht, 1975a, b). Another example is a cell based on a merocyanine dye absorber layer (Morel *et al.*, 1978), giving $V_{oc} = 1.2$ eV, ff = 0.25, $J_{sc} = 1.8$ mA cm^{-2}, and $\eta_s = 0.7\%$ in sunlight at 78 mW cm^{-2} intensity. The absorber layer is entirely depleted under normal illumination and the quantum efficiency appears to be bias voltage dependent, which may be indicative of very short diffusion lengths.

Note added in proof. Several significant improvements in CdTe and $CuInSe_2$ cells were reported in *Proc. 16th IEEE Photovoltaic Specialists Conf.* (1982).

1. A shallow diffused layer $p^+/n/n^+$ single-crystal CdTe-based homojunction with η_s = 13% (M. Barbé, F. Bailly, D. Lincot, and G. Cohen-Solal).
2. An all-thin-film CdTe/CdS heterojunction cell formed by close-spaced vapor transport with η_s = 10.5% (Y.-S. Tyan and E. A. Perez-Albuerne).
3. An all-screen-printed 64-cm^2 CdS/CdTe thin-film device with η_s = 5.8% (H. Uda, H. Matsumoto, Y. Komatsu, A. Nakano, and S. Ikegami).
4. An ITO/*p*-CdTe (single crystal) cell with η_s = 10.5% (active area) (J. G. Werthen, T. C. Anthony, A. L. Fahrenbruch, and R. H. Bube).
5. All-thin-film, vacuum evaporated $CdS/CuInSe_2$ devices with η_s = 10%. (R. A. Mickelsen and W. S. Chen).

REFERENCES

Anderson, W. A. (1977). "Si Schottky Photovoltaic Diodes for Solar Energy Conversion." Rutgers Univ., New Brunswick, New Jersey, Quarterly Progress Report, NSF/RANN/SE/AER 73-03197/PR/77/3.

Anderson, W. A., Delahoy, A. E., and Milano, R. A. (1974). *J. Appl. Phys.* **45,** 3913.

Anderson, W. A., Kim, J. K., and Delahoy, A. E. (1977). *IEEE Trans. Electron Devices* **ED-24,** 453.

Anderson, D. A., and Spear, W. E. (1977). *Phil. Mag.* **35,** 1.

Antypas, G. A., and Moon, R. L. (1974). *J. Electrochem. Soc.* **120,** 1574.

Aranovich, J., Golmayo, D., Fahrenbruch, A. L., and Bube, R. H. (1980). *J. Appl. Phys.* **51,** 4260.

Ashok, S., Fonash, S. J., Singh, R., and Wiley, P. (1981). *IEEE Electron Device Lett.* **EDL-2,** 184.

Bachmann, K. J., Buehler, E., Shay, J. L., and Wagner, S. (1976). *Appl. Phys. Lett.* **29,** 121.

Bachmann, K. J., Schreiber, H., Jr., Sinclair, W. R., Schmidt, P. H., Thiel, F. A., Spencer, E. G., Pasteur, G., Feldmann, W. L., and Sree Harsha, K. (1979). *J. Appl. Phys.* **50,** 3441.

Best, J. S., McCaldin, J. O., McGill, T. C., Mead, C. A., and Mooney, J. B. (1976). *Appl. Phys. Lett.* **29,** 433.

Bhattacharyya, A. B., Nahar, R. K., Nagchoudhuri, D., and Vasi, J. (1979). *J. Appl. Phys.* **50,** 390.

Bhushan, M. (1982). *Appl. Phys. Lett.* **40,** 51.

Bhushan, M., and Catalano, A. (1981). *Appl. Phys. Lett.* **38,** 39.

Blakers, A. W., and Green, M. A. (1981). *Appl. Phys. Lett.* **39,** 483.

Blakeslee, A. E., and Vernon, S. M. (1979). *Solar Cells* **1,** 81.

Bonnet, D., and Rabinhorst, H. (1972), *Proc. 9th IEEE Photovoltaic Specialists Conf.*, p. 129.

Bube, R. H. (1977). *Proc. Soc. Photo-Opt. Instrum. Eng.* **114,** 7.

Bube, R. H., Buch, F., Fahrenbruch, A. L., Ma, Y. Y., and Mitchell, K. W. (1977). *IEEE Trans. Electron. Devices* **ED-24,** 487.

Bucher, E. (1978). *Appl. Phys.* **17,** 1.

Card, H. C., and Rhoderick, E. H. (1971). *J. Phys. D* **4,** 1589.

Carlson, D. E. (1977). *IEEE Trans. Electron Devices* **ED-24,** 449.

Carlson, D. E. (1980). *Sol. Energy Mater.* **3,** 503.

Carlson, D. E., and Magee, C. W. (1979). *Proc. 2nd EC Photovoltaic Sol. Energy Conf., Berlin,* p. 312. D. Reidel Publ., Dordrecht, Holland.

Carlson, D. E., and Wronski, C. R. (1976). *Appl. Phys. Lett.* **28,** 671.

Catalano, A., Dalal, V., Fagen, E. A., Hall, R. B., Masi, J. V., Meakin, J. D., Warfield, G., and Barnett, A. M. (1978). *Proc. 1st EC Photovoltaic Sol. Energy Conf., Luxembourg, 1977,* p. 644. D. Reidel Publ., Dordrecht, Holland.

Catalano, A., D'Aiello, R., Dresner, J., Faughnan, B., Firester, A., Kane, J., Smith, Z. E., Schade, H., Swartz, G., and Triano, A. (1982). *Proc. 16th IEEE Photovoltaic Specialists Conf.,* to be published.

Chapman, G. H., Shewchun, J., Garside, B. K., Loferski, J. J., and Beaulieu, R. (1979). *Sol. Energy Mater.* **1,** 451.

Charlson, E. J., and Lien, J. C. (1975). *J. Appl. Phys.* **46,** 3982.

Charlson, E. J., and Richardson, W. F. (1978). *Proc. 13th IEEE Photovoltaic Specialists Conf.,* p. 656.

Cheek, G., and Mertens, R. (1980). *Solar Cells* **1,** 405.

Chiang, C. K., Cohen, M. J., Garito, A. F., Heeger, A. J., Mikulski, C. M., and MacDiarmid, A. G. (1976). *Solid State Commun.* **18,** 1451.
Chittick, R. C., Alexander, J. H., and Sterling, H. F. (1969). *J. Electrochem. Soc.* **116,** 77.
Chu, M., Fahrenbruch, A. L., Bube, R. H., and Gibbons, J. F. (1978). *J. Appl. Phys.* **49,** 322.
Chu, S. S., Chu, T. L., and Yang, H. T. (1978). *Proc. 13th IEEE Photovoltaic Specialists Conf.,* p. 956.
Clemen, C., and Bucher, E. (1978). *Proc. 13th IEEE Photovoltaics Specialists Conf.,* p. 1255.
Cohen, M. J., and Harris, J. S., Jr. (1978). *Tech. Dig. Int. Electron Devices Meet.,* p. 247.
Courreges, F. G., Fahrenbruch, A. L., and Bube, R. H. (1980). *J. Appl. Phys.* **51,** 2175.
Cusano, D. (1963). *Solid-State Electron.* **6,** 217.
D'Aiello, R. V., Robinson, P. H., and Cressel, H. (1976). *Appl. Phys. Lett.* **28,** 231.
Dalal, V. (1980). *Solar Cells* **2,** 261.
De Visschere, P., and Pauwels, H. (1978). *Proc. 1st EC Photovoltaic Sol. Energy Conf., Luxembourg, 1977,* p. 330. D. Reidel Publ., Dordrecht, Holland.
Dubow, J. B., Burk, D. E., and Sites, J. R. (1976). *Appl. Phys. Lett.* **29,** 494.
Engemann, D., and Fischer, R. (1974). *Proc. 12th Int. Conf. Phys. Semicond.,* Teubner, Stuttgart.
Fabre, E., Baudet, Y., and Makram-Ebied, S. (1978). *Proc. 13th IEEE Photovoltaic Specialists Conf.,* p. 1101.
Feng, T., Fishman, C., and Ghosh, A. K. (1979a). *J. Appl. Phys.* **50,** 8070.
Feng, T., Ghosh, A. K., and Fishman, C. (1979b). *J. Appl. Phys.* **50,** 4972.
Fishman, C., Feng, T., and Ghosh, A. K. (1978). *Tech. Dig. Int. Electron Devices Meet.,* p. 93.
Fishman, C., Ghosh, A. K., and Feng, T. (1979). *Sol. Energy Mater.* **1,** 181.
Fossum, J. G., and Burgess, E. L. (1978). *Appl. Phys. Lett.* **33,** 238.
Franz, S. L., Kent, G., and Anderson, R. L. (1977). *J. Electron. Mater.* **6,** 107.
Fritzche, H. (1980). *Sol. Energy Mater.* **3,** 447.
Ghosh, A. K., Fishman, C., and Feng, T. (1978b). *Proc. 13th IEEE Photovoltaic Specialists Conf.,* p. 519.
Ghosh, A. K., Fishman, C., and Feng, T. (1978a). *J. Appl. Phys.* **49,** 3490.
Godfrey, R. B., and Green, M. A. (1979). *Appl. Phys. Lett.* **34,** 860.
Goodnik, S. M., Wilmsen, C. W., and Wager, J. F. (1980). *Proc. 14th IEEE Photovoltaic Specialists Conf.,* p. 1175.
Green, M. A., King, F. D., and Shewchun, J. (1974). *Solid-State Electron.* **17,** 551.
Green, M. A., Godfrey, R. B., and Davies, L. W. (1978). *Proc. 13th IEEE Photovoltaic Specialists Conf.,* p. 651.
Green, M. A., Godfrey, R. B., Willison, M. R., and Blakers, A. W. (1980). *Proc. 14th IEEE Photovoltaic Specialists Conf.,* p. 684.
Gu, J., Kitahara, T., Fujita, S., and Sakaguchi, T. (1975). *Jpn. J. Appl. Phys.* **14,** 499.
Hsu, L., and Wang, E. Y. (1978). *Proc. 13th IEEE Photovoltaics Specialists Conf.,* p. 536.
Jan, Z. S., Bube, R. H., and Knights, J. C. (1979). *J. Elect. Mater.* **8,** 47.
Jan, Z. S., Bube, R. H., and Knights, J. C. (1980). *J. Appl. Phys.* **51,** 3278.
Justi, E. W., Schneider, G., and Seredynski, J. (1973). *Energy Convers.* **13,** 53.
Kamimura, K., Suzuki, T., and Kunioka, A. (1981). *Appl. Phys. Lett.* **38,** 259.
Kazmerski, L. L. (1977). Univ. Maine, Orono, Second Quarter Report to NSF-RANN and ERDA, NSF/AER 75-19576/PR/77/2 (1977).
Kazmerski, L. L., and Sanborn, G. A. (1977). *J. Appl. Phys.* **48,** 3178.

Kazmerski, L. L., and Sheldon, P. (1978). *Proc. 13th IEEE Photovoltaic Specialists Conf.*, p. 541.
Kazmerski, L. L., Ayyagari, M. S., and Sanborn, G. A. (1975). *J. Appl. Phys.* **46,** 4865.
Kazmerski, L. L., White, F. R., and Morgan, G. K. (1976). *Appl. Phys. Lett.* **29,** 268.
Kleta, J. K., and Pulfrey, D. L. (1980). *IEEE Electron Device Lett.* **EDL-1,** 107.
Kipperman, A. H. M., Backerra, S. C. M., and Maaskamp, H. J. (1978a). *Proc. 1st EC Photovoltaic Sol. Energy Conf., Luxembourg, 1977,* p. 956. D. Reidel Publ., Dordrecht, Holland.
Kipperman, A. H. M., Backerra, S. C. M., Maaskamp, H. J., and van Zolingen (1978b). *Proc. 1st EC Photovoltaic Sol. Energy Conf., Luxembourg, 1977,* p. 961. D. Reidel Publ., Dordrecht, Holland.
Knights, J. C. (1976). *Phil. Mag.* **34,** 663.
Lai, S. W., Franz, S. L., Kent, G., Anderson, R. L., Clifton, J. K., and Masi, J. V. (1975). *Proc. 11th IEEE Photovoltaic Specialists Conf.*, p. 398.
Lanza, C., and Hovel, H. J. (1976). *Proc. 12th IEEE Photovoltaic Specialists Conf.*, p. 96.
LeComber, P. G., and Mort, J., eds. (1973). "Electronic and Structural Properties of Amorphous Semiconductors." Academic Press, New York.
LeComber, P. G., and Spear, W. E. (1970). *Phys. Rev. Lett.* **25,** 509.
Lincot, D., Mimila-Arroya, J., Triboulet, R., Marfaing, Y., and Barbe, M. (1979). *Proc. 2nd EC Photovoltaic Sol. Energy Conf., Berlin,* p. 424. D. Reidel Publ., Dordrecht, Holland.
Loferski, J. J., Crisman, E. E., Armitage, W., and Chen, L. Y. (1973). *Proc. 10th IEEE Photovoltaic Specialists Conf.*, p. 58.
Loferski, J. J., Kwietniak, M., Piekoszewski, J., Spitzer, M., Arya, R., Roessler, B., Beaulieu, R., Vera, E., Shewchun, J., and Kazmerski, L. L. (1981). *Proc. 15th IEEE Photovoltaic Specialists Conf.*, p. 1056.
Ma, Y. Y., Fahrenbruch, A. L., and Bube, R. H. (1977). *Appl. Phys. Lett.* **30,** 423.
Madan, A., and Ovshinsky, S. R. (1980). *J. Non-Cryst. Solids* **35,** 171.
Madan, A., McGill, J., Czubatyj, W., Yang, J., and Ovshinsky, S. R. (1980). *Appl. Phys. Lett.* **37,** 826.
Manifacier, J. C., and Szepessy, L. (1977). *Appl. Phys. Lett.* **31,** 459.
Marfaing, Y., and Chevallier, J. (1971). *IEEE Trans. Electron Devices* **ED-18,** 465.
Maruska, H. P., Feng, T., Ghosh, A. K., and Eustace, D. J. (1981). *Proc. 15th IEEE Photovoltaic Specialists Conf.*, p. 1412.
McGill, J., Wilson, J. I. B., and Kinmond, S. (1979). *J. Appl. Phys.* **50,** 548.
Mickelsen, R. A., and Chen, W. S. (1981). *Proc. 15th IEEE Photovoltaics Specialists Conf.*, p. 800.
Miller, D. (1981). SERI Emerging Materials Contract Review Meeting, Golden, Colorado, Feb. 26.
Mimila-Arroyo, J., Marfaing, Y., Cohen-Solal, G., and Triboulet, R. (1979). *Sol. Energy Mater.* **1,** 171.
Mitchell, K. W., Fahrenbruch, A. L., and Bube, R. H. (1977). *J. Appl. Phys.* **48,** 4365.
Mizrah, T., and Adler, D. (1976). *Appl. Phys. Lett.* **29,** 682.
Molhotra, A. K., and Neudeck, G. W. (1976). *Appl. Phys. Lett.* **28,** 47.
Moon, R. L., James, L. W., Vander Plas, H. A., and Nelson, N. J. (1978). *Appl. Phys. Lett.* **33,** 196.
Morel, D. L., Ghosh, A. K., Feng, T., Stogryn, E. L., Purwin, P. E., Shaw, R. F., and Fishman, C. (1978). *Appl. Phys. Lett.* **32,** 495.
Mott, N. F., and Davis, E. A. (1971). "Electronic Processes in Non-Crystalline Materials." Oxford Univ. Press (Clarendon), London and New York.

Nakayama, N., Matsumoto, H., Yamaguchi, K., Ikegami, S., and Hioki, Y. (1976). *Jpn. J. Appl. Phys.* **15,** 2281. Also see Nakayama *et al., Jpn. J. Appl. Phys.* **19,** 703 (1980).
Nash, T. R., and Anderson, R. L. (1977). *IEEE Trans. Electron Devices* **ED-24,** 468.
Neugroschel, A., Lindholm, F. A., and Pao, S. C. (1978). *Appl. Phys. Lett.* **33,** 168.
Ng, K. K., and Card, H. C. (1980). *IEEE Trans. Electron Devices* **ED-27,** 716.
Pankove, J. I. (1978). *Appl. Phys. Lett.* **32,** 812.
Pankove, J. I., and Carlson, D. E. (1980). *Annu. Rev. Mater. Sci.* **10,** 43.
Pawlewicz, W. T. (1978). *J. Appl. Phys.* **49,** 5595.
Ponpon, J. P., and Siffert, P. (1976). *J. Appl. Phys.* **47,** 3248.
Ponpon, J. P., and Siffert, P. (1977). *Rev. Phys. Appl.* **12,** 427.
Ponpon, J. P., and Siffert, P. (1978). *Proc. 13th IEEE Photovoltaics Specialists Conf.,* p. 639.
Pulfrey, D. L. (1978). *IEEE Trans. Electron Devices* **ED-25,** 1308.
Rothwarf, A. (1981). *Proc. 15th IEEE Photovoltaic Specialists Conf.,* p. 690.
Salter, G. C., and Thomas, R. E. (1975). *Proc. 11th IEEE Photovoltaic Specialists Conf.,* p. 364.
Schoijet, M. (1979). *Sol. Energy Mater.* **1,** 43.
Schwartz, G. P., Griffiths, J. E., and Schwartz, B. (1979). *J. Vac. Sci. Technol.* **16,** 1383.
Scranton, R. A., Mooney, J. B., McCaldin, J. O., McGill, T. C., and Mead, C. A. (1976). *Appl. Phys. Lett.* **29,** 47.
SERI/DOE Photovoltaic Material and Device Measurements Workshop, *Focus on Amorphous Silicon,* San Diego (1980). Reprinted in Solar Cells **2,** No. 3 and 4.
Shay, J. L., Wagner, S., and Kasper, H. M. (1975). *Appl. Phys. Lett.* **27,** 89.
Shay, J. L., Wagner, S., Bettini, M., Bachmann, K. J., and Buehler, E. (1977). *IEEE Trans. Electron Devices* **ED-24,** 483.
Shewchun, J., Dubow, J., Myszkowski, A., and Singh, R. (1978a). *J. Appl. Phys.* **49,** 855.
Shewchun, J., Singh, R., Burk, D., Spitzer, M., Loferski, J., and Dubow, J. (1978b). *Proc. 13th IEEE Photovoltaic Specialists Conf.,* p. 528.
Shewchun, J., Dubow, J., Wilmsen, C. W., Singh, R., Burk, D., and Wager, J. F. (1979). *J. Appl. Phys.* **50,** 2832.
Shewchun, J., Burk, D., and Spitzer, M. B. (1980). *IEEE Trans. Electron Devices* **ED-27,** 705.
Singh, R., Green, M. A., and Rajkanan, K. (1981). *Solar Cells* **3,** 95.
Spear, W. E., and LeComber, P. G. (1972). *J. Non-Cryst. Solids* **8–10,** 727.
Spear, W. E., and Lecomber, P. G. (1975). *Solid State Commun.* **17,** 1193.
Spear, W. E., and LeComber, P. G. (1976). *Phil. Mag.* **33,** 935.
Spear, W. E., LeComber, P. G., Kinmond, S., and Brodsky, M. H. (1976). *Appl. Phys. Lett.* **28,** 105.
Sree Harsha, K. S., Bachmann, K. J., Schmidt, P. H., Spencer, E. G., and Theil, F. A. (1977). *Appl. Phys. Lett.* **30,** 645.
Staebler, D. L., and Wronski, C. R. (1977). *Appl. Phys. Lett.* **31,** 292.
Staebler, D. L., and Wronski, C. R. (1980). *J. Appl. Phys.* **51,** 32.
Staebler, D. L., Crandall, R. S., and Williams, R. (1981). *Proc. 15th IEEE Photovoltaic Specialists Conf.,* p. 249.
Sterling, H. F., and Swann, R. C. (1965). *Solid-State Electron.* **8,** 653.
Stirn, R. J., and Yeh, Y. C. M. (1975). *Appl. Phys. Lett.* **27,** 95.
Stirn, R. J., and Yeh, Y. C. M. (1977). *IEEE Trans. Electron Devices* **ED-24,** 476.
Stirn, R. J., Yeh, Y. C. M., Wang, E. Y., Ernest, F. P., and Wu, C. J. (1977). *Tech. Dig. Int. Electron Devices Meet.,* p. 48.
Street, R. A., Knights, J. C., and Biegelsen, D. K. (1978). *Phys. Rev. B* **18,** 1880.

Tang, C. W., and Albrecht, A. C. (1975a). *Nature* (*London*) **254,** 507.

Tang, C. W., and Albrecht, A. C. (1975b). *J. Chem. Phys.* **63,** 953.

Tarr, N. G., and Pulfrey, D. L. (1979). *Proc. 2nd EC Photovoltaic Sol. Energy Conf., Berlin,* p. 58. D. Reidel Publ. Dordrecht, Holland.

Tawada, Y., Kondo, M., Okamoto, H., and Hamakawa, Y. (1981). *Proc. 15th IEEE Photovoltaic Specialists Conf.,* p. 245.

Thomas, R. E., North, R. B., and Norman, C. E. (1980). *IEEE Electron Device Lett.* **EDL-1,** 79.

Thompson, W. G., and Anderson, R. L. (1978). *Solid-State Electron.* **21,** 603.

Tsai, M.-J., Fahrenbruch, A. L., and Bube, R. H. (1980). *J. Appl. Phys.* **51,** 2696.

Turner, M. J., and Rhoderick, E. H. (1968). *Solid-State Electron.* **11,** 291.

Tyan, Y.-S. (1980). U.S. Patent No. 4,207,119, issued June 9. (Tyan was with Eastman-Kodak Co., Rochester, N.Y.)

Van Halen, P., Thomas, R. E., Mertens, R., and Van Overstraeten, R. *Proc. 12th IEEE Photovoltaic Specialists Conf.,* p. 907.

Wager, J. F., and Wilmsen, C. W. (1979). *J. Appl. Phys.* **50,** 4172.

Wagner, S., and Bridenbaugh, P. M. (1977). *J. Cryst. Growth,* **39,** 151.

Wagner, S., Shay, J. L., Migliorato, P., and Kasper, H. M. (1974). *Appl. Phys. Lett.* **25,** 434.

Waldrop, J. R., Cohen, M. J., Heeger, A. J., and MacDiarmid, A. G. (1981). *Appl. Phys. Lett.* **38,** 53.

Wang, E. Y., and Legge, R. N. (1978). *IEEE Trans. Electron Devices* **ED-25,** 800.

Wilson, J. I. B., McGill, J., and Kinmond, S. (1978). *Nature* (*London*) **272,** 153.

Wronski, C. R. (1977). *IEEE Trans. Electron Devices* **ED-24,** 351.

Wronski, C. R., Carlson, D. E., and Daniel, R. E. (1976). *Appl. Phys. Lett.* **29,** 602.

Yamaguchi, K., Matsumoto, H., Nakayama, N., and Ikegami, S. (1976). *Jpn. J. Appl. Phys.* **15,** 1575.

Yamaguchi, K., Nakayama, N., Matsumoto, H., and Ikegami, S. (1977). *Jpn. J. Appl. Phys.* **16,** 1203.

Yeh, Y. C. M., Ernest, F. P., and Stirn, R. J. (1978). *Proc. 13th IEEE Photovoltaic Specialists Conf.,* p. 966.

Yin, S.-Y., Fahrenbruch, A. L., and Bube, R. H. (1978). *J. Appl. Phys.* **49,** 1294.

CHAPTER 12

CONCENTRATORS, CONCENTRATOR SYSTEMS, AND PHOTOELECTROCHEMICAL CELLS

In this chapter several systems for direct conversion of solar energy into electricity that involve photovoltaics are discussed. These include concentrator systems and photoelectrochemical cells.

Concentration of solar energy serves several purposes:

1. The solar efficiency of photovoltaic cells generally increases with illumination intensity (e.g., Fig. 8.11).

2. For higher concentration ratios C, the concentrator itself begins to dominate the system cost, thus allowing expensive, high-efficiency cells to be used economically to increase system efficiency.

3. Higher-energy densities obtained with concentration raise the temperature of the heat flow that can be extracted from hybrid photovoltaic-thermal systems and make the collection of that heat more efficient.†

† Over one-third of the industrial process heat in the United States is used at $T \lesssim 127°C$. [“Analysis of the Economic Potential of Solar Thermal Energy to Provide Process Heat,” Final Rep. COO-2829-1,2,3; ERDA Contract No. ET-76-C-02,28929.000 (1977). InterTechnology Corp., Warrenton, Virginia.]

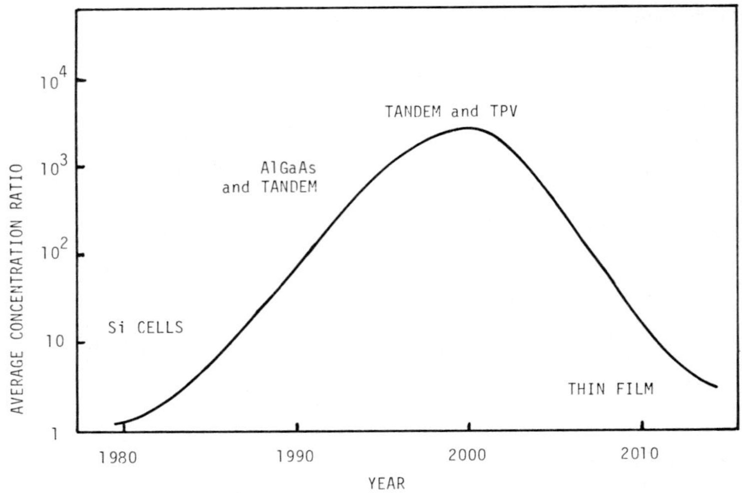

Fig. 12.1. Scenario: average concentration ratio of commercial photovoltaic systems versus time.

A variety of concentration schemes exist, some of which form an image of the sun (such as parabolic mirrors and Fresnel lenses), whereas others concentrate without imaging (heliostat fields and compound parabolic collectors). These are discussed briefly to elucidate basic principles and to illustrate the wealth of options available. The elementary analysis of cost per watt for a concentrator/solar-cell system is considered to illustrate some of the factors involved in optimization. Carrier generation and transport in photovoltaic cells at high intensities have been discussed in Chapters 4 and 5. The problems of current collection and R_s losses were treated in Chapter 6. These topics are mentioned only briefly here. We complete the first section of this chapter by considering the photovoltaic converters used in concentration systems, particularly the high-intensity Si, AlGaAs, and tandem cell devices.

A more complex photovoltaic system is considered next: a thermophotovoltaic converter that involves conversion of photon energy to lower energies and photon trapping to increase efficiency.

Finally some other novel devices that do not rely on concentration *per se* are mentioned, such as pyroelectric and photoelectrolytic converters.

An interesting perspective on concentration is given by Fig. 12.1, which illustrates a *possible* scenario for photovoltaics extending past the turn of the century.† The average concentration ratio of the photovoltaic

† K. W. Mitchell, personal communication (ca. 1977).

systems in commercial use is plotted versus the year. In the 1970s and 1980s moderate concentration systems utilizing rather conventional Si cells can make an economic impact at \$2 to \$3 per peak watt using available technology. Sometime in the late 1980s through the 1990s, large, high-concentration systems designed around very high-efficiency heteroface and tandem cells could provide electrical energy at \$0.40 to \$1 per peak watt as well as process heat. Judging from the current state of research, only in the late 1990s and early 2000s will thin-film cells make a large contribution, after the necessary increases in cell efficiency and stability have been demonstrated, and fabrication plants have been built on the large scale necessary for cost effectiveness.

An overview of concentrator systems utilizing both photovoltaic and thermal conversion is offered by Backus (1975).

12.1 ELEMENTARY CONCENTRATOR SYSTEMS

Besides the wide diversity of types, sizes, and concentration ratios of the concentrators *per se,* the large number of trade-offs possible in concentrator systems (e.g., thermal energy usage, fabrication costs) multiplies the options—and the complexity of optimization—tremendously. To be meaningful, a broad picture requires input from developmental endeavors (see e.g., Edenburn *et al.*, 1978). We therefore attempt only a simple approach to bring out fundamentals.

12.1.1 About Concentrators

A book by Meinel and Meinel (1976) (dealing mainly with devices of ratio $\lesssim 40$) gives a feeling for the surprising number of concentrator types and variations, and a very readable book by Welford and Winston (1978) discusses nonimaging systems. Because of the diversity of the devices available, it seems appropriate to present here only an overview of the general principles as they apply to photovoltaics.

Concentrators, a few of which are shown in Fig. 12.2, can be divided into two major categories: imaging and nonimaging. Examples of imaging concentrators are simple lenses, Fresnel lenses and reflectors, spherical and parabolic reflectors, and cylindrical reflectors and lenses. The second, nonimaging category includes parabolic mirrors operated out of focus, heliostat fields, compound parabolic collectors, and planar concentrators.

Whatever the concentration mechanism [excepting planar concentrators (Section 12.1.4)], a theoretical maximum concentration ratio can be calculated from any of several different fundamental considerations. For example, if the power density at the sun's surface is P_Σ and the sun's

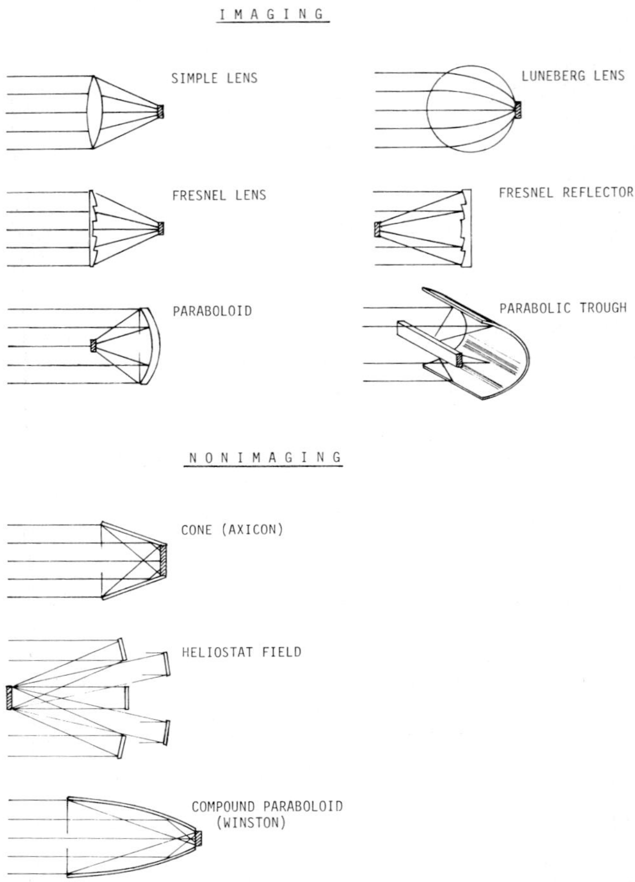

Fig. 12.2. A few of the wide variety of imaging and nonimaging concentrators. [The Luneberg lens is filled with a medium of variable refractive index $n_r(r)$.]

radius is r_Σ, then the total power radiated by the sun is $4\pi P_\Sigma r_\Sigma^2$. Ideally the concentrator aperture (Fig. 12.3) sees a flux $P_x = (4\pi P_\Sigma r_\Sigma^2)/4\pi r_e^2$, where r_e is the sun–earth distance, giving the ratio of flux at the concentrator aperture to that at the sun's surface as $P_x/P_\Sigma \lesssim (r_\Sigma^2/r_e^2)$. The flux at the cell P_s is not greater than that at the sun's surface; otherwise the equilibrium temperature of a body at the focus would be greater than the sun, violating the second law of thermodynamics. Therefore,

$$C_{max} = P_x/P_s \lesssim P_\Sigma(r_\Sigma^2/r_e^2)/P_s \lesssim r_\Sigma^2/r_e^2 = 1/\sin^2(\alpha_\Sigma/2) = 4.6 \times 10^4, \tag{12.1}$$

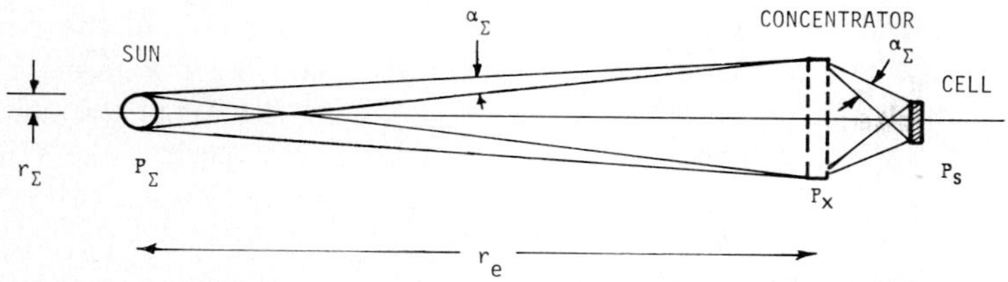

Fig. 12.3. Concentrator geometry (distances and angles exaggerated for clarity).

where $\alpha_\Sigma = 0.533°$ is the minimum angle subtended by the sun as viewed from the earth. The effective angle subtended by the sun is increased somewhat by circumsolar radiation, as shown in Fig. 2.6.

Likewise, Welford and Winston (1978) and others have shown, using geometric arguments, that the maximum concentration ratio for any concentrator can be expressed in terms of the acceptance half-angle of the concentrator θ_c and that the maximum ideal ratio is

$$C_{max} \leq (n_r/\sin\,\theta_c)^2 \tag{12.2}$$

for a concentrator with radial symmetry such as a paraboloid, where n_r is the index of refraction at the point where the flux is measured. If the acceptance half-angle is equal to $\alpha_\Sigma/2$ and $n_r = 1$, the same result as in Eq. (12.1) is obtained. For a concentrator with symmetry about a line perpendicular to the direction of incidence (a trough)

$$C_{max} = 1/\sin\,\theta_c. \tag{12.3}$$

The maximum ratio is reduced considerably by optical aberrations, especially for imaging concentrators.

In popular usage there are at least four different meanings of the term *concentration ratio C:* (1) the *theoretical maximum* ratio [e.g., Eq. (12.2)], (2) a *geometrical* concentration ratio defined by the ratio of the area of the input aperture to that of the output aperture, (3) an *optical* concentration ratio that includes aberration losses (i.e., not all incoming rays reach the output aperture), and (4) an *effective* ratio that accounts for all losses (i.e., the ratio of input to output power density).

In any real concentrator there is, of course, a complex variation of intensity across the plane of the collector (cell) that may also vary with wavelength. The collector area corresponding to the maximum practical efficiency for a given system may be smaller or larger than the simple geometrically projected focal spot area. The manner in which the intensity falls off at the edges of the focal spot must be taken into account, along with other factors such as tracking accuracy, wavelength, variations in the reflecting surface of the concentrating mirror, and the specific response of the cell to nonuniform illumination. The maximum *usable* concentration ratio may also be limited by the behavior of the cell's antireflection coating at high incidence angles [e.g., James and Williams (1978) give a limit of 11.5° from the normal for a GaAs-based cell]. These factors are well illustrated by a consideration of Fresnel lens optics by James and Williams (1978).

A useful parameter for discussion of concentrators is the ratio of focal length f to diameter D called the "f-number" of the device. For an ideal, aberration-free imaging system, the image diameter d is given by $d \simeq \alpha f$, where α is the angle subtended by the object (for object distance much greater than image distance). In these terms the concentration ratio is

$$C \simeq D^2/d^2 \simeq 1/(f/D)^2\alpha_\Sigma^2 \tag{12.4}$$

(about 1.13×10^4 for $f/D = 1$).† In some sense, the ratio (f/D) expresses the limitation on concentration ratio imposed by aberrations, since each type of concentrator has a minimum f-number beyond which aberrations are unacceptably large.

Of the refracting concentrators, simple lenses appear to be impractically heavy and expensive. Fresnel lenses are promising, however, since they are cheaply pressed from acrylic once a mold is made, and f/D ratios $\simeq 4$ are practical. Since they are made by a molding process, aspheric shapes can be made economically. The angle of each facet of the Fresnel lens is independent, and spherical aberration can be eliminated, although chromatic aberration cannot be. For a conventional imaging configuration, James and Williams (1978) found a practical limit of $C \simeq 700$ for flat, spherical-aberration-corrected, plastic Fresnels after accounting for losses. Using nonimaging facet angles, they obtained considerably higher concentration ratios.

Reflecting imagers can be made in many forms. Spherical mirrors are limited by aberration to $f/D \lesssim 5$, whereas good paraboloidal mirrors can maintain small focal spots to $f/D \simeq 1.5$, although $f/D \lesssim 2$ is probably a more common limit.

The compound parabolic (Winston) collector shown in Fig. 12.2 is an example of a nonimaging concentrator; the theory of such nonimaging devices is discussed by Welford and Winston (1974, 1978). Although this system is capable of extremely large concentration ratios, approaching the maximum possible, the area of the reflecting surface for a given entrance aperture area becomes unacceptably large at high ratios.

There is also a novel concentrator utilizing diffraction by a volume ho-

† Meinel and Meinel (1976) give a more general expression for the f-number that is appropriate for "fast" (large f-number) optical systems.

lographic grating. The holographic concentrator has the additional advantage of being able to separate portions of the solar spectrum spatially, thus facilitating the use of multicolor cells (Bloss *et al.*, 1980; *Optical Spectra*, 1981).

Increases in concentrator efficiency and reductions in cost can be obtained by the use of compound systems for high concentration ratios. For example, the light may be concentrated first by a paraboloidal mirror and then further concentrated by a nonimaging device with considerably better overall efficiency. This combination can achieve close to the maximum theoretical concentration ratio with the minimum surface area.

12.1.2 Cost of Concentrator Systems

Some elementary relations are considered here that show the forces at play in optimizing the cost effectiveness of a concentrator/solar-cell converter system such as is shown in Fig. 12.4.

The effective concentration ratio C is

$$C = P_s/P_x = \eta_x A_x/A_s, \tag{12.5}$$

where P_x is the incident, unconcentrated solar power density at the concentrator entrance aperture and P_s is the power density seen by the cell. The cell and concentrator areas are A_s and A_x, and η_x is the concentrator efficiency (typically $\geq 85\%$). The total power output of the system is

$$\text{Power Output} = \eta_s A_s P_s = \eta_s \eta_x A_x P_x. \tag{12.6}$$

If we represent the cost per unit area of the cell by ϕ_s and that of the concentrator by ϕ_x, then the cost per peak watt of the system in dollars per peak watt W_p is given by

$$\$/W_p = \frac{\text{cost/unit area}}{\text{power/unit area}} = \phi_s/C\eta_x\eta_x P_x + \phi_x/\eta_s\eta_x P_x,$$

$$\$/W_p = \frac{1}{\eta_x\eta_s P_x}[\phi_s/C + \phi_x]. \tag{12.7}$$

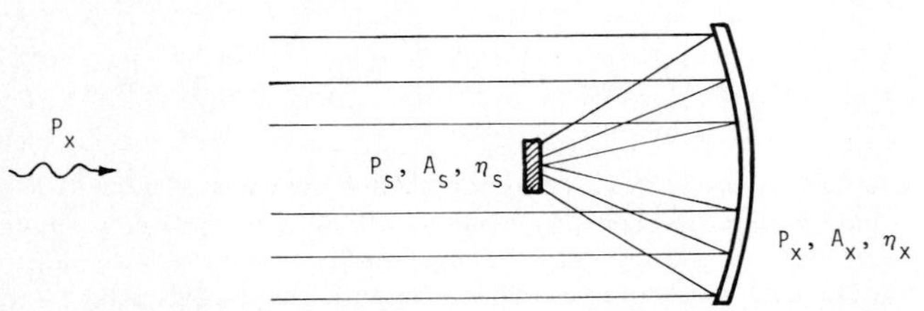

Fig. 12.4. Schematic of concentrator–solar-cell converter system.

It is clear that the highest possible cell efficiency must be used to minimize cost and that as C increases, the portion of the system cost due to the cell decreases. Equation (12.7) is plotted in Fig. 12.5 for a concentrator cost that increases with C as increasingly accurate fabrication be-

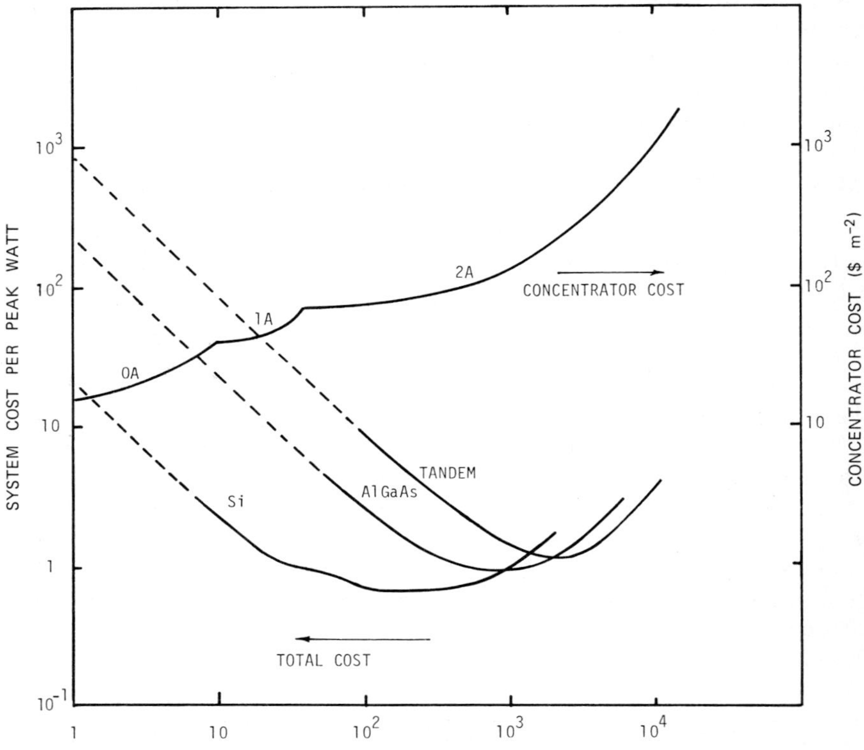

Fig. 12.5. Concentrator system cost per peak watt versus concentration ratio. Assumptions are (in 1982 dollars)

Cell	η_s	Cell cost (cm^{-2})	Cell cost (per peak watt)
Si	18%	\$0.30	\$16.70
AlGaAs	24%	\$5.00	\$200.00
Tandem	34%	\$25.00	\$735.00

Concentrator efficiency $\eta_x = 85\%$. The concentrator cost curve shown (as well as the AlGaAs and tandem cell costs) is highly hypothetical and no conclusions should be drawn about the relative merits of the systems. The concentrator cost curve is divided into sections representing no tracking 0A (with seasonal adjustment), one-axis tracking 1A (e.g., a parabolic trough tracking east–west), and two-axis tracking 2A. Thermal collection is disregarded here. The concentrator cost includes structural support and tracking.

comes necessary and more sophisticated tracking devices are needed. A constant approximation for cell cost is used in Fig. 12.5, since there are two concentration-dependent cell cost factors that tend to offset each other: cell efficiency increases with illumination intensity and cell fabrication cost increases with C. As C approaches 1 the concentrator cost becomes equal to the structure cost used to support flat plate arrays at the proper inclination. The curves show a minimum cost per peak watt at a particular value of C that is increasingly well defined for higher concentration ratios. The figure demonstrates the sensitivity to concentrator cost, a function that is largely unknown at this time.

12.1.3 Practical Aspects of Concentrator Systems

Different ranges of concentration ratio require somewhat different design considerations, as indicated by Table 12.I, where various aspects of the systems are divided (rather arbitrarily) into three ranges. The table demonstrates the increasing complexity as concentration ratio increases. At the low ratios the systems can be small and without significant operator control or maintenance, whereas at high ratios installations must be very large, with around-the-clock operators required.

For very low concentration ratios, tracking the sun may not be necessary. For trough concentrators that self-track on one axis to some degree, ratios up to 10 or 20 may be used with only seasonal adjustment to the proper ecliptic angle. For low to intermediate ratios at least one axis tracking is necessary with seasonal adjustments on the ecliptic axis. For high concentration ratios, continuous two-axis tracking is required with a

Table 12.I
Ranges of Concentration

Concentration ratio	2–20	20–200	200 and above
Tracking	None or 1-D (with possible 2-D seasonal adjustment)	1-D or 2-D	2-D required
Wind loading effect on stability	Mild	Low	Heavy
Thermal system	Passive air	Forced air or convective H_2O	Forced H_2O
Thermal cell design	Little modification (good heat sinking)	Some modification	Special design
Cell type	Si (slightly modified C = 1 type cell)	Special Si or AlGaAs	AlGaAs, tandem, or TPV converter
Cell efficiency	16–19%	20–24%	≃35%
Series resistance (total), Ω cm²	0.1	0.01	0.001

pointing accuracy of tenths of degrees at extremely high ratios. A variety of tracking options is compared by Meinel and Meinel (1976, Chap. 3).

In the temperate climates the diffuse component of solar insolation can amount to more than 40% of the total. The lowest concentration ratios (~10) can make use of a portion of this diffuse radiation, but at larger ratios virtually all of it is lost. Palz (1978)† discusses the effect of the direct/diffuse ratio on collection for various concentrator systems.

Environmental conditions that affect the concentrator system structurally (wind, snow, temperature variations) or degrade reflectivity (dust, rain, snow, hail, bird droppings) are an important concern considering the requirement of 20-or-more-year expected lifetimes for concentrator systems. Preliminary qualitative data suggest that dust may not be a severe problem. Cleaning and protection from the elements are certainly important considerations for cost analyses, however (Edenburn *et al.*, 1978).

The effect of large solar collectors on the local microclimate is an obvious environmental concern. Although conventional fossil and nuclear plants eject large amounts of heat and, when used on a large scale, cause local microclimate changes, solar collectors do just the opposite and could carry off up to ≃20% of the incident energy to other locales. These effects would be expected to be small unless areas of many square miles are involved.

12.1.4 Planar Concentrators

In northern Europe up to 60% of the average incident solar illumination is diffuse, and much of the incident energy is shifted toward the blue, predominantly between 0.32 and 0.70 μm (Goetzberger, 1978). Conventional concentrators are not effective for the diffuse component. An interesting device that promises to cope with this situation is the planar concentrator shown in Fig. 12.6, which utilizes fluorescent energy conversion of photons to lower energies and internal-reflection light trapping. Incident sunlight is absorbed by fluorescent dyes that are dissolved in a large acrylic or glass sheet. The dyes then emit photons with energies lying in a relatively narrow band centered in the 1–2-eV range. The dyes are chosen so that the emission spectrum is displaced from the absorption spectrum to reduce readsorption losses. Most of the photons emitted within the absorber suffer total internal reflection at the surfaces and are collected by solar cells at the edges of the sheet. Such planar concen-

† Palz describes concentrator systems that have been in use for thermal applications for many years and also gives considerable perspective on the future role of concentrators in the total world's energy supply.

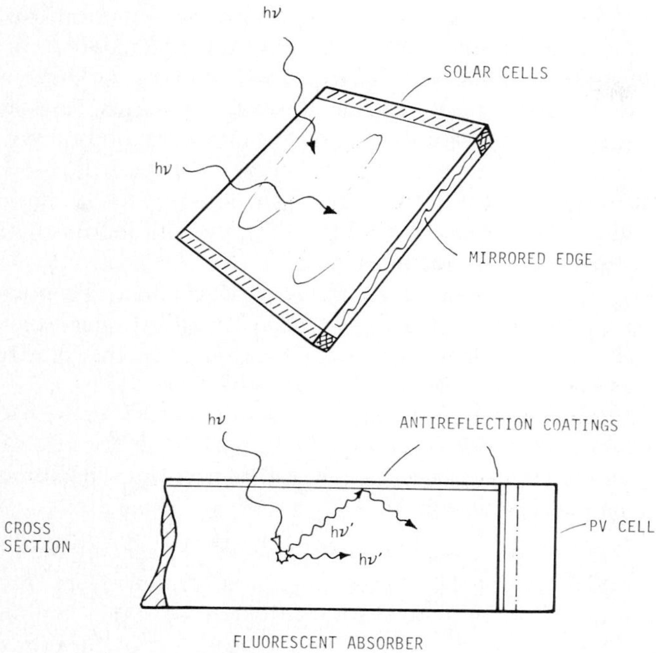

Fig. 12.6. Schematic of planar concentrator.

trators can have large geometrical concentration ratios (~200) and yet accept light from a wide range of incidence angles.

The device was apparently first reported by Garwin (1960). Goetzberger (1977, 1978) and Wittwer *et al.* (1980) given an extended consideration of the device, obtaining a theoretical maximum efficiency of 34% when used with tandem (spectrally split, Section 12.2.4) photovoltaic cells. Other work has been reported by Swartz *et al.* (1977) and by Rapp and Boling (1978), who utilized thin fluorescent films on a glass substrate to obtain preliminary measurements of concentrator efficiency indicating 2.3% overall system efficiency.

12.2 PHOTOVOLTAIC CELLS USED IN CONCENTRATOR SYSTEMS

12.2.1 Electronic Considerations

The electronic considerations for the use of photovoltaic cells under high-intensity illumination have been discussed: changes in carrier transport in the quasi-neutral regions in Section 4.5.3, junction transport in Section 5.2.5, and the changes in the photovoltaic parameters and the influence of series resistance in Sections 6.4.3 and 8.2.4. These modifications are summarized here. In the absence of series resistance, the zero-order theory shows that increasing the concentration ratio C causes: (1) the value of J_{sc} to increase linearly with C, (2) the value of V_{oc} to increase logarithmically, (3) the value of ff to increase slowly, and (4) J_{sc} to increase slightly and V_{oc} and ff to drop strongly with increasing cell temperature (as discussed in Section 6.4.2).

However, series resistance can have a large effect. To preserve the solar efficiency the effective R_s must be reduced to values of less than about $(0.5/C)\ \Omega\ \mathrm{cm}^2$. Thus the bulk resistivities of the semiconductor layers as well as the contacts and grids become crucial.

The resulting almost logarithmic variation of the η_s with C for an AlGaAs/GaAs cell model is shown in Fig. 8.11, with the effects of R_s.

The following effects can be considered as first-order perturbations on the variations just discussed:

1. As the minority carrier density becomes larger at higher C, the dominant junction transport mechanism is driven from recombination/generation ($A \simeq 2$) to injection/diffusion ($A = 1$), and finally, to high-injection behavior when $n \simeq p \gtrsim N_A$ (or N_D) in the quasi-neutral region QNR that controls the transport.

2. In the high-injection mode both *majority* and minority carrier densities increase as the injection level is increased, because of the strong charge neutrality condition. This increase in majority carrier density (either by photogeneration or forward-bias injection or both) results in an increase in conductivity, called *conductivity modulation.*† If the photon flux and/or bias is high enough to induce conductivity modulation in a large portion of the base layer of a solar cell, then the contribution to R_s from that portion can be substantially reduced. Because the maximum minority carrier density in an illuminated solar cell is increased strongly by forward bias (by factors of 10–20 in going from J_{sc} to V_{oc} conditions), the base layer can be driven from the low-injection to the high-injection mode by forward bias. For a backwall absorber, the region of high injection germinates at the light-incident side of the QNR and, with increasing photon flux Γ or bias, can expand to fill the entire base region (Schwartz and Lundstorm, 1980). The onset of high injection is given *roughly* by $\Gamma \gtrsim 10N_D(kT\mu_p/q\tau_p)^{1/2}$ (for an n-type base).‡ Thus, assuming $N_D \simeq$

† For photogeneration, this corresponds to photoconductivity.

‡ High injection for a n-QNR dominated homojunction occurs when $p_n \gtrsim N_D$ at the depletion layer edge x_n. From Eqs. (5.10) and (5.12) we have $p_n(x_n) = J_pL_p/qD_p$, where J_p is the diode current. The diode current at the maximum power point is $\simeq 0.1J_L \simeq 0.1q\Gamma$.

10^{15} cm^{-3}, conductivity modulation effects could be obtained if $C \lesssim 40-100$ for n-Si, whereas for n-GaAs, $C \lesssim 600-1000$ is required.

3. At high injection levels a Dember voltage is generated that can either add to or subtract from V_{oc}, depending on the incident light direction with respect to the n- and p-layers. Thus p^+/n cells with the light incident on the p^+-side would be preferred if R_s and the fact that L_n is generally larger than L_p were not considerations. Dember voltages of $\simeq 0.025$ V at $C = 10^2$ and $\simeq 0.05$ V to $C = 10^3$ are calculated for p^+/n Si cells by Lindholm and Fossum (1977).

4. Under concentrated illumination, the high photogenerated carrier densities can saturate the recombination centers, causing the minority carrier lifetime to increase. This causes J_{sc} to increase with Γ superlinearly, resulting in substantial increases in quantum efficiency at high concentration ratios. Nasby *et al.* (1981) report increases in the ratio of J_{sc} to C of $>8\%$ at $C = 10^3$ for Si-based cells.

For concentrator systems, particularly those with small f-numbers, a large portion of the illumination is not directed normal to the cell surface. The range of the angle of incidence must be considered in the design of antireflection coatings and in the evaluation of an effective optical absorption coefficient. Such designs must also consider the electrical effects of the nonuniform illumination intensity generally obtained with high-ratio concentrators (Sanderson *et al.*, 1980).

The electronic aspects of high-intensity solar cells have been considered by Dalal and Moore (1977) and by Weaver and Nasby (1981). Spaderna and Navon (1978) give a detailed theoretical analysis of Si concentrator cells including the effects of increased temperature and the two-dimensional aspect of current flow to the grid lines. Electronic transport at high injection levels in solar cells has been analyzed by Wolf and Wolf (1981) and by Grung (1981).

12.2.2 Thermal Considerations

In high-concentration cells, thermal transport is important; the following simple example gives some insight. Consider a simple n^+/p Si cell of thickness d mounted on a heat sink at temperature T_0. Thermal energy is deposited within the semiconductor by thermalization of hot photoexcited carriers at a rate

$$P_{th} \simeq \int_{E_g}^{\infty} h\nu[d\Gamma(h\nu)/d(h\nu)]\, d(h\nu) - \eta_s P_s. \tag{12.8}$$

Of the photons with energies below E_g, some are reflected at the front surface, some are absorbed within the cell, principally by free-carrier absorption, and some are reflected from the back contact, making a second pass before leaving the cell. The free-carrier absorption is generally on the order of 5% or less when integrated over the solar spectrum and contributions from $h\nu < E_g$ are neglected in this example. Since the thermalization energy is largely deposited near the top of the cell where the higher-energy photons are absorbed, we make the assumption that it is all deposited at the light-incident surface. The energy balance is

$$P_{th} \simeq \mathcal{K}(T - T_0)/d,$$

where $\mathcal{K}$ is the thermal conductivity and d is effective thickness of the cell. The temperature at the surface of the cell (and at the junction) is then

$$T \simeq T_0 + P_{th}(d/\mathcal{K}) \simeq T_0 + [(0.052 \text{ W cm}^{-2})Cd/\mathcal{K}] \qquad (12.9)$$

for this example, with $E_g = 1.1$ eV, $\eta_s = 17\%$ and AM1.5. In Fig. 12.7 the solar efficiency of such a cell is plotted versus concentration ratio, assuming J_{sc} is not dependent on cell temperature, that $J_{sc} = CJ_{sc0}$ with J_{sc0} the value at $C = 1$ sun, and that J_0 scales with temperature as $J_{00} \exp(-E_g/kT)$ with $J_{00} \neq J_{00}(T)$. For an excellent thermal bond with an infinite heat sink $\mathcal{K}$ can be taken as the thermal conductivity of Si (1.41 W cm^{-1} °C^{-1}), giving a conductance of 70.5 W cm^{-2} °C^{-1}. Effective thermal conductances of $\mathcal{K}/d \simeq 1$ and $\mathcal{K}/d \simeq 0.1$ W cm^{-2} °C^{-1} are *roughly* characteristic of forced H_2O (Chaffin, 1981) and of passive (convection) air cooling with efficient cooling fins, respectively. Given excellent heat sinking, η_s is not appreciably reduced by temperature rise (in this simple example) until $C \simeq 10^3$.

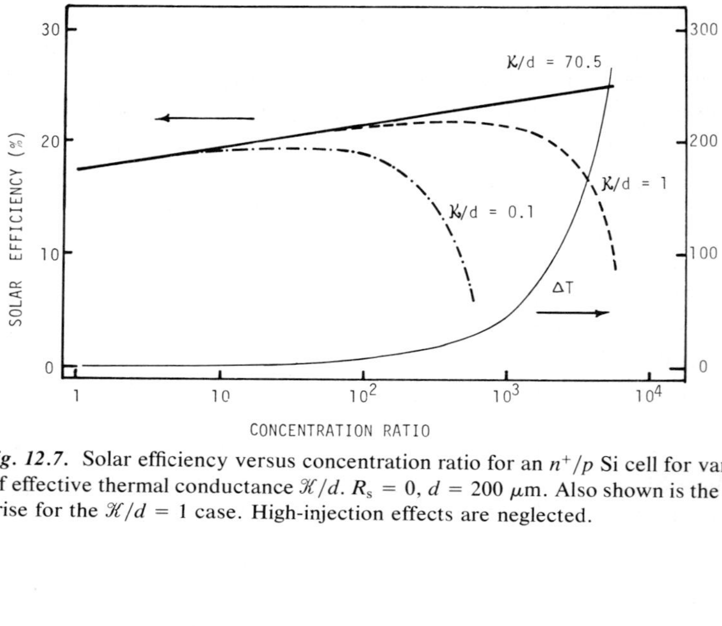

Fig. 12.7. Solar efficiency versus concentration ratio for an n^+/p Si cell for various values of effective thermal conductance $\mathcal{K}/d$. $R_s = 0$, $d = 200$ μm. Also shown is the temperature rise for the $\mathcal{K}/d = 1$ case. High-injection effects are neglected.

A thermal analysis for a 1 kW, $C = 50$ system is given by Burgess and Edenburn (1976) that includes consideration of both H_2O and active and passive air cooling.

12.2.3 Practical Concentrator Cells

Silicon cells are generally considered for low to moderate concentration ratios, whereas GaAs-based and tandem cells are considered for high ratios. The design of Si cells to operate at higher ratios and temperatures is usually a rather straightforward modification of the AM1 cells discussed in Chapter 7. However, series resistance considerations and the need for a back-surface field (BSF) to confine the carriers are critical. Grid spacing and contact resistance are reduced and the grid pattern must be optimized.

Solar efficiencies of greater than 17% at $C = 20-400$ (O'Donnell *et al.*, 1978) and greater than 20% at $C = 20-100$ (Nasby *et al.*, 1981) have been obtained with Si cells of fairly conventional design. The latter cell was a $p^+/n/n^+$ structure with a n-base resistivity of 0.3 Ω cm ($N_D \simeq 2 \times 10^{16}$ cm^{-3}). For this cell the ratio of J_{sc} to Γ was superlinear for $C \lesssim 10$ because of high-injection effects.

An excellent report on the design of a 1 kW Fresnel lens concentrator system, operating at $C \simeq 50$ and using Si cells is given by Burgess and Edenburn (1976). This paper includes practical cost projections for the system: \$216 m^{-2} for the concentrator, \$3289 m^{-2} for the Si cells, and \$270 m^{-2} for the total system (active area basis), all estimated quite conservatively.

Several special Si-based designs for even higher concentration ratios include the vertical multijunction cells (Section 7.6.1; also see Frank and Goodrich, 1980) and the V-groove multijunction cell, that utilizes $p^+/i/n^+$ cells in a configuration of tiny triangular prisms to aid photon collection by internal reflection (Chappell, 1978). A theoretical study of the conversion efficiency limit of planar $p^+/i/n^+$ Si cells under concentration was reported by Chappell (1980).

The use of AlGaAs/GaAs cells in concentrator systems has been largely discussed in Section 8.2.4. A design by James and Moon (see Vander Plas *et al.*, 1978) for such a system is shown in Fig. 12.8. The efficiency of the AlGaAs/GaAs cells (grown by liquid phase epitaxy) used with the system is 23% at $C = 945$. The grid design shown has an average spacing of 62 μm and the cell operates at $\simeq 50°C$.

With the use of other semiconductors than Si the possibilities for special cell designs is multiplied; a summary of such designs is given by Mitchell (1981). Hovel (1978) presents a good overview of the use of photo-

(a)

Fig. 12.8a. Concentrator array using AlGaAs/GaAs LPE solar cells and Fresnel lens concentrators (concentration ratio $\simeq 10^3$). Shown in front of the array are (left to right) R. L. Moon, L. W. James, and H. A. Vander Plas.

voltaic conversion under concentrated sunlight. A report on the status of the Department of Energy sponsored photovoltaic concentrator research and development (including cost estimates) by Boes and Shafer (1981) is also recommended for perspective.

12.2.4 Tandem Cells

Since concentrator systems emphasize solar cell efficiency rather than cell cost, novel designs for even higher efficiencies can be used. One such

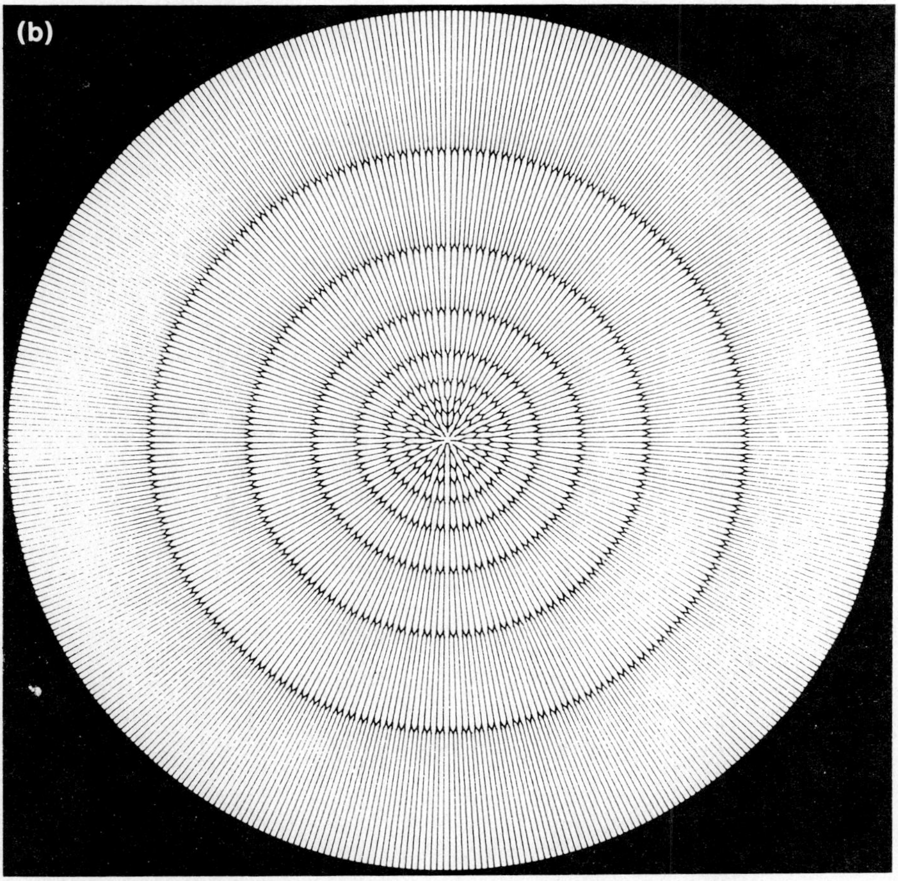

Fig. 12.8b. Grid pattern for a concentrator cell such as that used in the array. [Courtesy of Varian Associates, Palo Alto, California (1982).]

is the tandem (or multicolor) cell. The two largest fractions of the energy loss in conventional cells are by thermalization of photogenerated carriers with energy $h\nu > E_g$ and from photons with $h\nu < E_g$ passing through the cell (Fig. 6.25). By splitting the solar spectrum into several broad bands that are routed to cells with different absorber band gaps and connected in tandem, both these losses can be reduced. In the limit of a large number of cells operated in tandem, the system becomes a series of nearly monochromatic cells operated at photon energies slightly above E_g for each cell. For example, the calculated efficiency of the Si cell of Fig. 7.17 (with a relatively low $\eta_s = 13.7\%$ at AM1) is $\simeq 28\%$ when operated with photons of energies between 1.3 and 1.5 eV. Considerably higher efficiencies could be obtained for cells specifically designed for nearly monochromatic use. Two such schemes are shown in Fig. 12.9. A band diagram for a triple heterojunction converter is also shown.

The tandem cell concept appears to have been first reported by Jackson in 1955 (see Jackson, 1958), one year after the invention of the solar cell. Gokcen and Loferski (1979; see also Loferski, 1956; Loferski, 1976) calculated the maximum theoretical efficiency of the system for different numbers of component cells N, as shown in Fig. 12.10. His calculations indicate that η_s as large as 60% could be obtained by stacking enough cells (for somewhat optimistic conditions: $\eta_Q = 1$, $R_s = 1$, etc.). In reality the almost exponentially increasing complexity of fabrication would put an upper limit on N of 3 or 4.

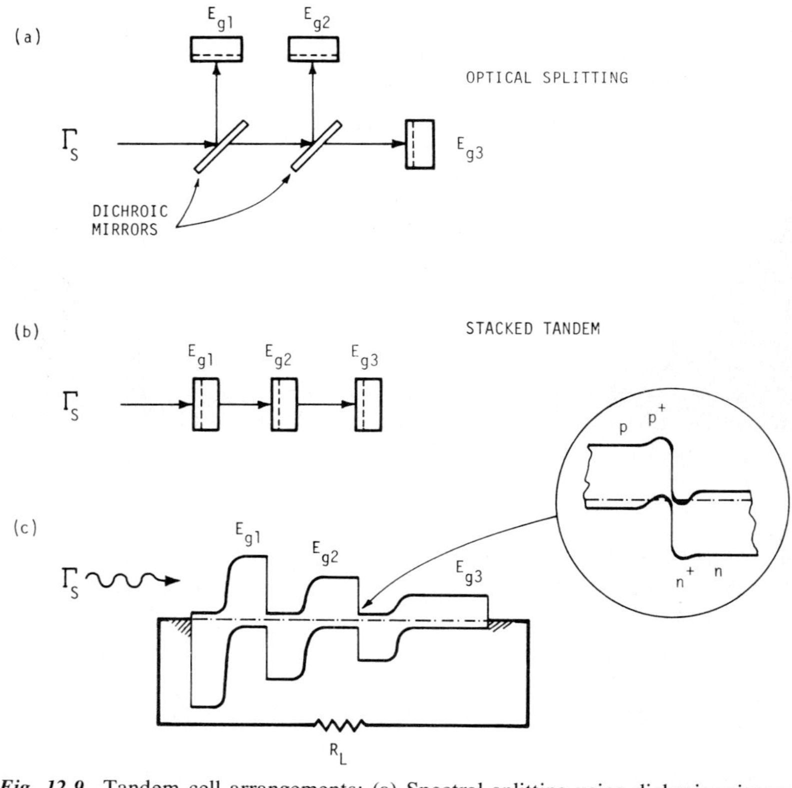

Fig. 12.9. Tandem cell arrangements: (a) Spectral splitting using dichroic mirrors. (b) In-line (stacked) tandem cells using inherent spectral splitting. (c) Band diagram of stacked tandem device with nonindependent component cells.

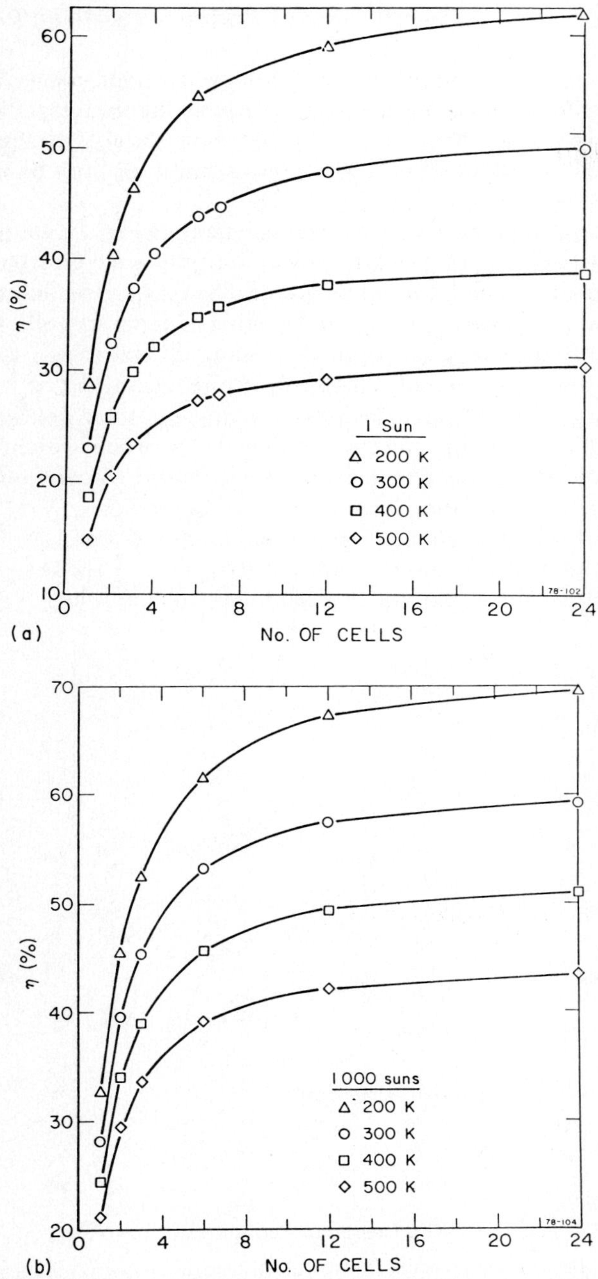

Fig. 12.10. Solar efficiency of ideal tandem cell converters versus number of cells. Band gaps are distributed so that J_L is nearly equal in all cells. AM0 spectrum: (a) 1 sun, (b) 1000 suns. [From N. A. Gokcen and J. J. Loferski, *Sol. Energy Mater.* **1,** 271 (1979).]

Masden and Backus (1978), using a more involved computer analysis and quite realistic material properties, found theoretical efficiencies of 31% for a two-cell system using Si and a material with $E_g = 1.75$ eV ($C = 125$). Iso-efficiency curves for the variation of both band gaps of a two-cell system are shown in Fig. 12.11.

In the dichroic mirror-splitting arrangement of Fig. 12.9a, the cells are electrically independent and the power outputs sum directly. The efficiency of the dichroic filters can be greater than 90%. The major limitation for such a system appears to be the mounting of separate cells—each with its electrical connections, optical alignment fixturing, and cooling facilities—near the focus of a high-ratio concentrator.

Moon *et al.* (1978) have measured a tandem solar cell efficiency of 28.5% at $C = 145$ for an unoptimized, two-cell, dichroic-mirror, spectrum-splitting arrangement using AlGaAs and Si cells (efficiency of cells plus filter, discounting lens losses). They predict $\eta_s > 33\%$ could be obtained with an optimum band-gap combination. Borden *et al.* (1981) report the practical realization of this system with a record module efficiency of 20.5% for a ten-unit module.

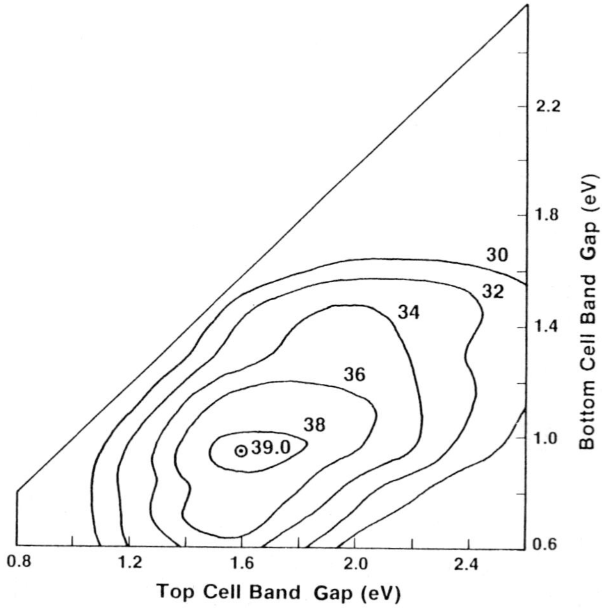

Fig. 12.11. Iso-efficiency curves for a two-cell tandem device versus the energy band gaps of the cells. The cells are electrically independent. $T = 300°$K, $C = 1000$ suns, and AM2. [After K. W. Mitchell, *Proc. 15th IEEE Photovoltaic Specialists Conf.* (1981), p. 142. © 1981 IEEE.]

For the stacked arrangement of Fig. 12.9c there are two obvious considerations. The ohmic contacts between cells must be optically transparent but of very low resistance. The use of p^+/n^+ tunnel junction contacts for this purpose has the additional advantage of providing front- and back-surface fields to confine the photogenerated carriers. The second consideration is that for most designs that are practical from a fabrication standpoint, the cells are not electrically independent and the total J_L can be no larger than the smallest J_L of any of the cells. This introduces a constraint in design: the band gaps must be adjusted so that J_L is the same for all cells, and of course this can be optimized for only a single air-mass spectrum input. Figure 12.12 illustrates this situation where the efficiency of an idealized tandem cell (with $\eta_Q = 1$ and $R_s = 1$) is plotted versus E_{g1} for AM1.5 with and without the restriction on J_L. Because of the integration involved, the change in η_s with air-mass number is probably somewhat smaller than one would infer from this simple example, however.

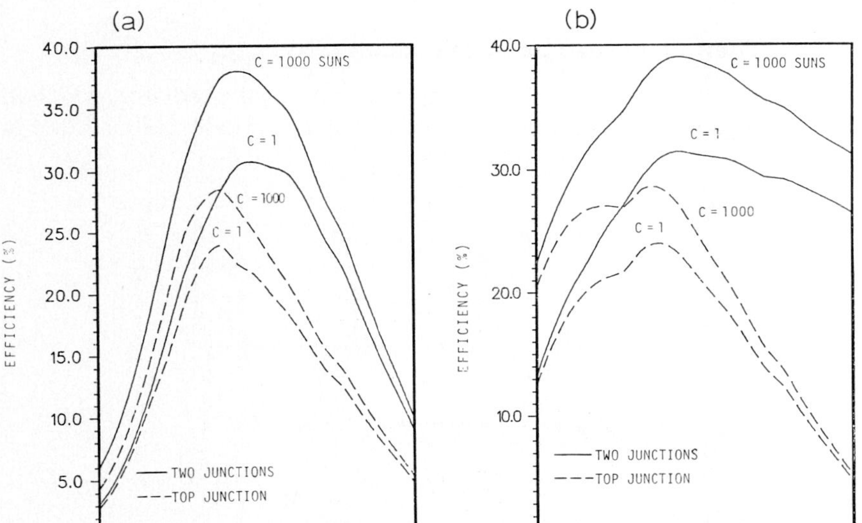

Fig. 12.12. Calculated solar efficiency of a two-cell tandem stack versus band-gap energy E_{g1} of the top cell (light-incident side). The second (bottom cell) band-gap energy E_{g2} is fixed at 0.95 eV. T = 300°K. (a) Cells are connected in series. (b) Cells are electrically independent. The maximum η_s in both cases is at a slightly higher energy than that which would give equal J_{sc} values for the two cells connected independently. [After K. W. Mitchell, *Tech. Dig.—Inter. Electron Devices Meet.*, p. 254 (1978). © 1978 IEEE.]

Preliminary experimental data by Moon *et al.* (1978) indicate that the loss due to series connection is quite small except near sunset and sunrise. A computer analysis by Lamorte and Abbott (1978) for AM0 through AM5 supports this conclusion.

12.3 COMPLEX SYSTEMS

Many more complex systems have been designed. These include hybrid systems where the photovoltaic cells are operated at higher temperatures and the excess heat is extracted to drive electrical generators or to be used as process heat. The hybrid system has the advantage that heat can be stored more conveniently than electrical energy, allowing continued operation of the plant at night.

Conversion of high-temperature solar-generated heat directly from the concentrator into electricity can be accomplished by standard heat engine methods or by more exotic means: thermophotovoltaic conversion, thermionic emission conversion, and pyroelectric conversion.

12.3.1 Thermophotovoltaic Conversion (TPV)

In TPV, concentrated solar energy is used to heat a radiator, which in turn illuminates a specially constructed photovoltaic cell, as illustrated in

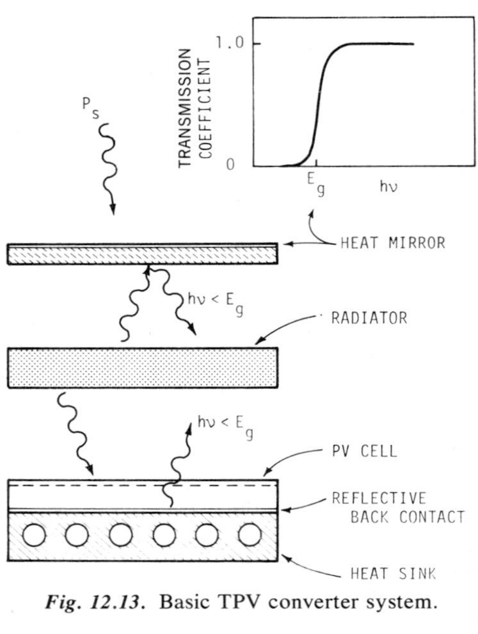

Fig. 12.13. Basic TPV converter system.

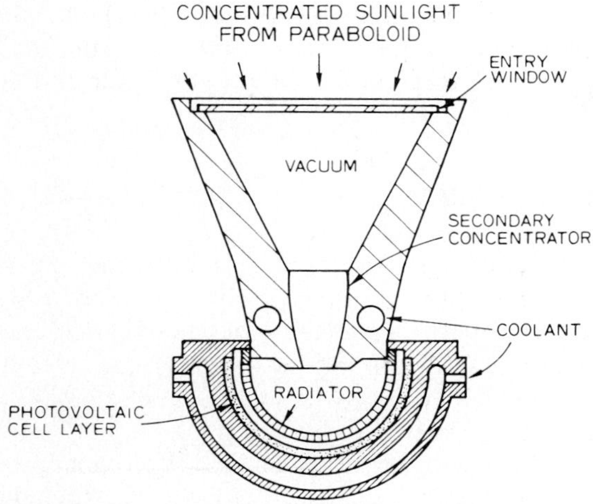

Fig. 12.14. TPV module design. [After R. Swanson, ''Silicon Photovoltaic Cells in TPV Conversion,'' Interim Report EPRI ER-633 February (1978). Electric Power Res. Inst., Palo Alto, California.]

Fig. 12.13. Two efficiency-increasing factors result: (1) the spectral output of the radiator (T_R = 2000–2400°K) is a better match to the photovoltaic cell, and (2) photons with energy $h\nu < E_g$ can be trapped to help heat the radiator. The schematic of Fig. 12.13 shows a ''heat mirror'' with a spectral transmission, shown in the inset, that serves to trap photons with $h\nu < E_g$. The cell is largely transparent to these photons; they pass through the cell and are returned to the radiator by reflection from the back mirror, which may also serve as the back electrical contact.

A more sophisticated design by Swanson (1978, 1979) is shown in Fig. 12.14. The sunlight is concentrated to $C = 5000$ by a first-stage paraboloidal reflector and enters the converter module, where it is further concentrated to $C = 20{,}000$ by a nonimaging concentrator. The light then diverges to the radiator. The heat mirror is in essence replaced by the narrow throat of the device, i.e., the exit hole for $h\nu < E_g$ is small in comparison to the radiator area. The cell operates at currents equivalent to a concentration ratio $C = 300$–500.

The energy usage of a conventional photovoltaic cell and that of the TPV converter are compared in Fig. 12.15. In a PV cell the thermalization energy ($h\nu - E_g$) plus the energy of the photons that pass through the cell ($h\nu < E_g$) amount to some 56% of the input. If such a cell has a monochromatic ($h\nu \simeq E_g$) efficiency η_M of 50%, only 22% of the input can be used effectively. In contrast, the TPV converter has much lower losses for $h\nu > E_g$ because the radiator temperature is much lower and photons with $h\nu < E_g$ are recycled; ideally up to $\simeq$55% of the input can be used.

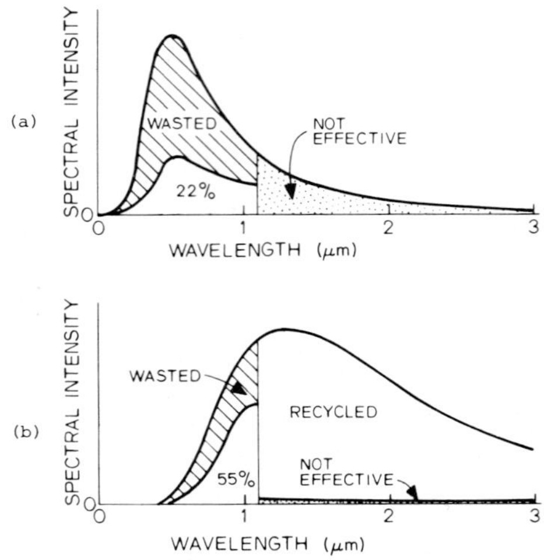

Fig. 12.15. (a) Solar cell efficiency with sunlight spectrum. (b) TPV cell efficiency with incandescent spectrum. [After R. Swanson, "Silicon Photovoltaic Cells in TPV Conversion," Interim Report EPRI ER-633 February (1978). Electric Power Res. Inst., Palo Alto, California.]

Using an idealized treatment, following that of Swanson, the TPV efficiency η_{TPV} is given by

$$\eta_{TPV} = P_{out}/P_{abs} = J_L(E_g/q)\eta_M/P_{abs}, \tag{12.10}$$

where $\eta_M = V_m J_m / J_L(E_g/q) = V_{oc} J_{sc} \mathrm{ff} / J_L(E_g/q)$ is a "monochromatic" efficiency that expresses the inherent losses in the photovoltaic cell (i.e., fill factor, $V_{oc} < E_g$, R_s, etc.).† Note that η_M is independent, to a first approximation, of the input spectrum to the photovoltaic cell, depending only on how much that spectrum affects J_L. The absorbed power is given by

$$P_{abs} = \int_0^\infty h\nu[A_{ph}(h\nu) + A_{par}(h\nu)]\frac{d\Gamma}{d(h\nu)}\, d(h\nu) = \overline{A}_{ph}P_s + \overline{A}_{par}P_s, \tag{12.11}$$

where A_{ph} is the absorptance resulting in photogeneration and A_{par} is the parasitic absorptance from free-carrier absorption and at the contacts. The quantities $\overline{A}_{ph}$ and $\overline{A}_{par}$ are weighted absorptances given by

$$\overline{A} = \int_0^\infty h\nu\, A(h\nu)\frac{d\Gamma}{d(h\nu)}\, d(h\nu) \bigg/ \int_0^\infty h\nu \frac{d\Gamma}{d(h\nu)}\, d(h\nu).$$

† For a state-of-the-art Si p/n junction cell, $\eta_M \simeq 35\%$ at AM0, $C = 1$. For a GaAs-based high-efficiency heteroface cells $\eta_M \simeq 40\%$ at AM0, $C = 1$ and 50% at AM1.5, $C = 10^3$.

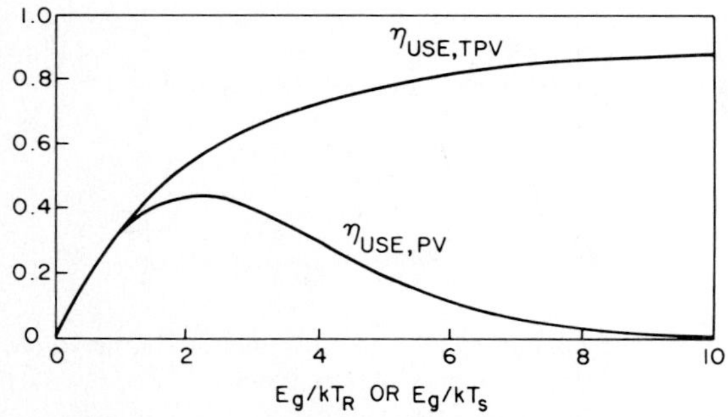

Fig. 12.16. Thermophotovoltaic and photovoltaic use factors. [After R. Swanson, "Silicon Photovoltaic Cells in TPV Conversion," Interim Report EPRI ER-633 February (1978). Electric Power Res. Inst., Palo Alto, California.]

It is assumed here that all power not absorbed is recycled.† Equation (12.10) can be rewritten

$$\eta_{TPV} = \eta_{USE,TPV}\eta_M/\{1 + [\overline{A}_{par}/\overline{A}_{ph}]\}, \tag{12.12}$$

where

$$\eta_{USE,TPV} = J_L E_g/q\overline{A}_{pH}P_s \tag{12.13}$$

is a factor such that $(1 - \eta_{USE,TPV})$ represents the excess kinetic energy loss due to thermalization of photogenerated carriers with initial energies greater than E_g. The factor η_{USE} is specific to the input spectrum, whereas η_M is not. Equation (12.10) can also be applied to a conventional photovoltaic cell, in which case

$$\eta_{USE,PV} = J_L E_g/qP_s. \tag{12.14}$$

Both are compared in Fig. 12.16, assuming that $A_{ph} = 1$ for $h\nu < E_g$ and $A_{ph} = 0$ for $h\nu < E_g$, and that blackbody radiation distributions obtain with the sun's temperature $T_s = 6500°K$. Note that the plot of $\eta_{USE,PV}$ is quite similar to the plot of η_s versus E_g of Fig. 6.7 with a maximum at approximately the same energy (1.32 eV here for $T_s = 6500°K$). These curves show the advantage of TPV and also show that E_g/kT_R (where T_R is the radiator temperature) should be as large as possible, suggesting *lower* radiator temperatures. Intuition, however, tells us that η_{TPV} must somehow decrease for decreasing T_R. Since η_M decreases quite slowly with T_R, the principal factor that reduces η_{TPV} (and provides a maximum) is the

† This treatment neglects power reradiated from the throat of the converter, estimated at 7% for Swanson's system.

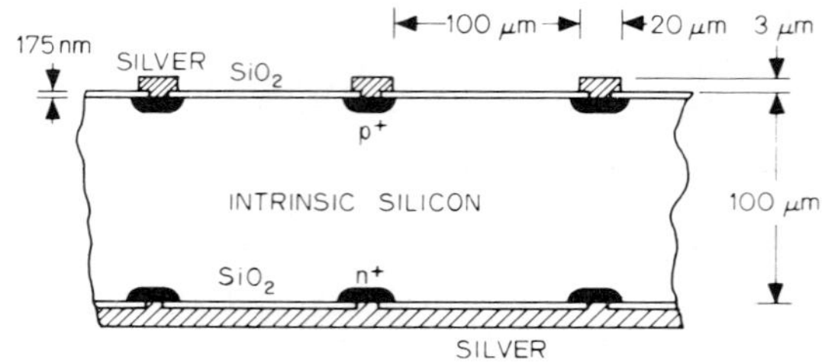

Fig. 12.17. TPV $p^+/i/n^+$ junction photocell cross section. [After R. Swanson, "Silicon Photovoltaic Cells in TPV Conversion," Interim Report EPRI ER-633 February (1978). Electric Power Res. Inst., Palo Alto, California.]

increasing effect of $\overline{A}_{par}$ for lower $h\nu$. Coupled with an ideal Si cell model, these results suggest a theoretical maximum $\eta_{TPV} \simeq 50\%$ with $T_R \simeq 2200°K$ and 63% for GaAs with $T_R = 2400°K$ (Yeargan *et al.*, 1976). Limiting efficiency calculations for TPV have also been done by Würfel and Ruppel (1980) and by Edenburn (1980).

Swanson has tested the TPV concept using a resistance-heated cavity radiator at 2300°K furnishing a J_L equivalent to about $C = 200$ AM1. To reduce free-carrier absorption, a Si $p^+/i/n^+$ junction cell was used in which most of the useful absorption occurs in a 100-μm-thick intrinsic layer and only the photogenerated carriers are involved.† The cell, shown in Fig. 12.17, has no continuous highly doped n or p layer; instead it has a grid of p^+ and n^+ material on either side of the intrinsic layer. The grid lines are less than L_n or L_p apart, and current collection is by diffusion rather than drift, obviating thin-layer R_s considerations (cf. Section 7.6.2). Conductivity modulation by photoexcitation plays an essential role in reducing resistive losses there and places a premium on achieving a long recombination lifetime in this region. Experimental results indicate that parasitic absorption is indeed a critical design factor, but efficiencies of 26% have been achieved (Swanson, 1979).

12.3.2 Thermionic, Thermoelectric, and Pyroelectric Conversion

Several more unusual means for conversion of sunlight to electricity via heat are mentioned briefly here. Thermionic conversion (Brosens, 1965)‡ utilizes concentrated sunlight to heat a thermionic emitter to a temperature of 1000–2700°C. The electrons at the high-energy end of the energy distribution in the emitter have enough energy to escape from the

† The electrical characteristics of $p/i/n$ junctions are described by Berz (1977, 1979).
‡ Also see the *Proc. IEEE Thermionics Specialists Conf.* (1964–1971).

surface into a vacuum and be collected a short distance away. Although the currents developed can be very large, only small potentials can be achieved because of the buildup of a retarding space charge in the vacuum gap between the emitter and collector.

Thermoelectric conversion (Gaibnazarov *et al.*, 1969) converts the temperature difference generated by concentrated sunlight directly to electrical energy using arrays of efficient, large-area thermocouples. Again currents are large but voltages are small. A compromise must be made between series-resistance loss and the considerable heat flow through large-area conductors. PbSe, SnTe, Bi_2Te_3, and PbTe are some of the semiconductors used that have large (often degenerate) electronic conduction but relatively small thermal conduction. Present efficiencies are below 5% but efficiencies of $\simeq$14% might be reached with several different couples in series.

Pyroelectric conversion (Sher, 1976) utilizes the piezoelectric charge transfer generated by a transient heat pulse in a material such as $BaTiO_3$ or LaF_3. The transient can be generated by rotating the absorbing piezoelectric absorber in a concentrated light beam, thus giving a roughly sinousoidal current. In contrast to the devices just mentioned, the pyroelectric generator is a high-impedance device, giving high voltage but low current. Efficiencies of up to 25% are projected (Sher, 1976).

12.4 PHOTOELECTROCHEMICAL CONVERSION

12.4.1 Overview

Photoelectrochemical (PEC) cells utilize the junction between a semiconductor and an electrolyte (usually liquid) to convert light to electrical and/or chemical energy. The production of an emf at an illuminated semiconductor–electrolyte interface—the first photovoltaic cell—was reported by Becquerel in 1839, long before similar observations in solid-state junctions. Indeed Gerischer (1979) points out that all our traditional energy resources, with the exception of wind and water power, are the result of the photoelectrochemical process known as photosynthesis. The efficiency of devices using reduction–oxidation (redox) reactions similar to photosynthesis, as imitated by man, appears quite limited; nature is much more proficient at the large-scale complexity required (Katz and Norris, 1973). But other photoelectrochemical devices exist that have demonstrated solar efficiencies above 12%.

Most PEC devices are quite analogous to Schottky barrier solar cells, with the metal layer being replaced by a highly conductive electrolyte, often aqueous, and a counter electrode. Equalization of the electrochemical potential across the built-in potential barrier causes the formation of a depletion layer within the semiconductor and an oppositely charged layer within the electrolyte. Like the situation in the Schottky barrier, the magnitude of the potential barrier is determined by the bulk properties of the semiconductor, the redox levels in the electrolyte, and by the interface properties at the junction.

Two of the major types of photoelectrochemical devices are *photovoltaic* cells, in which energy is withdrawn only in the form of electrical power and the chemical system is invariant, and *photoelectrolysis* devices, in which energy is withdrawn in the form of chemical redox reaction products (and possibly electrical power as well). In both these PEC devices electron–hole pairs are generated within the semiconductor and separated by the depletion layer field. The electrolyte is as transparent as possible. For an *n*-type material, electron flow is through the semiconductor (anode) and the external circuit, whereas holes are injected into "donor states" in the electrolyte, producing an oxidation reaction. Similarly, electrons injected into the electrolyte at the counter electrode (cathode) produce reduction reactions there. States must exist at the proper energy in the electrolyte for efficient injection of carriers; the redox potentials must be related to the semiconductor bands so that acceptor (donor) levels exist in the electrolyte at an energy equal to or below (above) the conduction (valence) band edge and the electron (hole) quasi-Fermi level, as shown in Fig. 12.18.

In the photovoltaic device, charge transport through the electrolyte is accompanied by various redox reactions that leave the electrolyte and, it is hoped, the semiconductor electrode unchanged. Band bending should be as large as possible to obtain the largest possible V_{oc}.

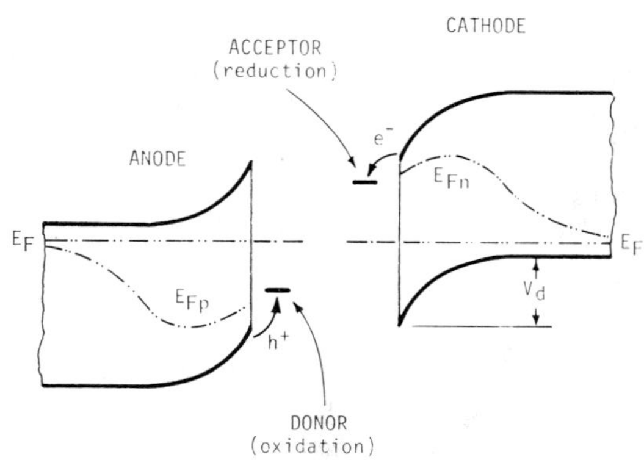

Fig. 12.18. Electrode reactions at semiconductor/electrolyte interfaces.

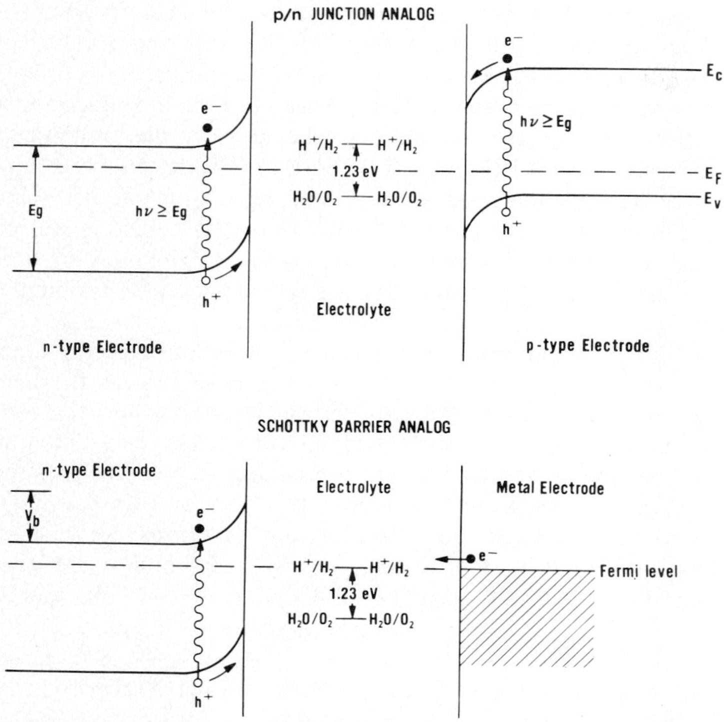

Fig. 12.19. Energy level diagram for photoelectrolysis cells. Reaction at *n*-type electrode: $2h^+ + H_2O \rightarrow \frac{1}{2}O_2 + 2H^+$; reaction at *p*-type electrode: $2e^- + 2H^+ \rightarrow H_2$. [A. J. Nozik, *J. Cryst. Growth* **39**, 200 (1977).]

In photoelectrolysis devices, reaction products are withdrawn from the cell to be "burned" elsewhere to provide heat or electrical power (e.g., by fuel cell). The source of these fuels might be H_2O or more complex chemicals with lower dissociation energies and more favorable kinetics.

Such a cell might be operated at short-circuit or with a small applied bias to provide the necessary potential within the electrolyte. Both classes of cells can also utilize a semiconductor of opposite type as the counter electrode, shown in Fig. 12.19, or two different semiconductor materials, in analogy to the heterojunction.

In a third class of devices, the *photogalvanic cell* (Clark and Eckert, 1975; Lichtin, 1976), light is absorbed within a highly absorbing electrolyte to provide energy for a redox reaction. Electrical power is generated by subsequent charge transfer to the electrode by a photo-oxidized or photoreduced molecule diffusing from the bulk of the electrolyte. This device is, in principle, perhaps closest of the PEC devices to photosynthesis. These devices show very low efficiencies at present, mostly because of back reactions (the analog of recombination in semiconductors), and our discussion centers around the other devices mentioned earlier.

The advantages of PEC devices are ease of fabrication (since no diffusion or critical metallization steps are needed), a transparent "window" with the absence of lattice mismatch considerations (which is particularly useful when polycrystalline materials are used), and the possibility of producing chemical fuels that could alleviate the considerable problem of energy storage.

The major problems to be overcome are dissolution of the electrode materials, encapsulation of cells, the complexity of the electrochemistry in the layers of electrolyte adjacent to the interface, and the low efficiencies obtained for the photoelectrolysis cells at present. The first consideration may be solved by the use of nonaqueous electrolytes and the addition of redox ions to the electrolyte, the electron transfer kinetics of which effectively compete with the corrosion reactions. Reviews on the theoretical and experimental aspects of PEC conversion are given by Gerischer (1970, 1979), Nozik (1977), Paleocrassas (1974), and Heller (1977).

12.4.2 Energy Balance in Photoelectrochemical Cells

On formation of a semiconductor/electrolyte junction, redistribution of charge occurs both because the electrochemical potentials in each part are initially different and because of surface charges. The formation of the depletion layer within the semiconductor electrode is the same as that in a Schottky barrier. In the electrolyte adjacent to the interface, however, redistribution of charge is accomplished by chemical changes within the electrolyte. Solvated ions of either sign can be attracted to the semiconductor surface, either by residual charge on the electrode or by image forces, and these form the outer Helmholtz layer, with the ions separated from the electrode by their solvation sheath. Ions that are involved in electron exchange with the semiconductor surface states or strong chemical interactions, leading to their adsorption, form the inner Helmholtz layer. Since the ionic concentration of typical electrolytes is high [of the order of $10^{21}/\text{cm}^3$ (about 1 molar)], the thickness of these layers is small (perhaps of the order of 10 Å or less), and carriers can tunnel through them. In analogy to the p/n^+ or p^+/n solid-state junctions (where one side is highly doped), the portion of the band bending occurring within the electrolyte is negligibly small. However, when current is passed through the cell, polarization voltages can be set up at both the semiconductor and the counter-electrode by the kinetics of ion transport associated with these or similar layers.

The energy levels of ions in the electrolyte can be defined by considering the effect of bringing a single electron from an infinite distance to the ion in question (or from the ion to ∞). This operation gives an electron affinity for the oxidized species (e.g., $2H^+ + 2e^- \rightarrow H_2$) and an ionization energy for the reduced species (e.g., $H_2 + 2h^+ \rightarrow 2H^+$), each of which is considerably broadened by thermal and positional fluctuations. The standard redox potential for a redox couple is the average of these two energies (and might be considered to correspond to the intrinsic level in a semiconductor). Thus the Fermi level (electrochemical potential) of a redox couple is equal to the standard redox potential for equal concentrations of the reduced and oxidized species. The dissolution of H_2O, for example, involves two redox couples ($2e^- + 2H^+ \rightarrow H_2$ and $2h^+ + H_2O \rightarrow \frac{1}{2}O_2 + 2H^+$) whose redox potentials are 1.23 eV apart.

In the electrolyte, the redox Fermi levels for each reaction are conventionally measured with respect to a normal hydrogen electrode (NHE) whose energy in turn has been determined to lie about 4.5 eV below the vacuum level, so that a quantity analogous to the semiconductor electron affinity can be measured. The factors determining the barrier potential at a semiconductor/electrolyte interface are similar, in principle, to those in a semiconductor/semiconductor junction: a difference in electron affinity coupled with the effects of interface states and dipole layers. In general, however, the band bending V_d associated with the semiconductor/electrolyte junction depends on the nature of the electrolyte and, with many semiconductors, its pH. Thus surface state pinning is of a different nature at these junctions and may be absent altogether in some cases.

The effect of these considerations is demonstrated by an energy balance for a Schottky analog photoelectrolysis cell given by Nozik (1975, 1976, 1977)

$$E_g - qV_d - \delta_n = \Delta G/m\mathcal{F} + q\eta_a + q\eta_c + qIR_s. \qquad (12.15)$$

The band bending within the semiconductor is V_d, the energy difference between the conduction band and the Fermi level ($E_c - E_F$) is δ_n, $\Delta G/m\mathcal{F}$ is the free energy per electron for the overall cell reaction (e.g., 1.23 eV for H_2O decomposition), η_a and η_c are the anode and cathode overpotentials,† and IR_s is the ohmic loss (the voltage drop across the Helmholtz

† The anode (cathode) overpotential is the energy difference between the valence (conduction) band edge and the oxidation (reduction) redox Fermi level at the semiconductor electrolyte interface.

layers is neglected here). The sum on the left is just the energy available from the photon flux for doing electrochemical work in the electrolyte. The band banding, although necessary for efficient charge separation in the semiconductor, subtracts from the energy available for electrochemical work in this case. In some cases an applied bias V_a must be added to drive the reaction (on the left side of the equation).

Photoelectrolytic cells can use a greater portion of the photon energy than do conventional solid-state cells; the energy lost by thermalization of photogenerated carriers over the potential barrier ($V_d > V_{oc}$) in solid-state cells need not be lost in their photoelectrolytic analogs, providing that the overpotentials are low in the latter devices.

Relations similar to Eq. (12.15) can be written for p/n homojunction and heterojunction analog cells, where both electrodes are semiconductors (Fig. 12.19). In the Schottky analog photoelectrolysis cell, the energy available for electrochemical work comes from single photons of energy $h\nu \geq E_g$, whereas in the homojunction and heterojunction types, the process involves two photons, each absorbed in different semiconductor layers. In the heterojunction case, the energy available for electrochemical work may be greater than that available from either of the single photons making up the process. Such a cell is analogous in a sense to the tandem photovoltaic cells of Section 12.2.4.

The complexity of the situation near the electrolyte/semiconductor interface makes it difficult at this time to predict the minimum usable semiconductor band gap, the corresponding current, or the overvoltages at the electrodes, so experimental "predictions" must be used. Gerischer (1979, p. 167) gives theoretical maximum efficiencies ranging from $\simeq 11\%$ for a cell utilizing one semiconductor electrode to $\simeq 24\%$ for a cell using two semiconductor layers with different band gaps. He discusses the factors that decrease these efficiencies to the rather small values actually observed.

The experimental solar efficiency of photoelectrolytic cells is given by

$$\eta_{PE} = (V_{H_2} - V_a)I/P_s, \tag{12.16}$$

where V_{H_2} is the voltage equivalent of the heat of combustion of H_2 (1.48 eV) and the remaining terms have their usual meaning. Another definition uses the dissociation energy of H_2O (1.23 eV) that would be appropriate if the hydrogen were burned in a fuel cell.

The alternative method of H_2 generation, against which PEC cells must compete, is conventional electrolysis equipment. Such equipment has efficiencies of $\gtrsim 70\%$ for conversion of H_2O to H_2 and O_2 and if driven by conventional Si cells, overall system efficiencies of 10–12% could result.

Variants of photoelectrolysis cells, called photocatalytic cells, employ the absorbed light energy to overcome an activation energy barrier to run a downhill (exoergic) chemical reaction rather than storing energy in an uphill reaction. An example is the exoergic reaction $N_2 + 3H_2 \rightarrow 2NH_3$. Photocatalytic cells may be useful in saving energy now expended by industries to run such reactions at high temperatures and/or pressures.

12.4.3 Practical Photoelectrolytic Devices

The objective of these devices is to produce significant quantities of fuel by dissociation of the electrolyte (e.g., $H_2O \rightarrow H_2 + \frac{1}{2}O_2$ or $CO_2 + 2H_2O \rightarrow CH_3OH + \frac{3}{2}O_2$) without dissociation of the electrodes. It now appears that the current produced by most early devices was mainly electrode dissolution current. Fujishima *et al.* (1969; Fujishima and Honda, 1972) were apparently the first to demonstrate photodissociation of H_2O without dissociation of the electrode (TiO_2, $E_g = 3.0$ eV, $V_a \simeq 0.3$ V). Many other photoelectrolytic cells have been studied involving, for example, $SrTiO_3$ (Mavroides *et al.*, 1976), $BaTiO_3$ (Kennedy and Frese, 1976), SnO_2 (Wrighton *et al.*, 1976), GaP (Nozik, 1976), CdS (Anderson and Chai, 1976), and other semiconductors.

The quantum efficiencies of many of these devices are large, 30–100%, at small light intensities. At normal AM1 intensities, however, polarization voltages, which increase with current, tend to decrease the quantum efficiency. Thus the overall solar efficiency of these devices is small, 0.2–0.8% at present, mainly because of the high semiconductor band gaps that are needed (because the oxidation of water at these materials generally goes through the high-energy intermediate, H_2O_2, or through a lower-energy surface state). Ghosh and Maruska (1977) have reported a solar efficiency of 0.8% using a TiO_2 electrode sensitized with a high doping concentration of Cr to respond to a larger wavelength range. Dye molecules adsorbed on the semiconductor surface may also extend the wavelength response in large-band-gap photoelectrolysis cells.

12.4.4 Chemically Invariant Cells

In photovoltaic-type cells there is no net chemical reaction; the anode reaction is reversed at the cathode, and only electrical energy is withdrawn. In such cells the band bending must be as large as possible to maximize V_{oc}. The cell is operated at the maximum power point and the calculation of theoretical maximum efficiency is similar to that for metal Schottky barrier cells. The optimal band gap for the semiconductor is in the 1.3–1.7 eV range. In contrast to the situation for the photoelectrolytic cell, the *p/n* junction analog appears to be of no great advantage to the basic operation of the chemically invariant cell, since no net electrochemical work is utilized in the latter case. However, *p/electrolyte/n* cells might be used in spectral splitting or tandem arrangements to increase system efficiency (Section 12.2.4). Photocorrosion of the electrode occurs when dissolution of the electrode material is thermodynamically *and* kinetically more favorable than the redox reaction in the electrolyte. In this system there is much greater freedom to choose the redox couples, however, since the ions are used only for charge transport within the electrolyte. Thus photocorrosion can be suppressed by proper choice of redox couples. For CdS anodes an electrolyte containing NaOH, Na_2S, and S is used, and the polysulfide redox reaction occurs rather than dissolution of CdS. The common ion effect probably also helps for this electrolyte choice. Efficiencies of up to 32% have been reported using monochromatic light with photon energy close to the band-gap energy of CdS.

Until recently, PECs made from Si have had a lifetime of minutes because of the formation on the Si surface of SiO_2, which blocks electron transfer. Experiments done at MIT have shown that this process may be averted by chemically attaching to the Si surface molecules that are much more reactive toward the redox couples in the electrolyte and thus bypass the passivation reaction. The combination of the advantages of PECs with the low cost and large technological base of Si may be important in the future.

A cell using *n*-GaAs/electrolyte/C with an aqueous electrolyte of K_2Se, K_2Se_2, and KOH was reported by Chang *et al.* (1977) to give a solar efficiency of 8.8%. The Fermi level at the semiconductor surface is apparently not pinned by interface states as it would be in a Schottky barrier (Heller, 1979). Parkinson *et al.* (1978) found that various cations absorbed on the surface of the *n*-GaAs produced appreciable variations in the $I-V$ characteristics and the V_{oc} of these cells, apparently changing the interfacial recombination velocity.† The value of J_{sc} remained constant. Ruthenium was strongly chemisorbed and, following absorption of a few atomic layers or less of Ru (and texturization of the surface to prevent reflection losses) the solar efficiency was increased to 12% (AM1). A Ru-treated polycrystalline cell using CVD-deposited GaAs with a grain size of 3–6 μm has also demonstrated a conversion efficiency of 7.3%. These results are particularly important since this is one of the first instances of the intentional use of a foreign interfacial dopant to reduce the interfacial recombination velocity.

† Also contributing to the increased efficiency was a "texturized" antireflection surface on the *n*-GaAs formed by chemical etching.

REFERENCES

Anderson, W. W., and Chai, Y. G. (1976). *Energy Convers.* **15,** 85.

Backus, C. E. (1975). *J. Vac. Sci. Technol.* **12,** 1032.

Berz, F. (1977). *Solid-State Electron.* **20,** 709.

Berz, F. (1979). *Solid-State Electron.* **22,** 293.

Bloss, W. H., Griesinger, M., and Reinhardt, E. R. (1980). *Proc. 3rd EC Photovoltaic Sol. Energy Conf., Cannes,* p. 401. D. Reidel Publ., Dordrecht, Holland.

Boes, E. C., and Shafer, B. D. (1981). *Proc. 15th IEEE Photovoltiacs Specialists Conf.,* p. 305.

Borden, P. G., Gregory, P. E., Moore, O. E., James, L. W., and Vander Plas, H. (1981). *Proc. 15th IEEE Photovoltiac Specialists Conf.,* p. 311.

Brosens, P. J. (1965). *J. Eng. Power* **87,** 281.

Burgess, E. L., and Edenburn, M. W. (1976). *Proc. 12th IEEE Photovoltiac Specialists Conf.,* p. 775.

Chaffin, R. J. (1981). *Proc. 15th IEEE Photovoltaic Specialists Conf.,* p. 173.

Chang, K. C., Heller, A., Schwartz, B., Menezes, S., and Miller, B. (1977). *Science* **196,** 1097.

Chappell, T. I. (1978). *Proc. 13th IEEE Photovoltaic Specialists Conf.,* p. 791.

Chappell, T. I. (1980). *IEEE Trans. Electron Devices* **ED-27,** 760.

Clark, W. D. K., and Eckert, J. A. (1975). *Sol. Energy* **17,** 147.

Dalal, V. K., and Moore, A. R. (1977). *J. Appl. Phys.* **48,** 1244.

Edenburn, M. W. (1980). *Sol. Energy* **24,** 367.

Edenburn, M. W., Schueler, D. G., and Boes, E. C. (1978). *Proc. 13th IEEE Photovoltaic Specialists Conf.,* p. 1028.

Frank, R. I., and Goodrich, J. L. (1980). *Proc. 14th IEEE Photovoltaic Specialists Conf.,* p. 423.

Fujishima, A., and Honda, K. (1972). *Nature (London)* **238,** 37.

Fujishima, A., Honda, K., and Kukuchi, S. (1969). *Kogyo Kagaku Zasshi* **72,** 108.

Gaibnazarov, M., Malevskii, Y. N., and Rezgol, I. A. (1969). *In* "Semiconductor Solar Energy Convertors" (V. A. Baum, ed.), p. 9. Consultants Bureau, New York.

Garwin, R. L. (1960). *Rev. Sci. Instrum.* **31,** 1010.

Gerischer, H. (1979). Solar photoelectrolysis with semiconductor electrodes. *In* "Solar Energy Conversion" (B. O. Seraphin, ed.), Topics in Applied Physics. Vol. 31, p. 115. Springer-Verlag, Berlin and New York.

Ghosh, A. K., and Maruska, H. P. (1977). *J. Electrochem. Soc.* **124,** 1516.

Goetzberger, A. (1977). *Appl. Phys.* **14,** 123.

Goetzberger, A. (1978). *Appl. Phys.* **16,** 399.

Gokcen, N. A., and Loferski, J. J. (1979). *Sol. Energy Mater.* **1,** 271. p. 24

Grung, B. L. (1981). *Proc. 15th IEEE Photovoltaics Specialists Conf.,* p. 394.

Heller, A., ed. (1977). "Semiconductor-Liquid Junction Solar Cells." Proceedings of the Airlie Conference, Proc. Vol. 77-3. Electrochem. Soc., Pennington, New Jersey.

Heller, A. (1979). Personal communication.

Hovel, H. J. (1978). *IBM J. Res. Develop.* **22,** 112.

Jackson, E. D. (1958). "Transactions on the Use of Solar Energy," Vol. 5, p. 122. Univ. of Arizona Press, Tucson.

James, L. W., and Williams, J. K. (1978). *Proc. 13th IEEE Photovoltaic Specialists Conf.,* p. 673.

Katz, J. J., and Norris, J. R. (1973). *Curr. Top. Bioenerg.* **5,** 41.

Kennedy, J. H., and Frese, K. W., Jr. (1976). *J. Electrochem. Soc.* **123,** 1683.

Lamorte, M. F., and Abbott, D. (1978). *Proc. 13th IEEE Photovoltiac Specialists Conf.*, p. 874.
Lichtin, N. N. (1976). NSF/RANN Rep. AER 72-03579/A03/75. Dept. Chem., Boston Univ., Massachusetts.
Lindholm, F. A., and Fossum, J. G. (1977). *IEEE Trans. Electron Devices* **ED-24,** 325.
Loferski, J. J. (1956). *J. Appl. Phys.* **27,** 777.
Loferski, J. J. (1976). *Proc. 12th IEEE Photovoltaic Specialists Conf.*, p. 957.
Masden, G. W., and Backus, C. E. (1978). *Proc. 13th IEEE Photovoltaic Specialists Conf.*, p. 853.
Mavroides, J. G., Kafaleas, J. A., and Kolesar, D. F. (1976). *Appl. Phys. Lett.* **28,** 241.
Meinel, A. B., and Meinel, M. P. (1976). "Applied Solar Energy—An Introduction." Addison-Wesley, Reading, Massachusetts.
Mitchell, K. W. (1981). *Proc. 15th IEEE Photovoltaic Specialists Conf.*, p. 142.
Moon, R. L., James, L. W., Vander Plas, H. A., Yep, T. O., Antypas, G. A., and Chai, Y. (1978). *Proc. 13th IEEE Photovoltiac Specialists Conf.*, p. 859.
Nasby, R. D., Garner, C. M., Weaver, H. T., Sexton, F. W., and Rodriguez, J. L. (1981). *Proc. 15th IEEE Photovoltaics Specialists Conf.*, p. 132.
Nozik, A. J. (1975). *Nature (London)* **257,** 383.
Nozik, A. J. (1976). *Appl. Phys. Lett.* **29,** 150.
Nozik, A. J. (1977). *J. Cryst. Growth* **39,** 200.
O'Donnell, D. T., Robb, S. P., Rule, T. T., Sanderson, R. W., and Backus, C. E. (1978). *Proc. 13th IEEE Photovoltaics Specialists Conf.*, p. 791.
Optical Spectra (1981). September, Vol. **15**(9), p. 20 (anon.).
Paleocrassas, S. N. (1974). *Sol. Energy* **16,** 45.
Palz, W. (1978). "Solar Electricity—an Economic Approach to Solar Energy." Butterworth–Unesco, London and Boston.
Parkinson, B. A., Heller, A., and Miller, B. (1978). *Appl. Phys. Lett.* **33,** 521.
Rapp, C. F., and Boling, N. L. (1978). *Proc. 13th IEEE Photovoltaic Specialists Conf.*, p. 690.
Sanderson, R. W., O'Donnell, D. T., and Backus, C. E. (1980). *Proc. 14th IEEE Photovoltaics Specialists Conf.*, p. 431.
Schwartz, R. J., and Lundstorm, M. S. (1980). *Proc. 14th IEEE Photovoltaic Specialists Conf.*, p. 1392.
Sher, A. (1976). *Proc. 12th IEEE Photovoltaics Specialists Conf.*, p. 1000.
Spaderna. D. W., and Navon, D. H. (1978). *IEEE Trans. Electron Devices* **ED-25,** 1290.
Swanson, R. M. (1978). *Tech. Dig.—Inter. Electron Devices Meet.*, p. 70.
Swanson, R. M. (1979). *Proc. IEEE* **67,** 446.
Swartz, B. A., Cole, T., and Zewail, A. H. (1977). *Opt. Lett.* **1,** 73.
Vander Plas, H. A., James, L. W., Moon, R. L., and Nelson, N. J. (1978). *Proc. 13th IEEE Photovoltaic Specialists Conf.*, p. 934.
Weaver, H. T., and Nasby, R. D. (1981). *IEEE Trans. Electron Devices* **ED-28,** 465.
Welford, W. T., and Winston, R. (1974). *Sol. Energy* **16,** 89.
Welford, W. T., and Winston, R. (1978). "The Optics of Nonimaging Concentrators: Light and Solar Energy." Academic Press, New York.
Wittwer, V., Heidler, K., Zastrow, A., and Goetzberger, A. (1980). *Proc. 14th IEEE Photovoltaics Specialists Conf.*, p. 760.
Wolf, M. A., and Wolf, M. (1981). *Proc. 15th IEEE Photovoltaics Specialists Conf.*, p. 387.
Wrighton, M. S., Morse, D. L., Ellis, A. B., Ginley, D. S., and Abrahamson, H. B. (1976). *J. Am. Chem. Soc.* **98,** 44.
Würfel, P., and Ruppel, W. (1980). *IEEE Trans. Electron Devices* **ED-27,** 745.
Yeargan, J. R., Cook, R. G., and Sexton, F. W. (1976). *Proc. 12th IEEE Photovoltaic Specialists Conf.*, p. 807.

APPENDIX 1

AIR-MASS 1.5 SOLAR INSOLATION DATA

The solar insolation data below are the "Revised Air-Mass 1.5 Spectral Distribution," reproduced from *Terrestrial Photovoltaic Measurement Procedures,* ERDA/NASA/1022-77/16, NASA TM 73702, NASA-Lewis Research Center, Cleveland, Ohio 44134 (1977). The total irradiance = 83.18 mW cm^{-2} for $0.295 \lessapprox \lambda \lessapprox 2.537$ μm (ref. Henry Curtis, NASA-Lewis). These data have been plotted in Figs. 2.5, 6.2, and 6.3.

Wave-length, μm	Irradiance, $W/(m^2\ \mu m)$	(Number of photons)/(sec cm^2)[a]	Wave-length, μm	Irradiance, $W/(m^2\ \mu m)$	(Number of photons)/(sec cm^2)[a]
0.295	0	-----------	0.455	1162.43	2.5396×10^{15}
.305	1.32	9.9792×10^{11}	.465	1180.61	2.7161×10^{15}
.315	20.96	1.7405×10^{13}	.475	1212.72	2.8347×10^{15}
.325	113.48	1.0841×10^{14}	.485	1180.43	2.8948×10^{15}
.335	182.23	2.4591×10^{14}	.495	1253.83	3.0058×10^{15}
.345	234.43	3.5699×10^{14}	.505	1242.28	3.1451×10^{15}
.355	286.01	4.5903×10^{14}	.515	1211.01	3.1530×10^{15}
.365	355.88	5.8232×10^{14}	.525	1244.87	3.2182×10^{15}
.375	386.80	6.9247×10^{14}	.535	1299.51	3.3983×10^{15}
.385	381.78	7.3599×10^{14}	.545	1273.47	3.5013×10^{15}
.395	492.18	8.5893×10^{14}	.555	1276.14	3.5338×10^{15}
.405	751.72	1.2539×10^{15}	.565	1277.74	3.6040×10^{15}
.415	822.45	1.6264×10^{15}	.575	1292.51	3.6919×10^{15}
.425	842.26	1.7619×10^{15}	.585	1284.55	3.7666×10^{15}
.435	890.55	1.8777×10^{15}	.595	1262.61	3.7871×10^{15}
.445	1077.07	2.1817×10^{15}	.605	1261.79	3.8169×10^{15}

Wave-length, μm	Irradiance, $W/(m^2\ \mu m)$	(Number of photons)/(sec cm^2)[a]	Wave-length, μm	Irradiance, $W/(m^2\ \mu m)$	(Number of photons)/(sec cm^2)[a]
0.615	1255.43	3.8695×10^{15}	1.128	100.86	4.5607×10^{15}
.625	1240.19	3.8992×10^{15}	1.131	116.87	1.8592×10^{14}
.635	1243.79	3.9436×10^{15}	1.137	108.68	3.8673×10^{14}
.645	1233.96	3.9961×10^{15}	1.144	155.44	5.3137×10^{14}
.655	1188.32	3.9677×10^{15}	1.147	139.19	2.5515×10^{14}
.665	1228.40	4.0195×10^{15}	1.178	374.29	4.6631×10^{15}
.675	1210.08	4.1171×10^{15}	1.189	383.37	2.4856×10^{15}
.685	1200.72	4.1311×10^{15}	1.193	424.85	9.7029×10^{14}
.695	1181.24	4.1418×10^{15}	1.222	382.57	7.1250×10^{15}
.6983	973.53	1.2483×10^{15}	1.236	383.81	3.3230×10^{15}
.700	1173.31	6.4301×10^{14}	1.264	323.88	6.2418×10^{15}
.710	1152.70	4.1324×10^{15}	1.276	344.11	2.5654×10^{15}
.720	1133.83	4.1199×10^{15}	1.288	345.69	2.6742×10^{15}
.7277	974.30	2.9610×10^{15}	1.314	284.24	5.3696×10^{15}
.730	1110.93	8.8089×10^{14}	1.335	175.28	3.2209×10^{15}
.740	1086.44	4.0700×10^{15}	1.384	2.42	2.9831×10^{15}
.750	1070.44	4.0493×10^{15}	1.432	30.06	5.5317×10^{14}
.7621	733.08	4.1577×10^{15}	1.457	67.14	8.8455×10^{14}
.770	1036.01	2.6980×10^{15}	1.472	59.89	7.0321×10^{14}
.780	1018.42	4.0123×10^{15}	1.542	240.85	7.9947×10^{15}
.790	1003.58	3.9999×10^{15}	1.572	226.14	5.4969×10^{15}
.800	988.11	3.9902×10^{15}	1.599	220.46	4.8178×10^{15}
.8059	860.28	2.2067×10^{15}	1.608	211.76	1.5719×10^{15}
.825	932.74	7.0375×10^{15}	1.626	211.26	3.1027×10^{15}
.830	923.87	1.9358×10^{15}	1.644	201.85	3.0638×10^{15}
.835	914.95	1.9288×10^{15}	1.650	199.68	9.9992×10^{14}
.8465	407.11	3.2212×10^{15}	1.676	180.50	4.1424×10^{15}
.860	857.46	3.6707×10^{15}	1.732	161.59	8.2262×10^{15}
.870	843.02	3.7067×10^{15}	1.782	136.65	6.6025×10^{15}
.875	835.10	1.8448×10^{15}	1.862	2.01	5.0932×10^{15}
.8875	817.12	4.5865×10^{15}	1.955	39.43	1.8535×10^{15}
.900	807.83	4.5747×10^{15}	2.008	72.58	2.9643×10^{15}
.9075	793.87	2.7358×10^{15}	2.014	80.01	4.6397×10^{14}
.915	778.97	2.7088×10^{15}	2.057	72.57	3.3654×10^{15}
.925	217.12	2.3093×10^{15}	2.124	70.29	5.0424×10^{15}
.930	163.72	4.4507×10^{14}	2.156	64.76	2.3306×10^{15}
.940	249.12	9.7273×10^{14}	2.201	68.29	3.2869×10^{15}
.950	231.30	1.1441×10^{15}	2.266	62.52	4.7856×10^{15}
.955	255.61	5.8437×10^{14}	2.320	57.03	3.7303×10^{15}
.965	279.69	1.2950×10^{15}	2.338	53.57	1.1684×10^{15}
.975	529.64	1.9783×10^{15}	2.356	50.01	1.1027×10^{15}
.985	496.64	2.5345×10^{15}	2.388	31.93	1.5673×10^{15}
1.018	585.03	9.0087×10^{15}	2.415	28.10	9.8088×10^{14}
1.082	486.20	1.8141×10^{16}	2.453	24.96	1.2367×10^{15}
1.094	448.74	3.0761×10^{15}	2.494	15.82	1.0422×10^{15}
1.098	486.72	1.0335×10^{15}	2.537	2.59	5.0182×10^{14}
1.101	500.57	8.2066×10^{14}			

[a]Number of photons/(sec-cm^2) in the wavelength interval between the corresponding wavelength and the one preceding it. Calculated using the average wavelength and irradiance for each wavelength interval.

APPENDIX 2

TRANSPORT EQUATION SOLUTIONS FOR DARK AND ILLUMINATED CASES

A.1 THE DARK FORWARD-BIAS CASE

Carrier densities and currents produced by dark, forward-bias injection into a quasi-neutral region (QNR) are given below. The configuration is shown in Fig. A.1. The QNR is assumed to be p-type and bounded by x_p and $x_{p'}$, with $y_0 = |x_p - x_{p'}|$ and $y = |x - x_p|$. For brevity $S_n = S$. The boundary conditions are $n_p = n_{p0}[\exp(qV/kT) - 1]$ at $y = 0$ and surface recombination velocity S at $y = y_0$. The carrier density at a point y is given by

$$n_p(y) = n_{p0}\left[\exp\left(\frac{qV}{kT}\right) - 1\right]\left\{\cosh\left(\frac{y}{L_n}\right) - \left\{\frac{(SL_n/D_n)\cosh(y_0/L_n) + \sinh(y_0/L_n)}{(SL_n/D_n)\sinh(y_0/L_n) + \cosh(y_0/L_n)}\right\}\sinh\left(\frac{y}{L_n}\right)\right\} + n_{p0}. \quad \text{(A.1)}$$

If $y_0 \gg L_n$, Eq. (A.1) reduces to

$$n_p(y) \simeq n_{p0}\left[\exp\left(\frac{qV}{kT}\right) - 1\right]\left[\exp\left(-\frac{y}{L_n}\right)\right] + n_{p0}. \quad \text{(A.2)}$$

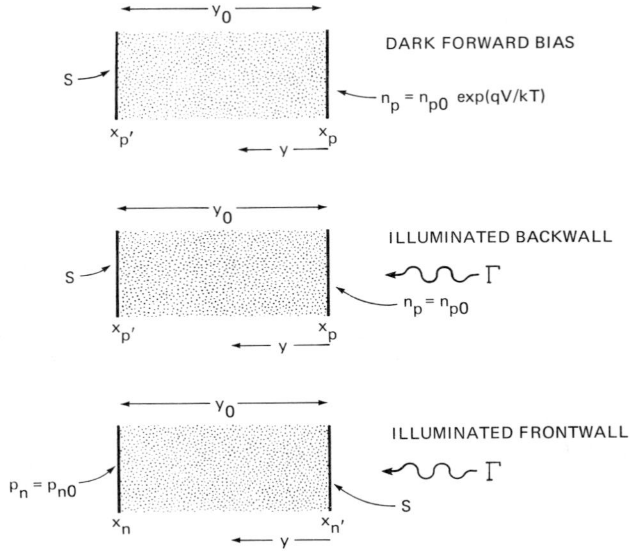

Fig. A.1. Configurations and boundary conditions.

The current density at a point y is given by

$$J_p(y) = \left(\frac{qD_n}{L_n}\right)\left[\exp\left(\frac{qV}{kT}\right) - 1\right] \times \left\{\sinh\frac{y}{L_n} - \left\{\frac{(SL_n/D_n)\cosh(y_0/L_n) + \sinh(y_0/L_n)}{(SL_n/D_n)\sinh(y_0/L_n) + \cosh(y_0/L_n)}\right\}\cosh\left(\frac{y}{L_n}\right)\right\}. \tag{A.3}$$

If $y_0 \gg L_n$, Eq. (A.3) reduces to

$$J_p(y) \simeq -\left(\frac{qD_n}{L_n}\right) n_{p0}\left[\exp\left(\frac{qV}{kT}\right) - 1\right]\left\{\exp\left(-\frac{y}{L_n}\right)\right\}. \tag{A.4}$$

At the depletion layer edge, $y = 0$, the current density is given by

$$J_p(0) = -\left(\frac{qD_n}{L_n}\right) n_{p0}\left[\exp\left(\frac{qV}{kT}\right) - 1\right] \times \left\{\frac{(SL_n/D_n)\cosh(y_0/L_n) + \sinh(y_0/L_n)}{(SL_n/D_n)\sinh(y_0/L_n) + \cosh(y_0/L_n)}\right\}. \tag{A.5}$$

If $S \to \infty$, this reduces to

$$J_p(0) \simeq -\left(\frac{qD_n}{L_n}\right) n_{p0}\left[\exp\left(\frac{qV}{kT}\right) - 1\right] \coth\left(\frac{y_0}{L_n}\right). \tag{A.6}$$

If $y_0 \gg L_n$, Eq. (A.5) reduces to

$$J_p(0) \simeq -\left(\frac{qD_n}{L_n}\right) n_{p0}\left[\exp\left(\frac{qV}{kT}\right) - 1\right]. \tag{A.7}$$

A.2 THE ILLUMINATED CASE

Carrier densities and currents produced by monochromatic photogeneration in the QNRs are given below for the zero-bias voltage case. The electric field in the QNRs is assumed to be zero. We let $\alpha(\lambda) = \alpha$ and $\Gamma(\lambda) = \Gamma$ for brevity.

(a) For the *backwall* configuration shown in Fig. A.1, the QNR is bounded by x_p and $x_{p'}$, with $y = |x - x_p|$ and $y_0 = |x_p - x_{p'}|$. For brevity $S_n = S$. The boundary conditions are $n_p = n_{p0}$ at $y = 0$ and surface recombination velocity S at $y = y_0$. The carrier density at a point y is given by

$$n_p(y) = \left[\frac{\alpha\Gamma\tau_n}{(\alpha^2L_n^2 - 1)}\right]\left\{\cosh\left(\frac{y}{L_n}\right) - \exp(-\alpha y)\right.$$
$$\left. - \left\{\frac{(SL_n/D_n)[\cosh(y_0/L_n) - \exp(-\alpha y_0)] + \sinh(y_0/L_n) + \alpha L_n \exp(-\alpha y_0)}{(SL_n/D_n)\sinh(y_0/L_n) + \cosh(y_0/L_n)}\right\} \sinh\left(\frac{y}{L_n}\right)\right\} + n_{p0}. \tag{A.8}$$

If $y_0 \gg L_n$, Eq. (A.8) reduces to

$$n_p(y) \simeq [\alpha\Gamma\tau_n/(\alpha^2L_n^2 - 1)][\exp(-y/L_n) - \exp(-\alpha y)] + n_{p0}. \tag{A.9}$$

The current density a point y is given by

$$J_n(y) = \frac{q\Gamma}{(1 - \alpha^{-2}L_n^{-2})}\left\{\left(\frac{1}{\alpha L_n}\right)\sinh\left(\frac{y}{L_n}\right) + \exp(-\alpha y)\right.$$
$$\left. - \left(\frac{1}{\alpha L_n}\right)\left\{\frac{(SL_n/D_n)[\cosh(y_0/L_n) - \exp(-\alpha y_0)] + \sinh(y_0/L_n) + \alpha L_n \exp(-\alpha y_0)}{(SL_n/D_n)\sinh(y_0/L_n) + \cosh(y_0/L_n)}\right\}\cosh\left(\frac{y}{L_n}\right)\right\}. \tag{A.10}$$

The current density at the depletion layer edge $y = 0$ is given by

$$J_n(0) = \frac{q\Gamma}{(1 - \alpha^{-2}L_n^{-2})} \left\{1 - \left(\frac{1}{\alpha L_n}\right) \times \left\{\frac{(SL_n/D_n)\{\cosh(y_0/L_n) - \exp(-\alpha y_0)\} + \sinh(y_0/L_n) + \alpha L_n \exp(-\alpha y_0)}{(SL_n/D_n)\sinh(y_0/L_n) + \cosh(y_0/L_n)}\right\}\right\}. \tag{A.11}$$

If $y_0 \gg L_n$, then Eq. (A.11) reduces to the familiar form

$$J_n(0) \simeq \frac{q\Gamma}{(1 - \alpha^{-2}L_n^{-2})}\left[1 - \frac{1}{\alpha L_n}\right] = \frac{q\Gamma}{1 + (1/\alpha L_n)}. \tag{A.12}$$

(b) For the *frontwall* configuration shown in Fig. A.1, the QNR is bounded by x_n and $x_{n'}$, with y measured from the *front* surface $y = |x - x_{n'}|$ and $y_0 = |x_n - x_{n'}|$. For brevity $S_p = S$. The boundary conditions are $p_n = p_{n0}$ at $y = y_0$ and surface recombination velocity S at $y = 0$. The carrier density at a point y is given by

$$p_n(y) = \frac{\alpha\Gamma\tau_p}{(\alpha^2L_p^2 - 1)} \times \left\{\frac{[(SL_p/D_p) + \alpha L_p]\sinh[(y_0 - y)/L_p] + [\exp(-\alpha y_0)][(SL_p/D_p)\sinh(y/L_p) + \cosh(y/L_p)]}{(SL_p/D_p)\sinh(y_0/L_p) + \cosh(y_0/L_p)} - \exp(-\alpha y)\right\} + p_{n0}. \tag{A.13}$$

If $y_0 \gg L_p$, Eq. (A.13) reduces to

$$p_n(y) \simeq \frac{\alpha\Gamma\tau_p}{(\alpha^2L_p^2 - 1)}\left\{\left[\frac{L_p(S + \alpha D_p)}{SL_p + D_p}\right]\exp\left[-\frac{y_0}{L_n}\right] + [\exp(-\alpha y_0)]\exp\left(-\frac{y_0 - y}{L_p}\right) - \exp(-\alpha y)\right\} + p_{n0}. \tag{A.14}$$

The current density at a point y is given by

$$J_p(y) = \frac{q\Gamma}{(1 - \alpha^{-2}L_p^{-2})} \times \left\{\frac{[(S/\alpha D_p) + 1]\cosh[y_0 - y)/L_p] - [\exp(-\alpha y_0)] \times \{(S/\alpha D_p)\cosh(y/L_p) + (1/\alpha L_p)\sinh(y/L_p)\}}{(SL_p/D_p)\sinh(y_0/L_p) + \cosh(y_0/L_p)} - \exp(-\alpha y)\right\}. \tag{A.15}$$

The current density at the depletion layer edge $y = y_0$ is given by

$$J_p(y_0) = \frac{q\Gamma}{(1 - \alpha^{-2}L_p^{-2})} \times \left\{ \frac{[(S/\alpha D_p) + 1] - [\exp(-\alpha y_0)] \times \{(S/\alpha D_p)\cosh(y_0/L_p) + (1/\alpha L_p)\sinh(y_0/L_p)\}}{(SL_p/D_p)\sinh(y_0/L_p) + \cosh(y_0/L_p)} - \exp(-\alpha y_0) \right\}. \tag{A.16}$$

For $S \to \infty$, this reduces to

$$J_p(y_0) \simeq \frac{q\Gamma}{(1 - \alpha^{-2}L_p^{-2})} \left\{ \frac{[1 - \exp(-\alpha y_0)\cosh(y_0/L_p)]}{(\alpha L_p)\sinh(y_0/L_p)} - \exp(-\alpha y_0) \right\}. \tag{A.17}$$

INDEX

Q

R

S

T